Edited by
Pierre Pichat

Photocatalysis and Water Purification

Related Titles

Duke, M., Zhao, D., Semiat, R. (eds.)

Functional Nanostructured Materials and Membranes for Water Treatment

2013

ISBN: 978-3-527-32987-8

Cecen, F., Aktas, Ö.

Activated Carbon for Water and Wastewater Treatment

Integration of Adsorption and Biological Treatment

2012

ISBN: 978-3-527-32471-2

Gottschalk, C., Libra, J. A., Saupe, A.

Ozonation of Water and Waste Water

A Practical Guide to Understanding Ozone and its Applications

2010

ISBN: 978-3-527-31962-6

Ganoulis, J.

Risk Analysis of Water Pollution

2009

ISBN: 978-3-527-32173-5

Parmon, V. N., Vorontsov, A. V., Kozlov, D., Smirniotis, P.

Photocatalysis

Catalysts, Kinetics and Reactors

2008

ISBN: 978-3-527-31784-4

Wiesmann, U., Choi, I. S., Dombrowski, E.-M.

Biological Wastewater Treatment

Fundamentals, Microbiology, Industrial Process Integration

2007

ISBN: 978-3-527-31219-1

Turovskiy, I. S., Mathai, P. K.

Wastewater Sludge Processing

2006

ISBN: 978-0-471-70054-8

Edited by Pierre Pichat

Photocatalysis and Water Purification

From Fundamentals to Recent Applications

WILEY-VCH

WILEY-VCH Verlag GmbH & Co. KGaA

The Editor

Prof. Dr. Pierre Pichat
CNRS, Ecole Centrale de Lyon
(STMS), Photocatalyse et
Environnement
69134 Ecully CEDEX
France

Library of Congress Card No.: applied for

British Library Cataloguing-in-Publication Data
A catalogue record for this book is available from the British Library.

Bibliographic information published by the Deutsche Nationalbibliothek
The Deutsche Nationalbibliothek lists this publication in the Deutsche Nationalbibliografie; detailed bibliographic data are available on the Internet at <http://dnb.d-nb.de>.

Print ISBN: 978-3-527-33187-1
ePDF ISBN: 978-3-527-64542-8
ePub ISBN: 978-3-527-64541-1
mobi ISBN: 978-3-527-64543-5
oBook ISBN: 978-3-527-64540-4

Materials for sustainable energy and development (Print) ISSN: 2194-7813

Materials for sustainable energy and development (Internet) ISSN: 2194-7821

Cover Design Simone Benjamin, McLeese Lake, Canada
Typesetting Laserwords Private Limited, Chennai, India
Printing and Binding Markono Print Media Pte Ltd, Singapore

Editorial Board

Contents

Series Editor Preface

The Wiley Series on New Materials for Sustainable Energy and Development

Sustainable energy and development is attracting increasing attention from the scientific research communities and industries alike, with an international race to develop technologies for clean fossil energy, hydrogen and renewable energy as well as water reuse and recycling. According to the REN21 (Renewables Global Status Report 2012 p. 17) total investment in renewable energy reached $257 billion in 2011, up from $211 billion in 2010. The top countries for investment in 2011 were China, Germany, the United States, Italy, and Brazil. In addressing the challenging issues of energy security, oil price rise, and climate change, innovative materials are essential enablers.

In this context, there is a need for an authoritative source of information, presented in a systematic manner, on the latest scientific breakthroughs and knowledge advancement in materials science and engineering as they pertain to energy and the environment. The aim of the *Wiley Series on New Materials for Sustainable Energy and Development* is to serve the community in this respect. This has been an ambitious publication project on materials science for energy applications. Each volume of the series will include high-quality contributions from top international researchers, and is expected to become the standard reference for many years to come.

This book series covers advances in materials science and innovation for renewable energy, clean use of fossil energy, and greenhouse gas mitigation and associated environmental technologies. Current volumes in the series are:

Supercapacitors. Materials, Systems, and Applications
Functional Nanostructured Materials and Membranes for Water Treatment
Materials for High-Temperature Fuel Cells
Materials for Low-Temperature Fuel Cells
Advanced Thermoelectric Materials. Fundamentals and Applications
Advanced Lithium-Ion Batteries. Recent Trends and Perspectives
Photocatalysis and Water Purification. From Fundamentals to Recent Applications

In presenting this volume on Supercapacitors, I would like to thank the authors and editors of this important book, for their tremendous effort and hard work in completing the manuscript in a timely manner. The quality of the chapters reflects well the caliber of the contributing authors to this book, and will no doubt be recognized and valued by readers.

Finally, I would like to thank the editorial board members. I am grateful to their excellent advice and help in terms of examining coverage of topics and suggesting authors, and evaluating book proposals.

I would also like to thank the editors from the publisher Wiley-VCH with whom I have worked since 2008, Dr Esther Levy, Dr Gudrun Walter, and Dr Bente Flier for their professional assistance and strong support during this project.

I hope you will find this book interesting, informative and valuable as a reference in your work. We will endeavour to bring to you further volumes in this series or update you on the future book plans in this growing field.

Brisbane, Australia
31 July 2012

G.Q. Max Lu

Preface

Thousands of articles, patents, and reviews have appeared about heterogeneous photocatalysis using semiconductors, mainly TiO_2. Among them, many deal with the purification of water, which is obviously a very important problem for our environment and poses a challenge. Because of this abundance of publications, it is very difficult even for the experts to get a clear view of "photocatalysis and water purification". In my opinion, this book – which gathers reviews from eminent scientists with a long experience in the field – will allow the readers to gain the general view they expect. The knowledge on the fundamentals is integrated with the potential implementation of photocatalytic water purification using UV lamps or solar light. To that end, this book is structured in four sections with a total of 14 chapters.

The three opening papers explore the fundamentals. Chapter 1 focuses on the active species, especially the oxygen species, which follow the excitation of the photocatalyst by band gap irradiation, and provides information on the roles of these species. Chapter 2 identifies the degradation pathways of organic pollutants and indicates how these pathways are influenced by the pollutant molecular structure, the photocatalyst, and irradiation characteristics. Dealing also with pathways, Chapter 3 presents a synthetic and critical view of the kinetic models and shows how the effects of various parameters can be rationalized.

Improving the photocatalytic efficacy is obviously an extremely important question that is addressed in Chapters 4–9. Understanding how the diverse characteristics of the photocatalyst intervene and whether these characteristics can be tailored to increase the efficacy is considered in Chapters 4 and 5. On the basis of a critical appraisal of the present knowledge it is concluded in Chapter 4 that it is, unfortunately, premature to indicate uncontestable guidance for photocatalyst design. Chapter 5 discusses in detail the various ways by which the photocatalyst can be modified. The use of composite materials is also taken into account. The conclusion is similar to that of Chapter 4. As the photocatalyst is in practice rarely dispersed in the water to be treated, Chapter 6 covers the main methods used to immobilize the photocatalyst and details the nature, shapes, and advantages and disadvantages of the supporting materials. Chapter 7 focuses on the interest of TiO_2 thin films and gives examples of the types of reactors in which they can be employed; some of the means allowing one to make these films sensitive in the visible region of the solar spectrum are also discussed. Chapter 8 provides a detailed and critical review of all the methodologies utilized for visible-light sensitization of photocatalysts and indicates how to properly evaluate the results. Finally, as the recombination of the photoproduced charge carriers is the main flaw of heterogeneous photocatalysis, there is also the possibility to effect the separation of these charge carriers by use of an applied potential. Chapter 9 reports on this technique, called photoelectrocatalysis, and concludes that the design of reactors needs further progress to achieve effective water purification.

Most studies regarding water photocatalytic purification refer to both organic pollutants representative of the different categories of organic compounds and to great classes of exogenous pollutants such as pesticides, dyes, and pharmaceuticals. It was therefore deemed necessary to also include in this book a section on the effects of photocatalysis on natural organic matter (NOM) and bacteria which are found in many waters. In some cases eliminating NOM in the course of the purification is desirable; NOM can also hinder the removal of pollutants. Consequently, Chapter 10 is devoted to photocatalysis and NOM; it is concluded that, in general, photocatalysis cannot be applied as a stand-alone process, but its coupling with another technique can be beneficial. Chapter 11 covers the photocatalytic inactivation of bacteria and explains in depth how the active species issued from excitation of the photocatalyst can damage the membrane and degrade the organic constituents of bacteria.

The last section deals with modeling and reactors. Chapter 12 focuses on the influence of irradiation, which is the initiation step of photocatalysis and stresses that irradiance measurement is essential; however, the observed kinetics is also influenced by other factors whose relative importance must be determined for reactor design. Chapter 13 demonstrates that rigorous mathematical modeling based on data collected at the laboratory and bench scales enables one to simulate the performance of photocatalytic slurry reactors of different shapes, sizes, irradiation systems, and configurations. Chapter 14 includes examples of reactors using photocatalysis as a stand-alone process or combined with another AOP; it also emphasizes that immobilization of the photocatalyst must be carefully studied and that further progress in photocatalyst sensitization to visible light is highly desirable. Chapter 15 reviews the characteristics of diverse types of solar reactors including full-size demonstration plants with a detailed presentation regarding the utilization of sun light; it is concluded that appraisal of the possibilities for both pollutants removal and disinfection requires further comparison with other technologies.

I think this book covers all the main topics of "photocatalysis and water purification". It will be helpful for students beginning research in the field and for technicians and scientists involved in the treatment of water by other technologies, who would like to learn about photocatalysis and its potentialities. Even senior scientists in the field will certainly be interested in the opinion of colleagues about some of the issues.

I sincerely thank the contributors for their response to my solicitation, their time, and efforts. I am also very grateful to the reviewers and the Wiley staff.

Pierre Pichat

About the Series Editor

Professor Max Lu
Editor, New Materials for Sustainable Energy and Development Series

Professor Lu's research expertise is in the areas of materials chemistry and nanotechnology. He is known for his work on nanoparticles and nanoporous materials for clean energy and environmental technologies. With over 500 journal publications in high-impact journals, including Nature, *Journal of the American Chemical Society, Angewandte Chemie, and Advanced Materials,* he is also coinventor of 20 international patents. Professor Lu is an Institute for Scientific Information (ISI) Highly Cited Author in Materials Science with over 17 500 citations (h-index of 63). He has received numerous prestigious awards nationally and internationally, including the Chinese Academy of Sciences International Cooperation Award (2011), the Orica Award, the RK Murphy Medal, the Le Fevre Prize, the ExxonMobil Award, the Chemeca Medal, the Top 100 Most Influential Engineers in Australia (2004, 2010, and 2012), and the Top 50 Most Influential Chinese in the World (2006). He won the Australian Research Council Federation Fellowship twice (2003 and 2008). He is an elected Fellow of the Australian Academy of Technological Sciences and Engineering (ATSE) and Fellow of Institution of Chemical Engineers (IChemE). He is editor and editorial board member of 12 major international journals including *Journal of Colloid and Interface Science and Carbon.*

Max Lu has been Deputy Vice-Chancellor and Vice-President (Research) since 2009. He previously held positions of acting Senior Deputy Vice-Chancellor (2012), acting Deputy Vice-Chancellor (Research), and Pro-Vice-Chancellor (Research

Linkages) from October 2008 to June 2009. He was also the Foundation Director of the ARC Centre of Excellence for Functional Nanomaterials from 2003 to 2009.

Professor Lu had formerly served on many government committees and advisory groups including the Prime Minister's Science, Engineering and Innovation Council (2004, 2005, and 2009) and the ARC College of Experts (2002–2004). He is the past Chairman of the IChemE Australia Board and former Director of the Board of ATSE. His other previous board memberships include Uniseed Pty Ltd., ARC Nanotechnology Network, and Queensland China Council. He is currently Board member of the Australian Synchrotron, National eResearch Collaboration Tools and Resources, and Research Data Storage Infrastructure. He also holds a ministerial appointment as member of the National Emerging Technologies Forum.

About the Volume Editor

Professor Pierre Pichat

Pierre Pichat, as "Directeur de Recherche", with the CNRS (National Center for Scientific Research, France), has been active in heterogeneous photocatalysis for many years and has founded a laboratory principally based on this field in Lyon, France. His research activity has concerned both basic investigations and applications regarding not only the photocatalytic treatment of water but also self-cleaning materials and the purification of indoor air. He has been invited to publish reviews or these domains. He has received an Appreciation Award acknowledging his pioneering contributions.

List of Contributors

Orlando M. Alfano
Universidad Nacional del Litoral and CONICET
Instituto de Desarrollo Tecnológico para la Industria Química (INTEC)
Paraje El Pozo. Colectora Ruta Nacional N° 168
3000, Santa Fe
Argentina

Rossano Amadelli
University of Ferrara
CNR-ISOF, U.O.S. of Ferrara
Department of Chemistry
Via L. Borsari 46
I-44121 Ferrara
Italy

Rose Amal
The University of New South Wales
School of Chemical Engineering
ARC Centre of Excellence for Functional Nanomaterials
Tyree Energy Technologies Building
Sydney NSW 2052
Australia

Masakazu Anpo
Osaka Prefecture University
(Vice President & Executive Director)
Research Organization for the 21st Century
1-1 Gakuencho, Naka-ku, Sakai
599-8531 Osaka
Japan

Rodolfo J. Brandi
Universidad Nacional del Litoral and CONICET
Instituto de Desarrollo Tecnológico para la Industria Química (INTEC)
Paraje El Pozo. Colectora Ruta Nacional N° 168
3000, Santa Fe
Argentina

Alberto E. Cassano
Universidad Nacional del Litoral and CONICET
Instituto de Desarrollo Tecnológico para la Industria Química (INTEC)
Paraje El Pozo. Colectora Ruta Nacional N° 168
3000, Santa Fe
Argentina

Chuncheng Chen
Institute of Chemistry
Chinese Academy of Sciences
(CAS)
Beijing National Laboratory for
Molecular Sciences
Key Laboratory of Photochemistry
No. 2, 1st North Street,
Zhongguancun
Beijing 100190
China

Pilar Fernández-Ibáñez
Plataforma Solar de Almería
(CIEMAT)
Carretera Senés, km 4
04200, Tabernas (Almería)
Spain

Akira Fujishima
Kanagawa Academy of Science
and Technology
Photocatalyst Group
KSP East 421, 3-2-1 Sakado,
Kawasaki
Kanagawa 213-0012
Japan

and

Tokyo University of Science
Division of Photocatalyst for
Energy and Environment
Research Institute for Science
and Technology
1–3 Kagurazaka, Shinjuku-ku
Tokyo 162-8601
Japan

Yu Horiuchi
Osaka Prefecture University
Department of Applied Chemistry
1-1 Gakuencho, Naka-ku, Sakai
599-8531 Osaka
Japan

William S. Jenks
Iowa State University
Department of Chemistry
3760 Gillman Hall
Ames, IA, 50011-3111
USA

Nicolas Keller
Laboratoire des Matériaux
Surfaces et Procédés pour la
Catalyse
Ecole Européenne de Chimie
Polymères et Matériaux
CNRS and Strasbourg University
25 rue Becquerel
67087, Strasbourg Cedex
France

Valérie Keller
Laboratoire des Matériaux
Surfaces et Procédés pour la
Catalyse
Ecole Européenne de Chimie
Polymères et Matériaux
CNRS and Strasbourg University
25 rue Becquerel
67087, Strasbourg Cedex
France

May Lim
The University of New South
Wales
School of Chemical Engineering
ARC Centre of Excellence for
Functional Nanomaterials
Tyree Energy Technologies
Building
Sydney NSW 2052
Australia

Sanly Liu
The University of New South Wales
School of Chemical Engineering
ARC Centre of Excellence for Functional Nanomaterials
Tyree Energy Technologies Building
Sydney NSW 2052
Australia

Wanhong Ma
Institute of Chemistry
Chinese Academy of Sciences (CAS)
Beijing National Laboratory for Molecular Sciences
Key Laboratory of Photochemistry
No. 2, 1st North Street, Zhongguancun
Beijing 100190
China

Sixto Malato
Plataforma Solar de Almería (CIEMAT)
Carretera Senés, km 4, 04200
Tabernas (Almería)
Spain

Maneil Ignacio Maldonado
Plataforma Solar de Almería (CIEMAT)
Carretera Senés, km 4, 04200
Tabernas (Almería)
Spain

Masaya Matsuoka
Osaka Prefecture University
Department of Applied Chemistry
1-1 Gakuencho
Naka-ku, Sakai
599-8531 Osaka
Japan

Valter Maurino
Univ. degli Studi di Torino
Dipt. di Chimica, Analitica
Via Pietro Giuria 5
10125 Torino
Italy

Claudio Minero
Univ. degli Studi di Torino
Dipt. di Chimica, Analitica
Via Pietro Giuria 5
10125 Torino
Italy

Atsuko Y. Nosaka
Nagaoka University of Technology
Department of Materials Science and Technology
Kamitomioka-cho 1603-1
Nagaoka, 940-2188
Japan

Yoshio Nosaka
Nagaoka University of Technology
Department of Materials Science and Technology
Kamitomioka-cho 1603-1
Nagaoka, 940-2188
Japan

Tsuyoshi Ochiai
Kanagawa Academy of Science and Technology
Photocatalyst Group
KSP East 421, 3-2-1 Sakado
Kawasaki
Kanagawa 213-0012
Japan

and

Tokyo University of Science
Division of Photocatalyst for Energy and Environment
Research Institute for Science and Technology
1–3 Kagurazaka
Shinjuku-ku
Tokyo 162-8601
Japan

Bunsho Ohtani
Catalysis Research Center
Hokkaido University
North 21, West 10
Sapporo 001-0021
Japan

Isabel Oller
Plataforma Solar de Almería (CIEMAT)
Carretera Senés, km 4, 04200
Tabernas (Almería)
Spain

David Ollis
North Carolina State University
Department of Chemical and Biomolecular Engineering
911 Partners Way
Raleigh, NC 27695-7905
USA

Yaron Paz
Technion-Israel Institute of Technology
Department of Chemical Engineering
3200, Haifa
Israel

Maria Inmaculada Polo-López
Plataforma Solar de Almería (CIEMAT)
Carretera Senés, km 4, 04200
Tabernas (Almeréa)
Spain

Cesar Pulgarin
Ecole Polytechnique Fédérale de Lausanne (EPFL)
Institute of Chemical Science and Engineering, GGEC 1015
Lausanne
Switzerland

Julián Andrés Rengifo-Herrera
Universidad Nacional de La Plata-UNLP CCT La Plata-CONICET
Centro de Investigación y Desarrollo en Ciencias Aplicadas "Dr. J.J. Ronco"- CINDECA
Departamento de Química
Facultad de Ciencias Exactas
Calle 47 No. 257, 1900 La Plata
Buenos Aires
Argentina

Angela Giovana Rincón
University of Cambridge
Department of Chemistry
Lensfield Road
Cambridge, CB2, 1EW
UK

Didier Robert
Laboratoire des Matériaux Surfaces et Procédés pour la Catalyse
CNRS and Strasbourg University
Antenne de Saint-Avold
Université de Lorraine
12 Rue Victor Demange
57500, Saint-Avold
France

Luca Samiolo
University of Ferrara
CNR-ISOF
U.O.S. of Ferrara
Department of Chemistry
Via L. Borsari 46
I-44121 Ferrara
Italy

Mariia L. Satuf
Universidad Nacional del Litoral
and CONICET
Instituto de Desarrollo
Tecnológico para la Industria
Química (INTEC)
Paraje El Pozo. Colectora Ruta
Nacional N° 168
3000, Santa Fe
Argentina

Nurit Shaham-Waldmann
Technion-Israel Institute of
Technology
Department of Chemical
Engineering
3200, Haifa
Israel

Masato Takeuchi
Osaka Prefecture University
Department of Applied Chemistry
1-1 Gakuencho
Naka-ku, Sakai
599-8531 Osaka
Japan

Takashi Toyao
Osaka Prefecture University
Department of Applied Chemistry
1-1 Gakuencho
Naka-ku, Sakai
599-8531 Osaka
Japan

Davide Vione
Univ. degli Studi di Torino
Dipt. di Chimica Analitica
Via Pietro Giuria 5
10125 Torino
Italy

Zhaohui Wang
Institute of Chemistry
Chinese Academy of Sciences
(CAS)
Beijing National Laboratory for
Molecular Sciences
Key Laboratory of Photochemistry
No.2, 1st North Street,
Zhongguancun
Beijing 100190
China

and

Donghua University
College of Environmental Science
and Engineering
No.2999, North Renmin Road
Shanghai 201620
China

Jincai Zhao
Institute of Chemistry
Chinese Academy of Sciences
(CAS)
Beijing National Laboratory for
Molecular Sciences
Key Laboratory of Photochemistry
No. 2, 1st North Street,
Zhongguancun
Beijing 100190
China

Part I
Fundamentals: Active Species, Mechanisms, Reaction Pathways

Photocatalysis and Water Purification: From Fundamentals to Recent Applications, First Edition. P. Pichat.

1
Identification and Roles of the Active Species Generated on Various Photocatalysts

Yoshio Nosaka and Atsuko Y. Nosaka

TiO_2 photocatalysts have been utilized for the oxidation of organic pollutants [1–5]. For further practical applications, the improvement in the photocatalytic efficiency and the extension of the effective wavelength of the irradiation light are desired. From this point of view, better understanding of the primary steps in photocatalytic reactions is prerequisite to develop prominent photocatalysts. The properties of TiO_2 and the reaction mechanisms in molecular level have been reviewed recently [6]. Therefore, this chapter describes briefly active species involved in the photocatalytic reactions for bare TiO_2 and TiO_2 modified for visible-light response, that is, trapped electrons, superoxide radical ($O_2^{\bullet -}$), hydroxyl radical ($OH^{\bullet}$), hydrogen peroxide (H_2O_2), and singlet oxygen ($^1O_2$).

1.1
Key Species in Photocatalytic Reactions

Since the photocatalytic reactions proceed usually with oxygen molecules (O_2) in air, the reduction of oxygen would be the important process in photocatalytic reduction. On the other hand, taking into account that the surface of TiO_2 photocatalysts is covered with adsorbed water molecules in usual environments and that photocatalysts are often used to decompose pollutants in water, oxidation of water would be the important process in photocatalytic oxidation. As shown in Figure 1.1, when O_2 is reduced by one electron (Eq. (1.1)), it becomes a superoxide radical ($O_2^{\bullet -}$) that is further reduced by one electron (Eq. (1.2)) or reacts with a hydroperoxyl radical ($HO_2^{\bullet}$, i.e., protonated $O_2^{\bullet -}$) to form hydrogen peroxide (H_2O_2). The latter reaction is largely pH dependent because the amount of $HO_2^{\bullet}$, whose p*K*a is 4.8, changes largely at pH around neutral [7]. One-electron reduction of H_2O_2 (Eq. (1.3)) produces hydroxyl radical ($OH^{\bullet}$). In the field of radiation chemistry, it is well documented that $OH^{\bullet}$ is produced by one-electron oxidation of H_2O with ionization radiation. However, the formation of $OH^{\bullet}$ in the photocatalytic oxidation process has not been confirmed,

Photocatalysis and Water Purification: From Fundamentals to Recent Applications, First Edition. P. Pichat.

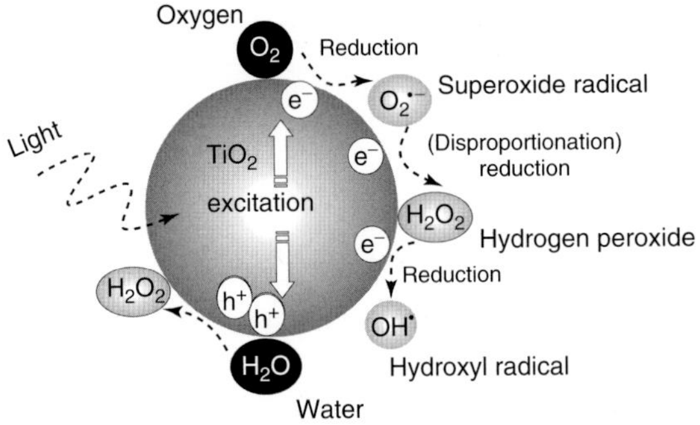

Figure 1.1 One-electron reduction steps of oxygen to OH radical and two-electron oxidation step of water to H_2O_2 observed in the TiO_2 photocatalyst.

as described later.

$$O_2 + e^- \rightarrow O_2^{\bullet -} \tag{1.1}$$

$$O_2^{\bullet -} + 2H^+ + e^- \rightarrow H_2O_2 \tag{1.2}$$

$$H_2O_2 + H^+ + e^- \rightarrow OH^\bullet + H_2O \tag{1.3}$$

Figure 1.2 shows the standard potentials [8] for the one-electron redox of active oxygen species as a function of pH of the solution. The conduction band (CB) bottom for anatase and rutile TiO_2 along with valence band (VB) top of TiO_2 is also depicted. The p*Ka* values for H_2O_2 and $OH^\bullet$ are 11.7 and 11.9, respectively [7]. Therefore, the linear lines showing pH dependence in Figure 1.2 change the inclination at the individual pH. It is notable that in the pH range between 10.6 and 12.3, one-electron reduction resulting in $OH^\bullet$ formation (Eq. (1.3)) occurs at a higher potential than that resulting in H_2O_2 formation (Eq. (1.2)). As commonly known, the potential of the VB of TiO_2 is low enough to oxidize H_2O, suggesting the possibility of the formation of $OH^\bullet$. However, the potentials in the figure are depicted based on the free energy change in a homogeneous aqueous solution. Therefore, it does not always mean that the one-electron oxidation of H_2O by VB holes at the surface of TiO_2 solid takes place in the heterogeneous system. Since the oxidation of H_2O to H_2O_2 and O_2 is also possible, only the potential difference between VB and $OH^\bullet$ should not be used easily for explaining the possibility of the formation of $OH^\bullet$. The competition between OH-radical-mediated reaction versus direct electron transfer has been studied as the effect of fluoride ions on the photocatalytic degradation of phenol in an aqueous TiO_2 suspension [9]. Under a helium atmosphere and in the presence of fluoride ions, phenol is significantly degraded, suggesting the occurrence of a photocatalytically induced hydrolysis [9].

Primary intermediates of water photocatalytic oxidation at the TiO_2 in aqueous solution were investigated by *in situ* multiple internal reflection infrared (MIRIR) absorption combined with the observation of photoluminescence from trapped holes [10]. The reaction is initiated by a nucleophilic attack of a H_2O molecule on a photogenerated hole at a surface two hold coordinated O site to form [TiO$^\bullet$HO–Ti].

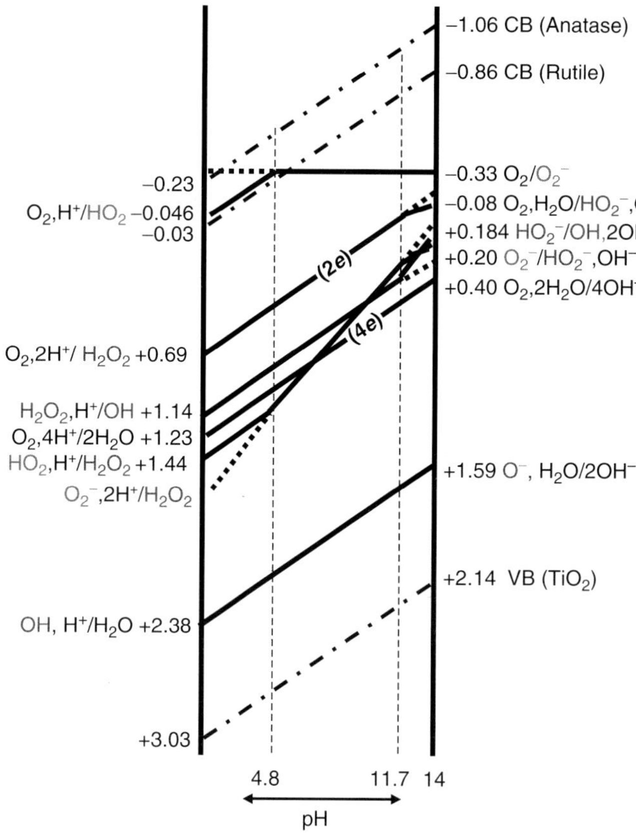

Figure 1.2 The standard potentials for the one-electron redox of active oxygen species along with the energy bands of TiO_2 as a function of pH of the solution. All redox couples are one-electron process except for those indicated with 2*e* and 4*e*.

A plausible reaction scheme is shown in Figure 1.3. Detailed investigations revealed the presence of TiOOH and TiOOTi as primary intermediates of the oxygen photoevolution reaction. This means that water is oxidized to form hydrogen peroxide adsorbed on TiO_2 surface, but the formation of OH radical in the oxidation process of water was denied.

Ultraviolet photoelectron spectroscopy (UPS) studies showed that the top of the O-2p levels for surface hydroxyl groups (Ti–OH) at the rutile TiO_2 (100) face is about 1.8 eV below the top of the VB at the surface [11]. This implies that surface hydroxyl groups cannot be oxidized by photogenerated holes in the VB. On the basis of the electronic structure of surface-bound water obtained from the data reported in the literature of X-ray photoelectron spectroscopy (XPS) study, it is evidenced that water species specifically adsorbed on terminal (surface) Ti atoms cannot be photooxidized under UV illumination [12]. The photogenerated VB free holes are favorably trapped at the terminal oxygen ions of the TiO_2 surface $(O^{2-})_s$

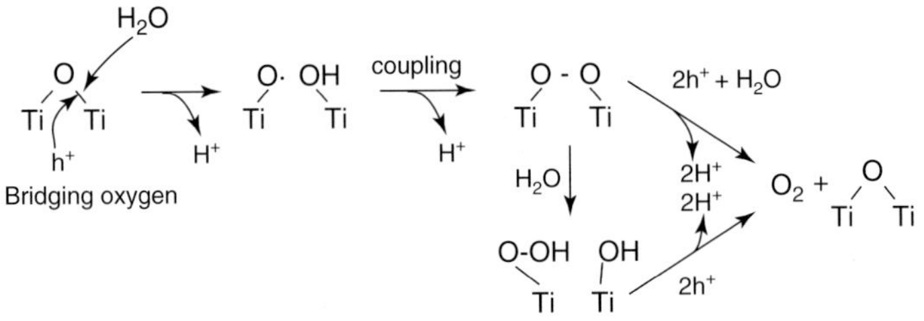

Figure 1.3 Reaction scheme for the oxygen photoevolution reaction on TiO_2 (rutile) in contact with an aqueous solution of pH 1–12. (Source: Reprinted with permission from Nakamura *et al.* [10]. © 2004 American Chemical Society.)

to generate terminal $(O^-)_s$ radicals, rather than being trapped at adsorbed water species to produce adsorbed $OH^\bullet$. As discussed later, when $OH^\bullet$ is detected in photocatalytic reactions, it should be formed by photocatalytic reduction of H_2O_2 (Eq. (1.3)).

1.2
Trapped Electron and Hole

Different from the semiconductor bulk, many electronic energy states may be formed within the band gap at the solid surface. These energy levels are capable of trapping VB holes and CB electrons. The trapped energy is considerably larger at the surface than in the bulk, indicating that it is energetically favorable for carriers to travel from the bulk to the surface [13]. At the surface, the trapping sites generally correspond to five-coordinated Ti^+ and two-coordinated O^- surface ions. When an appropriate acceptor (a scavenger), such as O_2 for electrons or methanol for holes, is adsorbed on the surface, it was suggested that the carriers should be preferentially transferred to the adsorbate rather than remain trapped at the surface sites [13].

When there are no molecules that can suffer the reaction, the existence of electrons and holes can be detected at a low temperature such as 77 K. To detect such paramagnetic species, electron spin resonance (ESR) spectroscopy is a valuable method [14, 15].

Holes and electrons could be observed by the absorption spectra just after the short pulse excitation under ambient temperature [16]. Trapped holes show that the absorption peaked at about 500 nm [17] and disappeared by the further reactions. On the other hand, trapped electrons show a broad absorption band that peaked at about 700 nm [18], which react mainly with oxygen molecules in air. Trapped electrons are so stable in the absence of O_2 that the kinetics can be explored by means of a stopped flow technique [19]. The reduction kinetics has been investigated through the electron acceptors such as O_2, H_2O_2, and NO_3^-, which are often present in photocatalytic systems. The experimental results clearly showed that the stored electrons reduce O_2 and H_2O_2 to water by multielectron transfer processes [19]. Moreover, NO_3^- is reduced via the transfer of eight electrons evidencing the formation of ammonium ions. On the other hand, in the reduction of toxic metal ions, such as Cu(II), two-electron transfer occurs, indicating the reduction of the copper metal ion into its nontoxic metallic form.

1.3 Superoxide Radical and Hydrogen Peroxide ($O_2^{\bullet-}$ and H_2O_2)

Since photocatalysts are usually used in air, photoexcited CB electrons transfer to the oxygen in air to form superoxide radical $O_2^{\bullet-}$. The highly sensitive MIRIR technique was applied and surface intermediates of the photocatalytic O_2 reduction were directly detected. Figure 1.4 shows the proposed mechanism of the reduction of molecular oxygen at the TiO_2 surface in aqueous solutions [20]. In neutral and acidic solutions, CB electrons reduce the surface Ti^{4+} that adsorbs H_2O, and then O_2 attacks it immediately to form superperoxo $TiOO^{\bullet}$ as shown in path A in Figure 1.4. This superperoxo is reduced to peroxo $Ti(O_2)$, which is equivalent to hydroperoxo TiOOH, when it is protonated (Figure 1.4). The hydroperoxo has the same structure with the hydrogen peroxide adsorbed on TiO_2 surface. On the other hand, in the alkaline solution, as shown in path B, the adsorbed O_2 receives a photogenerated CB electron to produce $O_2^{\bullet-}$. If it is not used for reactions or oxidized, the produced $O_2^{\bullet-}$ is converted to H_2O_2 by disproportionation with

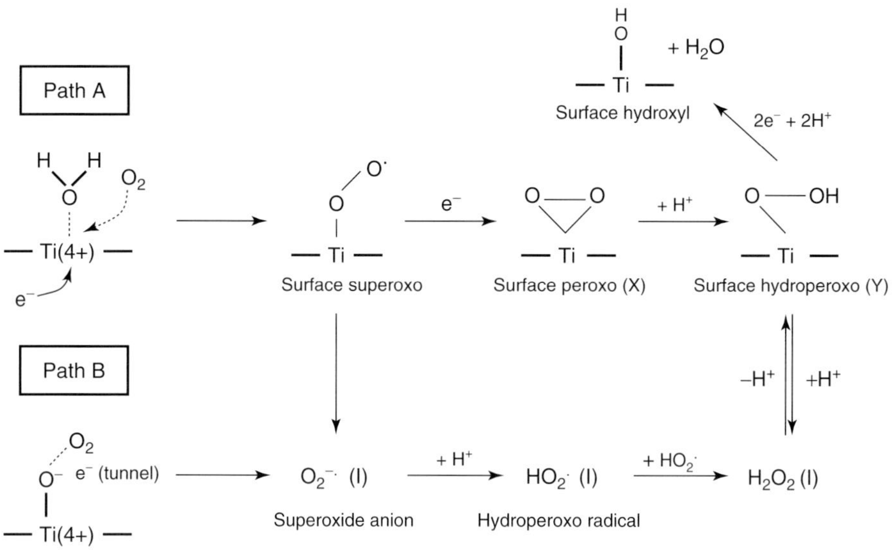

Figure 1.4 Reaction paths for the photocatalytic reduction of O_2 at the TiO_2 surface, suggested from IR measurements in neutral and acidic aqueous solutions (path A) and in an alkaline solution (path B). (Source: Reprinted with permission from Nakamura *et al.* [20]. © 2003 American Chemical Society.)

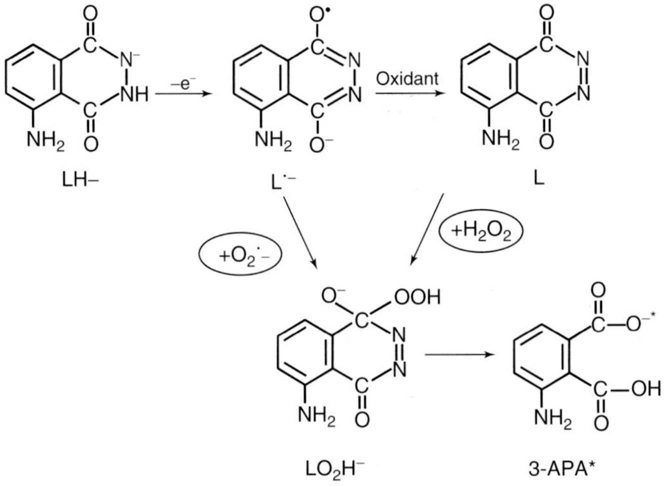

Figure 1.5 Chemiluminescence reactions for detecting $O_2^{\bullet-}$ and H_2O_2. The excited state of 3-aminophthalate (3-APA) is formed by two different reactions.

protons. Although the reaction rate for molecules having higher electron affinity is usually large, the reactivity of $O_2^{\bullet-}$ is generally weak. At pH lower than 4.8, it takes the form of $HO_2^{\bullet}$ by the protonation, whose lifetime is short owing to the rapid reaction with $O_2^{\bullet-}$ or $HO_2^{\bullet}$ to form stable H_2O_2 [7], as stated above.

Since the lifetime of $O_2^{\bullet-}$ is long in alkaline solution [21], it can be detected after stopping the irradiation. To detect $O_2^{\bullet-}$, a chemiluminescence method with luminol or luciferin analog (MCLA) has been used [22]. Figure 1.5 shows the reaction scheme for luminol chemiluminescence reactions. Luminol (LH^-) is easily oxidized in alkaline solution under air forming one-electron oxidized state ($L^{\bullet-}$), and reacts with $O_2^{\bullet-}$ to form unstable peroxide (LO_2H^-). This species releases N_2 to form the excited state of 3-aminophthalate (3-APA), which emits light at 430 nm. When L^- is oxidized further, a two-electron oxidation form of luminol (L), or a kind of diazo-naphthoquinones, is formed. It can react with H_2O_2 to form peroxides to proceed the same chemiluminescence reaction. Thus, using an oxidant, H_2O_2 could be separately detected by a luminol chemiluminescence method [23].

The decay profile of $O_2^{\bullet-}$ concentration does not obey first- or second-order kinetics, but obeys fractal-like kinetics, namely, the distribution of the distance between holes and adsorbed $O_2^{\bullet-}$ governs these decay kinetics [21]. For anatase thin film photocatalysts irradiated with very weak (1 μW cm^{-2}) UV light, the quantum yields of $O_2^{\bullet-}$ were reported to be 0.4 and 0.8 in air and water, respectively [24].

As suggested in Figure 1.2, $O_2^{\bullet-}$ may be produced by the photocatalytic oxidation of H_2O_2 (Eq. (1.4)).

$$HO_2^- + h^+ \rightarrow O_2^{\bullet-} + H^+ \tag{1.4}$$

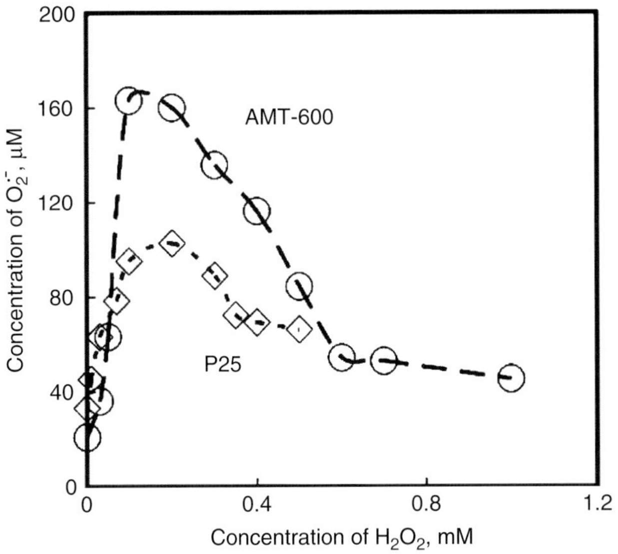

Figure 1.6 The effect of H_2O_2 on the concentration of $O_2^{\bullet-}$ measured after 100 s irradiation of the TiO_2 suspension of AMT-600 (TAYCA Corp.) and P25 (Nippon Aerosil Co., Ltd).

Figure 1.6 shows the amount of $O_2^{\bullet-}$ formed after 10 s in the presence of H_2O_2 of various concentrations. Increase in $O_2^{\bullet-}$ was observed with a small amount of H_2O_2, indicating the oxidation of H_2O_2 with photogenerated hole h^+ (Eq. (1.4)) or the increase in the reduction of O_2 owing to the suppression of photogenerated e^- from the recombination. When the amount of H_2O_2 was larger than 0.2 mmol l^{-1}, the formation of $O_2^{\bullet-}$ decreased, indicating that the adsorption of H_2O_2 on the whole surface blocks the access of O_2, which would increase the electron–hole recombination rate.

1.4 Hydroxyl Radical (OH•)

Although OH• has been usually recognized as the most important active species of the photocatalytic oxidation, recent reports confirmed that the contribution of OH• in the photocatalytic oxidation process is not usually dominant [6]. It should be emphasized that OH• has been referred too easily to be involved in the oxidation mechanism of photocatalytic reactions.

Several methods to detect OH• in photocatalytic reactions have been reported. Usually, the spin trapping reagents, such as DMPO (5,5-dimethyl-1-pyrroline-*N*-oxide), have been used to detect OH radicals (Figure 1.7a). However, it is not a molecule stable enough in aerated aqueous solutions and can be easily oxidized. In many reports, the possibility of the other reactions for DMPO than the OH radical adduction has not been anticipated. Based on the detailed study in [25],

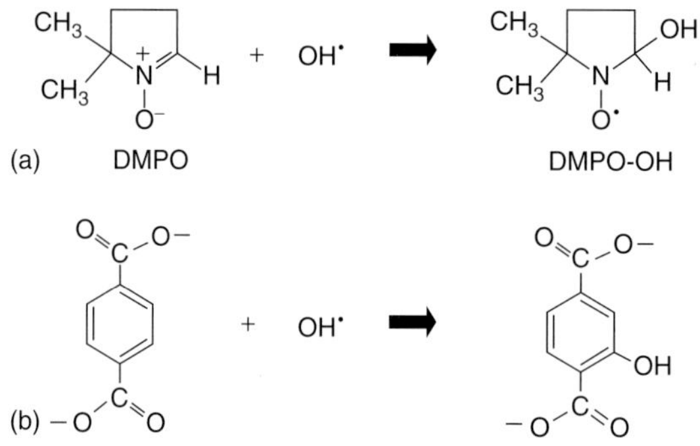

Figure 1.7 (a) The spin trapping reaction for OH• with DMPO. (b) The reaction of terephthalic acid with OH• forming fluorescent 2-hydroxy terephthalate.

it was indicated that the amount of radical adduct in the photocatalytic reaction was increased with DMPO concentration and that no saturation was observed, whereas OH• formed by photolysis of H_2O_2 could be trapped by excess amount of DMPO. This means that the OH radical adduct DMPO–OH• was formed by the photocatalytic reaction of DMPO itself and not through OH radicals. Thus, spin trapping experiments for detecting OH• must be carefully performed to prove the presence of OH• [25, 26].

A fluorescence probing method, based on the reaction of OH• with stable molecules seems more suitable than those with unstable spin trapping regents. In the field of radiation chemistry, the reactions of OH• with terephthalic acid (TA) and coumarin have been used because these products show strong fluorescence aiding in sensitive detection [27]. Therefore, this method has been adopted to detect OH• in photocatalytic reactions in aqueous suspension systems [28, 29]. The quantum yield of OH• in TiO_2 aqueous suspension was on the order of 10^{-5} [30]. Kinetic analysis for the formation rates of the OH• adduct (DMPO–OH•) along with the competitive adsorption of phosphate showed that, at a $pH = 4.25$, phthalic acid that was adsorbed on TiO_2 surface was oxidized directly by VB holes, with a quantum yield of 0.08 [31]. This high quantum yield could be attributed to the direct oxidation of adsorbed TA with VB holes.

Since radicals can be sensitively analyzed with ESR, nitroxide radical (3-carboxy-2,2,5,5-tetramethyl-1-pyrrolidine-1-oxy) has been used as a probe to detect OH radicals [32]. The quantum efficiencies of OH• for several TiO_2 photocatalysts were measured by the TA fluorescence method (Figure 1.7b) and compared with those obtained with the spin-trap and spin-probe ESR methods stated above [29]. The OH• yields measured by the TA fluorescence method were smaller by a factor of about 100, showing no correlation with those obtained by the DMPO spin trapping and the TA spin probing methods. Although the formation of OH• has been reported mainly using the spin trapping method, the contribution of the free OH• may be very small when the reactant is readily oxidized. Thus, the OH• should be distinguished from that generated by the trapped holes in photocatalytic reactions.

OH• was expected to be directly detected by means of ESR spectroscopy at low temperature. However, actually the OH• was not detected by ESR spectroscopy at 77 K, but only trapped holes were detected for hydrated TiO_2 particles [33]. Under hydrated conditions, when the frozen trapped holes were partly melted, they oxidized the adsorbed molecules [33]. Thus, the involvement of OH• in the oxidation process was not proved by direct detection with ESR.

Another definite method to confirm the presence of OH• is the observation of the optical absorption spectrum in gas phase. By scanning the excitation wavelength (282–284 nm) and monitoring the fluorescence at 310 nm, the spectrum could be identified as the absorption lines of OH•. This highly sensitive and selective technique is called as the laser-induced fluorescence (LIF) method. Using this method, the first direct observation of the presence of OH• in TiO_2 photocatalytic systems was reported [34]. The quantum yield of OH• calculated from the LIF intensity was about 5×10^{-5}. When the O_2 gas of low partial pressures was flowed, the formation of OH• was clearly enhanced. Since the addition of H_2O_2 on the TiO_2 surface increased the LIF intensity, H_2O_2 molecules were also considered to form by the reduction reactions of O_2. The addition of methanol (a scavenger of hole) decreased significantly the LIF signal intensity, suggesting the formation of H_2O_2 by the oxidation of surface OH groups by holes. This mechanism of OH• formation is illustrated in Figure 1.8 [34]. With a similar reaction system, the formation and diffusion of H_2O_2 have been reported using the LIF method [35]. Consequently, it was proved that OH radicals are mainly formed by the reduction of H_2O_2, which is formed by the two-electron reduction of O_2 and/or two-electron oxidation of H_2O.

Using a molecular fluorescence marker, the diffusion of OH• from TiO_2 surface during UV irradiation has been verified [36]. The detected amount of OH• decreased with decreasing the concentration of oxygen, that is, at $[O_2] = 0.2$ vol%, no significant amount of OH• was detected. This result indicates that the OH• formation is very sensitive to the oxygen concentration, and the reduction process of oxygen, which results in the formation of $O_2^{\bullet-}$ leading to H_2O_2, is a key process in the formation of OH•.

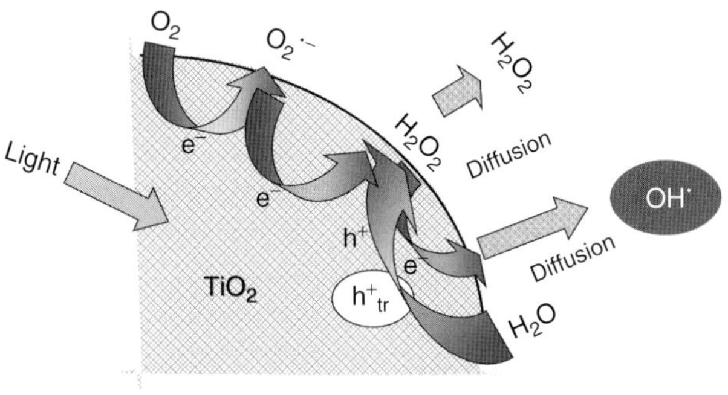

Figure 1.8 A plausible reaction scheme for the OH radical formation on the irradiated TiO_2 surface.

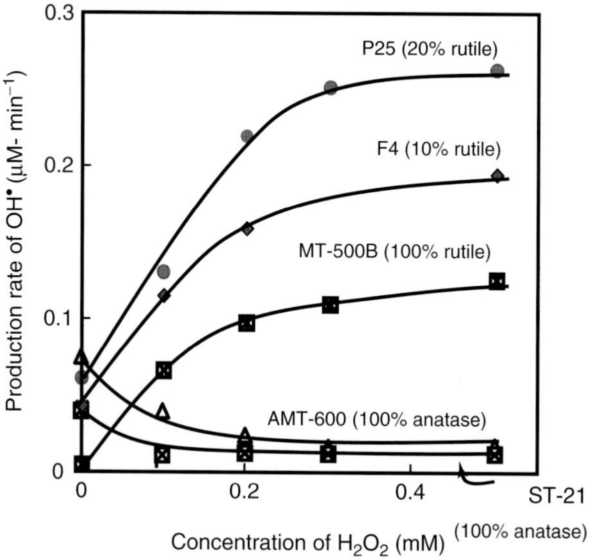

Figure 1.9 The formation rates of $OH^{\bullet}$ measured by a fluorescence probe method plotted for several TiO_2 photocatalysts as a function of the concentration of H_2O_2. (Source: Reprinted with permission from Hirakawa *et al.* [37]. © 2007 Elsevier.)

The effect of H_2O_2 addition on the rate of $OH^{\bullet}$ formation in aqueous suspension systems was measured for various TiO_2 [37]. As shown in Figure 1.9, the $OH^{\bullet}$ formation rates were increased with the addition of H_2O_2 for P25 (Nippon Aerosil Co, Ltd) and F4 (Showa Titanium Co., Ltd) TiO_2, which are rutile-containing anatase, and for rutile TiO_2 (MT-500B, TAYCA Corp.). The quite opposite tendency was observed for AMT-600 (TAYCA Corp.) and ST-21 (Ishihara Sangyo Co., Ltd), which consist of 100% anatase TiO_2, where the $OH^{\bullet}$ formation rate decreased on H_2O_2 addition. The increase of $OH^{\bullet}$ is attributable to the photocatalytic reduction of H_2O_2 (Eq. (1.3)). Since the rutile-containing anatase increased the $OH^{\bullet}$ generation, the structure of H_2O_2 adsorbed on the rutile TiO_2 surface is likely preferable to produce $OH^{\bullet}$.

1.5
Singlet Molecular Oxygen ($^1O_2$)

To explain the formation of singlet oxygen, the disproportionation of $O_2^{\bullet-}$ was proposed through the intermediate formation of $HO_2^{\bullet}$ as shown by Eq. (1.5) [38]. Since the energy difference of $HO_2^{\bullet} \rightarrow O_2$ from $HO_2^{\bullet} \rightarrow H_2O_2$ at a $pH = 0$ is calculated to be $+1.49$ V from Figure 1.2, O_2 may be excited to $^1O_2$. But, it becomes 0.53 V at $pH = 14$, which is smaller than the excitation energy of 0.98 eV (or 1270 nm in wavelength).

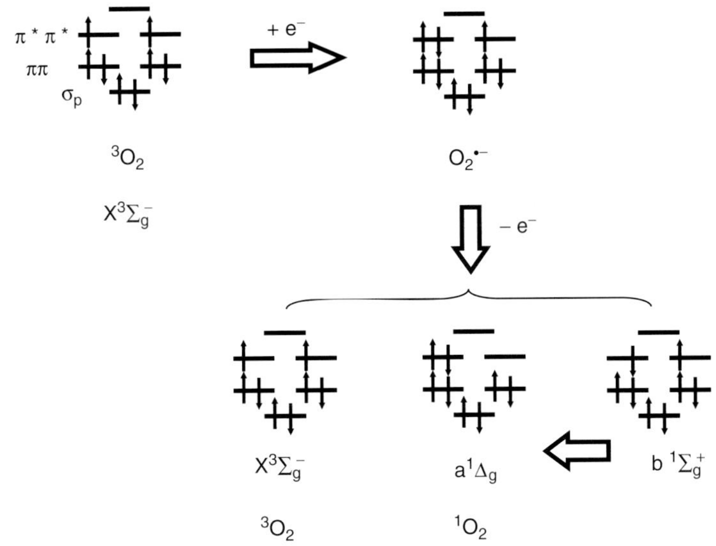

Figure 1.10 The spin states in the process of singlet molecular oxygen formation via the oxidation of O_2^{-}.

Alternatively, the oxidation of $O_2^{\bullet-}$ as indicated by Eq. (1.6) has been proposed as the formation mechanism [39]. Since $O_2^{\bullet-}$ is formed by the electron transfer of photoexcited CB electrons at the surface, it may be easily oxidized.

$$2HO_2^{\bullet} \rightarrow (HOOOOH) \rightarrow {}^1O_2 + H_2O_2 \quad (1.5)$$

$$O_2^{\bullet-} + h^+ \rightarrow {}^1O_2 \quad (1.6)$$

Figure 1.10 shows the plausible pathways for the consecutive reduction and oxidation of O_2. Since three electrons in the π^* state of $O_2^{\bullet-}$ cannot be distinguished from one another, three electronic states may be produced depending on the removed electron. These are ${}^3\Sigma_g^-$, ${}^1\Delta_g$, and ${}^1\Sigma_g^+$ states in the order from the lower energy. The last two states are electronic excited states of molecular oxygen and named as *singlet oxygen*. The lifetime of the ${}^1\Sigma_g^+$ state is very short and immediately transfers to the ${}^1\Delta_g$ state of singlet molecular oxygen (${}^1O_2$). The lifetime of the ${}^1\Delta_g$ state depends largely on its environment, ranging from a few microseconds in H_2O to a few milliseconds in air.

Among the detection methods to verify the formation of ${}^1O_2$, one of the most established methods is to observe the phosphorescence at 1270 nm, which is the radiative transition from the $a^1\Delta_g$ state to the $X^3\Sigma_g^-$ state of molecular oxygen. The phosphorescence at 1270 nm has been detected in a TiO_2 aqueous suspension system [39]. Quantum yields for ${}^1O_2$ generation measured for 10 commercial TiO_2 photocatalysts in air ranged from 0.12 to 0.38, while the lifetimes ranged from 2.0 to 2.5 μs [40]. The production and decay of ${}^1O_2$ in TiO_2 photocatalysis were investigated by monitoring the phosphorescence under various reaction

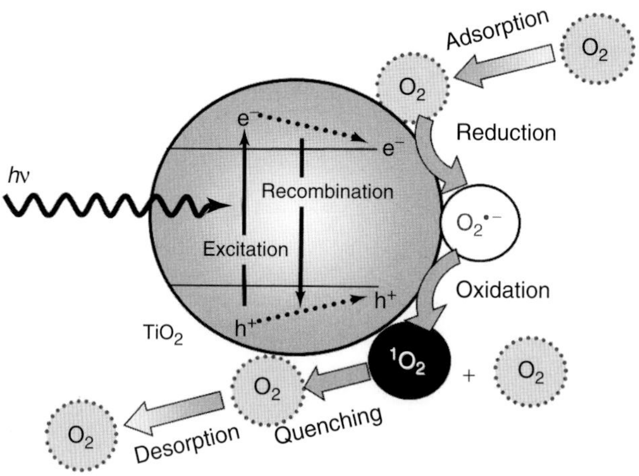

Figure 1.11 Photocatalytic processes of molecular oxygen on the TiO_2 surface. (Source: Reprinted with permission from Daimon *et al.* [40]. © 2007 American Chemical Society.)

conditions. The comparison among the effects of additives such as KBr, KSCN, KI, and H_2O_2 on the formation of $^1O_2$ and $O_2^{\bullet-}$ suggested that $^1O_2$ should be formed by the electron transfer mechanism (Eq. (1.6)), as illustrated in Figure 1.11. The formation of $^1O_2$ decreased at pH < 5 and pH > 11, indicating that the intermediate $O_2^{\bullet-}$ is stabilized at the terminal OH site of the TiO_2 surface in this pH range. Eighteen commercially available TiO_2 photocatalysts were compared on the formation of $^1O_2$ and $O_2^{\bullet-}$ in an aqueous suspension system. The formation of $^1O_2$ was increased with decreasing the size of TiO_2 particles, indicating that a large specific surface area causes a higher possibility of reduction producing $O_2^{\bullet-}$ and therefore, a large amount of $^1O_2$ is formed. The difference in the crystal phase (rutile and anatase) does not seem to affect the formation of $^1O_2$ [41].

Singlet oxygen is known to be reactive with some organic compounds such as olefines and amines. Therefore, in the presence of four kinds of organic molecules, methionine, pyrrole, collagen, and folic acid (pteroyl-L-glutamic acid), the decay of $^1O_2$ was measured [42]. Figure 1.12a represents the total decay of $^1O_2$, and Figure 1.12b shows the partial decay obtained after subtraction of the intrinsic exponential decay. The observed decay rates of $^1O_2$ with these organic molecules are significantly higher than those expected from the bimolecular rate constant reported for the reaction in homogeneous solution. By assuming pseudo-first-order reaction, the virtual concentrations of the reactant are in the vicinity of 0.01 mol l^{-1}. Since the concentration of the solution used in the experiments was 0.01 mmol l^{-1}, the organic reactants must be concentrated at the surface of TiO_2 by adsorption. These observations suggest that the reactant molecules should be adsorbed on the TiO_2 surface [42]. Although the 40% of $^1O_2$ was deactivated with folic acid, this deactivation process includes thermal deactivation besides the chemical reactions.

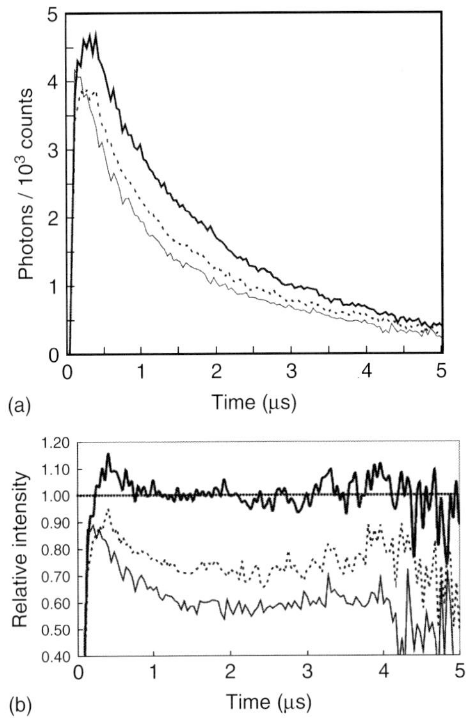

Figure 1.12 The time dependence of phosphorescence intensity for singlet molecular oxygen monitored at 1250 nm after the pulsed excitation on P25 TiO_2 aqueous suspension. Without additives (heavy line), with methionine (dashed line), and folic acid (fine line). (a) Original observation of emission intensity as the time profile. (b) The partial decay obtained after subtraction of the intrinsic exponential decay showing the fast decay by the reaction. (Source: Reprinted from Daimon et al. [42]. copyright 2008 Electrochemical Society of Japan.)

1.6 Reaction Mechanisms for Bare TiO_2

There are many reaction pathways in any photocatalytic reaction system. Whenever a certain pathway in question is discussed, the other pathways should also be considered simultaneously. To detect $^1O_2$ in the reaction system, sterically hindered cyclic amines, such as HTMP (4-hydroxy-2,2,6,6-tetramethylpiperidine), have been used as probe molecules [43]. Such amines are converted to the corresponding stable aminoxyl radical (nitroxide radical) which can be sensitively detected by ESR spectroscopy. In the case of HTMP, TEMPOL radical (4-hydroxy-2,2,6,6-tetramethyl piperidine 1-oxyl) is formed as a result of a photocatalytic reaction in a TiO_2 aqueous suspension. The time profiles of the radical formation and the effect of additives, such as SCN^-, I^-, methanol, and H_2O_2, on the initial formation rates were measured in order to elucidate the photocatalytic reaction mechanism for HTMP [44]. By assuming possible key reactants for the oxidation as shown in Figure 1.13,

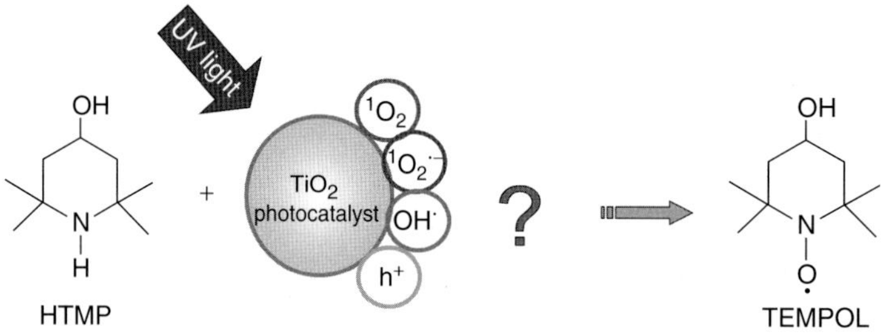

Figure 1.13 Key reactants considered for the kinetic analysis of the photocatalytic formation of nitroxide radicals.

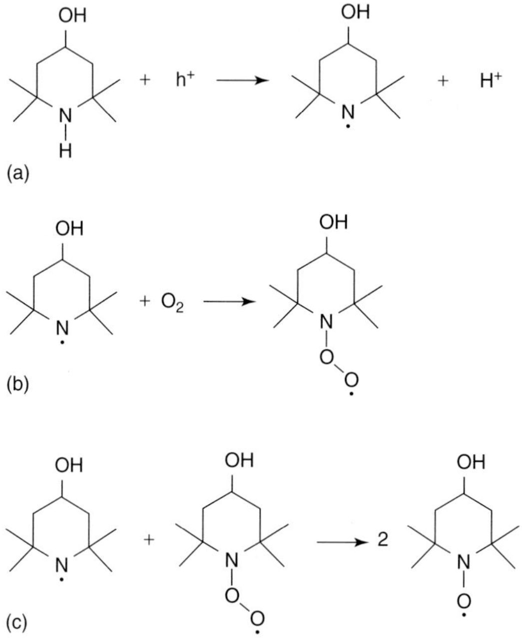

Figure 1.14 (a–c) Plausible photocatalytic reaction processes of sterically hindered cyclic amine.

the kinetics was analyzed to elucidate the reaction process. The experimental observations indicated that the direct photocatalytic oxidation of HTMP followed by reaction with O_2 is the dominant process in the formation of TEMPOL radicals (Figure 1.14). The possibility of the other processes, involving reactions with $^1O_2$, $O_2^{\bullet -}$, and $OH^\bullet$, was excluded from the reaction mechanism.

As stated above, $OH^\bullet$ is not produced through a main oxidation process even in the absence of organic compounds. However, in most of the research papers on photocatalysis published so far, $OH^\bullet$ has often been regarded to be involved in the actual oxidation mechanism of photocatalytic reactions. However, actually the primary reaction pathway for the oxidation is the direct reaction at the surface

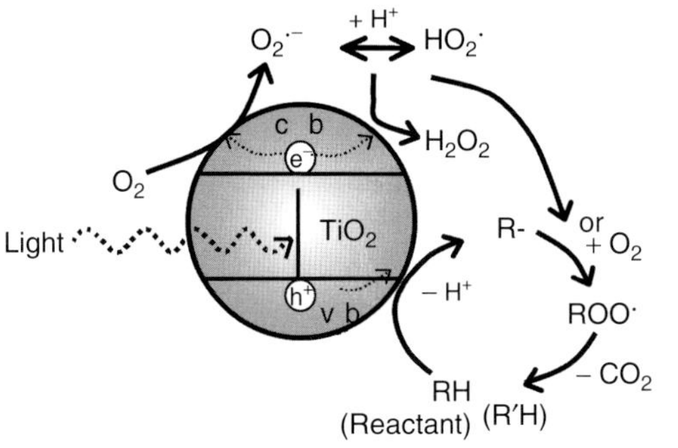

Figure 1.15 General reaction processes for the photocatalytic oxidation of organic molecules.

of TiO_2 with VB holes or trapped holes. Since it is generally known that the photocatalytic oxidation of organic compounds is accelerated by oxygen [45], the produced radical may react with the reduction products of O_2, namely, $O_2^{\bullet-}$ and H_2O_2. But the O_2 in air may directly react with the radical produced by the photocatalytic oxidation, because auto-oxidation, a kind of chain reaction with O_2 starting from organic free radicals, is well known [46]. The consumption of O_2 at the oxidation site of the photocatalyst has been suggested from the experiment of electrochemical probe reactions at the surface of illuminated TiO_2 photoelectrode [47]; the generalized reaction mechanism of the photocatalytic oxidation of organic molecules (RH) is illustrated in Figure 1.15. RH will degrade by losing one carbon atom by releasing CO_2, but the intermediates may be aldehyde R′CHO or carboxylate R′COO$^-$.

1.7
Reaction Mechanisms of Visible-Light-Responsive Photocatalysts

As promising practical applications of photocatalysts, the utilization of visible light has been promoted. Figure 1.16 shows the energy levels of some visible-light-responsive photocatalysts. Since the energy level of VB for metal oxides is governed by that of the O-2p orbital, a narrow-band-gap metal oxide semiconductor, such as WO_3, possesses CB energy lower than that of TiO_2. Since the one-electron reduction potential of O_2 is very close to that of the CB of TiO_2, as shown in Figure 1.2, WO_3 is unable to form $O_2^{\bullet-}$, as shown in Figure 1.16a. In this case, using a promoter such as deposited Pt [48], electrons could be stored to enable two-electron reduction of O_2 to H_2O_2. Doping to produce the mid-gap level has been proposed as the other visible-light-responsive photocatalysis. Since the energy level of VB has sufficient oxidation ability, shifting the VB by doping the N or S anion has been attempted (Figure 1.16b). In this case, photogenerated holes produced on the donor level are expected to have oxidation ability similar to that of bare TiO_2 [23]. As shown in Figure 1.16c, photocatalysts of the photosensitization type were

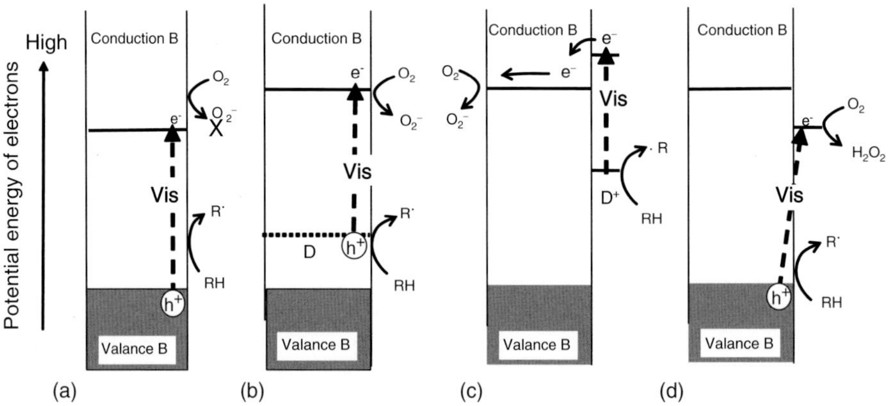

Figure 1.16 Classification of visible-light-responsive photocatalysts by the mechanism of reaction. (a) Narrow-band-gap semiconductor such as WO_3, (b) anion-doped TiO_2 such as nitrogen-doped TiO_2, (c) sensitizer-deposited TiO_2 such as $PtCl_6^{2-}$-deposited TiO_2, and (d) interfacial-charge-transfer-type TiO_2 such as Cu(II)-grafted TiO_2.

also proposed. The deposited compound absorbs the visible light and transfers the excited electron to produce a cation radical, which can oxidize organic pollutant molecules. In this case, enough oxidation power with good stability is required as an oxidized sensitizer [49]. Recently, interfacial charge transfer (IFCT)-type absorption originating from the excitation of VB electrons to deposited (or grafted) metal ions has been proposed (Figure 1.16d). In this case, if the deposited compound has a catalytic ability of O_2 reduction, the efficiency is expected to be increased [50].

The observation of active species such as $O_2^{\bullet -}$and H_2O_2 is inevitable to confirm the reaction mechanism proposed. Figure 1.17 shows the formation of $O_2^{\bullet -}$ as a function of irradiation time of 442 nm light for several visible-light-responsive photocatalysts. For $PtCl_6^{2-}$-modified TiO_2, a large amount of $O_2^{\bullet -}$ was observed immediately after the excitation, in concord with the electron transfer to the CB (Figure 1.16c). On the other hand, Fe-complex-deposited TiO_2 generated a small amount of $O_2^{\bullet -}$ probably because the excitation takes place from VB to Fe ions at the surface (Figure 1.16d). For the S-doped TiO_2, the amount of $O_2^{\bullet -}$ increased as the irradiation period increased, and reached a steady value in 30 s, while for the N-doped TiO_2, it gradually increased up to about 180 s. In a control experiment, P25 TiO_2 did not produce $O_2^{\bullet -}$ by visible-light irradiation at 442 nm, whereas on UV irradiation at 325 nm (with the similar number of photons to 442 nm) for 180 s, 20 nmol l^{-1} of $O_2^{\bullet -}$ was produced, indicating that the steady-state concentrations of $O_2^{\bullet -}$ for the N- and S-doped TiO_2 are higher than those for the undoped TiO_2 (P25) [23].

Figure 1.18 schematically shows the dominant reaction processes in the absence of organic substrates, which are deduced from the observation of $O_2^{\bullet -}$and H_2O_2 [23]. The S-doped TiO_2 surpassed the N-doped TiO_2 in the production of $O_2^{\bullet -}$, while the N-doped TiO_2 surpassed the S-doped TiO_2 in the production of H_2O_2. Since $O_2^{\bullet -}$ decays obeying the second-order kinetics, H_2O_2 is mainly produced from the disproportionation of $O_2^{\bullet -}$. The production of $O_2^{\bullet -}$ decreased by adding

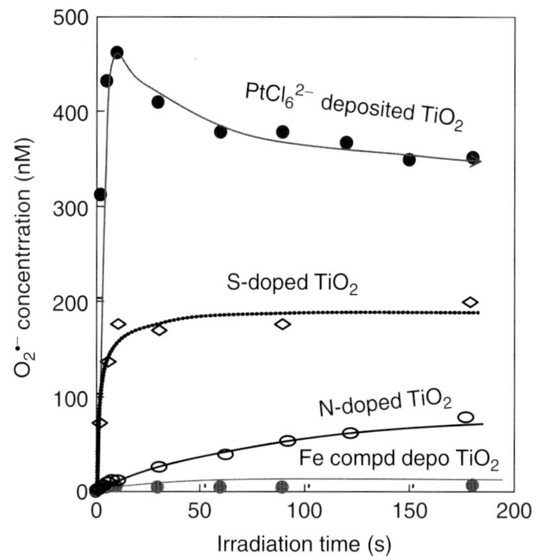

Figure 1.17 The concentration of $O_2^{\bullet -}$ produced in the suspension (15 mg per 3.5 ml) of the modified TiO_2 photocatalyst powders as a function of the irradiation time of 442 nm light.

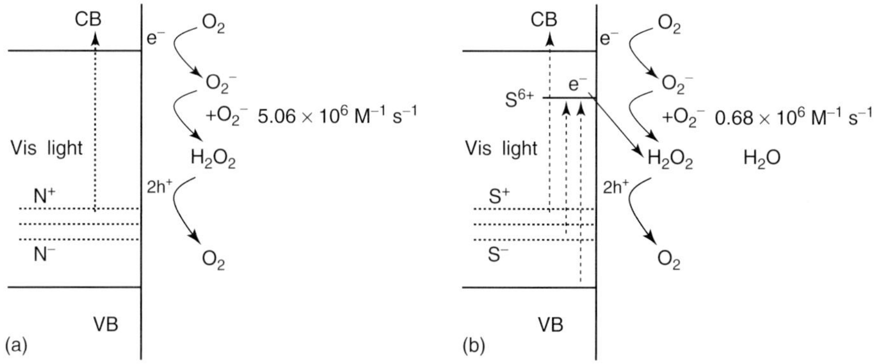

Figure 1.18 Proposed photocatalytic reaction processes of $O_2^{\bullet -}$ and H_2O_2 on the N- and S-doped TiO_2 in the absence of organic substrates. (a) The N-doped TiO_2 selectively produces H_2O_2, while (b) the S-doped TiO_2 produces $O_2^{\bullet -}$ and reduces H_2O_2 to water. (Source: Reprinted with permission from Hirakawa *et al.* [18]. © 2008 American Chemical Society.)

H_2O_2 to both N- and S-doped TiO_2. Therefore, H_2O_2 would not be oxidized in both N- and S-doped TiO_2, which is in prominent contrast to the undoped TiO_2 (P25). The H_2O_2 produced by the S-doped TiO_2 might be quickly reduced to H_2O via some intermediate states; the reactive oxygen species produced by the reduction of H_2O_2 may play an important role in the decomposition of organic molecules, and the S-doped TiO_2 may surpass N-doped TiO_2 in this ability.

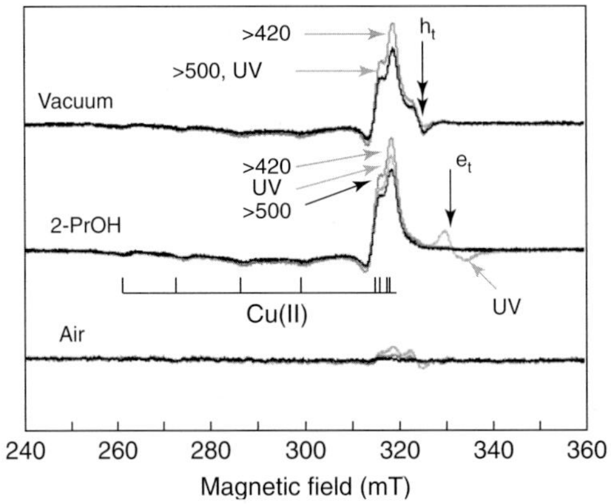

Figure 1.19 ESR difference spectra for Cu(II)/TiO_2 measured under vacuum, with 2-propanol, and air showing the effects of light irradiation at the wavelengths around 360 nm, longer than 420 and 500 nm, respectively. (Source: Reprinted with permission from Nosaka *et al.* [51]. © 2011 American Chemical Society.)

For visible-light-responsive photocatalysts of IFCT type with metal ions (Figure 1.16d), ESR spectroscopy could be utilized to analyze the state of the metal ions. Figure 1.19 shows the difference in the ESR spectra caused by light irradiation for Cu(II)-deposited TiO_2 [51]. The decrease of the large signal, characterized by a hyperfine splitting by Cu nuclear spin, indicates the decrease of Cu^{2+} ions on visible-light irradiation under vacuum. In the presence of air, the signal of Cu^{2+} did not change with the irradiation, indicating the reduced Cu^{+} reacts with O_2 to reproduce Cu^{2+}. This fact and the observation of $O_2^{\bullet-}$ formation clearly supported the IFCT mechanism for the Cu(II)-deposited TiO_2 photocatalysts. For WO_3 photocatalyst, which is classified to Figure 1.16a, the formation of $O_2^{\bullet-}$ was not observed on 442 nm excitation [51]. When it was grafted with Cu(II), the reduction of Cu^{2+} in ESR spectrum and the formation of H_2O_2 were observed. The formation of H_2O_2 indicated the function of Cu(II) as a promoter for two-electron reduction of O_2. The reaction mechanisms of Cu(II)/TiO_2 and Cu(II)/WO_3 photocatalysts are illustrated in Figure 1.20 [51]. Thus, the reaction pathways for different types of visible-light-responsive photocatalysts could be confirmed by the detection of primarily produced active species.

1.8 Conclusion

In order to explore the reaction mechanism of bare TiO_2 and TiO_2 photocatalysts modified for visible-light response, the detection and the behaviors of key species, such as trapped electrons, superoxide radical ($O_2^{\bullet-}$), hydroxyl radical ($OH^{\bullet}$), hydrogen peroxide (H_2O_2), and singlet oxygen ($^1O_2$), were discussed. Trapped electrons, which have been analyzed at 77 K with ESR spectroscopy, are so stable in the

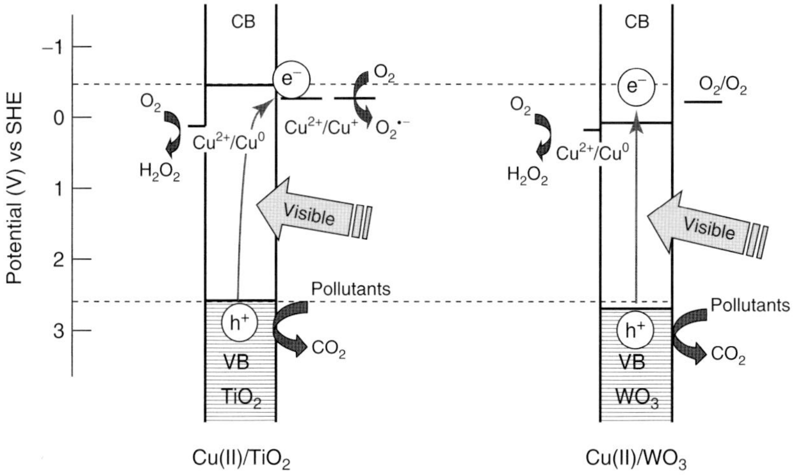

Figure 1.20 Energy diagrams of Cu(II)-grafted TiO_2 (rutile) and WO_3 photocatalysts at pH 7 showing the photocatalytic reaction mechanisms under visible-light irradiation. (Source: Reprinted with permission from Nosaka *et al.* [51]. © 2011 American Chemical Society.)

absence of O_2 that the kinetics can be investigated by means of a stopped flow technique. O_2 in air receives a photogenerated CB electron or trapped electron to produce $O_2^{\bullet-}$, which is converted to more stable H_2O_2. Since the rutile-containing anatase increased the $OH^{\bullet}$ generation, the structure of H_2O_2 adsorbed on the rutile TiO_2 surface might be preferable to produce $OH^{\bullet}$. Only partial decay of $^1O_2$ with folic acid, which is a well-known reactant of $^1O_2$, was observed, suggesting that the role of $^1O_2$ in photocatalysis is not major. A general reaction pathway for bare TiO_2, in which organic compounds are oxidized directly to form organic radicals followed by the auto-oxidation with O_2 and release of CO_2, has been proposed. Furthermore, for some types of visible-light-responsive photocatalysts, the reaction mechanisms were compared by the detection of the primarily produced key species and the reaction pathways could be proposed.

References

1. Fujishima, A., Zhang, X., and Tryk, D. A. (2008) TiO_2 Photocatalysis and related surface phenomena, *Surf. Sci. Rep.* **63**, 515–582.
2. Kaneko, M., Ohkura, I., (eds) (2002) *Photocatalysis*, Kodansha-Springer, Ltd, Tokyo.
3. Zhang, H., Chen, G., and Bahnemann, D.W. (2009) Photoelectrocatalytic materials for environmental applications, *J. Mater. Chem.* **19**, 5089–5121.
4. Y. Paz, (2010) Composite titanium dioxide photocatalysts and the "adsorb & shuttle" approach: a review, *Solid State Phenom.*, **162**, 135–162.
5. P. Pichat (2010), Some views about indoor air photocatalytic treatment using TiO_2: conceptualization of humidity effects, active oxygen species, problem of C_1–C_3 carbonyl pollutants, *Appl. Catal., B* **99** 428–434.
6. Henderson, M. A. (2011) A surface science perspective on TiO_2

photocatalysis, *Surf. Sci. Rep.* **66**, 185–297.

7. Bielski, B. H. J., Cabelli, D. E., and Arudi, R. L. (1985) Reactivity of HO_2/O_2^- radicals in aqueous solution, *J. Phys. Chem. Ref. Data*, **14**(4), 1041–1100.
8. Bard, A. J., Parsons, R., and Jordan, J., (eds) (1985) *Standard Potentials in Aqueous Solution*, Marcel Dekker, New York.
9. Minero, C., Mariella, G., Maurino, V., and Pelizzetti, E. (2000), Photocatalytic transformation of organic compounds in the presence of inorganic anions. 1. Hydroxyl-mediated and direct electron-transfer reactions of phenol on a titanium dioxide-fluoride system, *Langmuir*, **16**, 2632–2641.
10. Nakamura, R., and Nakato, Y. (2004) Primary intermediates of oxygen photoevolution reaction on TiO_2 (rutile) particles, revealed by in situ FTIR absorption and photoluminescence measurements, *J. Am. Chem. Soc.* **126**(4), 1290–1298.
11. Muryn, C. A., Hardman, P. J., Crouch, J. J., Raiker, G. N., Thornton, G. D., and Law, S. L. (1991) Step and point defect effects on TiO_2(100) reactivity, *Surf. Sci.*, **251–252**, 747–752.
12. Salvador, P. (2007) On the nature of photogenerated radical species active in the oxidative degradation of dissolved pollutants with TiO_2 aqueous suspensions: a revision in the light of the electronic structure of adsorbed water, *J. Phys. Chem. C*, **111**(45), 17038–17043.
13. Valentin, C. D. and Selloni, A. (2011) Bulk and surface polarons in photoexcited anatase TiO_2, *J. Phys. Chem. Lett.*, **2**, 2223–2228.
14. Formenti, M., Juillet, F., Meriaudeau, P., and Teichner, S. J. (1971) Heterogeneous photocatalysis for partial oxidation of paraffins. *Chem. Technol.*, **1**, 680–686.
15. Nakaoka, Y., and Nosaka, Y. (1997) ESR investigation into the effects of heat treatment and crystal structure on radicals produced over irradiated TiO_2 powder. *J. Photochem. Photobiol., A*, **110**, 299–307.
16. Murakami, Y., Nishino, J., Mesaki, T., and Nosaka, Y. (2011) Femtosecond diffuse reflectance spectroscopy of various commercially available TiO_2 powders, *Spectrosc. Lett.*, **44**(2) 88–94.
17. Yang, X., and Tamai, N. (2001) How fast is interfacial hole transfer? *In situ* monitoring of carrier dynamics in anatase TiO_2 nanoparticles by femtosecond laser spectroscopy, *Phys. Chem. Chem. Phys.* **3**, 3393–3398.
18. Kamat, P. V., Bedja, I., and Hotchandani, S. (1994) Photoinduced charge transfer between carbon and semiconductor clusters. one-electron reduction of C_{60} in colloidal TiO_2 semiconductor suspensions. *J. Phys. Chem.* **98** (37), 9137–9142.
19. Mohamed, H., Mendive, C. B., Dillert, R., and Bahnemann, D. W. (2011) Kinetic and mechanistic investigations of multielectron transfer reactions induced by stored electrons in TiO_2 nanoparticles: a stopped flow study, *J. Phys. Chem. A*, **115**, 2139–2147.
20. Nakamura, R., Imanishi, A., Murakoshi, K., and Nakato, Y. (2003) In situ FTIR studies of primary intermediates of photocatalytic reactions on nanocrystalline TiO_2 films in contact with aqueous solutions, *J. Am. Chem. Soc.* **125**(24), 7443–7450.
21. Hirakawa, T., and Nosaka, Y. (2002) Properties of $\cdot O_2^-$ and $\cdot OH$ formed in TiO_2 aqueous suspensions by photocatalytic reaction and the influence of H_2O_2 and some ions, *Langmuir*, **18**(8), 3247–3254.
22. Nosaka, Y., Yamashita, Y., and Fukuyama, H. (1997) Application of chemiluminescent probe to monitoring superoxide radicals and hydrogen peroxide in TiO_2 photocatalysis, *J. Phys. Chem. B*, **101**(30), 5822–5827.
23. Hirakawa, T., and Nosaka, Y. (2008) Selective production of superoxide ions and hydrogen peroxide over nitrogen- and sulfur-doped TiO_2 photocatalysts with visible light in aqueous suspension systems, *J. Phys. Chem. C*, **112**(40), 15818–15823.
24. Ishibashi, K., Fujishima, A., Watanabe, T., and Hashimoto, K. (2000) Generation and deactivation processes of superoxide formed on TiO_2 film illuminated by very weak UV light in air

or water, *J. Phys. Chem. B*, **104**(20), 4934–4938.

25. Grela, M. A., Coronel, M. E. J., and Colussi, A. J. (1996) Quantitative spin-trapping studies of weakly illuminated titanium dioxide sols. Implications for the mechanism of photocatalysis, *J. Phys. Chem.* **100**(42), 16940–16946.
26. Brezova, V., Stasko, A., Biskupic, S., Blazkova, A., and Havlinova, B. (1993) Kinetics of hydroxyl radical spin trapping in photoactivated homogeneous (H_2O_2) and heterogeneous (TiO_2, O_2) aqueous systems. *J. Phys. Chem.*, **98**, 8977–8984.
27. Matthews, R. W. (1980) The radiation chemistry of the terephthalic dosimeter, *Radiat. Res.*, **83**(1), 27–41.
28. Ishibashi, K., Fujishima, A., Watanabe, T., and Hashimoto, K. (2000) Detection of active oxidative species in TiO_2 photocatalysis using the fluorescence technique. *Electrochem. Commun.* **2**, 207–210.
29. Nosaka, Y., Komori, S., Yawata, K., Hirakawa, T., and Nosaka, A. Y. (2003) Photocatalytic OH radical formation in TiO_2 aqueous suspension studied by several detection methods, *Phys. Chem. Chem. Phys.* **5**, 4731–4735.
30. Ishibashi, K., Fujishima, A., Watanabe, T., and Hashimoto, K. (2000) Quantum yields of active oxidative species formed on TiO_2 photocatalyst, *J. Photochem. Photobiol., A*, **134**, 139–142.
31. Taborda, A. V., Brusa, M. A., and Grela, M. A. (2001) Photocatalytic degradation of phthalic acid on TiO_2 nanoparticles, *Appl. Catal., A*, **208**, 419–426.
32. Schwarz, P. F., Turro, N. J., Bossmann, S. H., Braun, A. M., Wahab, A.-M. A. A., and Durr, H. (1997) A new method to determine the generation of hydroxyl radicals in illuminated TiO_2 suspensions, *J. Phys. Chem. B*, **101**(36), 7127–7134.
33. Micic, O. I., Zhang, Y., Cromack, K. R., Trifunac, A. D., and Thurnauer M. C. (1993) Trapped holes on TiO_2 colloids studied by electron paramagnetic resonance, *J. Phys. Chem.*, **97**(28), 7277–7283.
34. Murakami, Y., Endo, K., Ohta, I., Nosaka, A. Y., and Nosaka, Y. (2007) Can OH radicals diffuse from the UV-irradiated photocatalytic TiO_2 surfaces? laser-induced-fluorescence study, *J. Phys. Chem. C*, **111**(30), 11339–11346.
35. Thiebaud, J., Thevenet, F., and Fittschen, C. (2010) OH radicals and H_2O_2 molecules in the gas phase near to TiO_2 surfaces, *J. Phys. Chem. C*, **114**, 3082–3088.
36. Naito, K., Tachikawa, T., Fujitsuka, M., and Majima, T. (2008) Real-time single-molecule imaging of the spatial and temporal distribution of reactive oxygen species with fluorescent probes: applications to TiO_2 photocatalysts, *J. Phys. Chem. C*, **112**(4), 1048–1059.
37. Hirakawa, T., Yawata, K., and Nosaka, Y. (2007) Photocatalytic reactivity for $\cdot O_2^-$ and •OH radical formation in anatase and rutile TiO_2 suspension as the effect of H_2O_2 addition, *Appl. Catal., A*, **325**(1), 105–111.
38. Konovalova, T.A., Lawrence, J., and Kispert, L. D. (2004) Generation of superoxide anion and most likely singlet oxygen in irradiated TiO_2 nanoparticles modified by carotenoids, *J. Photochem. Photobiol., A*, **162**, 1–8.
39. Nosaka, Y., Daimon, T., Nosaka, A.Y., and Murakami, Y. (2004) Singlet oxygen formation in photocatalytic TiO_2 aqueous suspension, *Phys. Chem. Chem. Phys.* **6**, 2917–2918.
40. Daimon, T. and Nosaka, Y. (2007) Formation and behavior of singlet molecular oxygen in TiO_2 photocatalysis studied by detection of near-infrared phosphorescence, *J. Phys. Chem. C*, **111**(11), 4420–4424.
41. Daimon, T., Hirakawa, T., Kitazawa, M., Suetake, J., and Nosaka, Y. (2008) Formation of singlet molecular oxygen associated with the formation of superoxide radicals in aqueous suspensions of TiO_2 photocatalysts, *Appl. Catal., A*, **340**, 169–175.
42. Daimon, T., Hirakawa, T., and Nosaka, Y. (2008) Monitoring the formation and decay of singlet molecular oxygen in TiO_2 photocatalytic systems and the reaction with organic molecules, *Electrochemistry*, **76**(2), 136–139.
43. Konaka, R., Kasahara, E., Dunlap, W. C., Yamamoto, Y., Chien, K. C., and

Inoue, M. (1999) Irradiation of titanium dioxide generates both singlet oxygen and superoxide anion. *Free Radical Biol. Med.*, **27**(3/4), 294–300.

44. Nosaka, Y., Natsui, H., Sasagawa, M., and Nosaka, A. Y. (2006) ESR studies on the oxidation mechanism of sterically hindered cyclic amines in TiO_2 photocatalytic systems, *J. Phys. Chem. B*, **110**(26),12993–12999.

45. Maldotti, A., Molinari, A., and Amadelli, R. (2002) Photocatalysis with organized systems for the oxofunctionalization of hydrocarbons by O_2. *Chem. Rev.* **102** (10), 3811–3836.

46. Clinton, N. A. Kenley, R. A., Traylor, T. G. (1975) Autoxidation of acetaldehyde. III. Oxygen-labeling studies, *J. Am. Chem. Soc.* **97**(13), 3757–3762.

47. Ikeda, K., Sakai, H., Ryo, R., Hashimoto, K., Fujishima, A. (1997) Photocatalytic reactions involving radical chain reactions using microelectrodes, *J. Phys. Chem. B*, **101**(14), 2617–2620.

48. Abe, R., Takami, H., Murakami, N., Ohtani, B. (2008) Pristine simple oxides as visible light driven photocatalysts: highly efficient decomposition of organic compounds over platinum-loaded tungsten oxide, *J. Am. Chem. Soc.*, **130** (25), 7780–7781.

49. Ishibai, Y., Sato, J., Nishikawa, T., Miyagishi, S. (2008) Synthesis of visible-light active TiO_2 photocatalyst with Pt-modification: role of TiO_2 substrate for high photocatalytic activity, *Appl. Catal., B*, **79**, 117–121.

50. Irie, H., Miura, S., Kamiya, K., Hashimoto, K. (2008) Efficient visible light-sensitive photocatalysts: grafting Cu(II) ions onto TiO_2 and WO_3 photocatalysts. *Chem. Phys. Lett.*, **457**, 202–205.

51. Nosaka, Y., Takahashi, S., Sakamoto, H., Nosaka, A. Y. (2011) Reaction mechanism of Cu(II)-grafted visible-light responsive TiO_2 and WO_3 photocatalysts studied by means of ESR spectroscopy and chemiluminescence photometry, *J. Phys. Chem. C* **115**(43), 21283–21290.

2
Photocatalytic Reaction Pathways – Effects of Molecular Structure, Catalyst, and Wavelength

William S. Jenks

2.1
Introduction

Exhaustive oxidation of organic impurities in aerated water can be achieved – for an overwhelming majority of organic compounds – through irradiation of TiO_2 and related catalysts. However, just as with ordinary combustion of hydrocarbons, an understanding of the chemical pathways is extremely valuable. Among the reasons are these: (i) operational or experimental conditions can lead to incomplete oxidation/combustion and therefore products of partial oxidation. In real systems, small compounds like 4-chlorophenol may be important, but so are much more complicated compounds. A thorough understanding of pathways can lead to predictability for the intermediate products of polyfunctional compounds. This concept applies well, for example, to the destruction of microcystins [1, 2], important drinking water contaminants released by cyanobacteria blooms; (ii) understanding of the chemical processes involved can lead to optimization of the operational or experimental conditions; and importantly; and (iii) the use of well-characterized partial oxidation reactions can lead to characterization of the catalyst or conditions themselves.

Toward that end, this chapter reviews some of the chemical pathways that have been elucidated for photocatalytic degradations in water and the mechanisms proposed behind them. A few key structural relationships are pointed out in which chemically simple modifications of a substrate/pollutant can grossly change the observed chemistry. Finally, the utility of certain molecules as interrogators of catalyst action is pointed out. It should also be noted that, due to the limitations of space and scope, the references in this review are intended to be pertinent and important examples, rather than exhaustive. Literally hundreds of papers purport to show at least partial degradation pathways in photocatalytic degradation, so the examples selected for this chapter are meant solely to illustrate the points being made and apologies are given in advance to authors whose work is uncited.

The emphasis in this chapter is on the chemical transformations mediated by the photocatalyst. Detailed discussions of the primary photochemical events can be found elsewhere in this volume and in other primary and review literature [3–8].

Photocatalysis and Water Purification: From Fundamentals to Recent Applications, First Edition. P. Pichat.

Nonetheless, a brief reiteration is worthwhile to make an important point about the understanding of degradation mechanisms. In the vast majority of studies on chemical pathways of photocatalysis, it is stable intermediate products that are analyzed, and the chemical mechanisms must be inferred; in studies focusing on the direct observation of the active oxidizing species, it is most often the case that only limited application to relevant molecules of interest for degradation is reported. Thus, most studies make mechanistic conclusions by extrapolation, inference, or both – and the conditions ongoing in photocatalytic degradations are sufficiently complex that care must be exercised when approaching the problem from either direction.

Initial light absorption by the photocatalyst produces a conduction band electron and a valence band hole. Though not literally separate "particles," it is a worthwhile shorthand to consider them as such. In order for the material to act as a catalyst, both the hole and electron must be consumed. It is generally believed that the electron and hole find low energy trap sites on a sub-picosecond timescale in crystalline TiO_2 [5]. The excitation and relaxation is thus simplified as Eq. (2.1).

$$TiO_2 + h\nu \rightarrow e^-_{cb} + h^+_{vb} \quad (2.1)$$

$$e^-_{cb} + O_2 \rightarrow O_2^- \quad (2.2)$$

$$h^+_{vb} + H_2O \rightarrow H^+ + HO^{\bullet} \quad (2.3)$$

$$H_2O_2 + h\nu \rightarrow 2\ HO^{\bullet} \quad (2.4)$$

$$H_2O_2 + e^-_{cb} \rightarrow HO^{\bullet} + HO^- \quad (2.5)$$

The trapped electron is often envisioned as a Ti^{3+} center, which can be detected by EPR [9]. The reducing potential of the TiO_2 conduction band is sufficient to convert molecular oxygen to superoxide ($O_2^{\bullet-}$, −0.28 V vs NHE), therefore, O_2 fulfills the first of its two major roles by removing the electron from the TiO_2 particle and minimizing recombination. The reactivity of the h^+_{vb} is more complex, or at least more controversial. Residing near 2.7 V (vs NHE), it is hard to state that there is a consensus on the nature(s) of the state(s) of the trapped hole, as reviewed by Fujishima *et al.* [5]. Moreover, a growing consensus of authors agrees that the oxidative reactivity observed in organic compounds is attributable to chemistry that appears to derive from (i) direct single electron transfer (SET) to the particle and (ii) a species that either is hydroxyl radical ($HO^{\bullet}$) or at least mimics its reactivity. We shall represent this second process as oxidation of water to produce a proton and hydroxyl radical (Eq. (2.3)), but realize that this hydroxyl-like chemistry may be due to more complex processes.

Superoxide produces hydrogen peroxide (H_2O_2) through disproportionation or by receiving a second electron. Hydrogen peroxide can also be produced through coupling of hydroxyl radicals, though there is no wide belief that this is a major source of it. Hydrogen peroxide can produce hydroxyl radical, either through secondary photolysis if UV irradiation is applied (Eq. (2.4)), or by secondary reduction by photoexcited TiO_2 (Eq. (2.5)). As implied below, the latter of these is thought to be more important.

Finally, some authors invoke "singlet oxygen," that is, $^1O_2(^1\Delta_g)$, the lowest excited state of molecular oxygen, in the reactivity observed on excitation of TiO_2 and related photocatalysts [10–15]. This intermediate is presumably formed, at least primarily, by back electron transfer from superoxide, a process represented schematically by Eq. (2.6).

$$O_2^{\bullet-} + h_{vb}^{+}\ (\text{or other electron acceptor}) \rightarrow {}^1O_2\left({}^1\Delta_g\right) \qquad (2.6)$$

Although the direct observation of singlet oxygen by spectroscopic methods is firm proof that it can be formed and may very well participate as an oxidant under certain circumstances, there is no strong evidence that it is a major contributor to the oxidation of organic molecules in semiconductor photocatalysis. The reasons to rule it out as a predominant mechanism are (i) the chemical selectivity of $^1O_2(^1\Delta_g)$ is well known, and many of the compounds easily destroyed by TiO_2 are inert to $^1O_2(^1\Delta_g)$; (ii) water is a singularly poor solvent for $^1O_2(^1\Delta_g)$ [16], in that its lifetime is inversely proportional to the density of OH groups in solution; and (iii) several studies have been carried out using fairly selective $^1O_2(^1\Delta_g)$ quenchers, such as azide ion [17–19], without grossly affecting the outcome of degradation. Moreover, switching from H_2O to D_2O is expected to have the effect of *increasing* the rate of degradation due to longer inherent $^1O_2(^1\Delta_g)$ lifetimes, yet this is in contrast to observation, in at least one case [20]. As a result, for the remainder of this chapter, it will remain acknowledged that $^1O_2(^1\Delta_g)$ may play an occasional role in substrate oxidation in these complex systems, but the focus will be on the other pathways.

2.2 Methods for Pathway Determination

For determination of chemical pathways of degradation with clear parent–daughter relationships, there remains no substitute for the relatively tedious classic method of carrying out degradations to very low conversion (perhaps 10–15%), followed by product determination, and then secondary experiments in which the observed daughter compounds are used as the starting materials for subsequent experiments of the same type. This can require significant synthetic effort to prepare the intermediate compounds. However, without this level of care, jumping two or three steps will result in unnecessary and unwanted speculation. Illustrative of this are the compounds in Scheme 2.1, which shows only the first two steps of degradation pathways for 4-chlorophenol observed at neutral pH [21, 22].

The most practical method for product identification remains mass spectrometry coupled to chromatography. However, this also requires authentic samples, whether commercial or synthetically obtained, for unambiguous identification. *De novo* analysis of mass spectral fragmentation too often leads to irresolvable questions of stereochemistry and sometimes even regiochemistry, such as the position of a new hydroxyl group. In the experience of the author's group, GC is often preferred to HPLC for its greater resolution power. However, this is frequently counterbalanced by the need to functionalize (e.g., silylate) compounds

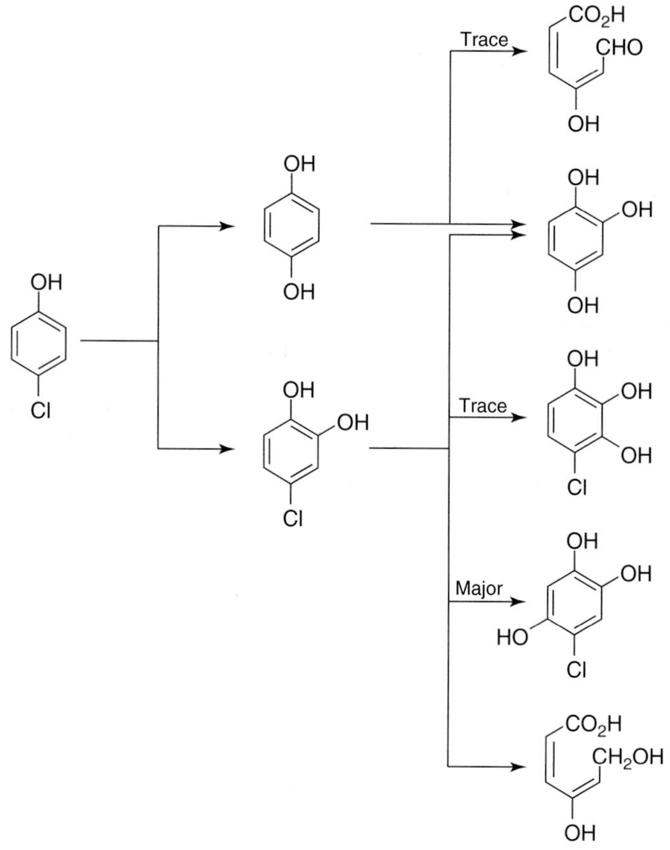

Scheme 2.1 Degradation tree for the first two transformations of 4-chlorophenol at neutral pH [21, 22].

for GC analysis, because many highly oxidized compounds do not otherwise survive gas chromatographic separation.

Many literature reports of the degradation of a particular pollutant or model do not give any information about degradation pathways. In most cases, this is due to a lack of investigation of the point. However, this should be distinguished from the fairly common case in which a few, if any, partial degradation intermediates are observed even when sensible experiments are performed. This result, although disappointing from a mechanistic investigation standpoint, is readily explainable in terms of the rates of consecutive reactions. In the sequence of A → B → C, if the pseudo-first-order rate constant k_{BC} exceeds k_{AB} by a substantial margin, then compound B does not accumulate and might not be detected in an ordinary kinetic run examining the transformation of A to C.

In the case of photocatalytic degradation, the early oxidative steps frequently produce compounds that (i) are themselves more vulnerable to oxidation than the original compound and (ii) are more highly functionalized and probably better adsorbers to the catalyst. This combination easily leads to the conditions where these intermediates will not be observed. As a result, it is recommended that authors who do not observe intermediates distinguish between cases in which no investigation was carried out and those in which none were observed despite looking. Although mechanistically less interesting, it is arguable that from a practical perspective, it is superior that no intermediates accumulate and the compound goes "directly" to full mineralization.

For more purely mechanistic investigations, additional tools are available even to those whose detection is limited to GC-MS type methods. Compounds whose chemistry is well-characterized can sometimes be used as "probes" for certain types of chemistry, much in the same spirit as in many other physical organic investigations that take advantage of certain rapid radical or cationic rearrangements, for example. Another important tool is isotopic labeling, which can be used to trace either the source of a new functionality or how one disappears. Examples have been published for clever use of labels in the molecular oxygen [23–25], water [20, 25, 26], and substrate [27, 28].

2.3
Prototypical Oxidative Reactivity in Photocatalytic Degradations

It is perhaps self-evident that to predict reactive pathways, an understanding of mechanism is required. However, it is just as important to understand that in the vast majority of cases – and particularly so in systems as complex as semiconductor-mediated photocatalytic degradation in aqueous systems – mechanism is mainly inferred, rather than directly demonstrated. Because there are so many potential reactive oxygen species (ROS), it is not surprising that different reaction mechanisms have been proposed throughout the literature to account for various oxidative steps. For successful extrapolation to new compounds, however, care must be taken to understand when each of the types of reactivity is to be expected, which ROS is likely to be involved, and the expected pathways for each of the ROS.

Accumulated evidence shows that the great majority of oxidative reactivity is due to one of the two major pathways: (i) direct SET from the organic substrate to the photocatalyst (i.e., reaction with the "hole") or (ii) reaction with hydroxyl radical, a surface-bound hydroxyl radical, or other surface species that mimics hydroxyl chemistry. This latter language is intentionally vague because there remains disagreement over the prevalence of free hydroxyl radicals in photocatalytic systems. Where there is hydrogen peroxide and ultraviolet light, or hydrogen peroxide and active reducing species, there will also be hydroxyl radicals, but the importance of the free versus solvated hydroxyl radical is not universally agreed. In addition to this difficulty is the fact that in many cases, SET and hydroxyl-like chemistry can lead to the same product, writing down ordinary and accepted mechanistic steps.

2.3.1
Oxidation of Arenes and the Importance of Adsorption

2.3.1.1 Hydroxylation and the Source of Oxygen

A prototypical example of the ambiguity of determining mechanism by examining products is the hydroxylation of arenes, for which multiple mechanisms can be written, as illustrated in Scheme 2.2.

SET from the arene results in the formation of an arene radical cation, which is trapped immediately by water. (Trapping of the radical center by O_2 is also plausible, but the water concentration is much higher than the concentration of O_2.) Deprotonation of the resulting adduct results in the same hydroxycyclohexadienyl radical formed by addition of $HO^•$ to the arene ring directly. The hydroxycyclohexadienyl radical itself can be converted to phenol (in this illustration) either by stepwise oxidation and deprotonation or by reversible [29] addition of O_2 and concerted loss of $HOO^•$, as observed in combustion and radiation chemistry [30, 31].

Regiochemistry for hydroxylation of arenes (irrelevant for benzene) qualitatively follows expectations for an electrophilic reaction; electron donating substituents generally induce ortho- and para-hydroxylation, whereas electron withdrawing groups (e.g., carboxylic acid groups, see earlier discussion) produce greater amounts of meta substitution. Precise ratios of the regioisomers have sometimes been invoked to distinguish between $HO^•$ addition and SET. Only in exceptional cases (e.g., quinoline, see earlier discussion) is such distinction *a priori* intuitive, given that both mechanisms are expected generally to follow these patterns. It is observed that the ratios are sensitive to conditions, catalyst identity, and so on, so it is

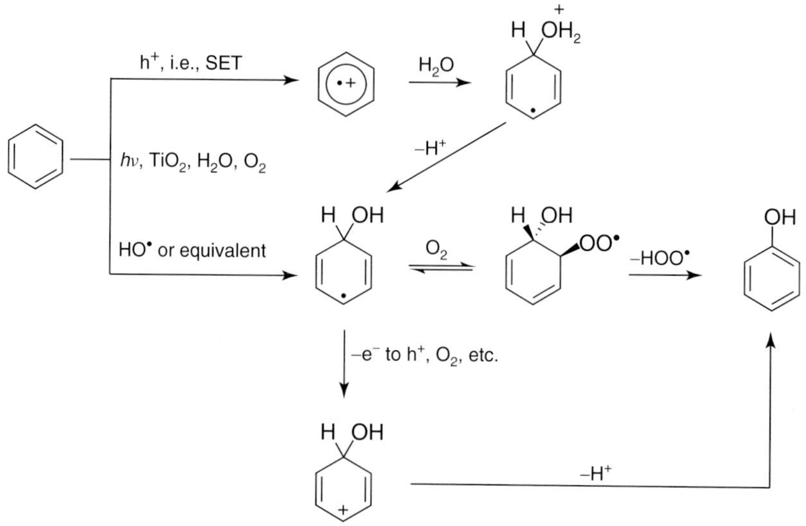

Scheme 2.2 Plausible mechanisms for hydroxylation of arenes under photocatalytic conditions.

meritorious to consider the mechanistic insights substitution regiochemistry can make possible.

A clever use of O_2 partial pressure to the exclusion of other variables reveals another potential complication [32]. Li and coworkers observed that the regiochemistry of benzoic acid hydroxylation was affected by the partial pressure of O_2. They asserted that addition of $HO^•$ to the aromatic ring (whether directly via $HO^•$ or by a stepwise SET mechanism) is reversible – that e^-_{cb} can reduce off the HO to give back the starting material. Thus the rate of reduction of the hydroxycyclohexadienyl isomers, if slightly variable, affects the observed regiochemical distribution of the hydroxylated benzoic acids.

If it is accepted that the hydroxyl radical is derived from solvent water or from rapidly exchangeable surface TiOH groups, then even the source of the HO of the hydroxylated product does not distinguish between SET and $HO^•$ chemistry, because in both cases the HO derives from solvent, rather than O_2. However, recent studies prove definitively that the situation is more complicated than once imagined. Bui and coworkers carried out the functionalization of benzene using water partially labeled with ^{18}O. Their results indicated that between 17 and 90% of the phenol oxygen atoms derived from the water, with anatase TiO_2 powders giving higher percentages than rutile ones. Control experiments showed that this lower than "expected" 100% value did not derive from unrelated exchange phenomena. The complementary experiments using ^{18}O labeled O_2 also resulted in partial labeling of the phenol. These authors speculated that the benzene radical cation was trapped directly by O_2 or superoxide, $O_2^{•-}$. However, they also acknowledged the possibility that hydroxyl radicals formed by way of superoxide and/or hydrogen peroxide (Eqs. (2.4) and (2.5)) could contribute to partial isotope scrambling.

In broader terms, these results were confirmed by a follow up study by Li and coworkers, who concluded that isotope scrambling was indeed due to *in situ* formation of H_2O_2. They examined a larger set of oxidation conditions, also using labeled water and O_2 and benzoic acid as the substrate [26]. An important result of their work was that scavenging of e^-_{cb} by benzoquinone changed the labeling results dramatically. By eliminating the reduction of O_2, the labeling experiments showed that the hydroxylation took place virtually entirely through water-derived oxygen atoms. Further control experiments using enzymes that selectively destroy superoxide or hydrogen peroxide suggest that the main source of O_2-derived hydroxyl oxygen atoms is single electron reduction hydrogen peroxide, via Eq. (2.5). In other words, the build up of H_2O_2, which occurs naturally because of the reduction of O_2, itself becomes a source for $HO^•$ by its secondary reduction.

A further important conclusion that derived from the work of Li *et al.* [26] is that SET reactions occurred on the surface of the photocatalyst particle, whereas hydroxylation could occur in the bulk phase. This is readily understandable if reduction of H_2O_2 is among the primary sources of the hydroxyl radical. Moreover, it is consistent with several previous studies, in which such a conclusion was inferred from chemical results.

2.3.1.2 **Ring-Opening Reactions**

On the basis of a variety of chemical inferences, the Jenks group proposed that the ring opening of hydroxylated aromatic compounds – particularly ortho-dihyroxylated – was initiated by SET chemistry, whereas HO substitution, ring hydroxylation, and alkyl group degradation was initiated by hydroxyl-like chemistry [21, 22], as illustrated in Scheme 2.3 for 4-chlorocatechol.

The mechanism for hydroxylation of aromatic rings is not expected to vary significantly, except in detail. As an arene is oxidized with additional HO or RO substitution, its oxidation/ionization potential naturally drops. Hydroxyl radical additions to alkenes and arenes, however, do not have a large range of rate constants. In contrast, there is evidence for a fairly wide range of time-scales (tens of picoseconds through microseconds) for oxidative reactions, as reviewed extensively by Fujishima and Serpone [4, 5, 33]. Ranges of this width are more reflective of multiple fundamental mechanisms, and, in particular, with electron transfer mechanisms in which the rate of the initial step will vary widely with the associated ΔG (and thus $\Delta G^{\ddagger}$). It must be recalled, however, that the ease of the oxidation of the organic may also be coupled to a greater stability of the resulting radical cation (or radical, after deprotonation). Thus, the more easily oxidized species, once formed, may be more selective in their reactivity.

Although the detailed mechanism of ring opening in these aromatic compounds is still at the level of informed speculation, there is a growing body of evidence in papers with definitive product identification [21, 22, 34–42] that the ring-opening step for many aromatic compounds follows the oxidation state pattern indicated in Scheme 2.3, that is, that the carbon-carbon cleavage occurs with net addition of O_2 across the broken bond. As reviewed elsewhere [43], one logical explanation for this reaction is the formation of a transient dioxetane across the double bond to be cleaved, but it must be emphasized that no dioxetane has been directly observed in any of these systems. (Dioxetanes, as a whole, are not particularly stable, and it is no surprise that they would not accumulate under such harsh conditions as in these experiments.)

2.3.1.3 **Indicators of SET versus Hydroxyl Chemistry in Aromatic Systems**

There is only a modest difference between the ionization potential of a phenol and a methoxybenzene because of the "methyl capping." However, that methyl cap radically changes the exchangeability of that oxygen for binding to Lewis

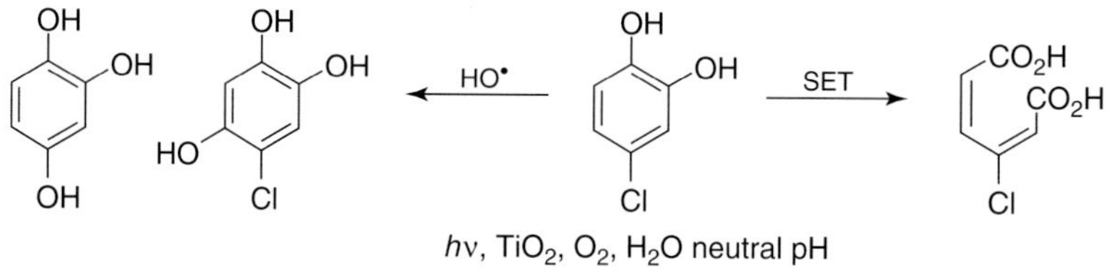

Scheme 2.3 Major primary products of photocatalytic degradation of 4-chlorocatechol at neutral pH. Ring-opening products are attributed to SET-initiated chemistry, while hydroxylation and substitution are attributed to hydroxyl-like chemistry.

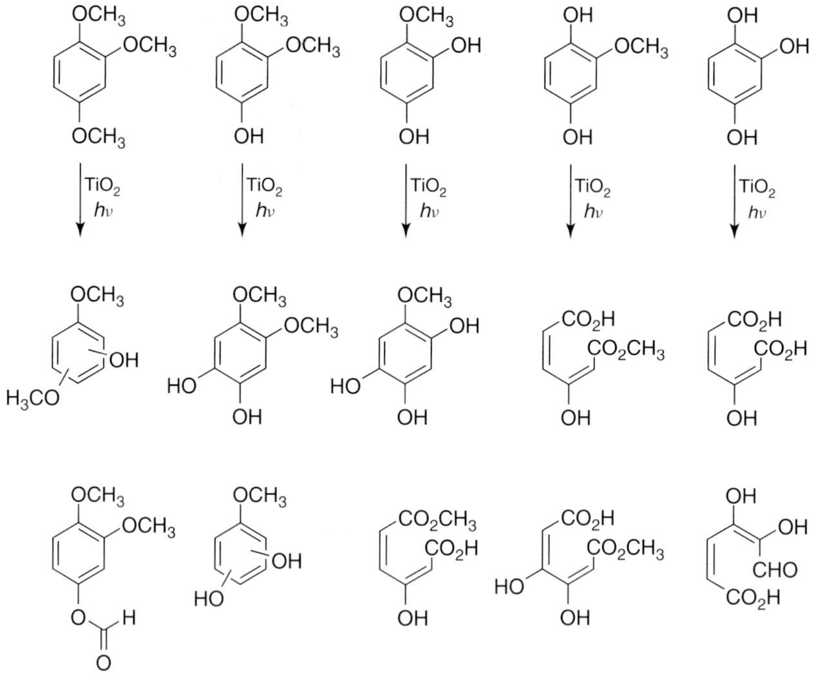

Scheme 2.4 The two most predominant early degradation products of a series of hydroxylated/methoxylated benzenes. Each product is illustrated in its most chemically suggestive tautomer, rather than its most predominant tautomer in solution. Ring-opened compounds are attributed to SET chemistry and other compounds are attributed predominantly to hydroxyl chemistry.

Acids, metal oxides, and so on, because the proton is easily removed in water whereas the methyl is not. A series of 1,2,4-trisubstituted benzenes with some combination of HO and CH_3O groups were oxidatively treated under similar conditions (Scheme 2.4) [38]. The hypothesis, verified by control experiments with additives and adsorption measurements, was that the hydroxyl-bearing compounds would adsorb more effectively to the TiO_2 catalyst (particularly the ones with *ortho*-dihydroxyl groups), whereas the "methyl-capped" groups would not adsorb as well. By exploring several different substitution patterns, it was hypothesized that variations in reactivity might be correlated with adsorption in this series of compounds with similar ionization potentials and substitution patterns.

At one extreme of the series, benzene-1,2,4-triol suffered almost entirely ring opening, predominantly between the two *ortho*-hydroxyls. When a single methyl group was capped, instead of a diacid, an acid-ester was observed as the ring-opened product. However, when both of the *ortho*-hydroxyl groups were methyl-capped, ring opening was replaced by either oxidation of one of the methyl groups or hydroxylation of the ring [38]. It is undoubtedly an oversimplification to suggest that all of the hydroxylation comes via hydroxyl radicals, but the clear inference to be drawn from this work is consistent with the more recent isotope studies: the

SET chemistry is occurring at the surface of the TiO_2 catalyst and the hydroxyl radical chemistry does *not* require any kind of specific chemisorption.

This conclusion was also reached by the Pichat group, who studied the rate of reaction of several substrates (4-chlorobenzoic acid, pyridine, anisole, among others) over a series of catalysts that varied only in their sintering time and temperature, and thus degree of crystallinity and surface defects. The pattern of rates for each substrate across the group of catalysts was interpreted in terms that hydroxyl radical chemistry could take place off the surface of the particles, while the SET required adsorption [44, 45].

In the context of using product distribution as an indicator of mechanism, it was important to verify that demethylation could be construed as a "hydroxyl" reaction, because (like hydroxylation of the ring), an electron transfer mechanism can be written. Regardless of the initial step, a carbon-oxygen bond must be broken. In the final part of the multistep mechanism, the removal of methyl group from any of these compounds can be viewed as any of the following: (i) a hydrolysis (e.g., nucleophilic attack at the methyl with an activated leaving group); (ii) oxidation of the methyl group followed by more facile hydrolysis; or (iii) substitution of an OH for an OCH_3. Although the latter is somewhat counterintuitive, isotopic labeling of the oxygen and hydrogens was used to demonstrate that both mechanisms (ii) and (iii) occurred in anisole (Scheme 2.5) [27]. The simplifying assumption was made that all chemistry was due to hydroxyl radicals because the results for TiO_2-mediated degradation were essentially identical to that for Fenton chemistry or hydrogen peroxide photolysis, but this assumption cannot be proven by simple chemical inference, and SET chemistry may contribute. Several products are observed in this reaction, including phenol. While phenol is not the major primary product of anisole degradation, of the anisole that is observed, only about a quarter of it retains the original oxygen atom!

The clear implication is that hydroxyl attack can occur at the *ipso* position even when an oxygen substituent is already there. The alternative hypothesis that the substitution is due to an SET reaction can be dismissed as unlikely, because virtually the same quantitative result is observed for direct photolysis of hydrogen peroxide. Moreover, comparable *ipso* attack of hydroxyl radical was clearly

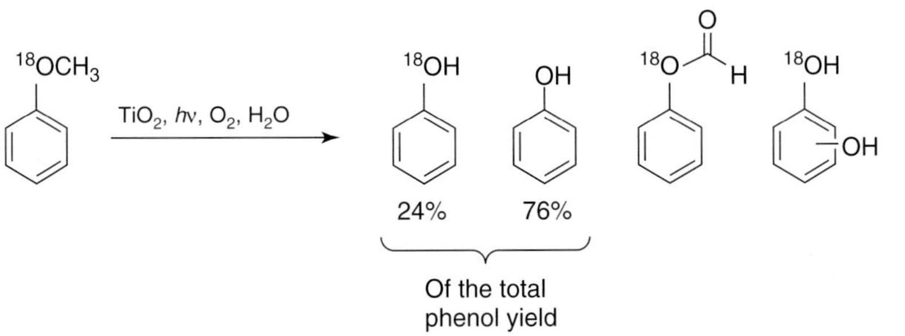

Scheme 2.5 Primary oxidation products of ^{18}O-labeled anisol.

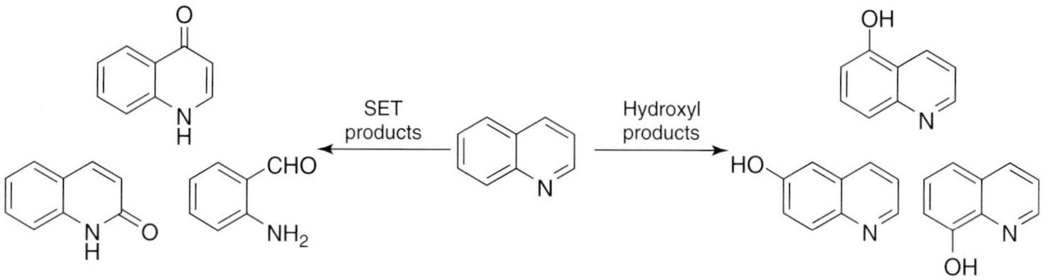

Scheme 2.6 Predominant early oxidation products of quinoline under TiO_2-mediated photocatalytic conditions.

implicated in pulse radiolysis studies (unambiguously making hydroxyl radicals) of 2,4-dichlorophenoxyacetic acid using similar labeling strategies [46].

Regioselectivity of hydroxylation can be indicative of the mechanism of functionalization in certain instances. The Pichat group noted that functionalization of quinoline could occur either on the "benzene half" or on the "pyridine half" of this compound and that the regioselectivity changed as a function of reaction conditions (Scheme 2.6) [47–49]. In practice, both primary and secondary oxidation products are observed, but they can still be traced back to the initial oxidation on one moiety or the other. In this case, *o*-aminobenzaldehyde, for example, is easily attributable to oxidation of the pyridine.

At low pH, where the quinoline is protonated to give the quinolinium ion, the electron density is clearly greater on the benzene side of the molecule. Under either acidic photocatalytic conditions or photo-Fenton conditions the functionalization occurs predominantly on that side. In contrast, at pH 6, 4-quinolone (tautomer of 4-hydroxyquinoline) and *ortho*-aminobenzaldehyde (and its *N*-formyl derivative) are predominant. This is attributed to the attack of water on the less electron dense pyridine half of the quinoline after removal of an electron. As elaborated below, this easily identifiable set of distinct products that report back alternative mechanisms can be useful in the chemical characterization of modified photocatalysts.

2.3.2
Carboxylic Acids

Certain well-defined functional groups have fairly generalizable reactivity. Carboxylic acids are relatively strong binding groups to TiO_2, particularly at relatively low pH. This makes them candidates for SET chemistry. Among their characteristic processes is the photo-Kolbe reaction [50, 51]. We use the term broadly to encompass either abstraction of a carboxylic acid hydrogen atom or direct one-electron oxidation of carboxylate anions. Either of these leads to rapid decarboxylation when the resulting radical is alkyl, as well as for formic acid. Most authors assume that the mechanism in photocatalytic degradation reactions takes place by direct SET. Formic acid is a substrate that is viewed as a "hole trap" in many studies of photocatalytic degradation. It also has the property (when in relatively high concentration and at the proper pH) of out-competing many other organic molecules for

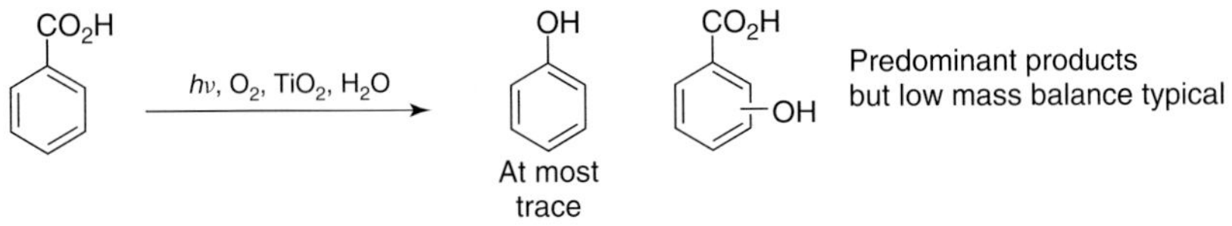

Scheme 2.7 Hydroxylation is the major product of TiO_2-mediated degradation of benzoic acid, not decarboxylation.

binding sites on the catalyst. See, for example, Refs. [26, 52–54]. The chemistry of formic, acetic, dimethylpropanoic, and other alkyl acids on both pristine TiO_2 and in the presence of water is thoroughly reviewed by Henderson [3], but for current purposes, it is sufficient to say that electron abstraction from the carboxylate and subsequent loss of CO_2 are rapid.

The same is not true of arenecarboxylic acids. Benzoic acid, salicylic acid, and various derivatives thereof, along with several biphenylcarboxylic acids have all been studied under photocatalytic degradation conditions [45, 55–63]. While detection of phenol has been claimed from benzoic acid, it is never in more than trace amounts, and the predominant products are salicylic acid and the other hydroxylated benzoic acids (Scheme 2.7). For the more highly functionalized 3,6-dichloropyridine-2-carboxylic acid (clopyralid), decarboxylation has been observed, but again, as a comparatively minor process [64]. Mass balances for arenecarboxylic acid partial degradations are typically low; this can be interpreted through the idea that the resulting products are even more strongly binding and easier to oxidize than the products, and thus disappear with greater rates without accumulating.

2.3.3
Alcohol Fragmentation and Oxidation

Alcohols of various sorts have been extensively studied, particularly on clean surfaces [3, 5]. It is commonly assumed that the initial step is reaction by SET, particularly for primary and secondary alcohols. In photoelectrochemical contexts, alcohols are known for their "current doubling," that is, injection of a second electron into the TiO_2 after the first has been removed. This amounts, of course, to standard two-electron oxidation of the alcohol to the corresponding ketone or aldehyde. However, it is again clear that, at least for more complex alcohols in aqueous solutions/suspensions, the reaction is stepwise. Strong inferential evidence for this comes from an example of 1-anisylneopentyl alcohol (AN), originally studied in nonaqueous (acetonitrile) TiO_2 suspensions [65], but more recently applied in aqueous systems [66–70], as illustrated in Scheme 2.8.

Among the two potential carbonyl products, the aldehyde is formed by loss of *tert*-butyl radical from $AN^{\bullet+}$, whereas the concerted two-electron oxidation product would produce anisyl *tert*-butyl ketone, which is only observed in trace quantities. With a discrete $AN^{\bullet+}$ intermediate, loss of the tertiary radical is expected to be favored, as is observed. If oxidation of the alcohol were to take place by hydrogen

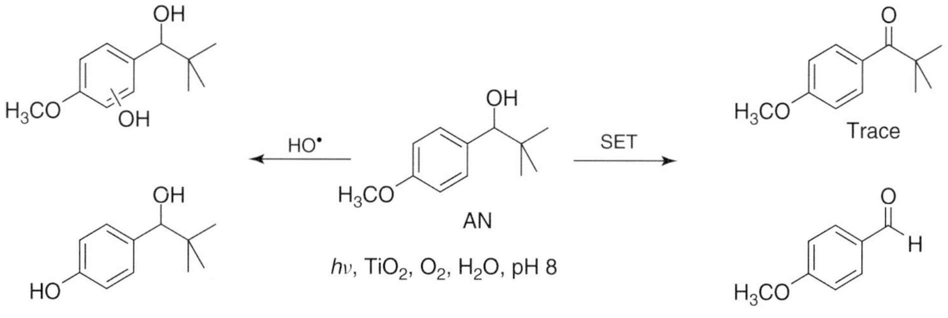

Scheme 2.8 The primary photodegradation products of 1-anisylneopentanol (AN). Only very small amounts of the anisyl *t*-butyl ketone are observed.

abstraction instead of SET, the first step would be to remove the carbinol hydrogen, and thus again, the ketone would be the predominant product. Thus, here, chemical inference does give a strong and singular indication of mechanism, that is, that SET is a primary and distinct step.

This particular alcohol, however, is quite hindered, given its phenyl and *tert*-butyl substituents. Furthermore, AN has only that one functional group by which to attach itself to the TiO_2. While quantitative measurements of adsorption compared to related structures have not been reported, it seems safe to speculate that it does not adsorb strongly to the TiO_2 surface. Thus, it is not surprising that, in addition to the SET-derived aldehyde, hydroxylation of the phenyl ring and degradation of the methyl are observed. These are attributed to hydroxyl chemistry.

There remains considerable mixed interpretation about how much chemistry of alcohols is derived from SET and how much from hydroxyl radicals. Even methanol has been used as a hydroxyl radical scavenger/actinometer [71]. Use of *tert*-butanol as a hydroxyl radical scavenger is fairly common and is justified on the basis that no alcohol oxidation products (e.g., acetone, due to fragmentation of a methyl) are observed. Hydroxyl radical attack at the methyls is instead the predominant path. It is worth noting, however, that not having a path by which to complete the classic alcohol oxidation does not mean that the first step of SET cannot occur. It is reasonable to speculate that *t*-BuOH does indeed quench a small fraction of h_{vb}^+ in addition to hydroxyl radical, but (i) due to its hindered nature, adsorption is weaker than for other alcohols, and the amount of h_{vb}^+ quenching is not large and (ii) the h_{vb}^+ quenching is chemically unproductive, that is, the surface-bound radical cation has a lifetime long enough that an electron is eventually restored to the alcohol.

2.3.4
Oxidation of Alkyl Substituents

The reactivity of alkanes and alkyl portions of larger organic molecules is not widely remarked on, though it is well documented that unfunctionalized compounds are mineralized under photocatalytic conditions. Because the alkanes or alkyl groups themselves have no specific "handle" by which to adsorb to the catalyst and are difficult to oxidize by SET, it is no surprise that their reactivity strongly resembles hydroxyl and autoxidation chemistry. It is to be expected that hydrogen abstraction by a hydroxyl-like species is the primary step. Under most circumstances, trapping of the alkyl radical by molecular oxygen will follow, giving a peroxyl radical ($ROO^•$). The peroxyl radicals will then go through ordinary reactivity cycles, depending on their structure and the concentrations of the species involved [30, 72].

Where interest in the degradation of alkyl carbons has been greater, if implicit, is by comparison to other functionality. As already noted for anisole, for example, the oxidation of the alkyl groups is a background against which chemistry of other functional groups is mapped.

2.3.5 Apparent Hydrolysis Reactions

One of these class of reactivity that alkyl oxidation can be compared to is the family of apparent hydrolysis reactions. The term apparent hydrolysis is used because ordinarily one "half" of an amide or ester is observed (e.g., the acid or amine portion), so the reaction looks superficially like an ordinary hydrolysis; however, given the available reactive species, it is clear that ordinary hydrolysis mechanisms do not apply. It is speculation here, but more likely in many cases, oxidation occurs on the alkyl portion (e.g., Scheme 2.5 for the "hydrolysis" of the ether in anisole), giving rise to a new functionality that is considerably easier to hydrolyze than the original.

Hydrolysis of amides under ordinary (nonphotocatalytic) conditions is typically quite difficult, requiring high temperatures and/or concentrations of acid or base. However, amide cleavages are commonly observed under photocatalytic conditions. Because of the paucity of real evidence on this, a larger number of leading references are provided to the reader who may be interested in elaborating this chemistry. Amide- and urea-containing herbicides and insecticides tend to suffer aromatic hydroxylation and alkyl oxidation competitively with apparent amide/urea hydrolysis [43, 73–83]. However, as yet, there is little direct information on the mechanism of the C–N cleavage. Similarly, hydrolysis of the CN group of benzonitrile is a minor process relative to aromatic hydroxylation but is observed [84]. (The detailed fate of nitrogen atoms in photocatalytic degradation, primarily as aromatic and aliphatic amines, is reviewed by other authors [85, 86]).

Isotopic labeling of the "hydrolysis" of phosphonate esters and other chemical evidence shows that it occurs by oxidation of the alcoholic carbon, rather than attack at the phosphorus center [28, 87–90]. At the time of this writing, the author is unaware of isotopic or other detailed mechanistic studies that clarify the mechanisms of "hydrolysis" of ordinary carboxylic esters or other related centers beyond the expectation that oxidation of the α-carbon will lead to the acid anhydride of a hemiacetal, which is surely quite sensitive to ordinary acid hydrolysis.

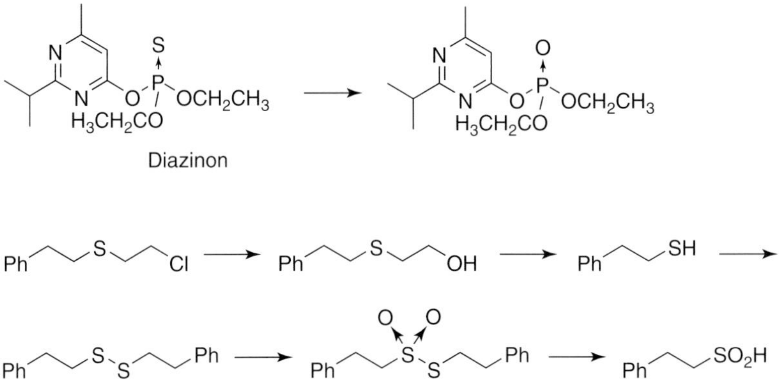

Scheme 2.9 Reported initial transformations for photocatalytic degradation of diazinon and 2-phenylethyl-2-chloroethyl sulfide.

2.3.6
Sulfur-Bearing Compounds

Studies of sulfur-bearing compounds have focused mainly on thionophosphates (e.g., diazinon, Scheme 2.9) as agricultural chemicals and on models for chemical warfare agents, such as the mustard model 2-phenylethyl-2-chloroethyl sulfide (Scheme 2.9) [87, 91–94]. For the thionophosphonates, the conversion of the sulfur back to the more stable phosphate is one of the predominant reactions early in the conversion. Multiple proposals are given for how the substitution occurs, but experimental confirmation of those mechanisms remains thin. Perhaps counterintuitively, given how facile sulfides are oxidized in general, photocatalytic treatment of 2-phenylethyl-2-chloroethyl sulfide did not result in observation of *S*-oxidized substances until formation of the disulfide [87]. However, more recent work on sulfide bearing agricultural chemicals did identify the respective sulfoxides and sulfones [95, 96].

2.4
Prototypical Reductive Reactivity in Photocatalytic Degradations

Though ordinarily one thinks of oxidative chemistry in the context of photocatalytic degradation, there are compounds whose mineralization must proceed through reductive pathways. One of these is certain haloalkanes. Carbon tetrachloride, for example, has its carbon at the same oxidation state as carbon dioxide, so oxidative attack is not possible. Bahnemann and others showed that injection of an electron results in formation of Cl^- and $CCl_3^{\bullet}$ with the rate highest at high pH. They hypothesized that surface-bound hydroxyls lead to formation of $HOCCl_3$. Under these conditions trichloromethanol will spontaneously eliminate to form phosgene ($OCCl_2$), which is then subject to hydrolysis to form CO_2 [97]. Hole scavengers increased the rate of reaction and increased O_2 concentration

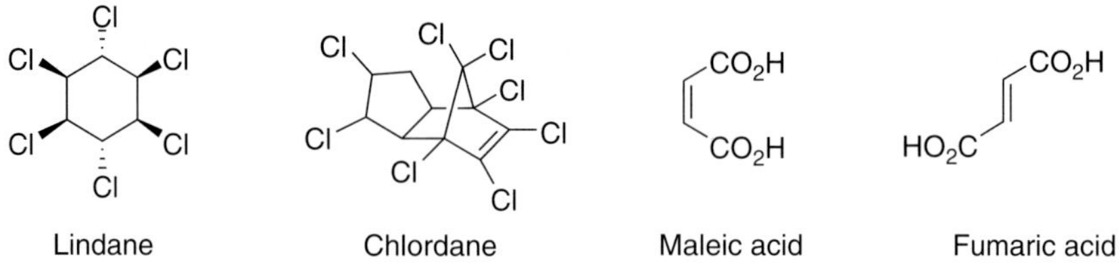

Scheme 2.10

decreased the efficiency, as expected for a reductive reaction. Formation of the trichloromethyl radical was later confirmed by Choi and Hoffman without detection of stable intermediates at high pH; they also trapped dichlorocarbene, indicative of a two-electron reduction [98]. Other compounds that are sufficiently saturated and polysubstituted, such as halothane (1-bromo-1-chloro-2,2,2-trichloroethane) can follow the same reductive initial step [99].

However, few compounds are so polysubstituted; even pesticides such as lindane and chlordane (Scheme 2.10) are heavily halogenated compared to most compounds. In this case, the mechanism is considerably more complex; Guillard and coworkers reported multiple chlorinated cyclohexanes, chlorinated cyclohexenes, and chloroaromatics [100]. This suggests that hydrogen abstraction, eliminations, and re-addition of HCl are all part of a complex network of reactions. The salient point is that once the compound is not fully saturated and perhalogenated, then the interaction of the adjacent hydrogen atoms, π-bonds, and other functionality become central in understanding the observed pathways.

In certain instances, more conventional organic structures also suffer reduction by TiO_2. As an example, hydroquinones are often intermediates in the degradation of aromatic compounds, as noted in Scheme 2.4. In fact, analysis of some of these compounds under realistic conditions provides the corresponding benzoquinone because of simple air oxidation. Richard, however, showed that benzoquinones are rapidly reduced to the corresponding hydroquinone under photocatalytic conditions [101]. As a result, benzoquinone had been used as an alternate oxygen acceptor for mechanistic experiments when the absence of molecular oxygen was required. For example see Ref [32].)

Another example of an organic electron acceptor is the four-carbon diacid pair of fumaric and maleic acid [102, 103]. In ordinary photocatalytic conditions (water, oxygen, TiO_2), fumaric acid and maleic acid are both mineralized, with some interconversion between the two that occurs in parallel. However, in the absence of molecular oxygen at low pH, a photostationary state of about 92% maleic acid and 7% fumaric acid was quickly obtained, regardless of which pure isomer was used as starting material. This was interpreted as deriving from reversible SET reduction to the radical anion, which could isomerize. The formation of the photostationary state could be blocked either by using the analogous esters (dimethyl maleate and dimethyl fumarate) or by fluorinating the TiO_2 and – in either case – blocking adsorption and SET processes [102].

2.5
The Use of Organic Molecules as Test Probes for Next-Generation Photocatalysts

The previous pages have described several effects of structure on the mechanisms and pathways of photocatalytic degradation, but have presented results obtained almost entirely with ordinary TiO_2: anatase, rutile, or mixed-phase as in P25. However, the great bulk of work in recent years has been to develop photocatalysts that are superior to ordinary TiO_2 in terms of absorption of visible light, while maintaining its other positive properties, such as its cost relative to other materials, robustness, and broad reactivity. Titanium dioxide is also absent of toxicity except for those related to the use of nanoparticles. Nearly uncountable papers have been published in which some form of doped or modified TiO_2 catalyst (or related oxide) has been prepared and evaluated for its photocatalytic efficiency. Often, and with good reason, P25 is used as the "industry standard" given its high reactivity and ready availability for purposes of reproducibility.

However, a flaw in many of these studies is the use of a single probe molecule to evaluate photocatalytic activity. Many papers rely on the decolorization of either methylene blue or an azo dye such as methyl orange (Scheme 2.11). The former has taken on such popularity that a standardized procedural method has been recommended [104, 105]. The use of these probes is very simple as they require only measurement of visible absorption, given that initial products are generally colorless. However, there are at least two problems: (i) these dyes naturally absorb the visible light that is being used and can self-sensitize their own destruction in the presence of oxygen [106] and (ii) decolorization tells us nothing about the mechanism of reaction. Addressing the first of these limitations, the Ohtani group has demonstrated that degradation of methylene blue by sulfur-doped TiO_2 (S-TiO_2) has an action spectrum that does not match the absorption spectrum of the catalyst [107]. Regarding the mechanistic information, obviously these large aromatic compounds are subject to hydroxylation, but they are also both subject to relatively easy reduction. Either oxidation or reduction results in discolorization.

These flaws limit the utility of methylene blue and azo dyes as testing agents. It cannot be overemphasized that multiple mechanisms must be probed with modified photocatalysts. As pointed out elsewhere in this volume, movement of the valence, conduction, or trap site energy levels (as is a necessity to change the absorption spectrum) must ultimately limit the ability of the catalyst to operate. Thus, probes must test for multiple forms of reactivity, and it is likely that multiple

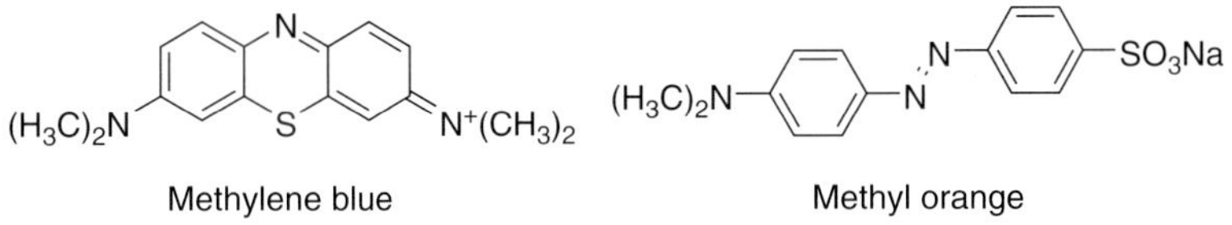

Scheme 2.11

substrates will be required. Certainly, knowing the chemical fate of the probe is essential.

The surface properties of the catalysts also affect the reactivity. The combination of electronic and physical properties means that relative reactivity will not necessarily correlate for one class of substrate versus another. Ryu and Choi took a standard set of eight undoped TiO_2 samples and tested them against a bank of 20 substrates in 8 different structure classes [108]. Each of the different catalysts was the most efficient catalyst for at least one of the substrates. This gives a second rationale for multiple test probes, that is, to give fair evaluation of rates of catalysis for the broad classes of structures that are of interest to the water purification community.

The approach taken in the author's laboratory is far from perfect, but has focused on the mechanistic end of these considerations. The rationale is that the first issue to settle with a new catalyst is whether it is truly functional in the visible range; definitive discussions of rate can wait until the basics of reactivity have been worked out. Toward that end, probes have been selected that exhibited both hydroxyl reactivity and SET reactivity in easily observable proportion, so that changes in reactivity are easily monitored. Probes could be as simple as hydroxybenzoic acid [109], but a priority was put on certainty of SET versus hydroxyl activity. The first probe used was 3-methoxyresorcinol (3-methoxybenzene-1,2-diol, Scheme 2.4), but it was observed that activity of this probe (i.e., its destruction) with visible irradiation was essentially universal, regardless of the absorption spectrum of the catalyst. This was attributed to formation of a charge-transfer complex (with visible absorption) between the very electron rich arene and the TiO_2 as previously documented in related cases [110–113].

Subsequently, quinoline and AN have been used, according to the schemes shown in Schemes 2.6 and 2.8. However, any substrate that exhibits the chemistry (or chemistries) of interest in a well-defined way could be chosen.

2.6
Modified Catalysts: Wavelength-Dependent Chemistry of Organic Probes

In recent years, many of the reported promising modified TiO_2 catalysts are based on impregnation of main group elements, such as nitrogen and sulfur. These are sometimes referred to as anion-doped, by comparison to the older transition-metal ("cation") doped catalysts. In this section, a few results from the author's labs are used as illustration of the types of experiments that can be done, followed by selected recent results that hint at explanations of wavelength dependencies by other authors.

In classic small molecule photochemistry, a substantial influence of excitation wavelength is rare, once the threshold for excitation is achieved. This is because relaxation of the excited state down to the lowest excited electronic configuration is generally faster than any other process.

With semiconductor photocatalysts, this is not necessarily the case. There are inherent reasons elaborated elsewhere that may result in variation of efficiency or selectivity [114, 115]. But when new elements are added, the electronic structure is changed. If, as seems most likely, this is because of the introduction of traps/defects/color centers, as opposed to a more generalized band gap narrowing [5, 116, 117], excitation in the visible is fundamentally a different transition than may occur on UV excitation. The kinetics of hole/electron trapping versus chemical reactivity may introduce the possibility that the reactivity may differ substantially depending on which absorption is stimulated. The most common assumption is that the reduction potential of the conduction band of TiO_2 remains relatively constant with doping/modification, and that the color change of the catalysts is due to higher energy occupied orbitals at the trap/defect/doping sites; these are sometimes referred to as mid-gap orbitals. These higher energy orbitals are thus inherently less oxidative than the valence band of pristine TiO_2, and at some energy level, must eventually affect the observable reactivity of the catalyst.

In the author's group's hands, several main-group-modified catalysts whose preparations were based on early literature reports turned out to have modest visible absorption, lower efficiency than the undoped analog when exposed to UV light, and little or no activity in the visible. However, sulfur-doped TiO_2 (S-TiO_2), prepared by the method of Ohno [118, 119], or by annealing S_8 directly into TiO_2 proved more interesting. With UV irradiation, the S-TiO_2 degraded both AN and quinoline slightly less efficiently than control TiO_2 (prepared identically, save for the sulfur). With visible irradiation (>495 nm), AN was not degraded by either catalyst and only the S-TiO_2 affected degradation of quinoline [69].

Product distributions, however, were also informative, particularly for quinoline degradation. While more SET products (pyridine side functionalization) than hydroxyl products (benzene side functionalization) were observed for both TiO_2 and S-TiO_2 with UV irradiation, under visible irradiation the SET products were observed exclusively, and only with S-TiO_2. With AN, no degradation was observed under visible irradiation. These results suggest that the S-TiO_2 was unable to generate hydroxyl chemistry, and was only causing degradation when the compound was adsorbed to the catalyst surface.

Another notable catalyst was a selenium-doped material (Se-TiO_2), prepared in a similar fashion as the S-TiO_2, save for using SeO_2 as the doping agent [67]. In this case, using UV excitation, Se-TiO_2 was equally efficient at degradation of AN, quinoline, or 3-methoxyresorcinol as undoped TiO_2. Again, it was active with visible irradiation (>435 nm), this time effecting degradation of both AN and quinoline with only observation of the SET-induced products. It is obvious that the utility of catalysts whose activity in the visible is limited to certain kinds of reactivity is less general than a catalyst capable of generating its full repertoire of reactivity throughout its excitation range.

Results from the Jenks group are hardly the only ones to suggest that wavelength dependence exists in the reactivity of certain photocatalysts. Mrowetz investigated the activity of N-TiO_2 with respect to formate oxidation and observed a lack of reactivity with irradiation >400 nm [106]. This was interpreted as being due to the oxidation potential of hydroxide being more positive than the holes that were formed on excitation in the visible. A similar conclusion about the lack of ability of N-TiO_2 to produce hydroxyl radical on visible irradiation (by oxidation of water;

hydroxyl radical formation through indirect reaction of superoxide by reductive pathways remains possible) was reached by Sakthivel and Kisch [120]. Tachikawa and coworkers carried out transient absorption experiments in nonaqueous systems using *p*-methylthiobenzyl alcohol as a one-electron donor and spectroscopic marker [121]. Although the conditions used for this experiment differ from those discussed throughout this chapter, it is noteworthy that the authors did not observe one-electron oxidation of their probe using visible excitation, nor did they with N-TiO_2 using a different probe, but similar conditions [122].

With I-TiO_2, chlorophenol was used as a probe and reactivity was reported with both UV and visible light, despite recognition that the trapped holes would be of low oxidizing power, given the absorption spectrum of the catalyst. These workers invoked hydroxyl radical chemistry throughout the excitation range, but argued that the hydroxyl radicals must come from reduction of peroxide [123]. Nakamura concludes that hydroxyl radicals can be made in the visible (by N-TiO_2), despite the low oxidizing power of the holes, but invokes a different mechanism not involving reduction of peroxide [124].

Emeline also used chlorophenol, this time with N-TiO_2, and observed a subtle wavelength dependence on the regiochemistry of hydroxylation. Arguing that the substitution of HO for Cl is fundamentally based on 4-chlorophenol acting as a direct electron acceptor, they noted the shutdown of 4-chlorocatechol formation at longer wavelengths and attributed this to the lack of sufficiently oxidizing holes, while reducing electrons remained [125].

2.7 Conclusions

From even this brief review, it is evident that there is room – in fact demand – for properly designed experiments using degradation substrates to explore various facets of catalyst behavior and properties. Compounds with varying adsorption capability can report on the locations of reactivity and sometimes on the nature of the reactive species. Compounds with multiple reactivity modes can report on chemoselectivity of the catalysts. Compounds with varying oxidation potentials can report on the ability of various catalysts to oxidize water and/or the substrates. Wavelength dependences will vary with the structure, adsorption, and reactivity of the probes. Subtle use of oxygen labeling is clearly critical in demonstrating the mechanisms of catalyst action.

Although this chapter has focused largely on TiO_2, this concept remains entirely valid as the photocatalysis community explores new materials. Most of the reactivity described herein will have its strong parallel with any photocatalyst that operates in a fundamentally similar manner, and the work that has been done to establish differences between SET and various ROS oxidations will surely find application for any relevant material. The concept of a bank of standard test molecules to evaluate new catalysts – in the spirit of Choi [108] but modified to include a mechanistic components – remains a strong one.

References

1. Antoniou, M.G., Shoemaker, J.A., de la Cruz, A.A., and Dionysiou, D.D. (2008) Unveiling New degradation intermediates/pathways from the photocatalytic degradation of microcystin-LR. *Environ. Sci. Technol.*, **42**, 8877.
2. Sharma, V.K., Triantis, T.M., Antoniou, M.G., He, X., Pelaez, M., Han, C., Song, W., O'Shea, K.E., de la Cruz, A.A., Kaloudis, T., Hiskia, A., and Dionysiou, D.D. (2012) Destruction of microcystins by conventional and advanced oxidation processes: a review. *Sep. Purif. Technol.*, **91**, 3.
3. Henderson, M.A. (2011) A surface science perspective on TiO_2 photocatalysis. *Surf. Sci. Rep.*, **66**, 185.
4. Serpone, N. and Emeline, A.V. (2008) *Fundamentals in Metal-Oxide Heterogeneous Photocatalysis*, Series on Photoconversion of Solar Energy, World Scientific Publishing, Vol. **3**, p. 275.
5. Fujishima, A., Zhang, X., and Tryk, D.A. (2008) TiO_2 Photocatalysis and related surface phenomena. *Surf. Sci. Rep.*, **63**, 515.
6. Linsebigler, A.L., Lu, G., and Yates, J.T. Jr., (1995) Photocatalysis on TiO_2 surfaces: principles, mechanisms, and selected results. *Chem. Rev.*, **95**, 735.
7. Tachikawa, T., Fujitsuka, M., and Majima, T. (2007) Mechanistic insight into the TiO_2 photocatalytic reactions: design of New photocatalysts. *J. Phys. Chem. C*, **111**, 5259.
8. Robertson, P.K.J., Bahnemann, D.W., Robertson, J.M.C., and Wood, F. (2005) *Handbook of Environmental Chemistry*, Springer-Verlag, Vol. **2**, p. 367.
9. Howe, R.F. and Gratzel, M. (1985) EPR observation of trapped electrons in colloidal titanium dioxide. *J. Phys. Chem.*, **89**, 4495.
10. Yamamoto, Y., Imai, N., Mashima, R., Konaka, R., Inoue, M., and Dunlap, W.C. (2000) Singlet oxygen from irradiated titanium dioxide and zinc oxide. *Methods Enzymol.*, **319**, 29.
11. Jańczyk, A., Krakowska, E.B., Stochel, G.Y., and Macyk, W. (2006) Singlet oxygen photogeneration at surface modified titanium dioxide. *J. Am. Chem. Soc.*, **128**, 15574.
12. Daimon, T. and Nosaka, Y. (2007) Formation and behavior of singlet molecular oxygen in TiO_2 photocatalysis studied by detection of near-infrared phosphorescence. *J. Phys. Chem. C*, **111**, 4420.
13. Nosaka, Y., Daimon, T., Nosaka, A.Y., and Murakami, Y. (2004) Singlet oxygen formation in photocatalytic TiO_2 aqueous suspension. *Phys. Chem. Chem. Phys.*, **6**, 2917.
14. Rengifo-Herrera, J.A., Pierzchala, K., Sienkiewicz, A., Forro, L., Kiwi, J., Moser, J.E., and Pulgarin, C. (2010) Synthesis, characterization, and photocatalytic activities of nanoparticulate N, S-codoped TiO_2 having different surface-to-volume ratios. *J. Phys. Chem. C*, **114**, 2717.
15. Barclay, L.R.C., Basque, M.C., and Vinqvist, M.R. (2003) Singlet-oxygen reactions sensitized on solid surfaces of lignin or titanium dioxide: product studies from hindered secondary amines and from lipid peroxidation. *Can. J. Chem.*, **81**, 457.
16. Murov, S.L., Carmichael, I., and Hug, G.L. (1993) *Handbook of Photochemistry*, 2nd edn, Marcel Dekker, Inc., New York.
17. Xu, Z., Jing, C., Li, F., and Meng, X. (2008) Mechanisms of photocatalytical degradation of monomethylarsonic and dimethylarsinic acids using nanocrystalline titanium dioxide. *Environ. Sci. Technol.*, **42**, 2349.
18. Karunakaran, C., Senthilvelan, S., and Karuthapandian, S. (2005) TiO_2-Photocatalyzed oxidation of aniline. *J. Photochem. Photobiol., A*, **172**, 207.
19. Zheng, S., Cai, Y., and O'Shea, K.E. (2010) TiO_2 Photocatalytic degradation of phenylarsonic acid. *J. Photochem. Photobiol., A*, **210**, 61.
20. Robertson, P.K.J., Bahnemann, D.W., Lawton, L.A., and Bellu, E. (2011) A study of the kinetic solvent isotope effect on the destruction of microcystin-LR and geosmin using

TiO2 photocatalysis. *Appl. Catal., B*, **108–109**, 1.

21. Li, X., Cubbage, J.W., and Jenks, W.S. (1999) Photocatalytic degradation of 4-chlorophenol. 2. The 4-chlorocatechol pathway. *J. Org. Chem.*, **64**, 8525.
22. Li, X., Cubbage, J.W., Tetzlaff, T.A., and Jenks, W.S. (1999) Photocatalytic degradation of 4-chlorophenol. 1. The hydroquinone pathway. *J. Org. Chem.*, **64**, 8509.
23. Pichat, P. (1988) Powder photocatalysts: characterization by isotopic exchanges and photoconductivity; potentialities for metal recovery, catalyst preparation and water pollutant removal. *NATO ASI Ser., Ser. C*, **237**, 399.
24. Pichat, P., Courbon, H., Enriquez, R., Tan, T.T.Y., and Amal, R. (2007) Light-induced isotopic exchange between O_2 and semiconductor oxides, a characterization method that deserves not to be overlooked. *Res. Chem. Intermed.*, **33**, 239.
25. Bui, T.D., Kimura, A., Ikeda, S., and Matsumura, M. (2010) Determination of oxygen sources for oxidation of benzene on TiO_2 photocatalysts in aqueous solutions containing molecular oxygen. *J. Am. Chem. Soc.*, **132**, 8453.
26. Li, Y., Wen, B., Yu, C., Chen, C., Ji, H., Ma, W., and Zhao, J. (2012) Pathway of oxygen incorporation from O_2 in TiO_2 photocatalytic hydroxylation of aromatics: oxygen isotope labeling studies. *Chem. Eur. J.*, **18**, 2030.
27. Li, X. and Jenks, W.S. (2000) Isotope studies of photocatalysis: dual mechanisms in the conversion of anisole to phenol. *J. Am. Chem. Soc.*, **122**, 11864.
28. Oh, Y.-C., Bao, Y., and Jenks, W.S. (2003) Isotope studies of photocatalysis. TiO_2-Mediated degradation of dimethyl phenylphosphonate. *J. Photochem. Photobiol., A*, **160**, 69.
29. Fang, X., Pan, X., Rahmann, A., Schuchmann, H.-P., and von Sonntag, C. (1995) Reversibility in the reaction of cyclohexadienyl radicals with oxygen in aqueous solution. *Chem. Eur. J.*, **1**, 423.
30. von Sonntag, C. and Schuchmann, H.-P. (1991) The elucidation of peroxyl radical reactions in aqueous solution with the help of radiation-chemical methods. *Angew. Chem., Int. Ed. Engl.*, **30**, 1229.
31. Pan, X.-M., Schuchmann, M.N., and von Sonntag, C. (1993) Oxidation of benzene by the OH radical. A product and pulse radiolysis study in oxygenated aqueous solution. *J. Chem. Soc., Perkin Trans. 2*, 289.
32. Li, Y., Wen, B., Ma, W., Chen, C., and Zhao, J. (2012) Photocatalytic degradation of aromatic pollutants: a pivotal role of conduction band electron in distribution of hydroxylated intermediates. *Environ. Sci. Technol.*, **46**, 5093.
33. Serpone, N., Lawless, D., and Pelizzetti, E. (1996) Subnanosecond characteristics and photophysics of nanosized TiO_2 particulates from R_{part} = 10 a to 134 a: meaning for heterogeneous photocatalysis. *NATO ASI Ser., Ser. III*, **12**, 657.
34. Matthews, R.W. (1988) Kinetics and photocatalytic oxidation of organic solutes over titanium dioxide. *J. Catal.*, **111**, 264.
35. Soana, F., Sturini, M., Cermenati, L., and Albini, A. (2000) Titanium dioxide photocatalyzed oxygenation of naphthalene and some of its derivatives. *J. Chem. Soc., Perkin Trans. 2*, 699.
36. Das, S., Muneer, M., and Gopidas, K.R. (1994) Photocatalytic degradation of wastewater pollutants. Titanium-dioxide-mediated oxidation of polynuclear aromatic hydrocarbons. *J. Photochem. Photobiol., A*, **77**, 83.
37. Ohno, T., Tokieda, K., Higashida, S., and Matsumura, M. (2003) Synergism between rutile and anatase TiO_2 particles in photocatalytic oxidation of naphthalene. *Appl. Catal., A*, **244**, 383.
38. Li, X., Cubbage, J.W., and Jenks, W.S. (2001) Variation in the chemistry of the TiO_2-mediated degradation of hydroxy- and methoxybenzenes: electron transfer and $HO•_{ads}$ initiated chemistry. *J. Photochem. Photobiol., A*, **143**, 69.
39. Amalric, L., Guillard, C., and Pichat, P. (1994) Use of catalase and superoxide dismutase to assess the roles of hydrogen peroxide and superoxide in the TiO_2 of ZnO photocatalytic destruction

of 1,2-dimethoxybenzene in water. *Res. Chem. Int.*, **20**, 579.
40. Amalric, L., Guillard, C., and Pichat, P. (1995) The GC-MS identification of some aliphatic intermediates from the TiO_2 photocatalytic degradation of dimethoxybenzenes in water. *Res. Chem. Intermed.*, **21**, 33.
41. Muneer, M., Qamar, M., and Bahnemann, D. (2005) Photoinduced electron transfer reaction of few selected organic systems in presence of titanium dioxide. *J. Mol. Catal. A*, **234**, 151.
42. Muneer, M., Bahnemann, D., Qamar, M., Tariq, M.A., and Faisal, M. (2005) Photocatalysed reaction of few selected organic systems in presence of titanium dioxide. *Appl. Catal., A*, **289**, 224.
43. Jenks, W.S. (2005) in *Environmental Catalysis* (ed V.H. Grassian), CRC Press, Boca Raton, FL, p. 307.
44. Agrios, A.G. and Pichat, P. (2006) Recombination rate of photogenerated charges versus surface area: opposing effects of TiO_2 sintering temperature on photocatalytic removal of phenol, anisole, and pyridine in water. *J. Photochem. Photobiol., A*, **180**, 130.
45. Enriquez, R., Agrios, A.G., and Pichat, P. (2007) Probing multiple effects of TiO_2 sintering temperature on photocatalytic activity in water by use of a series of organic pollutant molecules. *Catal. Today*, **120**, 196.
46. Peller, J., Wiest, O., and Kamat, P.V. (2003) Mechanism of hydroxyl radical-induced breakdown of the herbicide 2,4-dichlorophenoxyacetic acid (2,4-D). *Chem. Eur. J.*, **9**, 5379.
47. Cermenati, L., Albini, A., Pichat, P., and Guillard, C. (2000) TiO_2 Photocatalytic degradation of haloquinolines in water: aromatic products GM-MS identification. Role of electron transfer and superoxide. *Res. Chem. Intermed.*, **26**, 221.
48. Cermenati, L., Pichat, P., Guillard, C., and Albini, A. (1997) Probing the TiO_2 photocatalytic mechanisms in water purification by Use of quinoline, photo-fenton generated $OH^•$ radicals and superoxide dismutase. *J. Phys. Chem. B*, **101**, 2650.
49. Enriquez, R. and Pichat, P. (2001) Interactions of humic acid, quinoline, and TiO_2 in water in relation to quinoline photocatalytic removal. *Langmuir*, **17**, 6132.
50. Kraeutler, B. and Bard, A.J. (1978) Heterogeneous photocatalytic synthesis of methane from acetic acid - new kolbe reaction pathway. *J. Am. Chem. Soc.*, **100**, 2239.
51. Irawaty, W., Friedmann, D., Scott, J., and Amal, R. (2011) Relationship between mineralization kinetics and mechanistic pathway during malic acid photodegradation. *J. Mol. Catal. A: Chem.*, **335**, 151.
52. Imamura, K., Iwasaki, S.-I., Maeda, T., Hashimoto, K., Ohtani, B., and Kominami, H. (2011) Photocatalytic reduction of nitrobenzenes to aminobenzenes in aqueous suspensions of titanium(iv) oxide in the presence of hole scavengers under deaerated and aerated conditions. *Phys. Chem. Chem. Phys.*, **13**, 5114.
53. Liu, H., Imanishi, A., and Nakato, Y. (2007) Mechanisms for photooxidation reactions of water and organic compounds on carbon-doped titanium dioxide, as studied by photocurrent measurements. *J. Phys. Chem. C*, **111**, 8603.
54. Yoon, S.-H., Oh, S.-E., Yang, J.E., Lee, J.H., Lee, M., Yu, S., and Pak, D. (2008) TiO_2 Photocatalytic oxidation mechanism of As(III). *Environ. Sci. Technol.*, **43**, 864.
55. Hathway, T., Chernyshov, D.L., and Jenks, W.S. (2011) Selectivity in the photo-Fenton and photocatalytic hydroxylation of biphenyl-4-carboxylic acid and derivatives (viz. 4-Phenylsalicylic acid and 5-phenylsalicylic acid). *J. Phys. Org. Chem.*, **24**, (12) 1151–1156. ISSN: 0894-3230 DOI: 10.1002/poc.1839.
56. Matthews, R.W. (1987) Photooxidation of organic impurities in water using thin films of titanium dioxide. *J. Phys. Chem.*, **91**, 3328.

57. Mills, A., Holland, C.E., Davies, R.H., and Worsley, D. (1994) Photomineralization of salicylic acid: a kinetic study. *J. Photochem. Photobiol., A*, **83**, 257.
58. Richard, C. and Boule, P. (1994) Is the oxidation of salicylic acid to 2,5-dihydroxybenzoic acid a specific reaction of singlet oxygen? *J. Photochem. Photobiol., A*, **84**, 151.
59. Tunesi, S. and Anderson, M. (1991) Influence of chemlsorption on the photodecomposition of salicylic acid and related compounds using suspended TiO, ceramic membranes. *J. Phys. Chem.*, **95**, 3399.
60. Brezova, V., Ceppan, M., Brandstetrova, E., Breza, M., and Lapcik, L. (1991) Photocatalytic hydroxylation of benzoic acid in aqueous titanium dioxide suspension. *J. Photochem. Photobiol., A*, **59**, 385.
61. Izumi, I., Fan, F.-R.F., and Bard, A.J. (1981) Heterogeneous photocatalytic decomposition of benzoic acid and adipic acid on platinized titanium dioxide powder. The photo-kolbe decarboxylative route to the breakdown of the benzene ring and to the production of butane. *J. Phys. Chem.*, **85**, 218.
62. Qamar, M., Muneer, M., and Bahnemann, D. (2005) Titanium-dioxide-mediated photocatalyzed reaction of selected organic systems. *Res. Chem. Int.*, **31**, 807.
63. Velegraki, T. and Mantzavinos, D. (2008) Conversion of benzoic acid during TiO_2-mediated photocatalytic degradation in water. *Chem. Eng. J.*, **140**, 15.
64. Abramovic, B., Kler, S., Sojic, D., Lausevic, M., Radovic, T., and Vione, D. (2011) Photocatalytic degradation of metoprolol tartrate in suspensions of two TiO_2-based photocatalysts with different surface area. Identification of intermediates and proposal of degradation pathways. *J. Hazard. Mater.*, **198**, 123.
65. Baciocchi, E., Bietti, M., Ferrero, M.I., Rol, C., and Sebastiani, G.V. (1998) Photo-oxidative fragmentation of some a-alkyl substituted 4-methoxybenzyl alcohols and methyl ethers sensitized by TiO_2. *Acta Chem. Scand.*, **52**, 160.
66. Ranchella, M., Rol, C., and Sebastiani, G.V. (2000) Some evidence in favor of an electron transfer mechanism in the TiO_2 photosensitized oxidation of benzyl derivatives in aqueous media. *J. Chem. Soc., Perkin Trans. 2*, 311.
67. Rockafellow, E.M., Haywood, J.M., Witte, T., Houk, R.S., and Jenks, W.S. (2010) Selenium-modified TiO_2 and its impact on photocatalysis. *Langmuir*, **26**, 19052.
68. Hathway, T. and Jenks, W.S. (2008) Effects of sintering of TiO_2 particles on the mechanisms of photocatalytic degradation of organic molecules in water. *J. Photochem. Photobiol., A*, **200**, 216.
69. Rockafellow, E.M., Stewart, L.K., and Jenks, W.S. (2009) Is sulfur-doped TiO2 an effective visible light photocatalyst for remediation? *Appl. Catal., B*, **91**, 554.
70. Hathway, T., Rockafellow, E.M., Oh, Y.-C., and Jenks, W.S. (2009) Photocatalytic degradation using tungsten-modified TiO_2 and visible light: kinetic and mechanistic effects using multiple catalyst doping strategies. *J. Photochem. Photobiol., A*, **207**, 197.
71. Marugan, J., Hufschmidt, D., Lopez-Munoz, M.-J., Selzer, V., and Bahnemann, D. (2006) Photonic efficiency for methanol photooxidation and hydroxyl radical generation on silica-supported TiO_2 photocatalysts. *Appl. Catal., B*, **62**, 201.
72. Russell, G.A. (1957) Deuterium-isotope effects in the autoxidation of aralkyl hydrocarbons. Mechanism of the interaction of peroxy radicals. *J. Am. Chem. Soc.*, **79**, 3871.
73. Vulliet, E., Chovelon, J.-M., Guillard, C., and Herrmann, J.-M. (2003) Factors influencing the photocatalytic degradation of sulfonylurea herbicides by TiO_2 aqueous suspension. *J. Photochem. Photobiol., A*, **159**, 71.
74. Vulliet, E., Emmelin, C., Chovelon, J.-M., Guillard, C., and Herrmann, J.-M. (2002) Photocatalytic degradation of sulfonylurea herbicides in aqueous TiO_2. *Appl. Catal., B*, **38**, 127.

75. Pramauro, E., Prevot, A.B., Vincenti, M., and Brizzolesi, G. (1997) Photocatalytic degradation of carbaryl in aqueous solutions containing TiO_2 suspensions. *Environ. Sci. Technol.*, **31**, 3126.
76. Pramauro, E., Vincenti, M., Augugliaro, V., and Palmisano, L. (1993) Photocatalytic degradation of monuron in aqueous titanium dioxide dispersions. *Environ. Sci. Technol.*, **27**, 1790.
77. Macounová, K., Krysová, H., Ludvík, J., and Jirkovsky, J. (2002) Kinetics of photocatalytic degradation of diuron in aqueous colloidal solutions of Q-TiO_2 particles. *J. Photochem. Photobiol., A*, **156**, 273.
78. Maurino, V., Minero, C., Pelizzetti, E., and Vincenti, M. (1999) Photocatalytic transformation of sulfonylurea herbicides over irradiated titanium dioxide particles. *Colloids Surf., A*, **151**, 329.
79. Bahnemann, W., Muneer, M., and Haque, M.M. (2007) Titanium dioxide-mediated photocatalyzed degradation of few selected organic pollutants in aqueous suspensions. *Catal. Today*, **124**, 133.
80. Lambropoulou, D.A., Hernando, M.D., Konstantinou, I.K., Thurman, E.M., Ferrer, I., Albanis, T.A., and Fernandez-Alba, A.R. (2008) Identification of photocatalytic degradation products of bezafibrate in TiO_2 aqueous suspensions by liquid and gas chromatography. *J. Chromatogr. A*, **1183**, 38.
81. Abellán, M.N., Bayarri, B., Giménez, J., and Costa, J. (2007) Photocatalytic degradation of sulfamethoxazole in aqueous suspension of TiO_2. *Appl. Catal., B*, **74**, 233.
82. Guzsvany, V., Rajic, L., Jovic, B., Orcic, D., Csanadi, J., Lazic, S., and Abramovic, B. (2012) Spectroscopic monitoring of photocatalytic degradation of the insecticide acetamiprid and its degradation product 6-chloronicotinic acid on TiO_2 catalyst. *J. Environ. Sci. Health. Part A Toxic/Hazard. Subst. Environ. Eng.*, **47**, 1919.
83. Fukahori, S., Fujiwara, T., Ito, R., and Funamizu, N. (2012) Photocatalytic decomposition of crotamiton over aqueous TiO_2 suspensions: determination of intermediates and the reaction pathway. *Chemosphere*, **89**, 213.
84. Marci, G., Di Paola, A., Garcia-Lopez, E., and Palmisano, L. (2007) Photocatalytic oxidation mechanism of benzonitrile in aqueous suspensions of titanium dioxide. *Catal. Today*, **129**, 16.
85. Hidaka, H., Garcia-Lopez, E., Palmisano, L., and Serpone, N. (2008) Photoassisted mineralization of aromatic and aliphatic N-heterocycles in aqueous titanium dioxide suspensions and the fate of the nitrogen heteroatoms. *Appl. Catal., B*, **78**, 139.
86. Maurino, V., Minero, C., Pelizzetti, E., Piccinini, P., Serpone, N., and Hidaka, H. (1997) The fate of organic nitrogen under photocatalytic conditions: degradation of nitrophenols and aminophenols on irradiated TiO_2. *J. Photochem. Photobiol., A*, **109**, 171.
87. Vorontsov, A.V., Davydov, L., Reddy, E.P., Lion, C., Savinov, E.N., and Smirniotis, P.G. (2002) Routes of photocatalytic destruction of chemical warfare agent simulants. *New J. Chem.*, **26**, 732.
88. Vorontsov, A.V., Lion, C., Savinov, E.N., and Smirniotis, P.G. (2003) Pathways of photocatalytic gas phase destruction of HD simulant 2-chloroethyl ethyl sulfide. *J. Catal.*, **220**, 414.
89. O'Shea, K.E. (2003) Titanium dioxide-photocatalyzed reactions of organophosphorus compounds in aqueous media. *Mol. Supramol. Photochem.*, **10**, 231.
90. Oncescu, T., Stefan, M.I., and Oancea, P. (2010) Photocatalytic degradation of dichlorvos in aqueous TiO2 suspensions. *Environ. Sci. Pollut. Res.*, **17**, 1158.
91. Vorontsov, A.V., Kozlov, D.V., Smirniotis, P.G., and Parmon, V.N. (2005) TiO2 Photocatalytic oxidation. I. Photocatalysts for liquid-phase and gas-phase processes and the photocatalytic degradation of chemical warfare agent simulants in a liquid phase. *Kinet. Catal.*, **46**, 189.

92. Evgenidou, E., Konstantinou, I., Fytianos, K., and Albanis, T. (2006) Study of the removal of dichlorvos and dimethoate in a titanium dioxide mediated photocatalytic process through the examination of intermediates and the reaction mechanism. *J. Hazard. Mater.*, **137**, 1056.

93. Aungpradit, T., Sutthivaiyakit, P., Martens, D., Sutthivaiyakit, S., and Kettrup, A.A.F. (2007) Photocatalytic degradation of triazophos in aqueous titanium dioxide suspension: identification of intermediates and degradation pathways. *J. Hazard. Mater.*, **146**, 204.

94. Moctezuma, E., Leyva, E., Palestino, G., and de Lasa, H. (2007) Photocatalytic degradation of methyl parathion: reaction pathways and intermediate reaction products. *J. Photochem. Photobiol., A*, **186**, 71.

95. Dixit, V., Tewari, J.C., and Obendorf, S.K. (2009) Identification of degraded products of aldicarb due to the catalytic behavior of titanium dioxide/polyacrylonitrile nanofiber. *J. Chromatogr. A*, **1216**, 6394.

96. Fenoll, J., Hellin, P., Martinez, C.M., Flores, P., and Navarro, S. (2012) Semiconductor oxides-sensitized photodegradation of fenamiphos in leaching water under natural sunlight. *Appl. Catal., B*, **115-116**, 31.

97. Bahnemann, D. (1999) in *Handbook of Environmental Chemistry*, Vol. **2 L** (ed P. Boule), Springer, Berlin, p. 285.

98. Choi, W. and Hoffmann, M.R. (1996) Kinetics and mechanism of CCl_4 photoreductive degradation on TiO_2: the role of trichloromethyl radical and dichlorocarbene. *J. Phys. Chem.*, **100**, 2161.

99. Bahnemann, D.W., Moenig, J., and Chapman, R. (1987) Efficient photocatalysis of the irreversible one-electron and two-electron reduction of halothane on platinized colloidal titanium dioxide in aqueous suspension. *J. Phys. Chem.*, **91**, 3782.

100. Guillard, C., Pichat, P., Huber, G., and Hoang-Van, C. (1996) The GC-MS analysis of organic intermediates from the TiO_2 photocatalytic treatment of water contaminated by lindane (1.α., 2.α., 3.β., 4.α., 5.α., 6.β.-hexachlorocyclohexane). *J. Adv. Oxid. Technol.*, **1**, 53.

101. Richard, C. (1994) Photocatalytic reduction of benzoquinone in aqueous ZnO or TiO_2 suspensions. *New J. Chem.*, **18**, 443.

102. Oh, Y.-C., Li, X., Cubbage, J.W., and Jenks William, S. (2004) Mechanisms of catalyst action in the TiO_2-mediated photocatalytic degradation of maleic and fumaric acid. *Appl. Catal., B*, **54**, 105.

103. Franch, M.I., Ayllon, J.A., Peral, J., and Domenech, X. (2002) Photocatalytic degradation of short-chain organic diacids. *Catal. Today*, **76**, 221.

104. Tschirch, J., Dillert, R., Bahnemann, D., Proft, B., Biedermann, A., and Goer, B. (2008) Photodegradation of methylene blue in water, a standard method to determine the activity of photocatalytic coatings? *Res. Chem. Int.*, **34**, 381.

105. Tschirch, J., Dillert, R., and Bahnemann, D. (2008) Photocatalytic degradation of methylene blue on fixed powder layers: which limitations are to be considered? *J. Adv. Oxid. Technol.*, **11**, 193.

106. Mrowetz, M., Balcerski, W., Colussi, A.J., and Hoffmann, M.R. (2004) Oxidative power of nitrogen-doped TiO_2 photocatalysts under visible illumination. *J. Phys. Chem. B*, **108**, 17269.

107. Yan, X., Ohno, T., Nishijima, K., Abe, R., and Ohtani, B. (2006) Is methylene blue an appropriate substrate for a photocatalytic activity test? a study with visible-light responsive titania. *Chem. Phys. Lett.*, **429**, 606.

108. Ryu, J. and Choi, W. (2008) Substrate-specific photocatalytic activities of TiO_2 and multiactivity test for water treatment application. *Environ. Sci. Technol.*, **42**, 294.

109. Richard, C. (1993) Regioselectivity of oxidation by positive holes (h+) in photocatalytic aqueous transformations. *J. Photochem. Photobiol., A*, **72**, 179.

110. Kim, S. and Choi, W. (2005) Visible-light-induced photocatalytic degradation of 4-chlorophenol and phenolic compounds in aqueous suspensions of pure titanium: demonstrating the existence of a surface-complex-mediated path. *J. Phys. Chem. B*, **109**, 5143.

111. Agrios, A.G., Gray, K.A., and Weitz, E. (2003) Photocatalytic transformation of 2,4,5-trichlorophenol on TiO2 under Sub-band-Gap illumination. *Langmuir*, **19**, 1402.

112. Agrios, A.G., Gray, K.A., and Weitz, E. (2004) Narrow-band irradiation of a homologous series of chlorophenols on TiO_2: charge-transfer complex formation and reactivity. *Langmuir*, **20**, 5911.

113. Hurum, D.C., Gray, K.A., Rajh, T., and Thurnauer, M.C. (2004) Photoinitiated reactions of 2,4,6 TCP on degussa P25 formulation TiO_2: wavelength-sensitive decomposition. *J. Phys. Chem. B*, **108**, 16483.

114. Emeline, A., Salinaro, A., and Serpone, N. (2000) Spectral dependence and wavelength selectivity in heterogeneous photocatalysis. I. Experimental evidence from the photocatalyzed transformation of phenols. *J. Phys. Chem. B*, **104**, 11202.

115. Emeline, A.V., Lobyntseva, E.V., Ryabchuk, V.K., and Serpone, N. (1999) Spectral dependencies of the quantum yield of photochemical processes on the surface of wide-band-Gap metal oxides. 2. Gas/solid system involving scandia (Sc_2O_3) particles. *J. Phys. Chem. B*, **103**, 1325.

116. Serpone, N. (2006) Is the band Gap of pristine TiO_2 narrowed by anion- and cation-doping of titanium dioxide in second-generation photocatalysts? *J. Phys. Chem. B*, **110**, 24287.

117. Kuznetsov, V.N. and Serpone, N. (2009) On the origin of the spectral bands in the visible absorption spectra of visible-light-active TiO_2 specimens analysis and assignments. *J. Phys. Chem. C*, **113**, 15110.

118. Ohno, T., Mitsui, T., and Matsumura, M. (2003) Photocatalytic activity of S-doped TiO_2 photocatalyst under visible light. *Chem. Lett.*, **32**, 364.

119. Ohno, T., Akiyoshi, M., Umebayashi, T., Asai, K., Mitsui, T., and Matsumura, M. (2004) Preparation of S-doped TiO_2 photocatalysts and their photocatalytic activities under visible light. *Appl. Catal., A*, **265**, 115.

120. Sakthivel, S. and Kisch, H. (2003) Photocatalytic and photoelectrochemical properties of nitrogen-doped titanium dioxide. *ChemPhysChem*, **4**, 487.

121. Tachikawa, T., Tojo, S., Kawai, K., Endo, M., Fujitsuka, M., Ohno, T., Nishijima, K., Miyamoto, Z., and Majima, T. (2004) Photocatalytic oxidation reactivity of holes in the sulfur- and carbon-doped TiO_2 powders studied by time-resolved diffuse reflectance spectroscopy. *J. Phys. Chem. B*, **108**, 19299.

122. Tachikawa, T., Takai, Y., Tojo, S., Fujitsuka, M., Irie, H., Hashimoto, K., and Majima, T. (2006) Visible light-induced degradation of ethylene glycol on nitrogen-doped TiO_2 powders. *J. Phys. Chem. B*, **110**, 13158.

123. Tojo, S., Tachikawa, T., Fujitsuka, M., and Majima, T. (2008) Iodine-Doped TiO_2 photocatalysts: correlation between band structure and mechanism. *J. Phys. Chem. C*, **112**, 14948.

124. Nakamura, R., Tanaka, T., and Nakato, Y. (2004) Mechanism for visible light responses in anodic photocurrents at N-doped TiO_2 film electrodes. *J. Phys. Chem. B*, **108**, 10617.

125. Emeline, A.V., Zhang, X., Jin, M., Murakami, T., and Fujishima, A. (2009) Spectral dependences of the activity and selectivity of N-doped TiO_2 in photodegradation of phenols. *J. Photochem. Photobiol., A*, **207**, 13.

3
Photocatalytic Mechanisms and Reaction Pathways Drawn from Kinetic and Probe Molecules

Claudio Minero, Valter Maurino, and Davide Vione

3.1
The Photocatalyic Rate

The intrinsic complexity of the photocatalytic process comprises multiple elementary steps. Their analysis and kinetic relevance have been recognized in the former research on photocatalysis. Recently, the key role of surface species on the electron-transfer mechanism and the role of different crystal surfaces emerged as it had been the case in gas-phase thermal catalysis for a long time. Their intimate role is relevant for the whole process efficiency, eventual product selectivity, and applications.

The analysis of the photocatalytic rate as a function of substrate concentration, light intensity, and catalyst loading, and type of the produced intermediates could give some insight into the mechanism of electron transfer to/from the solution species. After an electron–hole pair is produced in the bulk TiO_2 by absorption of a photon with energy higher than its bandgap, the photogenerated hole and the electron diffuse to the surface accompanied by bulk recombination. Electron and hole polarons are localized at Ti^{3+} and O^- lattice sites, respectively. At the surface, the trapping sites generally correspond to undercoordinated ${Ti^{3+}}_{5c}$ and ${O^-}_{2c}$ surface atoms or to isolated OH species in the case of a hydroxylated surface. Using hybrid functional electronic structure calculations, at the (101) surface of anatase TiO_2, the polaron trapping energy is considerably larger at the surface than in the bulk, indicating that it is energetically favorable for the polarons to travel from the bulk to the surface [1]. On the surface, these holes and electrons will be consumed by surface recombination and interfacial charge transfer. The surface-continuity condition is that the hole–electron generation rate on the surface is equal to the consumption rate. Neglecting time transients, this corresponds to the steady-state assumption.

A basic kinetic model based on the primary events occurring in the photocatalytic transformation of the substrate, such as (i) light absorption, (ii) charge carrier recombination, (iii) interfacial hole transfer, (iv) electron transfer, and (v) an eventual further oxidation of the produced oxidized radical via a second oxidation or electron injection in the semiconductor conduction band was able to express the degradation rate of a reduced substrate $r = -\mathrm{d}\left\{\mathrm{Red}_1\right\}/\mathrm{d}t = k_3\left\{\mathrm{h}_\mathrm{s}^+\right\}\left\{\mathrm{Red}_1\right\}$,

Photocatalysis and Water Purification: From Fundamentals to Recent Applications, First Edition. P. Pichat.

where the curly brackets denotes surface concentrations, or the quantum yield ($\eta = \text{rate}/\varphi$, where φ is the volumetric rate of absorption of photons) as a function of microscopic kinetic rate constants [2].

$$\text{Light absorption:}\quad TiO_2 \rightarrow e_b^- + h_b^+ \tag{3.1}$$

Charge carrier migration to the surface:

$$e - b/h + b \rightarrow e - s/h + s \qquad (K\_1)(3.2)$$

Recombination at the surface :

$$e_s^- + h_s^+ \rightarrow \text{Heat, Light} \qquad (K\_2)(3.3)$$

$$\text{Interfacial charge transfer:}\quad h_s^+ + Red_{1s} \rightarrow Ox_{1s}^{\bullet} \qquad (K\_3)(3.4)$$

$$e_s^- + Ox_{2s} \rightarrow Red_{2s}^{\bullet} \qquad (K\_4)(3.5)$$

$$\text{Second oxidation step: } Ox_{1s}^{\bullet} + h_s^+ \rightarrow Ox_{1s} \qquad (K\_5a)(3.6)$$

$$\text{Current doubling: } Ox_{1s}^{\bullet} \rightarrow Ox_{1s} + e_{CB}^- \qquad (K\_5b)(3.7)$$

$$Ox_{1s}^{\bullet} + Red_{2s}^{\bullet} \rightarrow P \qquad (K\_6)(3.8)$$

In the above reaction framework, charges of radicals and eventual acid/base equilibrium have been intentionally omitted to simplify the managing without sacrificing generality. The resulting expressions (see Ref. [2] for details) for the quantum yield were expressed as a function of the a-dimensional variable γ (defined below):

$$\eta = -\gamma + \sqrt{\gamma(\gamma + 2)}, \text{considering Eqs. (3.1) – (3.5)} \\ \left(\text{and optionally Eq. (3.8)}\right), \text{or} \tag{3.9}$$

$$\eta = -\gamma + \sqrt{\gamma(\gamma + 1)}, \text{considering Eqs. (3.1) – (3.6) , or} \tag{3.10}$$

$$\eta = -\frac{\gamma}{2} + \sqrt{\frac{\gamma}{2}\left(\frac{\gamma}{2} + 2\right)}, \text{considering Eqs. (3.1) – (3.5) and (3.7)} \\ \left(\text{case of current doubling}\right) \tag{3.11}$$

In the case of Eqs. (3.1)–(3.5) and 3.8 the photocatalytic system is truly photocatalytic, as Ox_2 is added to Red_1 via the intervention of photons. The quantum yield depends just on the a-dimensional variable γ, which is then the master variable for the photocatalytic rate. It is here reported in a formulation slightly different from the before one [2] as

$$\gamma = k_o \frac{\theta_{Red}\theta_{ox} C_{cat}}{\varphi} \tag{3.12}$$

where $\theta_i = (K_i[I_{free}])/(1 + K_i[I_{free}])$, $k_o = k_{3,s}k_{4,s}S/2\ln(10)\ \ k_{2,s}a_s^2$, k_{is} are surface rate constants, a_s ($m^2\ mol^{-1}$) is the specific area of the adsorption site, and K_i is the adsorption constant. Often, in an open system in contact with air, θ_{ox} can be assumed constant. The rate is equal to $\eta\ \varphi$, where φ is the rate of absorption of photons in units consistent with those of the rate ($mol\ l^{-1}\ s^{-1}$).

The dependence of the rate on the substrate concentration is given in Figure 3.1 for the cases of negligible adsorption and significant adsorption, according to Eqs. (3.9)–(3.11).

Figure 3.1 shows the three cases under different conditions. The kinetic dependence of the rate on experimental parameters could hardly discriminate the three (a, b, c) mechanisms corresponding to Eqs. (3.9)–(3.11). Clearly, the quantum yield and the rate monotonically increase as γ increases. The increase of surface concentration of both species or the decrease of the absorbed light would increase the quantum yield. Accordingly, for a given substrate, the increase of the quantum yield can be achieved by reducing φ. The proper choice of the catalyst (with low recombination rate constant $k_{2,s}$, catalyst dependent, and large $k_{3,s}$, substrate and catalyst dependent) will increase both the rate and the quantum yield. They also increase with S, the specific surface area of the catalyst, or with the moles of adsorbing sites per gram of catalyst (S/a_s), with its loading C_{cat} (in a complex way because C_{cat} increases/decreases φ depending on the scattering properties of the catalyst particles [2]), and in an extent, depending on K_i with the solution concentration $[Red]_{free}$ (or $[Ox]_{free}$).

The rate depends always from the square root of the substrate concentration as long as it is far from the plateau, where it is independent. Conversely, the rate shows two different dependencies on the rate of light absorption φ as Figure 3.1 testifies: (i) in the case $k_o\theta_{Red}\theta_{Ox}\, C_{cat} \ll \varphi$, the rate is proportional to the square root of φ and (ii) in the opposite case for which $k_o\theta_{Red}\theta_{Ox}\, C_{cat} \gg \varphi$, the rate has a linear dependence on φ and the quantum yield η is constant. In the case of full coverage of the surface sites ($\theta_{Red} = 1$ if K_{Red} is sufficiently large or all available sites are saturated at a sufficiently large solution concentration), the rate will level off (because γ is constant) according to Eq. (3.13) at a value depending on φ, C_{cat}, the surface amount of oxidant, and obviously, the ratio k_o of surface kinetic constants.

$$\text{Rate}_{plateau} = k_o\theta_{ox}C_{cat}\left(\sqrt{1+\frac{2\varphi}{k_o\theta_{ox}C_{cat}}}-1\right) \tag{3.13}$$

For a given reactor, the rate(z) $= \eta\ \varphi(z)$, where $\varphi = \ln(10)\ \varepsilon\ C_{cat}\ I(z)$, ε is the catalyst absorption coefficient with dimension ($cm^2\ g^{-1}$), $I(z)$ is the actual incident photon irradiance (moles of photons $s^{-1}\ cm^{-2}$) at point z in the reactor, and C_{cat} is the catalyst concentration ($g\ l^{-1}$). A proper experimental evaluation of the kinetic model requires the integration of rate(z) over the reactor volume. Details of this procedure are given for a monodimensional reactor [2]. The model was successfully validated on kinetic data for phenol at low concentration [2].

3.1.1 Other Kinetic Models

The degradation rate is definitely proportional to the surface concentration of the substrate $\{Red_1\}$, as in the Langmuir–Hinshelwood (LH) approach still used today [3]. Under conditions for which $\{h_s^+\}$ (the lumped surface concentration of oxidizing species) is constant, the LH model would work correctly. As already

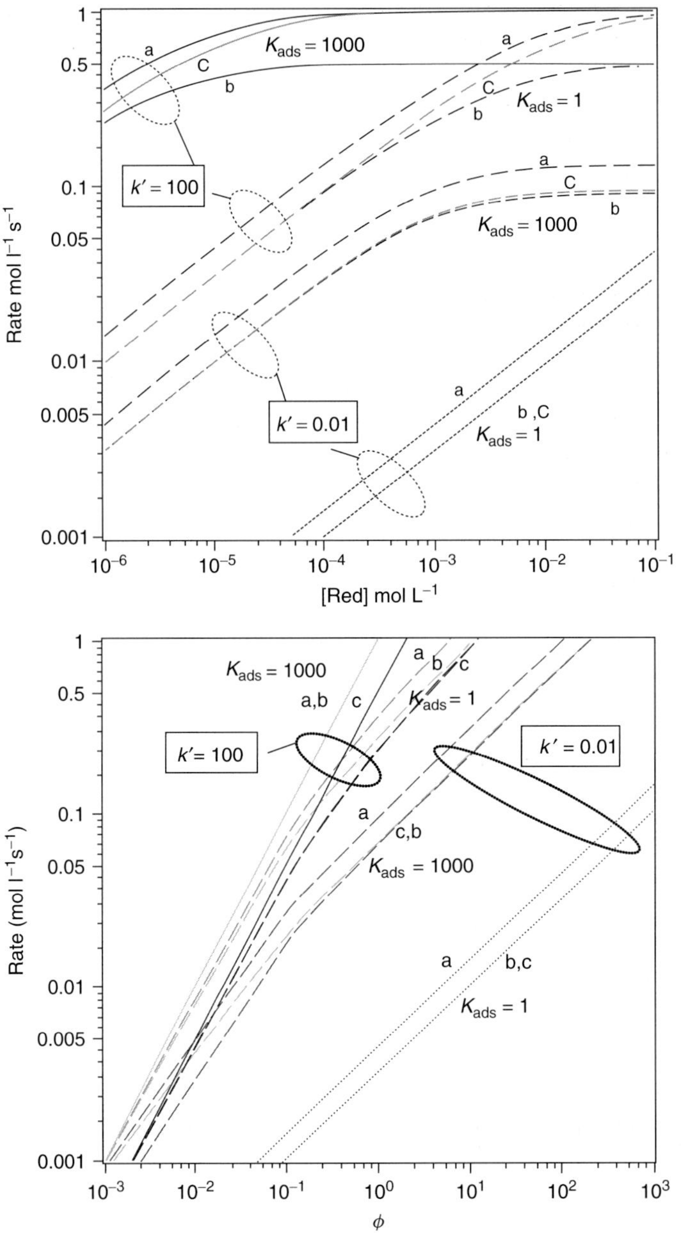

Figure 3.1 Top: The dependence of the rate according to Eqs. (3.9)–(3.11), indicated by letters a, b, c, respectively, on the substrate concentration (Red) and the adsorption constant K_{ads} ($K_{ads} = 1$ dot and long-dashed lines, $K_{ads} = 1000$ dashed and solid dot lines). The plot is on logarithmic scales. The actual (arbitrary) values $k' = k_o \theta_{ox} C_{cat}$ (Eq. (3.12)) are indicated, $\varphi = 1$. Bottom: The dependence of the rate according to Eqs. (3.9)–(3.11) on φ at constant (Red) = 0.001 mol l^{-1}. Symbols and condition as above.

criticized and pointed out [4], the actual concentration of the photogenerated reactive species $\{h_s^+\}$ is generally not constant as a function of the experimental conditions. This makes the adsorption constant in LH equation dependent on light intensity, which is obviously not compatible with the term constant and indicates that the LH model is inadequate.

The direct–indirect (D–I) model [5, 6] introduced in photocatalysis classical concepts such as direct, indirect, adiabatic, and inelastic electron-transfer mechanisms. The resulting kinetic set of equations and the rate equations are complex because of the detailing of factors that are lumped in the above kinetic equation. Two *alternative* types of interfacial hole-transfer mechanisms leading to different kinetic behavior are considered by the D–I model. For strong electronic interaction of dissolved substrate species with the semiconductor surface (specific adsorption), interfacial transfer of charge takes place via a mixture of adiabatic, indirect transfer (IT) of surface-trapped holes to dissolved substrate species and inelastic direct transfer (DT) of valence band (VB) free holes to substrate species bound to terminal Ti atoms. For weak electronic interaction of dissolved substrate species with the surface (total absence of specific adsorption), photocatalytic oxidation is assumed to take place exclusively via an IT mechanism involving surface-trapped holes. DT is assumed to coexist with IT in all cases where specific adsorption, even if small, takes place to some extent.

The model based on Eqs. (3.9)–(3.11) explicitly did not consider the distinct surface species that absorption of light would produce (deep or shallow surface-trapped hole and electrons, or free $OH^•$). For this reason, it can be applied to an indirect electron transfer (for physisorbed species, for which K_{Red} could be assumed low or null, with proper surface rate constants $k_{o,out}$ for outer sphere electron transfer) or to a direct electron transfer for chemisorbed species (with different surface rate constants $k_{o,in}$ for inner sphere electron transfer).

Conversely, the D–I model forecasts that the IT mechanism prevails on the DT mechanism for low enough photon flux, while DT prevails on IT for high enough illumination intensity. A photon-flux-independent quantum yield for the photocatalytic oxidation is predicted in the last case. Although it is difficult to understand on a chemical basis the intimate reason of the effect of light on a kinetic mechanism that depends on the way molecules are in contact with the surface, the constancy of the quantum yield with φ agrees with the guess of Eq. (3.9) for the case $k_o\theta_{Red}\theta_{Ox}\,C_{cat} \gg \varphi$. The conclusion of the D–I model thus applies for the case $k_{o,in} \gg k_{o,out}$ only for a given arbitrary value of φ as Figure 3.1 shows.

The D–I model belongs to the class of models depicted above in Eqs. (3.9)–(3.11) because the back reaction (see below) is explicitly neglected.

3.1.2
Substrate-Mediated Recombination

The (initial) rate of transformation levels off as a function of the substrate concentration at relatively low substrate concentration even for poor adsorbing substrates and at larger concentrations it often decreases (see a detailed discussion and older references in [4]). The typical case is phenol, which is reported to exhibit a weak interaction with the TiO_2 surface [7, 8], although its adsorption has been demonstrated from both aqueous [9–11] and gaseous phases [12]. In addition, in photoelectrochemical experiments, increasing the phenol concentration from 4 to 10 mM, the initial increasing photocurrent diminish, indicating that the phenol radical generated on oxidation, could act as a recombination center [8] or dimerize [13]. Consequently, the plateau of the rate and even more its further decrease could not be explained with the hypotheses used above.

The importance of adsorption is generally overestimated or misinterpreted. With regard to the adsorbed species, the possibility of adsorption rely on the functional groups, the orientation of charges or localized electronic densities, and hydrophobicity (chain length and molecular size). Because reactive species are present at the catalyst surface and do not migrate in solution [14], a positive effect of adsorption is expected as predicted by Eq. (3.9). However, there is evidence that a chemisorbed species seems less reactive than the physisorbed counterparts [15] and that the positive effect of adsorption is often not as much as that expected based on the surface concentration of species, as observed by comparing the reactivity of phenol and alkylated phenol [4].

The dimerization at the surface of the produced radical was demonstrated years ago in the degradation of various molecules [16], including chlorinated phenols [17]. The amount of detected intermediates was significant, suggesting that the local concentration of (very reactive) radicals is high. This imposes some caution to kinetic models development, mainly when, with the aim to reduce the complexity of resulting equations, some concentration is assumed negligible, as discussed below. If the surface concentration of produced radicals is large, and considering that they are very reactive, they could react not only with themselves but also with photogenerated surface species.

On the basis of the above concern, Minero [18] supposed, without explicitly affirming if they are free or chemisorbed, that the radical intermediates produced by oxidation and present at the surface would recombine with conduction band electrons (the so-called back reaction or substrate-mediated recombination), a concept that was invoked later by others [19] and also questioned (see below [20]).

$$\text{Back reactions } Ox_1^{\bullet} + e- \rightarrow Red_1 \qquad (K\_7)(3.14)$$

$$Red_2^{\bullet} + h^+ \rightarrow Ox_2 \qquad (K\_8)$$

The mediated recombination would reduce the rate $r = -d\{Red_1\}/dt = k_3\{h_s^+\}\{Red_1\} - k_7\{Ox_1^{\bullet}\}\{e_s^-\}$ by an amount proportional to the produced oxidized substrate. Thus, back reactions are favored by accumulation of the oxidized radical at the surface, because of the significant adsorption either for large K_i or for high substrate concentration. Conversely, the back reactions could be considered unimportant when the concentration of the substrate is low and it is poorly adsorbed. Application of Eq. (3.9) is thus limited far below the plateau appearing at larger substrate concentration. For other cases, a complex analytic solution forecasts that given three lumped kinetic parameters with physical significance, the rate can level off or also decrease after a maximum, as experimentally observed [18]. The leveling of the rate is mainly due to the balance of Eq. (K_7)(3.14) versus (3.4). Generally, analytical solutions for the rate are not possible, unless proper simplification/decoupling of the rate equations are carried out.

Recent literature reports a number of papers debating kinetic models able to take into account all the entangled dependences above discussed.

Valencia *et al.* [21] reported an implicit relationship of the rate considering the back reaction. Their approach considers the free electron/hole generation, the migration of the hole to a surface site, recombination of the trapped hole with free electrons, reaction of the trapped hole with Red_1, and the back reaction of the produced oxidized radical $Ox_1^{\bullet}$ with the free electron. They applied to the data of Emmeline *et al.* [22] their implicit equation for the rate and that obtained in the absence of back reaction (see Eq. (3.9) or the equivalent equation resulting from the IT mechanism of D–I model [20] in the absence of specific adsorption) showing that the disregard of the back reaction permits the fit only for low φ. Because a better fit is obtained using the back reaction concept, they criticized the assumption of Monllor-Satoca *et al.* [5, 6] that did not take into account the back reaction by assuming the concentration of $Ox_1^{\bullet}$ (incidentally in their notation $RH_{(aq)}^{\bullet}$) to be negligible.

In a recent reply to the paper [21], Montoya *et al.* [23] remarked that the set of equations used by Valencia *et al.* are a precise reproduction of the sequence of interfacial reactions sustaining the D–I model (see Section 3 in Ref. [20]), for which the rate equation given in Ref. [20] is similar to the rate equation derived from Eq. (3.9). The authors criticized the above approach because of the following.

1) The derived equations are somehow inconsistent and their derivation is doubtful.
2) The model was not based on the degree of interaction of dissolved substrate species with the semiconductor surface and on the nature of the interfacial charge transfer mechanisms, as performed by the D–I model. The critics is loose as in the master variable y the "degree of interaction" is accounted for by the adsorption constant. Although the D–I model does not refuse the existence of the "back reaction" (see, for instance, Eq. (3.13) in Ref. [20]), "it denies the actual influence of this step on the photocatalytic oxidation process of nonchemisorbed substrate species, since for $t \to 0$, the concentration of both $OH^{\bullet}_{aq}$ and $RH^{\bullet}_{aq}$ radicals is negligible with respect to the concentration of surface-trapped holes (h_s^+) [20]." Even though the first sentence has his valid grounds, the last sentence is contradictory because the model assumes the steady state, which is not valid at the transient $t \to 0$. The extrapolation to the initial rate is just a means to simplify the set of equations (assuming time-independent concentration of the substrate), which merely implies that $dC_i/dt = 0$ and not that $C_i = 0$ (where i is the transient intermediate). Finally, the sentence suffers the critics arose above on the experimental evidence that $RH^{\bullet}$ radicals are present at nonnegligible concentrations.
3) The "current doubling reaction" instead of the "back reaction" should be considered when analyzing the photocatalytic oxidation of nonchemisorbed

species. As evident from the previous section, the current doubling model (Eq. (3.11)) does not forecast a kinetic behavior different from that of the simpler model (Eq. (3.9) or (3.10)) because all of them are unable to predict a decrease in the rate as a function of the substrate concentration. In addition, it was never reported, in contrast with that observed for small alcohols and formic acid, that phenol radical cation is able to inject electron in the conduction band.

A numerical kinetic model that comprehensively considered the photoexcitation, bulk recombination, carrier diffusion, surface recombination, and interfacial hole transfer was developed to analyze the effects on the quantum yield of light intensity, grain size, carrier lifetime, and minority carrier diffusion coefficient for spherical TiO_2 nanoparticles [24]. Besides the analysis of charge carrier transport to the surface, the difference of the interfacial indirect charge transfer model from the D–I model of Villarreal *et al.* [5] is that electron trapping by the surface state is considered (see their ν_5) and this site could back react with surface-trapped holes (see their ν_{r3}). This model does not give the rate equation in analytical form.

3.2 Surface Speciation

3.2.1 Different Commercial Catalysts

It is not new that the titania catalysts prepared by different flame or solution strategies exhibit different chemical behavior. For example, hydroxyl groups on the surface of hydrated TiO_2 P25 are able to transform benzaldehyde molecules in hemiacetalic-like surface species, whereas C_6H_5CHO molecules are only weakly perturbed by interaction with the OH groups on TiO_2 Merck [25]. In the investigation on glycerol transformation [26], it was found that the selectivity among C_3 (glyceraldehyde (2,3-dihydroxypropanal) + dihydroxyacetone (1,3-dihydroxy-2-propanone)) and $C_2 + C_1$ products (formaldehyde and glycolaldehyde (2-hydroxyacetaldehyde)) was present only on P25. An important issue was that the selectivity was dependent on glycerol concentration. Other interesting features were the unusual rate dependence on glycerol concentration, with a sharp maximum, over Degussa P25 TiO_2, and the more usual increasing behavior over Merck TiO_2 and on both fluorinated TiO_2s.

Recently, a survey of the activity of different TiO_2 powders against a variety of substrates demonstrated a wide variation of substrate selectivity [27]. The ratio of phenol/formic acid removal rates varies from 0.8, for Anatase Aldrich, to 37, for Hombikat UV100. Augugliaro *et al.* [28] showed that the product selectivity in the transformation of *trans*-ferulic acid ((*E*)-3-(4-hydroxy-3-methoxy-phenyl)prop-2-enoic acid) to vanillin (4-hydroxy-3-methoxybenzaldehyde) is dependent on the nature of the anatase powder used. At similar substrate conversions, the vanillin yield on Merck TiO_2 is three- to fourfold higher with respect to the yield obtained over P25.

The product selectivity was observed in the photoconversion of NO depending on whether TiO_2 P25 was either prereduced or preoxidized [29] and in relation to the coordination type of surface Ti. The presence of TiO_2 containing tetrahedrally coordinated Ti(IV) ions plays a significant role in the decomposition of NO with a high selectivity for the formation of N_2 and O_2, whereas the catalysts involving the aggregated octahedrally coordinated titanium oxide species show a high selectivity to produce N_2O [30]. Pignatello *et al.* [31] first demonstrated a change of mechanism in the photocatalytic transformation of 2,4-D (2,4-dichlorophenoxyacetic acid) over P25 TiO_2 depending on the solution pH. Around pH 3, where adsorption of the substrate is favored through surface Ti(IV) by the carboxyl group, the phototransformation rate shows a maximum and the main products come from substrate decarboxylation. At pH above and below 3, the rate decreases and ring hydroxylation products prevail.

The published data on particles morphology suggest a good uniformity of the TiO_2 Merck particle surfaces and a more defective surface for the P25 particles. In a recent paper [32], we reported by a comparative FT-IR analysis under various conditions the presence of a variety of surface OH on P25, that is, at least three types of linear hydroxyl groups (in which OH is bound to a surface Ti, let say ≡Ti-OH) and three types of bridged hydroxyl groups (≡Ti-OH-Ti≡). There is a large consensus on the action of surface hydroxyls as surface hole traps, so their chemical nature is relevant to photocatalysis. The reaction of free carriers with surface chemisorbed molecules can be very rapid and efficient. Conversely, the trapped holes (and electrons) are less reactive than their bulk photogenerated precursors [33, 34].

3.2.2
Surface Manipulation

Drastic changes are caused on reactivity by manipulation of the surface texture chemical composition by exchange of surface groups. It is long known that fluoride adsorbs onto TiO_2 surfaces (see, for example, [35]) and the adsorption of fluoride inhibits the adsorption of other ligands, for example, catechol and hydrogen peroxide [[7, 36]].

Since the first reports in 2000 [7, 37], tens of papers have been published on this issue. Surface fluorination improves the photocatalytic degradation of a number of simple organic compounds, such as phenol [7], benzoic acid [38, 39], benzene [40], cyanide [41], *N*-nitrosodimethylamine [42], and a variety of organic dyes [43–47]. This positive effect on the photocatalytic degradation has been directly associated with the displacement of OH terminal groups from the TiO_2 surface, which would enhance the generation of free OH radicals [7, 48]. As a consequence, the oxidation of the organic substrates would occur in solution in which the probability of back reduction is reduced. A decrease in photocatalytic activity on illumination was observed for formic acid [39] and dichloroacetate [49], species that are strongly bound to the TiO_2 surface. In such cases, fluoride can displace them from the

Ti(IV) surface sites [50] hindering the direct hole transfer, which has been proposed to be the dominant mechanism [51, 52].

The importance of fluorination on the electron-transfer rate has been recognized [53]. By means of photopotential decay measurements, it was demonstrated that TiO_2 surface fluorination retards the reactivity of photogenerated electrons, both for recombination with surface-trapped holes and for transfer to oxygen, shifts upward the electronic levels in the potential energy scale, changes the mechanism from direct to indirect for strong adsorbing species, and inhibits the adsorption of intermediates that could serve as recombination centers [54, 55].

Fluorination of P25 greatly simplifies the surface IR spectrum [32], leaving only the component at 3674 cm^{-1} that was assigned to one type of the bridged hydroxyl groups. The ν_{OH} components removed by fluorination can be ascribed to hydroxyls sitting on defective sites, which interact more strongly with ligands. The surface of TiO_2 P25 is characterized by the presence of at least two different hydroxyls, with different coordination strength toward fluorides (and presumably to other ligands). The confirmation of this picture comes from the evolution of ν_{OH} patterns for TiO_2 Merck and their comparison with P25. Pristine and fluorinated Merck TiO_2 show similar ν_{OH} pattern, with a dominant spectroscopic feature at 3674 cm^{-1}. The effect of fluorination in this case is the decrease of the intensity at 3674 cm^{-1}, but the pattern does not' change. The spectra of pristine and fluorinated Merck TiO_2 are very similar to that of fluorinated P25. As demonstrated by Sun *et al.* [56], fluoride preferentially adsorbs on the {001} face of anatase phase. Surface Ti(IV) ions on these faces are more exposed, so more coordinatively unsaturated, with respect to the more stable {101} faces.

3.2.3
Crystal Faces

The morphology of TiO_2 nanoparticles, not only the size but also the shape and the exposed faces, can be important and significantly affect the photocatalytic activity [57]. Experimentally, Cho *et al.* [58], have shown that TiO_2 nanoparticles (3–33 nm) in aqueous solution possess similar UV light absorption and transmission spectra, indicating that nanoparticles have similar band structure, and the light transmission is not the cause for the difference in the photocatalytic activity.

The sensitivity to the surface structure is perhaps most appealing, as proposed by Ohno *et al.* [59], who first observed that the reductive deposition of Pt^{2+} to Pt occurs preferentially on the {101} face of anatase nanoparticle. An important finding [60] was that the energetically stable {101} face of the crystal (the surface energy is 0.44 J m^{-2}) showed a higher photocatalytic reduction activity than the {001} face with a higher surface energy (0.90 J m^{-2}). The (101) face yields a highly reactive surface for the reduction of O_2 molecules to superoxide radicals [61] and serves as possible reservoirs of the photogenerated electrons [62]. Murakami *et al.* [63] showed that the activity of acetaldehyde decomposition is higher in sharp nanocrystals dominated by {101} face, and the similar phenomenon was observed by Cho *et al.* [58] and Amano *et al.* [64] in chloroform and alcohol photodecomposition. By using a total scattering technique, Cernuto *et al.* [65] investigated methylene blue decomposition on TiO_2 nanoparticles from 4 to 10 nm and found that the photooxidation efficiency is sensitive to both the size and the shape .

Interestingly, on nanosized TiO_2 polyhedral particles exposing around 20% {001} faces, an inversion in the relative phototransformation rates of methylene blue and methyl orange is observed after fluorination. The conversion of methyl orange is faster on the hydroxylated catalyst, suggesting a key role of the surface complex of the anionic methyl orange on {001} face [66]. For a review of these effects, see also the Chapter 4.

In a recent paper [67], calculations by density functional theory (DFT) combined with the periodic continuum solvation model have been utilized to compute the electronic structure of anatase nanoparticles in aqueous solution. It was noted that the effective mass of the carriers and the calculated bandgap change converge rapidly for nanoparticles above 2 nm, which implies that the quantum size effect is important only for very small TiO_2 nanoparticles (<2 nm). Considering that the experimentally synthesized anatase particles in photocatalysis are generally above 2 nm, neither the carrier mobility nor the bandgap could be the key factors for determining the photocatalytic efficiency. Moreover, DFT calculations shows that the equilibrium shape of nanoparticles changes dramatically below ~30 nm, as small nanoparticles are flatter, and with the increase of the particle size, the particle are sharpened. In nanoparticles below 10 nm, the higher surface energy surface (001) is preferentially exposed compared to the lower surface energy surface {101}.

A recent experiment by Tachikawa *et al.* [62] by single-molecule imaging and kinetic analysis of the fluorescence from the products showed that reaction sites for the effective reduction of the probe molecules are preferentially located on the {101} face of the crystal rather than the {001} face with a higher surface energy. This can be explained in terms of face-specific electron-trapping probability [62] in line with the fact that {101} face exposes Ti5c sites where the adsorption of molecules can take place. Large differences in the OH adsorption energy on the surface of the same nanoparticle at different Ti5c sites have been calculated [67].

The production of adsorbed OH radicals on the surface is often assumed to be a key step for the photooxidation reactions in aqueous solution [68–70] (see also the Chapter 1). DFT calculations showed that the axial sites at the sharp ends are the most active sites for surface hole capture, while the edge sites at the conjunction between {101} faces are typically the least active, and that the terrace sites away from the edges are generally more active than the sites close to the edges [67]. Opposing to the case of the flat crystal, the highest occupied molecular orbital (HOMO) and lowest unoccupied molecular orbital (LUMO) are spatially separated in the sharp crystal. Because of the spatial separation, the sharper crystals are intrinsically more polarized and more photocatalytically active as the electron–hole pair separation at the early stage of photoexcitation process reduces the chance for the immediate electron–hole recombination. As the sharp crystals dominated by {101} faces can promote the photoreduction reaction and accelerate the consumption of photoelectrons, the chance of electron–hole recombination is further reduced. In addition, the sharp crystal can promote the photooxidation reaction by providing more active sites for the adsorption of key reaction intermediates. The optimum particle size was predicted around 15–20 nm.

A general consensus is that oxidative and reductive processes are spatially separated on different crystal faces, and the relative role of the two is by some means dependent on the relative ratio of the faces exposed.

3.2.4
Surface Traps for Holes

Some surface hole traps are shown in Scheme 3.1. A shallow surface hole trap can be schematized as in the central row of Scheme 3.1, where two limiting forms exist. The surface is considered hydrated to avoid explicitly write charges that depend on the type of crystal face, the type of bridging oxygen, the possible protonation, and electron density [71]. The actual hole localization is formally regulated by the concentration ratio K of the two forms. The actual electron density distribution of the left and right forms are $1/(1+K)$ and $K/(1+K)$, respectively. The species in the central row of Scheme 3.1 are referred as $(Ti(OH)(OH)) \rightarrow (Ti(O^{\bullet})(OH)) \rightleftarrows (Ti(OH)(O^{\bullet}))$, where the first (OH) refers to linear hydroxyl groups and the left (OH) to bridged hydroxyl groups. As three types of linear hydroxyl groups and three types of bridged hydroxyl groups have been detected [32] on P25, the notation Ti(OH)(OH) refers to all of them. It is important to remind that only one type of the bridging oxygen is present as Ti(F)(OH) on fluorinated P25 and as Ti(‡)(OH) on pristine and fluorinated Merck TiO_2.

The protonation of the linear hydroxyl at acidic pH disfavors the formation of $Ti(O^{\bullet})(OH)$. Conversely, the dissociation of the bridged hydroxyl occurring at

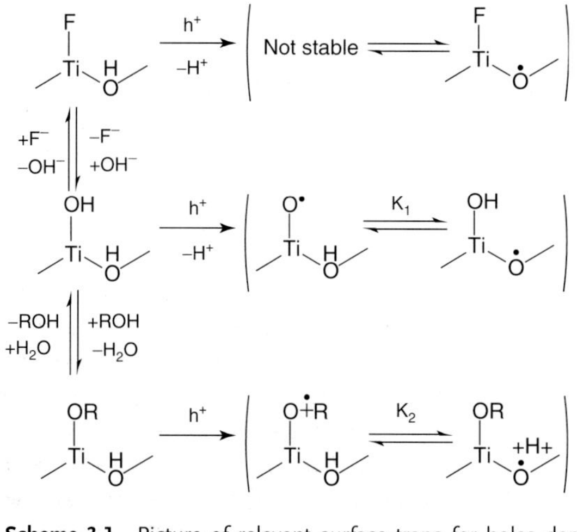

Scheme 3.1 Picture of relevant surface traps for holes depending on the surface speciation. Possibly, several types of these traps are present on the whole TiO_2 surface depending on faces, steps, and edges.

pH > 8, with creation of a net negative charge on the bridged oxygen, favors hole trapping on the oxygen. Then, K_1 will be >1 for naked titania. The most probable hole localization is then Ti(OH)(O•).

For fluorinated titania (Scheme 3.1, top row), the terminal hydroxyl is exchanged with fluoride and the left tautomeric form is impeded by the high fluoride electronegativity. Thus, $K/(1 + K) \approx 1$. In addition, as fluoride is more electronegative than -OH, the radical on bridging oxygen is less stable, that is, the surface trap Ti(F)(O•) is more shallow than Ti(OH)(O•) and its energy level is more resonant with free holes in the valence band. Since fluorination increases the rate for substrates that react with OH radicals and depresses the rate for substrates that react by direct electron transfer, the right form Ti(OH)(O•) performs as adsorbed OH radical. This view reconciles what stated in Ref. [7] where shallow traps were indicated with a concise formalism as (≡Ti–O•), neglecting the resonance form (Ti(‡)(O•)), and recent reconsideration of surface fluorination [72], where the decrease of the recombination rate concurrent with the increase of the electron-transfer rate with reduced dissolved species are invoked to explain the fluoride effect. Free electrons are more stabilized on -Ti(F)- than on -Ti(OH)-, and the surface-assisted recombination is reduced.

Alcohols, polyols, and carboxylic acids show good coordinative abilities toward Ti(IV) ions. At suitable concentration, these species are able to occupy surface sites (bottom row in Scheme 3.1). Surface complexation will form a surface deep trap for holes, as the oxidation potential of the surface complex Ti(OR)(OH) is lowered with respect to Ti(OH)(OH). Besides being an efficient recombination center, the oxidized surface complex (Ti(OR$^{\bullet+}$)(OH)) could have a chemical reactivity very different from the carbon-centered radical OR• generated via H-abstraction by the surface adsorbed OH•, namely, the Ti(OH)(O•) hole trap [73].

3.3
Multisite Kinetic Model

All the evidences reported above demonstrate the presence of at least two active sites with different reactivity and selectivity, whose ratio is related to the surface features of diverse catalysts (defects, type of exposed faces, surface charge, ion adsorption), which in turn depends on the catalyst synthesis and surface manipulation. In particular, it seems mandatory to postulate the presence of a population of strongly adsorbing surface sites, reasonably associated to coordinatively defective Ti(IV) centers on defects, edges, and low-stability faces such as the {001} and another population with low-coordinative ability.

Ollis and coworkers [74, 75] indicated the need of a two-site model to explain the partial deactivation observed in the gas-phase degradation of alcohols and aromatics over P25 TiO_2 and claimed the presence of a type I site, suitable for the adsorption of substrate (actually aromatics), water, and reaction intermediates, and a type II more hydrophilic (hydroxyl groups or bridging oxygen) site unsuitable for substrate adsorption.

To get a workable model in a recent paper on glycerol kinetic of transformation [73], only two different types of surface hole traps are considered. The sites A, namely, $(Ti(OH)(OH))_A$, are able to chemisorb the substrate, while sites B are not, namely, the sites $(Ti(OH)(OH))_B$ for P25 or $(Ti(\ddagger)(OH))_B$ for Merck TiO_2 or $(Ti(F)(OH))_B$ for fluorinated catalysts. A third site is considered as a trap for electrons because it seems located on a different crystal face as reported above. The kinetic framework based on this assumption is reported in Scheme 3.2.

The species $(Ti(\ddagger)(O^\bullet))_B$ or $(Ti(F)(O^\bullet))_B$ are similar to $(Ti(OH)(O^\bullet))_A$, but the depth of these shallow traps is manifestly different; this implies that the recombination reaction constants with free electrons (k_{AR} and k_{BR}) and those for substrate oxidation (k_{AO} and k_{BO}) are markedly different. In the presence of substrates that chemisorb, the site $(Ti(OR)(OH))_A$ is a deep trap for holes and the site $(Ti(OR^{\bullet+})(OH))_A$ must recombine free electrons more easily than $(Ti(O^\bullet)(OH))_A$. An oxidative attack to the nonchemisorbed substrate mediated by $Ti(OH)(O^\bullet)$ shallow surface hole traps and a direct hole transfer to the surface complex could produce very different intermediates (for example, a carbon-centered radical and an alkoxyl radical-like species) evolving to different stable products.

Accordingly, in the case of glycerol, path A(a) and path A(b) produce different detectable products (P1 = glyceraldehyde + dihydroxyacetone and P2 = formaldehyde + glycolaldehyde) [26]. A numerical simulation for the rate under steady-state

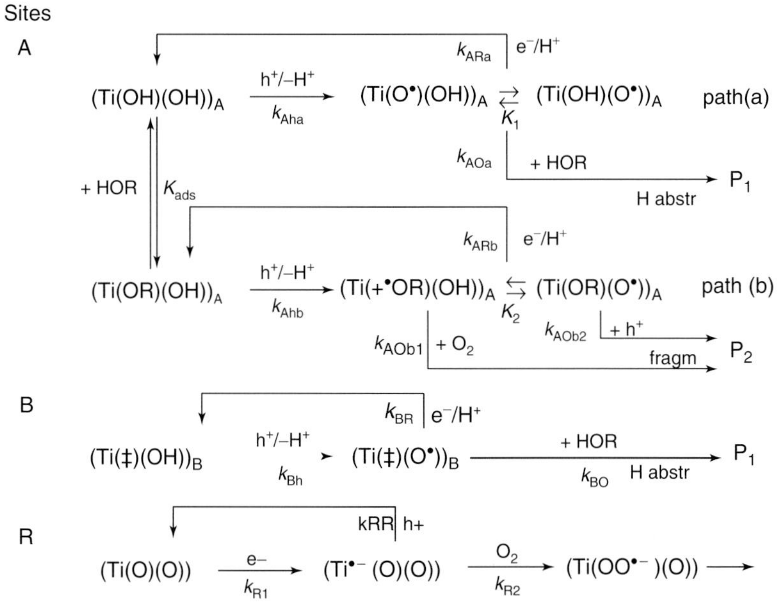

Scheme 3.2 The kinetic framework of the multisite kinetic model. The A sites are able to chemisorb the substrate and can perform the oxidation both by hole transfer from the shallow trap (path (a)) or on the deep trap (path (b)) depending on the surface coverage. Sites B that are not able to chemisorb the substrate are a second type of shallow trap. Sites R trap electrons. (Source: Reprinted with permission from Ref. [73].)

conditions for the shallow trap $(Ti(OH)(O^{\bullet}))_A$, deep trap $(Ti(OR)(O^{\bullet}))_A$, shallow trap $(Ti(⁝)(O^{\bullet}))_B$, electron trap $(Ti^{\bullet -}(O)(O))$, and free electrons and holes, together with the mass balance for sites of type A, B, and R, (C_{sA}, C_{sB}, and C_{sR}, respectively) is reported in Figure 3.2.

The rate simulation in Figure 3.2 reports the rate of substrate transformation together with the contributions of different paths to the overall rate. The plot is able to simulate quite well the strange rate dependence on the substrate concentration observed for glycerol [26]. It is remarkable that the contribution of the different paths that lead to different products (P1 and P2) is exactly predominant in the segment of glycerol concentration where they come out.

This model has only a numerical solution, and some efforts must be carried out to reduce it under some applicable hypotheses in analytical solutions [73]. However, it confirms that the kinetic pathways depicted in Scheme 3.2 are well founded. In particular, it is worth noting that the recombination constant with the electrons of $(Ti(OR)(O^{\bullet}))_A$ required to simulate the product distribution is very large, confirming that this site is a deep trap, whereas the others are not, as discussed in the preceding session.

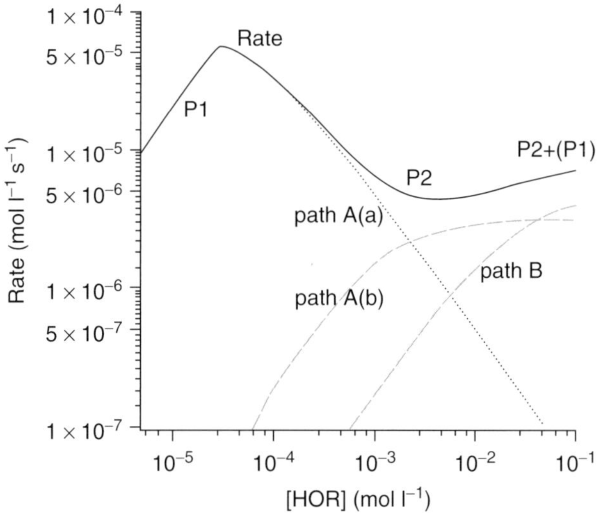

Figure 3.2 Numerical simulation of the MS kinetic model. The contribution to the rate of paths A(a), A(b), and B is shown together with the related formation of products P (P1 = glyceraldehyde + dihydroxyacetone and P2 = formaldehyde + glycolaldehyde). The following values have been used with reference to Scheme 3.2: $K_{ads} = 1 \times 10^4$, $\varphi = 1 \times 10^{-4}$ mol l^{-1} s^{-1}, number of sites C_{sA}, C_{sB}, C_{sR} = 1×10^{-4} mol l^{-1}; for path A: $k_{A1} = 5 \times 10^3$, $k_{AOa} = 2 \times 10^4$, $k_{ARa} = 1$, $k_{A2} = 1 \times 10^4$, $w_2 = 1$, $w_3 = 1 \times 10^4$, $k_{ARb} = 1 \times 10^8$; for path B: $k_B = 5 \times 10^2$, $k_{BO} = 2$, $K_{BR} = 1$; for reductive path: $k_{R1} = 10$, $k_{R2} = 10$, $k_{RR} = 1$. Units of kinetic constants are l mol^{-1} s^{-1} (second order) or s^{-1} (first order).

3.4 Conclusion

The multisite model model uses concepts of adsorption, direct and indirect electron transfer, and fundamental reaction steps developed for previous kinetic models in photocatalysis, for which a critical review is here reported. It assumes that *concurrent* direct and indirect electron transfer takes place on different sites of the same catalyst and that on sites able to adsorb the substrate, the relative weight of direct and indirect electron transfers is regulated by the adsorption constant. The MS kinetic model is here suggested as a new view to rationalize the set of different phenomena observed until now and in particular, the selectivity observed as a function of the substrate concentration for the same catalyst or that observed on different native or surface-manipulated catalysts.

References

1. Di Valentin, C. and Selloni, A. (2011) Bulk and surface polarons in photoexcited anatase TiO_2. *J. Phys. Chem. Lett.*, **2**(17), 2223–2228.
2. Minero, C. and Vione, D. (2006) A quantitative evalution of the photocatalytic performance of TiO_2 slurries. *Appl. Catal., B*, **67**, 257–269.
3. Herrmann, J.-M. (2010) Photocatalysis fundamentals revisited to avoid several misconceptions. *Appl. Catal., B*, **99**, 461–468.
4. Minero, C., Maurino, V., and Pelizzetti, E. (2003) in *Semiconductor Photochemistry and Photophysics*, Molecular and Supramolecular Photochemistry, Vol. **10** Chapter 6 (eds V. Ramamurthy and K.S. Schanze), Marcel Dekker, New York, p. 384. ISBN: 0824709586
5. Villarreal, T.L., Gómez, R., Gonzáles, M., and Salvador, P. (2004) A kinetic model for distinguishing between direct and indirect interfacial hole transfer in the heterogeneous photooxidation of dissolved organics on TiO_2 nanoparticle suspensions. *J. Phys. Chem. B*, **108**, 20278–20290.
6. Mora-Seró, I., Villareal, L., Bisquert, J., Pitarch, A., Gómez, R., and Salvador, P. (2005) Photoelectrochemical behavior of nanostructured TiO_2 thin-film electrodes in contact with aqueous electrolytes containing dissolved pollutants: a model for distinguishing between direct and indirect interfacial hole transfer from photocurrent measurements. *J. Phys. Chem. B*, **109**, 3371–3380.
7. Minero, C., Mariella, G., Maurino, V., and Pelizzetti, E. (2000) Photocatalytic transformation of organic compounds in the presence of inorganic anions. 1. Hydroxyl-mediated and direct electron-transfer reactions of phenol on a titanium dioxide–fluoride system. *Langmuir*, **16**, 2632–2641.
8. Monllor-Satoca, D. and Gómez, R. (2010) A photoelectrochemical and spectroscopic study of phenol and catechol oxidation on titanium dioxide nanoporous electrode. *Electrochim. Acta*, **55**, 4661–4668.
9. Palmisano, L., Schiavello, M., Sclafani, A., Martra, G., Borello, E., and Coluccia, S. (1994) Photocatalytic oxidation of phenol on TiO_2 powders. A Fourier transform infrared study. *Appl. Catal., B*, **3**, 117–132.
10. Horikoshi, S., Miura, T., Kajitani, M., Hidaka, H., and Serpone, N. (2008) A FT-IR (DRIFT) study of the influence of halogen substituents on the TiO_2-assisted photooxidation of phenol and p-halophenols under weak room light irradiance. *J. Photochem. Photobiol., A*, **194**, 189–199.
11. Agrios, A.G. and Pichat, P. (2006) Recombination rate of photogenerated charges versus surface area: opposing

effects of TiO_2 sintering temperature on photocatalytic removal of phenol, anisole, and pyridine in water. *J. Photochem. Photobiol., A*, **180**, 130–135.

12. Cropek, D., Kemme, P.A., Makarova, O.V., Chen, L.X., and Rajh, T. (2008) Selective photocatalytic decomposition of nitrobenzene using surface modified TiO_2 nanoparticles. *J. Phys. Chem. C*, **112**, 8311–8318.
13. Feng, Y.J. and Li, X.Y. (2003) Electrocatalytic oxidation of phenol on several metal-oxide electrodes in aqueous solution. *Water Res.*, **37**, 2399–2407.
14. Minero, C., Catozzo, F., and Pelizzetti, E. (1992) Role of adsorption in photocatalyzed reactions of organic molecules in aqueous titania suspensions. *Langmuir*, **8**, 481–486.
15. Tachikawa, T., Takai, Y., Tojo, S., Fujitsuka, M., and Majima, T. (2006) probing the surface adsorption and photocatalytic degradation of catechols on TiO_2 by solid-state NMR spectroscopy. *Langmuir*, **22**, 893–896.
16. Pichat, P. (2003) in *Chemical Degradation Methods for Wastes and Pollutants: Environmental and Industrial Applications* (ed M.A. Tarr), Marcel Dekker, New York, pp. 77–119.
17. Minero, C., Pelizzetti, E., Pichat, P., Sega, M., and Vincenti, M. (1995) Formation of condensation products in advanced oxidation technologies: the photocatalytic degradation of dichlorophenols on TiO_2. *Environ. Sci. Technol.*, **29**, 2226–2234.
18. Minero, C. (1999) Kinetic analysis of photoinduced reactions at the water semiconductor interface. *Catal. Today*, **54**, 205–216.
19. Tachikawa, T., Tojo, S., Fujitsuka, M., and Majima, T. (2004) Influence of metal ions on the charge recombination processes during TiO_2 photocatalytic One-electron oxidation reactions. *J. Phys. Chem. B*, **108**, 11054–11061.
20. Montoya, J.F., Velasquez, J.A., and Salvador, P. (2009) The direct–indirect kinetic model in photocatalysis: a reanalysis of phenol and formic acid degradation rate dependence on photon flow and concentration in TiO_2 aqueous dispersions. *Appl. Catal., B*, **88**, 50–58.
21. Valencia, S., Cataño, F., Rios, L., Restrepo, G., and Marín, J. (2011) A new kinetic model for heterogeneous photocatalysis with titanium dioxide: case of non-specific adsorption considering back reaction. *Appl. Catal., B*, **104**, 300–304.
22. Emeline, A., Ryabchuck, V., and Serpone, N. (2000) Factors affecting the efficiency of a photocatalyzed process in aqueous metal-oxide dispersions prospect of distinguishing between two kinetic models. *J. Photochem. Photobiol., A*, **133**, 89–97.
23. Montoya, J.F., Peral, J., and Salvador, P. (2012) Commentary on the article: "a new kinetic model for heterogeneous photocatalysis with titanium dioxide: case of non-specific adsorption considering back reaction, by S. Valencia, F. Cataño, L. Rios, G. Restrepo and J. Marín, published in applied catalysis B: environmental, 104 (2011) 300–304". *Appl. Catal., B*, **111–112**, 649–650.
24. Liu, B. and Zhao, X. (2010) A kinetic model for evaluating the dependence of the quantum yield of nano-TiO_2 based photocatalysis on light intensity, grain size, carrier lifetime, and minority carrier diffusion coefficient: indirect interfacial charge transfer. *Electrochim. Acta*, **55**, 4062–4070.
25. Martra, G. (2000) Lewis acid and base sites at the surface of microcrystalline TiO_2 anatase: relationships between surface morphology and chemical behaviour. *Appl. Catal., A*, **200**, 275–285.
26. Bedini, A., Maurino, V., Minella, M., Minero, C., and Rubertelli, F. (2008) Glycerol transformation through photocatalysis: a possible route to value added chemicals. *J. Adv. Oxid. Technol.*, **11**, 184–192.
27. Ryu, J. and Choi, W. (2008) Substrate-specific photocatalytic activities of TiO_2 and multiactivity test for water treatment application. *Environ. Sci. Technol.*, **42**, 294–300.
28. Augugliaro, V., Camera-Roda, G., Loddo, V., Palmisano, G., Palmisano, L., Parrino, F., and Puma, M.A. (2012) Synthesis of vanillin in water by TiO_2 photocatalysis. *Appl. Catal., B*, **111–112**, 555–561.

29. Courbon, H. and Pichat, P. (1984) Room temperature interaction of $N^{18}O$ with ultraviolet-illuminated titanium dioxide. *J. Chem. Soc., Faraday Trans. 1*, **80**, 3175–3185.
30. Zhang, J., Hu, Y., Matsuoka, M., Yamashita, H., Minagawa, M., Hidaka, H., and Anpo, M. (2001) Relationship between the local structures of titanium oxide photocatalysts and their reactivities in the decomposition of NO. *J. Phys. Chem. B*, **105**, 8395–8398.
31. Sun, Y. and Pignatello, J.J. (1995) Evidence for a surface dual hole-radical mechanism in the titanium dioxide photocatalytic oxidation of 2,4-D. *Environ. Sci. Technol.*, **29**, 2065–2072.
32. Minella, M., Faga, M.G., Maurino, V., Minero, C., Pelizzetti, E., Coluccia, S., and Martra, G. (2010) Effect of fluorination on the surface properties of titania P25 powder: an FTIR study. *Langmuir*, **26**, 2521–2527.
33. Rajh, T., Chen, L.X., Lukas, K., Liu, T., Thurnauer, M.C., and Tiede, D.M. (2002) Surface restructuring of nanoparticles: an efficient route for ligand–metal oxide crosstalk. *J. Phys. Chem. B*, **106**, 10543–10548.
34. Bahnemann, D.W., Hilgendorff, M., and Memming, R. (1997) Charge carrier dynamics at TiO_2 particles: reactivity of free and trapped holes. *J. Phys. Chem. B*, **101**, 4265–4275.
35. Vasudevan, D. and Stone, A.T. (1996) Adsorption of catechols, 2-aminophenols, and 1,2-phenylenediamines at the metal (hydr)oxide/water interface: effect of ring substituents on the adsorption onto TiO_2. *Environ. Sci. Technol.*, **30**, 1604–1613.
36. Maurino, V., Minero, C., Mariella, G., and Pelizzetti, E. (2005) Sustained production of H_2O_2 on irradiated TiO_2-fluoride systems. *Chem. Commun.*, 2627–2629.
37. Minero, C., Mariella, G., Maurino, V., Vione, D., and Pelizzetti, E. (2000) Photocatalytic transformation of organic compounds in the presence of inorganic ions. 2. Competitive reactions of phenol and alcohols on a titanium dioxide–fluoride system. *Langmuir*, **16**, 8964–8972.
38. Vione, D., Minero, C., Maurino, V., Carlotti, M.E., Picatonotto, T., and Pelizzetti, E. (2005) Degradation of phenol and benzoic acid in the presence of a TiO_2-based heterogeneous photocatalyst. *Appl. Catal., B*, **58**, 79–88.
39. Mrowetz, M. and Selli, E. (2006) H_2O_2 Evolution during the photocatalytic degradation of organic molecules on fluorinated TiO_2. *New J. Chem.*, **30**, 108–114.
40. Park, H. and Choi, W. (2005) Photocatalytic conversion of benzene to phenol using pure and modified TiO_2. *Catal. Today*, **101**, 291–297.
41. Chiang, K., Amal, R., and Tran, T. (2003) Photocatalytic oxidation of cyanide: kinetic and mechanistic studies. *J. Mol. Catal.*, **193**, 285–297.
42. Lee, J., Choi, W., and Yoon, J. (2005) Photocatalytic degradation of N-nitrosodimethylamine: mechanism, product distribution, and TiO_2 surface modification. *Environ. Sci. Technol.*, **39**, 6800–6807.
43. Yu, J. and Zhang, J. (2010) A simple template-free approach to TiO_2 hollow spheres with enhanced photocatalytic activity. *Dalton Trans.*, **39**, 5860–5867.
44. Dozzi, M.V., Schiavello, G.L., and Selli, E. (2010) Effects of surface modification on the photocatalytic activity of TiO_2. *J. Adv. Oxid. Technol.*, **13**, 305–312.
45. Chen, Y., Chen, F., and Zhang, J. (2009) Effect of surface fluorination on the photocatalytic and photo-induced hydrophilic properties of porous TiO_2 films. *App. Surf. Sci.*, **255**, 6290–6296.
46. Chen, K.T., Lu, C.S., Chang, T.H., Lai, Y.Y., Chang, T.H., Wu, C.W., and Chen, C.C. (2010) Comparison of photodegradative efficiencies and mechanisms of Victoria blue R assisted by nafion-coated and fluorinated TiO_2 photocatalysts. *J. Hazard. Mater.*, **174**, 598–609.
47. Lv, K., Li, X., Deng, K., Sun, J., Li, X., and Li, M. (2010) Effect of phase structures on the photocatalytic activity of surface fluorinated TiO_2. *Appl. Catal., B*, **95**, 383–392.

48. Mrowetz, M. and Selli, E. (2005) Enhanced photocatalytic formation of hydroxyl radicals on fluorinated TiO_2. *Phys. Chem. Chem. Phys.*, **7**, 1100–1102.
49. Park, H. and Choi, W. (2004) Effects of TiO_2 surface fluorination on photocatalytic reactions and photoelectrochemical behaviors. *J. Phys. Chem. B*, **108**, 4086–4093.
50. Lv, K. and Xu, Y. (2006) Effects of polyoxometalate and fluoride on adsorption and photocatalytic degradation of organic dye X3B on TiO_2: the difference in the production of reactive species. *J. Phys. Chem. B*, **110**, 6204–6212.
51. Monllor-Satoca, D., Gomez, R., Gonzalez-Hidalgo, M., and Salvador, P. (2007) The direct–indirect model: an alternative kinetic approach in heterogeneous photocatalysis based on the degree of interaction of dissolved pollutant species with the semiconductor surface. *Catal. Today*, **129**, 247–255.
52. Enriquez, R. and Pichat, P. (2006) Different Net effect of TiO_2 sintering temperature on the photocatalytic removal rates of 4-chlorophenol, 4-chlorobenzoic acid and dichloroacetic acid in water. *J. Environ. Sci. Health A*, **41**, 955–966.
53. Xu, Y., Lv, K., Xiong, Z., Leng, W., Du, W., Liu, D., and Xue, X. (2007) Rate enhancement and rate inhibition of phenol degradation over irradiated anatase and rutile TiO_2 on the addition of NaF: New insight into the mechanism. *J. Phys. Chem. C*, **111**, 19024–19032.
54. Monllor-Satoca, D. and Gomez, R. (2008) Electrochemical method for studying the kinetics of electron recombination and transfer reactions in heterogeneous photocatalysis: the effect of fluorination on TiO_2 nanoporous layers. *J. Phys. Chem. C*, **112**, 139–147.
55. Monllor-Satoca, D., Lana-Villarreal, T., and Gómez, R. (2011) Effect of surface fluorination on the electrochemical and photoelectrocatalytic properties of nanoporous titanium dioxide electrodes. *Langmuir*, **27**(24), 15312–15321.
56. Yang, H.G., Sun, C.H., Qiao, S.Z., Zou, J., Liu, G., Campbell Smith, S., Cheng, H.M., and Lu, G.Q. (2008) Anatase TiO_2 single crystals with a large percentage of reactive facets. *Nature*, **453**, 638–642.
57. Ohno, T., Sarukawa, K., and Matsumura, M. (2001) Photocatalytic activities of pure rutile particles isolated from TiO_2 powder by dissolving the anatase component in HF solution. *J. Phys. Chem. B*, **105**, 2417–2420.
58. Cho, C.H., Han, M.H., Kim, D.H., and Kim, D.K. (2005) Morphology evolution of anatase TiO_2 nanocrystals under a hydrothermal condition (pH=9.5) And their ultra-high photo-catalytic activity. *Mater. Chem. Phys.*, **92**, 104–111.
59. Ohno, T., Sarukawa, K., and Matsumura, M. (2002) Crystal faces of rutile and anatase TiO_2 particles and their roles in photocatalytic reactions. *New J. Chem.*, **26**, 1167–1170.
60. Tachikawa, T., Wang, N., Yamashita, S., Cui, S.-C., and Majima, T. (2010) Design of a highly sensitive fluorescent probe for interfacial electron transfer on a TiO_2 surface. *Angew. Chem. Int. Ed.*, **49**, 8593–8597.
61. Wu, N., Wang, J., Tafen, D.N., Wang, H., Zheng, J.-G., Lewis, J.P., Liu, X., Leonard, S.S., and Manivannan, A. (2010) Shape-enhanced photocatalytic activity of single-crystalline anatase TiO_2 (101) nanobelts. *J. Am. Chem. Soc.*, **132**, 6679–6685.
62. Tachikawa, T., Yamashita, S., and Majima, T. (2011) Evidence for crystal-face-dependent TiO_2 photocatalysis from single-molecule imaging and kinetic analysis. *J. Am. Chem. Soc.*, **133**, 7197–7204.
63. Murakami, N., Kurihara, Y., Tsubota, T., and Ohno, T. (2009) Shape-controlled anatase titanium(IV) oxide particles prepared by hydrothermal treatment of peroxo titanic acid in the presence of polyvinyl alcohol. *J. Phys. Chem. C*, **113**, 3062–3069.
64. Amano, F., Yasumoto, T., Prieto-Mahaney, O.-O., Uchida, S., Shibayama, T., and Ohtani, B. (2009) Photocatalytic activity of octahedral single-crystalline mesoparticles of anatase titanium(IV) oxide. *Chem. Commun.*, 2311–2313.
65. Cernuto, G., Masciocchi, N., Cervellino, A., Colonna, G.M., and

Guagliardi, A. (2011) Size and shape dependence of the photocatalytic activity of TiO_2 nanocrystals: a total scattering Debye function study. *J. Am. Chem. Soc.*, **133**, 3114–3119.

66. Li, Y.-F., Liu, Z.-P., Liu, L., and Gao, W. (2010) Mechanism and activity of photocatalytic oxygen evolution on titania anatase in aqueous surroundings. *J. Am. Chem. Soc.*, **132**, 13008–13015.

67. Li, Y.-F. and Liu, Z.-P. (2011) Particle size, shape and activity for photocatalysis on titania anatase nanoparticles in aqueous surroundings. *J. Am. Chem. Soc.*, **133**, 15743–15752.

68. Nosaka, Y., Komori, S., Yawata, K., Hirakawa, T., and Nosaka, A.Y. (2003) Photocatalytic OH radical formation in TiO_2 aqueous suspension studied by several detection methods. *Phys. Chem. Chem. Phys.*, **5**, 4731–4735.

69. Murakami, Y., Kenji, E., Nosaka, A.Y., and Nosaka, Y. (2006) Direct detection of OH radicals diffused to the gas phase from the UV-irradiated photocatalytic TiO_2 surfaces by means of laser-induced fluorescence spectroscopy. *J. Phys. Chem. B*, **110**, 16808–16811.

70. Hirakawa, T., Yawata, K., and Nosaka, Y. (2007) Photocatalytic reactivity for $O_2^{\bullet-}$ and $OH^{\bullet}$ radical formation in anatase and rutile TiO_2 suspension as the effect of H_2O_2 addition. *Appl. Catal., A*, **325**, 105–111.

71. Diebold, U. (2009) The surface science of titanium dioxide. *Surf. Sci. Rep.*, **48**, 53–229.

72. Montoya, J.F. and Salvador, P. (2010) The influence of surface fluorination in the photocatalytic behaviour of TiO_2 aqueous dispersions: an analysis in the light of the direct–indirect kinetic model. *Appl. Catal., B*, **94**, 97–107.

73. Minero, C., Bedini, A., and Maurino, V. (2012) Glycerol as a probe molecule to uncover oxidation mechanism in photocatalysis. *Appl. Catal., B*, **128**, 135–143.

74. Lewandowski, M. and Ollis, D.F. (2003) A Two-site kinetic model simulating apparent deactivation during photocatalytic oxidation of aromatics on titanium dioxide (TiO_2). *Appl. Catal., B*, **43**, 309–327.

75. Lewandowski, M. and Ollis, D.F. (2003) Extension of a Two-site transient kinetic model of TiO_2 deactivation during photocatalytic oxidation of aromatics: concentration variations and catalyst regeneration studies. *Appl. Catal., B*, **45**, 223–238.

Part II
Improving the Photocatalytic Efficacy

Photocatalysis and Water Purification: From Fundamentals to Recent Applications, First Edition. P. Pichat.

4
Design and Development of Active Titania and Related Photocatalysts

Bunsho Ohtani

4.1
Introduction – a Thermodynamic Aspect of Photocatalysis

Although it seems unnecessary to explain the importance of photocatalysis in both fundamental and application studies to the readers of this book and many reviews on photocatalysis have already been published [1, 2], a brief summary is presented here to start this chapter. A phenomenon that white pigments especially titanium(IV) oxide (titania) included in paints promotes, when exposed to sunlight, degradation of paint coatings had been recognized and was called "chalking" [3]. This might be the first example of titania photocatalysis, and efforts have been made to avoid this undesirable reaction by, for example, using less-active rutile (RUT) crystallites of titania or coating titania particle surfaces with an inert oxide layer. The mechanism of this "chalking" had been thought to be photoexcitation (photoabsorption) of titania without considering the electronic structure of photocatalysts, that is, being considered empirically or kinetically. Since the publication of a paper by Fujishima and Honda [4], photoabsorption has been described as band-to-band excitation, as described below, and possible difference in "photocatalytic activity" of different photocatalysts has been interpreted with the band structure of electronic energy in solid (semiconductor) materials, that is, photocatalysis has been understood thermodynamically.

The principle of photocatalysis by a solid material that absorbs light with its bulk is usually interpreted with a figure such as Figure 4.1. In a schematic representation of the electronic structures of semiconducting (or insulating) materials, a band model, an electron in an electron-filled valence band (VB) is excited by photoabsorption to a vacant conduction band (CB), which is separated by a bandgap (a forbidden band), from the VB, leaving a positive hole in the VB. One important point is that photoabsorption and electron–positive hole generation are inextricably linked; a VB electron is not excited after photoabsorption. These electrons and positive holes can induce reduction and oxidation, respectively, of chemical species adsorbed on the surface of a photocatalyst, unless they recombine with each other so as not to induce a redox reaction but to produce heat and/or photoemission. Such a mechanism accounts for the photocatalytic

Photocatalysis and Water Purification: From Fundamentals to Recent Applications, First Edition. P. Pichat.

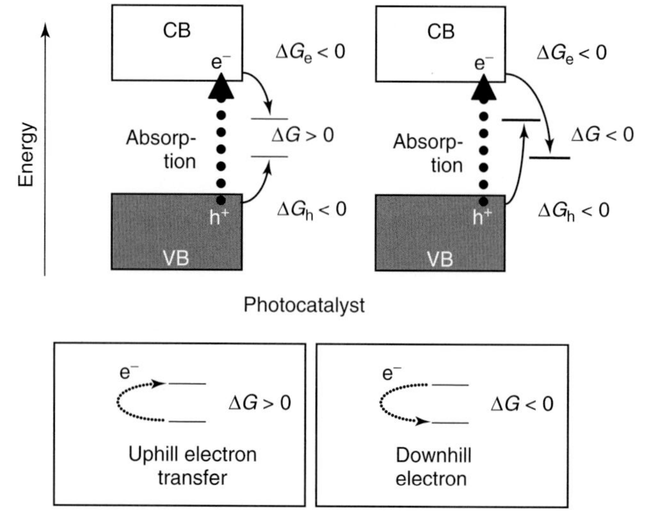

Figure 4.1 Gibbs energy change in photocatalytic reactions.

reactions of semiconducting (or insulating) materials absorbing photons by the bulk of materials.

What is shown in Figure 4.1 is, first, change in Gibbs energy (ΔG) of a given photocatalysis. As thermodynamics says, if ΔG is negative ($\Delta G < 0$ as is the case, for example, in oxidative decomposition of organic compounds under aerated conditions) and if ΔG is positive ($\Delta G > 0$, as in the case, for example, in splitting of water into hydrogen and oxygen), energy is released and stored, respectively. Therefore, if the standard electrode potential of the compound to be reduced by electrons is higher, that is, more negative (cathodic), than that of the compound to be oxidized by positive holes, ΔG is positive, that is, the reaction stores energy and vice versa. A notable point is that both situations, energy release and storage, are possible for photocatalytic reactions, while thermal catalyses are limited to reactions of negative ΔG, that is, spontaneous reactions. The reason why photocatalysts can drive even a reaction of positive ΔG, which does not proceed spontaneously, is that an overall redox reaction can proceed, even if its ΔG is positive, in a system in which reduction and oxidation steps are spatially or chemically separated (this is discussed later in relation to the enhancement of photocatalytic activity.); otherwise, the reaction between reduction and oxidation products proceeds to give no net products. Under these conditions, both Gibbs energy change for reactions of photoexcited electrons with oxidant (ΔG_e) and positive holes with reductant (ΔG_h) are required to be negative, that is, reactions by photoexcited electrons and positive holes proceed spontaneously after photoexcitation. In other words, the CB bottom and VB top positions must be higher (more cathodic) and lower (more anodic) than standard electrode potentials of an electron acceptor (oxidant) and an electron donor (reductant), respectively, to make Gibbs energy change of both half reactions negative, as has often been pointed out as a necessary condition for photocatalysis.

Discussion of the thermodynamics in photocatalysis as described above is valid when applicability of a given semiconducting (or insulating) material as a photocatalyst is examined; it has often been documented that the CB bottom and/or VB top of a given material do not satisfy the necessary conditions when negligible or poor photocatalytic activity of that material is found. This is natural considering that thermodynamics, in general, does not give information on reaction rate, that is, photocatalytic activity (discussed below), and the half reactions by electrons and positive holes, as well as their recombination, may govern the overall reaction rate. Although the design of photocatalysts includes, of course, the choice of material and the choice must be based on the above-mentioned thermodynamic discussion, further design to improve the photocatalytic activity of a chosen material should include such discussion on kinetics but not thermodynamics.

4.2 Photocatalytic Activity: Reexamination

The widely used scientific term "activity" often appears in papers on photocatalysis as "photocatalytic activity" Although the present author does not know who first started using this term in the field of photocatalysis, people involved in the field of conventional catalysis were using this term even before the 1980s, when photocatalysis studies had begun to be promoted by the famous work of the so-called "Honda–Fujishima effect" on photoelectrochemical decomposition of water into oxygen and hydrogen using a single-crystal titania electrode [4], as mentioned above. Most authors, including the present author, prefer to use the term "photocatalytic activity," but in almost all cases, the meaning seems to be the same as that of absolute or relative reaction rate. A possible reason why the term "photocatalytic activity" is preferably used is that the term may make readers think of "photocatalytic reaction rate" as a property or ability of a photocatalyst, that is, photocatalysts have their own activity, while "reaction rate" seems to be controlled by given reaction conditions including a photocatalyst. In the field of conventional catalysis, "catalytic activity" has been used to show a property or performance of a catalyst, since an "active site" (Figure 4.2), substantial or virtual, on a catalyst

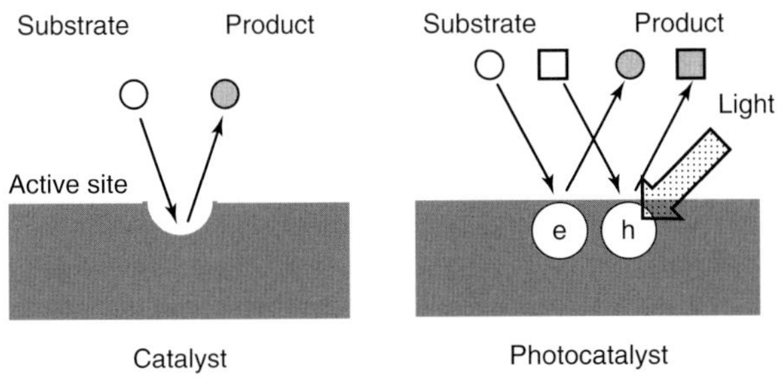

Figure 4.2 Schematic presentation of difference in mechanisms of catalytic and photocatalytic reactions: a catalyst contains active sites at which a substrate is converted into a product, while no such active sites are present on a photocatalyst.

accounts for the catalytic reaction. The estimated reaction rate per active site can be called "catalytic activity." In a similar sense, the term "turnover frequency," that is, number of turnovers per unit time of reaction, is sometimes used to show how many times one active site produces a reaction product(s) within unit time. On the other hand, it is clear that there are no active sites, as in the meaning used for conventional catalysis, in which rate of catalytic reaction is predominantly governed by the number or density of active sites, on a photocatalyst. The term "active site" is sometimes used for a photocatalytic reaction system with dispersed chemical species, for example, metal complexes or atomically adsorbed species, on support materials. However, even in this case, a photocatalytic reaction occurs only when the species absorb light, and thereby species not irradiated cannot be active sites. A possible mechanism of photoinduced reaction is the photoirradiation-induced production of stable "active sites" that work as reaction centers of conventional catalytic reactions, although this is different from the common mechanism of photocatalysis by electron–positive hole pairs. Anyway, photocatalytic reaction rate strongly depends on various factors such as the irradiance of irradiated light that initiates a photocatalytic reaction. Considering that the dark (nonirradiated) side of a photocatalyst or suspension does not work for the photocatalytic reaction, the use of the term "active site" seems inappropriate.

4.3 Design of Active Photocatalysts

An ordinary photocatalysis is induced by photoexcited electrons and positive holes, as has been described in the previous sections. Therefore, the rate of photocatalytic reaction must depend on photoirradiation irradiance (light flux) and efficiencies of both photoabsorption and electron–positive hole utilization. The efficiency of electron–positive hole utilization is called quantum efficiency, that is, the number (or rate) ratio of product(s) and absorbed photons, and even if quantum efficiency is high, the overall rate should be negligible when the photocatalyst does not absorb incident light. This is schematically represented as

$$(\text{Rate}) = (\text{Irradiance}) \times (\text{Photoabsorption efficiency}) \times (\text{Quantum efficiency}) . \tag{4.1}$$

Since all the parameters in Eq. (4.1) must be functions of light wavelength, the overall rate can be estimated by integration of a product of spectra of photoirradiation, photoabsorption, and quantum efficiencies. When we discuss the activity of a photocatalyst, it seems reasonable to evaluate a product of photoabsorption and quantum efficiencies, that is, apparent quantum efficiency. Assuming that quantum efficiency does not depend on the irradiation (absorption) irradiance, the actual reaction rate can be estimated by multiplying with the irradiance. On the basis of these considerations, enhancement of photocatalytic activity can be achieved by increase in both efficiencies. For example, preparing visible-light-absorbing photocatalysts, as a recent trend in the field of photocatalysis as discussed later, and depositing noble metal particles onto the surface of photocatalysts lead to the improvement of these efficiencies, respectively. In this sense, the design of active photocatalysts seems simple and feasible, but we encounter the problem that both efficiencies are related to each other and we do not know how we can improve the quantum efficiency because correlations between physical/structural properties and photocatalytic activity have only partly been clarified. In the following sections, the author will show what we know and do not know about photocatalytic activity.

4.4
A Conventional Kinetics in Photocatalysis: First-Order Kinetics

As is well known, first-order kinetics is commonly seen for reactions proceeding in homogeneous phases, that is, reactions in homogeneous solutions or gas phase. The rate of a monomolecular reaction, ideally, obeys a first-order rate expression that is explained by the assumptions: (i) the number (density) of molecules having kinetic energy larger than the activation energy, required to pass an intermediate state, is determined only by temperature of the reaction and (ii) the actual number (density) of molecules having energy for the activation is proportional to the concentration (or pressure) of molecules. On the basis of the assumption of this situation, kinetic data are analyzed by making a plot of the logarithm of concentration of a substrate or a product as a function of time [5] and, if a linear line is obtained, absolute value of the slope of the line is a rate constant, k (Figure 4.3). The rate (r) of consumption of a substrate (S) is shown by the following equation.

$$r = -\frac{\mathrm{d}[\mathrm{S}]}{\mathrm{d}t} = k[\mathrm{S}] \tag{4.2}$$

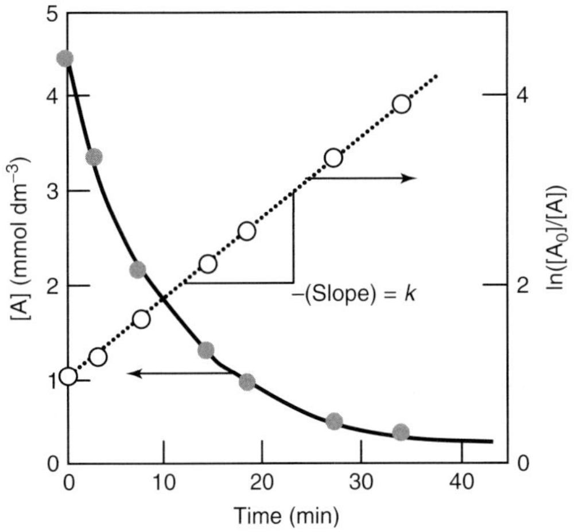

Figure 4.3 First-order kinetic analysis for a reaction consuming a substrate (A). Plot of logarithm of relative consumption (in the present plot, ratio of the initial concentration of substrate A and the concentration of A at a given time) against time of reaction.

On the other hand, kinetics of reactions occurring on a solid surface, that is, catalysis or photocatalysis, must be significantly different; such a monomolecular reaction cannot proceed on the surface of solids. However, there are at least two possible extreme cases in which the overall kinetics of photocatalysis (and catalysis) obeys a first-order rate expression. One is the so-called "diffusion-controlled" process, in which surface reactions and the following detachment process occur very rapidly to make a negligible surface concentration of adsorbed molecules, and the overall rate is the same as that of adsorption of substrate molecules. Under these conditions, the overall rate is proportional to concentration of the substrate in a solution or gas phase (bulk), that is, first-order kinetics may be observed. Assuming a very thin diffusion layer on the surface of a photocatalyst, the rate of diffusion may be proportional to surface area of the photocatalyst, since the rate of diffusion is proportional to the cross-sectional area of diffusion and the difference in concentration of a substrate. Therefore, the rate constant k calculated from the linear plot may be a product of surface area of a photocatalyst and true rate constant of reaction between electron–positive hole pairs and an adsorbed substrate.

The other extreme case is the so-called "surface-reaction-limited" process, in which surface adsorption is kept in equilibrium during the reaction and the overall rate is the same as that of reaction occurring on the surface, that is, reaction of photoexcited electrons and positive holes with a surface-adsorbed substrate. Under these conditions, the overall rate is not always proportional to concentration of the substrate in the bulk unless the adsorption isotherm obeys a Henry-type equation or in the lower concentration range of a Langmuir-type adsorption, in which the amount of adsorption is proportional to concentration in the bulk. The meaning of the obtained rate constant, k, is discussed in the next section.

In the former extreme case, the rate of photocatalytic reaction obeys the first-order rate law, while this is only formal and does not mean the mechanism of monomolecular reaction with activation energy. Thus, it seems that the above-mentioned kinetic analysis with the first-order rate law does not give valuable information for the design of highly active photocatalysts. This suggests that in scientific studies of not only photocatalysis but also the other areas, we must consider discrimination of "evidence" and "consistency" in experimental data. In other words, it is indispensable to recognize every fact to be a "necessary condition" but not a "sufficient condition" in a strict scientific sense. For example, even when the reaction rate obeys a first-order rate law giving a linear relation in a plot of data, that is only a necessary condition for a monomolecular reaction in homogeneous phase and also a necessary condition for heterogeneous photocatalytic reaction under diffusion-limited conditions or surface-reaction-limited conditions with a Henry-type adsorption or a lower-concentration-region Langmuir-type adsorption.

4.5
A Conventional Kinetics in Photocatalysis: Langmuir–Hinshelwood Mechanism

Although the term "Langmuir–Hinshelwood (LH) mechanism" [6] has commonly been employed for description of the mechanism of photocatalytic reaction in suspension systems, there has been no strict definition of the LH mechanism in photocatalytic reactions. In most reported cases, the authors claimed that the mechanism of a photocatalytic reaction proceeds through an LH mechanism when a reciprocal linear relation is obtained between reaction rate and concentration of the reaction substrate in a solution. The following equation may reproduce the experimental result of reaction rate, r:

$$r = \frac{kSKC}{KC + 1}, \tag{4.3}$$

where k, K, S, and C are rate constant of the reaction of the surface-adsorbed substrate with e^- (h^+), an equilibrium constant of adsorption, limiting amount of surface adsorption, and concentration of the substrate in the bulk in equilibrium, respectively [5]. This equation is derived from the assumption that the substrate is adsorbed by a photocatalyst obeying a Langmuir isotherm and the adsorption equilibrium is maintained during the photocatalytic reaction, that is, the rate of substrate adsorption is much faster than that of the reaction with electrons or holes, as has been discussed in the previous section. Such a situation is sometimes called "irradiance-limited" conditions, that is, photoabsorption is the rate-determining step (although the definition of "rate-determining step" in photocatalysis seems unclear). Two kinds of plots are often employed for analysis: one is the plot of reciprocal rate against reciprocal equilibrium concentration, and the other is the plot of ratio of concentration to rate as a function of equilibrium concentration. Theoretically, these plots give the same values of parameters, k, S, and K, while the former plot reflects mainly the lower concentration data with probable relatively large experimental errors (Figure 4.4).

Within the author's knowledge, the original definition, in the field of catalysis, of "LH mechanism" is a bimolecular (two kinds of molecules) reaction proceeding on a surface in which both the kinds of molecules are adsorbed at the same surface adsorption sites with the surface reaction being the rate-determining step. Thus, the general rate equation for the LH mechanism, not shown here, includes two sets of parameters for the two kinds of molecules, and, as an exception, when one set of parameters is neglected for a monomolecular reaction, the equation can be similar to the reaction of a substrate adsorbed in Langmuirian manner. At least in the field of catalysis, however, the term LH mechanism is seldom used for such monomolecular surface reactions, since the LH mechanism has been discussed for a bimolecular surface reaction for comparison with the Rideal–Eley mechanism, in which a surface-adsorbed substrate molecule reacts with a molecule coming from the bulk. Even if one defines the LH mechanism as a reaction of a surface-adsorbed substrate obeying a Langmuir isotherm determining the overall rate, the frequently reported reciprocal linear relation between the concentration of the substrate in solution and the photocatalytic reaction rate is not always proof of the above-mentioned mechanism. From the linear plot, two parameters are obtained [5]. One (kS, often shown as "k," not as "kS") is a limiting rate of the reaction at the infinite concentration giving maximum adsorption and is a product of rate constant and adsorption capacity of a photocatalyst, and this may

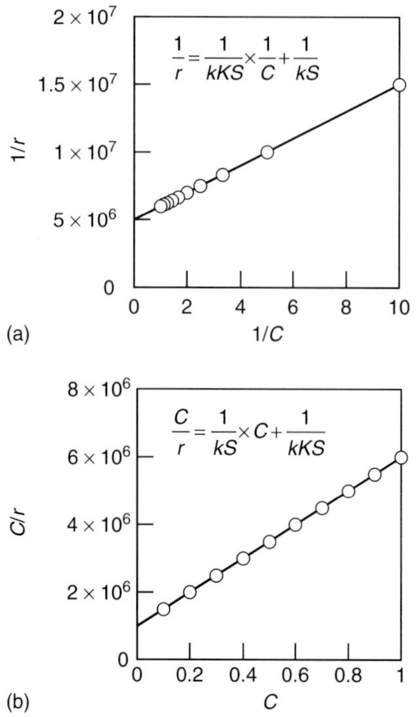

Figure 4.4 (a,b) Simulation of linearized plots for kinetics governed by surface concentration of substrates adsorbed on the photocatalyst surface in a Langmuirian fashion, where *r*, *C*, *k*, *K*, and *S* are the rate of reaction (mol s^{-1}), concentration of a substrate (mol L^{-1}), rate constant (10^{-4} s^{-1}), adsorption equilibrium constant (5 L mol^{-1}), and saturated amount of adsorption (2×10^{-3} mol).

be a photocatalytic activity. A notable point is that this parameter contains the term *S*, reflecting the surface area of a photocatalyst, similar to the term *k* in the first-order kinetics described in the preceding section. The other parameter, *K*, is the equilibrium constant for adsorption showing the strength of adsorption and must be the same as that estimated from an adsorption isotherm measured in the dark [7]. If this kinetically obtained *K* is found to be different from that obtained in dark adsorption measurement, the LH mechanism cannot be adopted. In other words, dark adsorption measurement is always required.

4.6
Topics and Problems Related to Particle Size of Photocatalysts

As described above, experimental results for photocatalysis, that is, rates obeying first-order kinetics or the so-called LH mechanism, often suggest the effect of surface area, that is, the larger the surface area, the higher is the activity of a photocatalyst, and this may be empirically correct in the field of photocatalysis.

Since surface area and particle size are closely related to each other, it can empirically be stated that the smaller the particle size is, the higher is the activity of a photocatalyst. For samples of titania, anatase (ANA), or RUT having a density of about 4 g cm^{-3}, the product of specific surface area in units of $m^2\ g^{-1}$ and diameter in units of nanometers is calculated to be 1500 on the basis of the assumption of a spherical shape of titania particles of monodispersed size distribution as represented by the following equation:

$$d\left(m^{-1}\right) = \frac{6}{\rho\left(kg\ m^{-3}\right) \cdot S\left(m^2 kg^{-1}\right)}, \quad (4.4)$$

where d, ρ, and S are the diameter, density, and specific surface area of particles. Thus, surface area should ideally be inversely proportional to particle size. In this section, the impact of and methods for estimation of particle size or surface area are discussed.

Particle size of a photocatalyst is often estimated by Scherrer's equation using data of powder X-ray diffraction pattern (XRD); the most intense XRD peaks appearing at about 25° and 27° (when a copper target is used for X-ray radiation) for ANA and RUT crystallites, respectively, are used for calculation. In Scherrer's equation, particle size L, expansion of crystallite in the direction vertical to the corresponding lattice plane in a strict scientific sense, is represented by

$$L = \frac{K\lambda}{\beta \cos\theta}, \quad (4.5)$$

where K, λ, β, and θ are a constant, wavelength of X-ray, corrected full width at half maximum (FWHM) of the XRD peak, and angle of diffraction, respectively. Since K, often called "shape factor," had been calculated in the derivation of this equation to be 0.891, K has its effective digits. Therefore, in order to show the size L, with, for example, three-digit accuracy, a value of "0.891" should be used as K. Another point to be noted is that FWHM should be corrected by two kinds of correction procedures: one is correction for broadening because of $K_{\alpha 2}$ radiation and the other is for broadening because of the optical path in the diffractometer. In general, the former and the latter corrections are made by assuming a radiation intensity ratio of $K_{\alpha 1}$ to $K_{\alpha 2}$ and by using FWHM of a standard large crystalline sample, respectively. There are at least three ways to perform the latter correction, but there seems to have been no discussion on the best way, and the simplest way, subtraction of FWHM of the standard, has often been employed. To the author's knowledge, both corrections yield appreciable difference in the size of particles, that is, a large error is expected without such corrections. Therefore, when the size of particles is demonstrated, the methods used for FWHM corrections should be described [5].

Thus, the size of particles in the direction vertical to the corresponding lattice plane, giving the XRD peak, can be estimated using Scherrer's equation. When compared with data obtained in other ways, for example, transmission electron microscopy (TEM) or scanning electron microscopy (SEM), we often observe difference because of this. Further difference may be caused by broadening of FWHM of XRD peaks because of distortion of the crystalline lattice. The Williamson–Hall equation includes this, as well as the effect of particle size [8]. In other words, Scherrer's equation neglects the effect of crystal lattice distortion. Consequently, for samples that are expected to have a large degree of distortion, analysis using the Williamson–Hall equation must be carried out.

Then, can we know particle size from the above-mentioned analysis? There still seems to be one problem, that is, distribution of size and its effect on evaluation of average size should be considered. Figure 4.5 shows hypothetic fractions of four kinds of spherical particles of 1, 10, 25, and 100 nm in diameter calculated on the basis of number, surface area, and volume (weight). Although number-based fractions are all the same, 25%, for the four kinds of particles, surface-area-based and volume-based fractions are 93% and 98%, respectively. Therefore, the average sizes of particles calculated on the basis of these fractions are 34, 93, and 99 nm when number, surface area, and volume fractions are employed. (In a similar sense, average molecular weight is calculated mainly in two modes, number average and weight average, in the field of polymer science.) Since particle size is measured by SEM or TEM images and particles are categorized by size and counted (making a histogram) from images in ordinary procedures, the average size is a number-based one unless volume fraction is considered. Then, what is the mode of average size of particles, with a wide size distribution, obtained by Scherrer's equation: average in number, surface area, or volume-based one? There seems to be no clear answer. The author's group has experienced that many commercial titania samples obey Eq. (4.4)) when (primary) particle size is measured using Scherrer's equation, presumably because average particle size obtained using Scherrer's equation corresponds to the surface-area-based or volume-based average [9].

Another important topic related to particle size of a photocatalyst is "quantum-size effect." This frequently used term explains that when the size (radius) of solid particles is smaller than their Bohr radii, the bottom of the CB and the top of the VB shift in negative (high electronic energy = cathodic) and positive (low electronic energy = anodic) directions, respectively, to result in expansion of the bandgap. The Bohr radii for ANA and RUT particles have been estimated to be 2.5 and 0.3 nm,

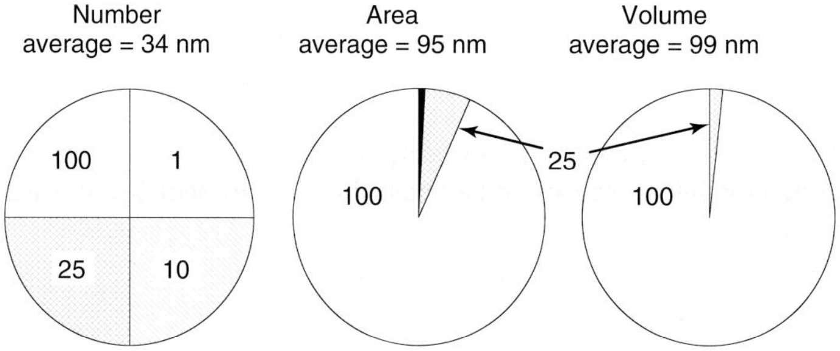

Figure 4.5 Example of fractions of four kinds of particles (25% each in number) of sizes 1, 10, 25, and 100 nm in number, area, and volume (weight). Figures in circle graphs show particle size. Average size of particles is calculated on the basis of fractions.

respectively [10]. Preparation of crystalline titania particles of such small size seems to be impossible, and titania particles claimed in papers to show a quantum-size effect might be larger than these sizes. Even if a blue (shorter wavelength) shift of the absorption edge of those samples is observed, it might be due to the amorphous part of titania, not due to the quantum-size effect. It can be said, at least for a titania photocatalyst, that use of quantum-size effect for interpreting results seems inappropriate.

4.7 Recombination of a Photoexcited Electron and a Positive Hole

When photocatalytic activity of a given photocatalyst seems less than that expected from its relatively large specific surface area or small particle size, decrease in the photocatalytic activity tends to be attributed to recombination of a photoexcited electron and a positive hole. This may be true, but it should be noted that direct evidence for enhanced recombination has rarely been shown in papers.

The electron–positive hole recombination kinetics may depend on its fashion; if one electron is excited and this is recombined with a positive hole, the rate of recombination obeys the first-order rate law, while if multiple electron–positive hole pairs appear at the same time within one photocatalyst particle, the rate obeys the second-order rate law. In fact, for titania particles, photoabsorption at 620 nm by trapped electrons showed second-order decay with a component of baseline in a femtosecond pump-probe (PP) diffuse reflection spectroscopic analysis of titania samples, as follows:

$$(\text{Absorption}) = \alpha \left\{ \frac{[\mathrm{e_0}]}{1 + k_\mathrm{r} [\mathrm{e_0}] \times t} + B \right\}, \tag{4.6}$$

where α, $[\mathrm{e_0}]$, k_r, t, and B are a constant, initial concentration of trapped electrons at time zero, second-order rate constant, time after pump (excitation) pulse (at 310 nm), and baseline component, respectively [11]. A baseline component might correspond to electrons trapped in deep traps that do not participate in the reaction within the measured time region. Different from kinetic analysis based on the first-order rate law, analysis based on the second-order rate law requires absolute values of concentration ($[\mathrm{e_0}]$ in Eq. (4.6)) and photoabsorption coefficient (α in Eq. (4.6)) of a photocatalyst, although these cannot be determined experimentally, at least when the analyses are performed and calculation is performed, for convenience, assuming α to be unity. An example of these kinetic analyses is shown in Figure 4.6 for Degussa (Evonik) P25 [12]. It should be noted that the thus-obtained second-order rate was relative, but k_r's of different titania samples in the form of powder were proportional to those in suspension systems, suggesting that k_r can be a measure of rate of recombination. At the same time, however, it must be noted that such a second-order recombination process cannot be reproduced under

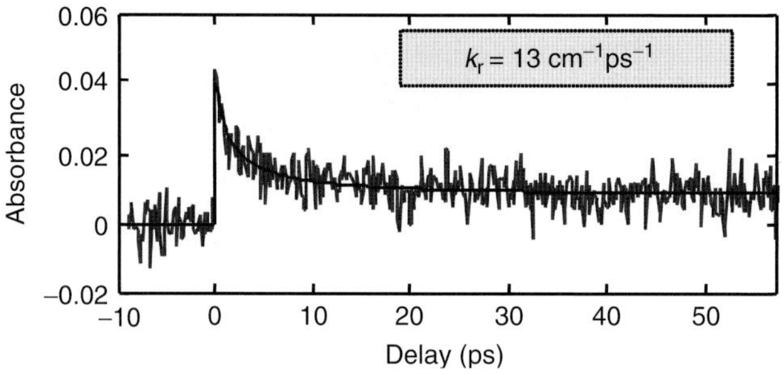

Figure 4.6 An example of picosecond-time-region decay of photoabsorption (620 nm) of trapped electrons in Degussa (Evonik) P25 particles after excitation by about 100-fs pump pulse (310 nm). The curve was analyzed by a second-order rate law (Eq. (4.6)) with a baseline component (B), and a second-order rate constant (k_r) was obtained to be 13 cm^{-1} ps^{-1}.

ordinary photoirradiation conditions in which lower irradiance induces single-electron photoexcitation and mutual recombination occurs, obeying the first-order rate law.

4.8 Evaluation of Crystallinity as a Property Affecting Photocatalytic Activity

Then, how do electron–positive hole pairs recombine in ordinary photocatalytic reaction systems? One of the most probable "recombination centers" is lattice defects that trap electrons (or positive holes) and it has often been claimed that high crystallinity, that is, low density of lattice defects, leads to high photocatalytic activity, but how can we evaluate the crystallinity? Considering inorganic solid materials as photocatalysts, they are usually crystals and the crystalline form is often determined by an XRD pattern. A serious problem in XRD analyses is that only crystals are detected, while an amorphous part, even if included in samples, exhibits no diffraction peaks; a hallow may appear for amorphous parts, but there has been no report showing quantitative determination of amorphous content. As a result, amorphous content must be determined as the rest of the crystalline part and therefore precise determination of crystalline content is needed. As a principle, XRD peak intensity is believed to be in proportion to content of corresponding crystallites, but a problem is how we can get widely acceptable standard samples of each crystal, because smaller crystallites may exhibit lower peak intensity [13]. A possible interpretation of this particle-size-dependent intensity problem is that the outermost surface of particles cannot be involved in a crystal, which can be defined only for bulk, and the ratio of surface/bulk becomes appreciable when the size of crystallites is small. Another possibility is that XRD peak intensity decreases because of the anisotropic shape of smaller particles reflecting preference of a lattice plane. Thus, precise analysis of crystalline content can be guaranteed only when pure crystalline particles included in a sample are extracted and used for making

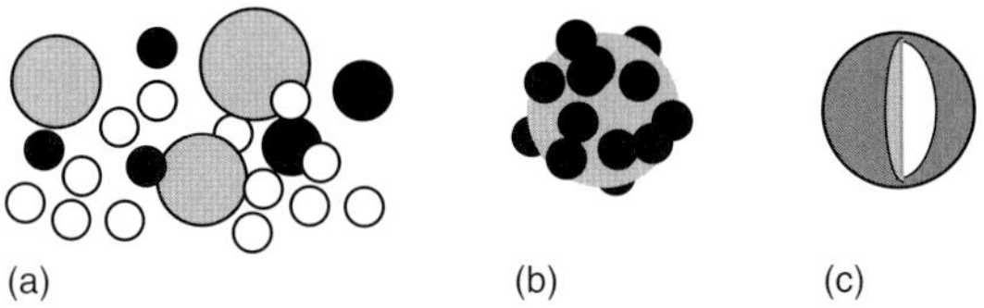

Figure 4.7 Representative patterns of mixture of crystallites (white and gray particles) and amorphous parts (black). (a) Simple mixture, (b) loaded, and (c) core–shell.

XRD calibration curves [14, 15] based on the assumption that crystallites and amorphous particles are separated but not in the form of, for example, a core–shell structure (Figure 4.7). In other words, if a sample particle is of a core–shell structure, precise determination of crystalline content may be very difficult.

Confusion related to the term "crystallinity" arises because the term has sometimes been discussed on the basis of sharpness of an XRD peak with description, for example, "sharpness of the peak indicated higher crystallinity of a photocatalyst." Since an XRD peak width reflects the size of a particle, that is, the depth of crystallites measured in the direction vertical to a corresponding lattice plane, as described in the preceding section [16], peak sharpness indicates the size of crystallites. In this sense, "crystallinity" based on XRD peak sharpness is used to show how crystallites grow to be larger-sized particles. Another meaning of the term "crystallinity" is to show perfectness of crystals, that is, higher crystallinity corresponds to lesser density of lattice defects. Assuming that density of lattice defects in larger crystallites is relatively low, sharpness of XRD peaks can also be a relative measure of "crystallinity." However, the crystallinity evaluated in this way seems not to be a general parameter in the discussion of photocatalytic activity.

4.9
Electron Traps as a Possible Candidate of a Recombination Center

It has often been shown that photoirradiation of titania samples leads to the formation of trivalent titanium species (Ti^{3+}). For example, for the detection of active species photogenerated in titania particles, electron paramagnetic (spin) resonance (EPR) spectroscopy has often been used and has revealed that electrons are trapped on the surfaces and/or titanium atoms in the bulk, thus forming Ti^{3+} [17], while positive holes might be trapped by oxygen atoms in the crystalline lattice and/or surface hydroxyl groups. This suggests that there might be some trapping sites for photoexcited electrons in titania particles. Similarly, electrons trapped in certain sites have generally been detected by laser spectroscopy. For example, as was discussed in Section 4.6, femtosecond PP spectroscopy for titania particles and colloids has shown that certain sites trap electrons within a few tens of femtoseconds after the photoexcitation, resulting in the appearance of a broad absorption at around 500–650 nm, and that their recombination with positive holes induces decay of absorption within 100 ps, obeying second-order kinetics. A major problem in the above-mentioned EPR and laser spectroscopic techniques is the complexity of the measurement procedures for determining kinetic parameters of electron–positive hole recombination. Another problem is that samples in those spectroscopic measurements might be under conditions different from those for ordinary photocatalytic reactions; samples are sometimes used in a film form or placed under vacuum for measurements.

On the basis of these facts, we tried to develop a method to measure the density of lattice defects working as electron traps. A working hypothesis is that density of electrons accumulated in titania particles as a form of Ti^{3+} under photoirradiation in the presence of a strong electron donor, but in the absence of an electron acceptor such as molecular oxygen, is a relative measure of the density of lattice defects, which work as electron traps. The first method developed was determination of Ti^{3+} by methyl viologen [18]. This enabled evaluation of the density of lattice defects M_d under experimental conditions similar to those of an ordinary photocatalytic reaction. Interestingly, a linear relation between M_d values and the second-order rate constant calculated for decay of trapped electrons in femtosecond laser spectroscopy described above was observed for commercial titania samples, suggesting that recombination of electron–positive hole pairs occurs at those electron traps. Another interesting feature of M_d is correlation with specific surface area of each sample. As shown in Figure 4.8, it seems that M_d was increased with increase in specific surface area, indicating that such electron traps are predominantly located on the surface, and its density reflected in the slope of plots depended on the kind of sample; it might be possible to reduce M_d while keeping a relatively large surface area.

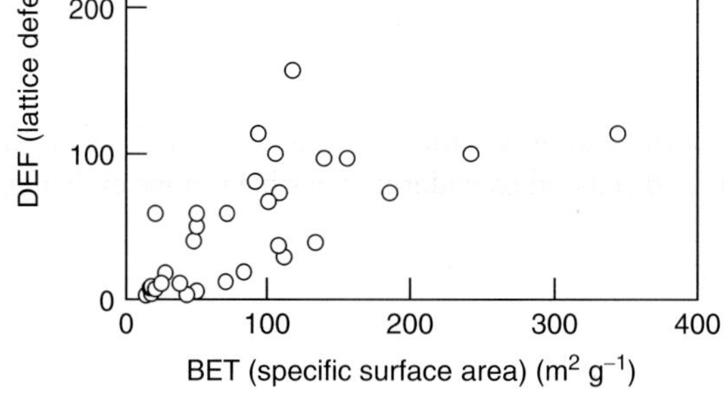

Figure 4.8 Correlation between density of lattice defects (DEF) and specific surface area. (Source: Data reported in Ref. [19b] are used for this plot.)

It was also shown that the electronic energy (potential) of those electron traps is located just below the CB edge of titania in ranges of 0–0.35 V for a typical ANA titania powder and 0–0.25 V for a typical RUT titania powder on the basis of results obtained by controlling the rate of reaction between Ti^{3+} and methyl viologen in various pH conditions [18]. Recently, we have also developed another method for determination of lattice defects using photoacoustic spectroscopy [20].

4.10
Donor Levels – a Meaning of n-Type Semiconductor

It is now possible to measure the density of lattice defects working as electron traps as a significant parameter governing the photocatalytic activity of titania photocatalysts by a photochemical or photoacoustic spectroscopic method. Then, what is the origin of these lattice defects? n-Type character of titania, and the most metal oxides, may interpret this.

In the early stage, in the 1980s, of studies on photocatalysis, a photocatalyst particle was often thought to be a "short-circuited photoelectrochemical cell" [21] because titania particles loaded with platinum were often employed as a photocatalyst working under deaerated conditions. This is similar to an ordinary photoelectrochemical cell in which titania and platinum work as an anode and a cathode, respectively, which are short circuited with each other. It was assumed that photoexcited electrons in titania migrate to platinum through a junction (border) between titania and platinum. This junction had been discussed and concluded to be an ohmic one, not a Schottky one, considering the possible n-type character of titania electrodes prepared by heat treatment at a high temperature under vacuum or hydrogen atmosphere (Figure 4.9a,b). There must be such a contact between titania and platinum. Then, do photoexcited electrons in titania actually migrate to platinum and reduce protons to liberate hydrogen in a photocatalytic reaction? At present, we have no experimental techniques to observe such migration through the junction directly and to check whether liberated hydrogen contains electrons originating from titania. As indirect evidence, that is, a supporting fact, Nakabayashi *et al.* [22] reported that the observed isotope distribution in hydrogen liberated by the photocatalytic reaction with platinized titania suspended in a mixture of water (H_2O) and heavy water (D_2O) was consistent with the ratio calculated on the basis of the assumption of hydrogen generation on the platinum surface, but not on the titania surface, although the interpretation of the results of this isotope experiment has to be discussed considering possible isotope exchange between water and surface hydroxyls. This suggests that titania particles act also as an n-type semiconductor, and when electrons in titania, but not in platinum, are excited by photoirradiation, migration of electrons from titania to platinum occurs. It has also been thought that there is a Schottky junction at an n-type semiconductor–electrolyte solution interface; a more cathodic position of a Fermi level of n-type semiconductors results in transfer of electrons in the donor levels of an n-type semiconductor to the electrolyte and a depletion (space charge) layer appears (Figure 4.9c,d). This is the reason why titania electrodes, not only single crystalline but also porous polycrystalline, show rectifying properties, that is, negligible dark anodic current even under anodically polarized conditions because of the so-called "Schottky barrier" inhibiting electron injection from an electrolyte to an n-type semiconductor electrode in the dark and under photoirradiation (Figure 4.9e,f); electron transfer occurs from an n-type semiconductor electrode to an electrolyte under cathodic polarization because no barrier exists for such electron flow, while under photoirradiation and anodic polarization, positive holes can oxidize redox species in the electrolyte and photoexcited electrons flow as an anodic current. The thickness of the depletion layer, governing the slope of the electric field, depends on the concentration of donors (impurities). For particles of titania that are not treated to produce lattice defects (donor levels), the donor concentration seems to be low and the depletion layer thickness might be larger than the particle size, resulting in a small potential slope in the particles. In the author's idea, however, particles may have a relatively large surface area and surface defects as electron donor levels, and the donor levels release electrons on contact with an electrolyte solution to leave vacant (electron-deficient) donor sites on and in titania particles. The electron traps detected and measured by the photochemical reaction introduced in the previous section might be these vacant electron donor sites. This seems consistent with the fact that the position of electron traps has been estimated to be just below the bottom of the CB.

4.11 Dependence of Photocatalytic Activities on Physical and Structural Properties

4.11.1 Correlation between Physical Properties and Photocatalytic Activities

The discussion described above suggested that surface area, providing adsorption sites for the substrate, and lattice defects, working as recombination centers (and at the same time as possible hopping sites for electron migration), may hold a key for photocatalytic activity of titania particles. It is easily predicted, however, that finding a correlation between those properties and photocatalytic activity is not straightforward because properties may be related to each other (e.g., the density of lattice defects depends strongly on the surface area) and there are possibly other properties affecting photocatalytic activity. In this section, our recent trial to find a correlation is introduced.

It seems that ordinary physical or structural properties measured for photocatalysts have not yet been proved to be decisive factors for photocatalytic activities. It must be true that activities of photocatalysts of given components prepared or treated in different ways or under different conditions are different and this is because physical and structural properties of those photocatalysts differ depending on the preparation/treatment conditions. Thus, we believe that physical and structural properties must control the photocatalytic activity [23]. A serious and

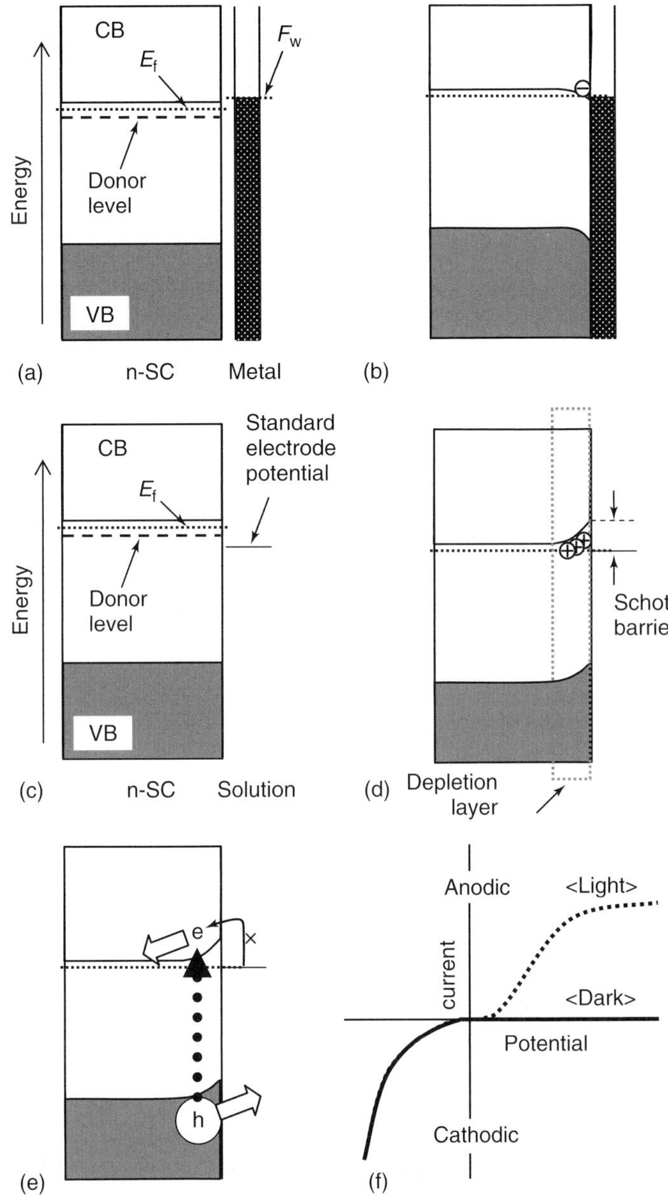

Figure 4.9 (a) n-Type semiconductor (n-SC) and metal with work function (F_w) and (b) ohmic contact between them. (c) n-SC and solution and (d) Schottky junction with a depletion layer where electrons in donor levels flowed out to the solution to lower the Fermi level (E_F). (e) n-SC–electrolyte interface under photoirraidiation and (f) ideal potential–current plots in the dark and under photoirradiation.

essential problem is that we, at least the author, do not know how properties affect photocatalytic activity. This is probably because those properties, although we do not know how many properties are required for analysis, are changed at the same time, that is, it is practically impossible to control only one property without changing other properties. As a representative example, when titania photocatalysts are prepared by hydrolysis of a titanium compound, titanium(IV) sulfate [24] or titanium(IV) tetra(2-propoxide) [25], followed by calcination in air, higher-temperature (>873 K) calcination leads to higher crystallinity, smaller specific surface area, and RUT crystallites, while lower-temperature (<673 K) calcination leads to lower crystallinity, larger specific surface area, and ANA crystallites. It has generally been reported that photocatalytic activity of titania particles prepared in such a way for photocatalytic hydrogen liberation from a deaerated aqueous 2-propanol solution in the presence of *in situ* photodeposited platinum decreased drastically at the temperature at which crystalline transformation from ANA to RUT occurred. Since both crystalline form and specific surface area were changed drastically at the same time and there have been no reported ways to extract (discriminate) the intrinsic effect of each property, it is rather difficult to determine which property (or both of them) governs the photocatalytic activity. Discussions on property–activity correlations reported so far, including those reported by a group of the author, may involve such an unsolved problem. It seems empirically correct that RUT titania samples with a small surface area that are prepared at a high temperature show relatively low photocatalytic activity, but it is impossible to state in strict scientific (logical) sense that the conversion of crystalline form or drastic reduction of specific surface area is the reason for such low photocatalytic activity.

4.11.2
Statistical Analysis of Correlation between Physical Properties and Photocatalytic Activities – a Trial

One plausible method for extraction of the intrinsic effect of each physical and structural property is statistical analysis of data on physical and structural properties and photocatalytic activities for samples of the same composition, such as titania. In a recent study by the author's research group, statistical analysis was performed for photocatalytic activities and physical and structural properties of 35 commercial titania powders to find the predominant property (properties) determining the activity of a given reaction system [26]: standardized photocatalytic activities for five kinds of reactions, **a–e** listed in Table 4.1, were fairly well reproduced by a linear combination of six kinds of physical and structural properties of photocatalysts, that is, specific surface area (BET) measured by nitrogen adsorption at 77 K using the Brunauer–Emmett-Teller (BET) equation, density of lattice defects (DEF), primary particle size (PPS), secondary particle size (SPS), and existence of ANA and RUT phases.

Matrix solutions (property–activity coefficients) are shown in Table 4.2 with squared regression coefficients (R^2) found between observed and predicted values of reaction rates for each system. It was found that reactions **a** and **e** gave relatively

Table 4.1 List of photocatalytic reactions for evaluation of activities.

Entry	Description	Stoichiometry	Conditions	Platinum loading
a	Silver deposition along with oxygen liberation	$4Ag^+ + 2H_2O \rightarrow 4Ag + O_2 + 4H^+$	Deaerated aqueous suspension	No
b	Dehydrogenation of methanol	$CH_3OH \rightarrow HCHO + H_2$	Deaerated aqueous suspension	Yes
c	Oxidative decomposition of acetic acid in water	$CH_3COOH + 2O_2 \rightarrow 2CO_2 + 2H_2O$	Aerated aqueous suspension	No
d	Oxidative decomposition of acetaldehyde in air	$CH_3CHO + 5/2O_2 \rightarrow 2CO_2 + 2H_2O$	Air	No
e	L-Pipecolinic acid synthesis from L-lysine	L-Lysine → L-Pipecolinic acid + NH_3	Deaerated aqueous suspension	Yes

larger R^2s, that is, higher reproducibility of the results fitting to a linear combination of properties, while those for the others were also fairly high. For the coefficients, k, if the value is positive, the larger is the parameter, the higher is the photocatalytic activity and vice versa. A significant feature found in Table 4.2 is that the coefficient of k_{ANA} has a large positive value in all cases, that is, ANA seems active, except for reaction **a**. This is the first example, as far as the author knows, of support for a general (common) understanding in this field of photocatalysis that ANA is more active than RUT, especially for photocatalytic reactions (**c** and **d**) in which molecular oxygen participates as an electron acceptor, since it can be said that previously reported data were for only a limited number of titania samples and neglected the influence of other properties. On the other hand, the absolute value of k_{RUT} was always small and it was negative in **d**, suggesting that RUT phase in the photocatalysts is rather inert compared with ANA; RUT seems not to disturb the reaction by ANA. An exception is k_{RUT} for reaction **b**, which was comparable to k_{ANA}. The value of k_{RUT} was larger than that of k_{ANA} only for reaction **a**, while the absolute value was small even in this reaction compared with those of k_{ANA} in the other reactions.

The observed trends in the effect of crystalline phase, ANA and RUT, can be explained by the position of the CB bottom of these crystals. Reaction **b** was performed using *in situ* platinized titania particles, because it is well known that bare titania samples show negligible photocatalytic activities. The positive values of k_{RUT} and k_{ANA} coefficients seem to be reasonable because the CB bottom of ANA (−0.20 V vs standard hydrogen electrode (SHE)) and RUT (+0.04 V) crystals has been reported [27] to be almost the same as or slightly more negative than the standard electrode potential of hydrogen evolution (H^+/H_2; 0 V at pH = 0). The value of k_{ANA} is slightly larger than that of k_{RUT}, and this might be related to the slightly more negative CB level of ANA. Such comparable activity of RUT phase in this reaction

Table 4.2 Squared multiple correlation coefficient (R^2) and partial regression coefficients (k).

Coefficient	a	b	c	d	e
R^2	0.86	0.52	0.58	0.60	0.85
k_{BET}	−0.01	0.43	−0.09	0.13	0.19
k_{DEF}	−0.15	−0.25	0.19	0.43	0.32
k_{PPS}	0.12	−0.20	−0.18	−0.20	−0.52
k_{SPS}	0.57	0.08	−0.20	−0.04	−0.07
k_{RUT}	0.14	0.28	0.11	−0.06	0.02
k_{ANA}	0.04	0.40	0.57	0.55	0.63

has been shown previously through action spectrum analysis. On the other hand, reactions **c** and **d**, exhibiting similar trends in their coefficients, were operated under aerated conditions with reduction of molecular oxygen by photogenerated electrons. The better activity of ANA crystallites (large positive k_{ANA}) may be caused by the above-mentioned CB bottom position. The potential of one-electron reduction of O_2, $O_2^{\bullet-}/O_2$ (−0.05 V) or $HO_2^{\bullet}/O_2$ (−0.28 V), is slightly more negative than that of hydrogen evolution and, therefore, a small difference (about 0.16 V) in the CB bottom positions of ANA and RUT may have a significant influence.

As described above, comparable positive k_{RUT} and k_{ANA} values are obtained for reaction **b**, and this is interpreted by the position of the CB bottom relative to the electrode potential of hydrogen evolution. Relatively large positive k_{BET}, as well as negative k_{PPS}, and large negative k_{DEF} suggested that this reaction requires both a large amount of adsorbed methanol and less probability of electron–positive hole recombination. These trends may be the reason for the decrease in photocatalytic activity, for 2-propanol dehydrogenation of titania samples calcined at a higher temperature to result in crystalline conversion from ANA and RUT and drastic decrease in the surface area, as described in the preceding section; the decrease in the activity is attributable to the drastic decrease in the surface area, not to the transformation of ANA to RUT, assuming similarity of reactions using methanol and 2-propanol.

4.11.3
Common Features of Titania Particles with Higher Photocatalytic Activity

Needless to say, enhancement of the photocatalytic activity of ANA titania for decomposition of water and degradation of organic compounds has been a challenging subject. One of the possible strategies for this is to find necessary properties by analyzing highly active photocatalysts among a large number of photocatalysts, and statistical analysis has been found to be valuable for finding the necessary properties for each photocatalytic reaction system. On the other hand, during the statistical analysis, we noticed some general trends in the photocatalytic activity data: samples showing higher activity for reaction **a** were less active in the other reactions and vice versa, reflecting the different tendency of regression coefficients between **a** and the others listed in Table 4.2. Comparing the rankings of higher activities for reaction **a** and reactions **b–e** (not shown here), three titania samples, P25 (Degussa–Nippon Aerosil (Evonik)), TIO-4 (Catalysis Society of Japan, Reference Titania Catalyst), and ST-F4 (Showa Titanium), appeared commonly in both rankings; they are the best photocatalysts applicable to a wide range of photocatalytic reactions. P25 and TIO-4 are originally the same and are provided by Nippon Aerosil. As common features, they consist of ANA–RUT mixed particles with a specific surface area of about 50 $m^2\ g^{-1}$, that is, PPS of about 30 nm. P25 is considered to be one of the best photocatalysts based on an accumulation of results of nonsystematic empirical studies. The present results support this conventional understanding, for the first time, in a scientific sense. In other words, a claim of superiority of a given titania photocatalyst requires that a given sample exhibits higher levels of photocatalytic activity for several reactions with different reactions conditions, such as those introduced here. Checking the physical and structural properties of these highly active titania photocatalysts, none of the properties seem to be extraordinary high or low (large or small), that is, "balance" is important.

4.11.4
Highly Active Mesoscopic Anatase Particles of Polyhedral Shape

Another finding through morphological studies by SEM of various titania photocatalysts was the presence of mesoscopic angular particles. This feature has been also found in SEM analysis of our developed samples [19] and suggests that a property (properties), related to particle morphology, other than the six properties used for the above-mentioned statistical analysis may influence photocatalytic activity. Mesoscopic decahedral titania particles were fabricated from titanium(IV) chloride ($TiCl_4$) through a gas-phase reaction process with rapid heating and quenching [28, 29]. On the other hand, mesoscopic octahedral ANA titania particles were prepared by hydrothermal reaction of titanate nanowires in a Milli-Q water at 443 K for 24 h [30]. ANA crystalline structure was confirmed by XRD patterns, and BET surface area in the range of 10–50 $m^2\ g^{-1}$ was found for both decahedral and octahedral particles. TEM analysis revealed single-crystalline particles exposing {101} and {001} facets and {101} facets for decahedral (Figure 4.10) and octahedral particles, respectively, from their electron diffraction patterns. Photocatalytic activities were examined in several reaction systems: reactions **a–d** in Table 4.1. In general, decahedral particles exhibited photocatalytic activities higher than or comparable to those of P25 in all the tested reaction systems, while octahedral particles showed high activity, especially in reaction **c**, presumably reflecting their crystalline shape. Within the author's experience, it is not easy to obtain such titania photocatalysts exhibiting higher photocatalytic activity in a wide range of photocatalytic reaction systems.

Then, what is the reason for the higher photocatalytic activity of decahedral (octahedral) titania particles? It is easy to claim that such polyhedral shape in mesoscopic scale enhances the photocatalytic activity, but there seems no scientific evidence supporting that morphology, but not other physical and structural properties, governs the photocatalytic activity. Then, how does the decahedral shape

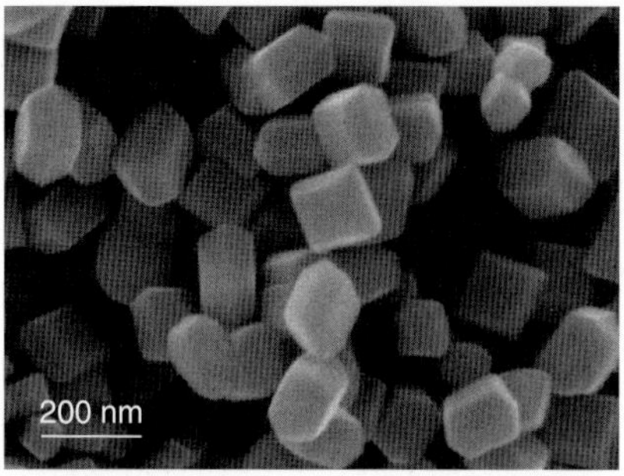

Figure 4.10 A representative SEM image of decahedral anatase titania particles prepared by controlled gas-phase reaction of titanium(IV) chloride and oxygen at 1473 K. Most particles expose two square (001) facets and eight trapezoidal (101) facets.

itself affect the photocatalytic activity? It was suggested that high levels of photocatalytic activity of decahedral ANA particles could not be reproduced by correlation equations derived in the above-mentioned multivariable analysis, especially for reactions **b** (methanol dehydrogenation) and **c** (acetic acid decomposition) [31], that is, the observed photocatalytic activities for these reactions were much higher than those calculated using the six physical and structural parameters, not representing their morphology, and coefficients listed in Table 4.2. This suggests that particle morphology may affect the overall photocatalytic activity, and a study along this line is now in progress.

4.12
Synergetic Effect

The term "synergetic effect," appearing frequently in papers on photocatalysis, can be defined as follows: when more than two kinds of photocatalysts are mixed and used as a photocatalyst, the overall photocatalytic activity is higher than the sum of activities of each photocatalyst (Figure 4.11). However, when a certain component alone is not a photocatalyst and a mixture with another photocatalyst that shows improved photocatalytic activity, that nonphotocatalyst component might be called "cocatalyst," and the improvement cannot be attributed to a synergetic effect. Ohno *et al.* [32] presented plots of photocatalytic activity of ANA–RUT mixtures showing that the mixture in a certain ratio had much higher activity for photocatalytic oxidation of naphthalene to dialdehyde than those of pure ANA and RUT, that is, a synergetic effect. Checking a synergetic effect by the use of plots such as Figure 4.11 seems reasonable, but in a strict scientific sense, there is still a room for discussion. Taking into consideration the fact that a photocatalytic reaction proceeds by photoabsorption of the photocatalyst and the fact that the total number of absorbed photons is not directly proportional to the mass (volume) of the photocatalyst, a control experiment using each component (or the other mixture in a given ratio) should be carried out with adjustment to make the flux of absorbed photons the same as, or at least similar to, that for the mixture.

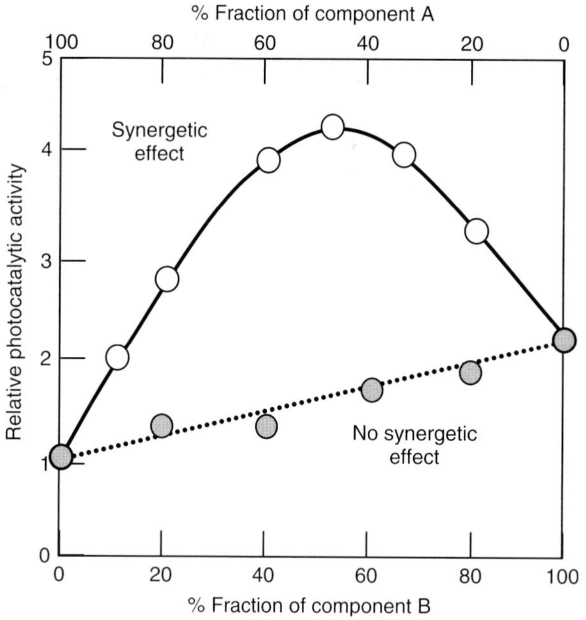

Figure 4.11 Hypothetical representation of relative photocatalytic activities of mixtures of components A and B. A synergetic effect can be suggested at least when a mixture shows better photocatalytic activity compared with those of pure components A and B.

One of the representative discussions on synergetic effect in photocatalysis is for ANA–RUT mixed crystalline (not an artificial mixture) photocatalysts, such as Degussa P25; it has been suggested that transfer of photoexcited electrons and positive holes between interconnecting ANA and RUT particles may enhance charge separation and hence improve the efficiency of utilization of electron–hole pairs [33]. However, within the author's knowledge, there have been no reports showing direct evidence of such charge migrations between particles and the expected lower level of activity of pure ANA or RUT particles alone. Thus a synergetic effect had not yet been proved for P25, and the effect seems to be speculation because each component, ANA and RUT, in P25 was not isolated to make a plot like that in Figure 4.11 before the isolation of ANA by the author's group [14]. Recently, the author's group found that experimental data obtained using ANA and RUT crystallites isolated (extracted) from P25 were consistent with the assumption of no synergetic effect of ANA and RUT in P25 [15].

4.13 Doping

Since the discovery of visible-light-driven photocatalysis of titania doped with nitrogen by Asahi *et al.* [34], "doping" has been a keyword for preparation of visible-light-active photocatalysts; any photocatalysts with poor visible light activity would be modified with metal or nonmetal elements to work under visible light irradiation. There appear to be two reasons for the explosive growth in the publication of papers on doped material: one reason is the lack of methodology to prove visible-light-induced photocatalysis, as discussed in the last part of this chapter and the other reason is the ambiguous definition of the term "doping." The original meaning of "doping" might be random incorporation of atoms or ions in a crystalline lattice, that is, modification of the bulk crystalline structure but not modification of surfaces. Such location of heteroatoms or ions has been rarely discussed. The effect of doping should be discussed on the basis of this structural information, for example, average density of heteroatoms/ions and their spatial distribution, although there have been few reports containing such discussion so far, because there are few reliable methods for obtaining such structural information. Related to this problem, it has been claimed in recent reports that heterocyclic compounds, heptazine derivatives, were produced on the surface of titania particles during the general procedure for nitrogen doping using urea as a nitrogen source and that the surface-adsorbed (or attached) derivatives work as a photosensitizer and/or photocatalyst [35]. In this case, nitrogen is not actually "doped" in the titania lattice but is included as a surface modifier.

It is useful to prepare modified photocatalysts with visible light absorption by introducing heteroatoms/ions, even though introduced heteroatoms/ions are not incorporated in crystalline lattices. However, a problem is that newly appearing visible light photoabsorption is sometimes not reflected in the photoinduced reaction under visible light photoirradiation. As has been reported previously, organic dyes are inappropriate as test compounds for photoinduced degradation under aerated conditions because those dyes might work as visible light photosensitizers when adsorbed on solid surfaces, and thereby, it is preferable to show resemblance of absorption (diffuse reflection) and action spectra. In other words, photocatalysis by doped (modified) photocatalysts can be proved, presumably only, through action spectrum analysis. It has been shown that through action spectrum analysis, showing the resemblance of a diffuse reflectance spectrum with an action spectrum for photocatalytic oxidative decomposition of acetic acid in an aerated aqueous solution, sulfur-doped titania was proved to work as a visible-light-sensitive photocatalyst [36] and that doping of (or at least modification with) sulfur induced photoabsorption and photocatalytic activity in the visible light region.

4.14 Conclusive Remarks

As has been discussed in this chapter and the author's recent reviews [5], our knowledge of photocatalysis seems still insufficient and therefore, it is impossible, in a strict scientific sense, to "design" a highly active photocatalyst, that is, what we, at least the present author, believe as requisites for active photocatalysts, for example, better photocatalytic activity of ANA compared with that of RUT or higher activity of smaller particles seems to be only an accumulation of information on empirical phenomena. In this sense, it can be said that all photocatalysts, especially those of titania and related compounds, with relatively high photocatalytic activity had been found only accidentally. What we have to do for "true" design or active photocatalysts is to obtain a correct understanding of those phenomena with "chemistry"; use of a chemical term such as "rate-determining step," "first-order kinetics," or "L–H mechanism" itself does not make sense. Of course, this chapter could not give readers a clear answer for active photocatalyst design, although the author has tried to explain photocatalysis and related phenomena from the standpoint of chemistry.

An important application of photocatalysis, as well as application to water/air purification, is splitting water into hydrogen and oxygen by a photocatalyst. Although the concept was shown by Fujishima and Honda more than 40 years ago [4], there have been only a few reported successful examples of photocatalytic water splitting, and it seems that we have no good strategy for preparation of photocatalysts working under visible light photoirradiation. As was discussed in the beginning of this chapter, the principle of photocatalysis, reduction and oxidation by photoexcited electrons and positive holes, respectively, is the same both for water splitting and water/air purification, that is, more cathodic and anodic positions of the CB bottom and the VB top compared with standard electrode potential of oxidants (electron acceptors) and reductants (electron donors). A possible reason for the unsuccessful development of active water-splitting photocatalysts is that we have focused only on those band position requirements. The original paper by Fujishima and Honda suggested implicitly, the present author thinks, that electrochemical or chemical bias is necessary for achievement of photoelectrochemical water splitting, and we have neglected this for more than 40 years [37]. This implies that understanding the phenomena and/or experimental results scientifically and finding those meanings are indispensable for the development of active photocatalysts. The author hopes that this review will help readers to understand photocatalysis and related phenomena and to design and develop active photocatalysts.

Acknowledgments

This work was partly supported by the Project to Create Photocatalyst Industry for Recycling-oriented Society supported by NEDO, the New Energy and Industrial Technology Development Organization and Grant-in-Aid for "Scientific Research (A) (General)" from the Ministry of Education, Culture, Sports, Science, and Technology (MEXT) of Japan.

References

1. (a) Hoffmann, M.R., Martin, S.T., and Choi, W.Y. (1995) Environmental applications of semiconductor photocatalysis. *Chem. Rev.*, **95**(1), 69–96; (b) Pichat, P. (2003) in *Chemical Degradation Methods for Wastes and Pollutants: Environmental and Industrial Applications* (ed M.A. Tarr), Marcel Dekker, Inc., New

York, Basel, pp. 77–119; (c) Zhang, H., Chen, G., and Bahnemann, D.W. (2009) Photoelectrocatalytic materials for environmental applications. *J. Mater. Chem.*, **19**(29), 5089–5121.

2. (a) Fujishima, A., Zhang, X.T., and Tryk, D.A. (2008) TiO_2 Photocatalysis and related surface phenomena. *Surf. Sci. Rep.*, **63**(12), 515–582; (b) Fujishima, A., Zhang, X., and Tryk, D.A. (2007) Heterogeneous photocatalysis: from water photolysis to applications in environmental cleanup. *Int. J. Hydrogen Energy*, **32**(14), 2664–2672.
3. For example: Jacobsen, A.E. (1949) Titanium dioxide pigments. Correlation between photochemical reactivity and chalking. *Ind. Eng. Chem.*, **41**(3), 523–526.
4. Fujishima, A. and Honda, K. (1972) Electrochemical photolysis of water at a semiconductor electrode. *Nature*, **238**, 37–39.
5. Problems found in papers on photocatalysis were reviewed in: (a) Ohtani, B. (2008) Preparing articles on photocatalysis – beyond the illusions, misconceptions and speculation. *Chem. Lett.*, **37**, 216–229; (b) Ohtani, B. (2010) Photocatalysis A to Z – What We know and What We Don't know. *J. Photochem. Photobiol., C*, **11**(4), 157–178.
6. Jenny, B. and Pichat, P. (1991) Determination of the actual photocatalytic rate of H_2O_2 decomposition over suspended TiO_2—fitting to the Langmuir–Hinshelwood form. *Langmuir*, **7**(5), 947–954.
7. As an example of papers showing the coincidence of equilibrium adsorption constants obtained from the photocatalytic reaction rate and adsorption in the dark, see: Amano, F., Nogami, K., and Ohtani, B. (2010) Correlation between surface area and photocatalytic activity for acetaldehyde decomposition over metal oxide photocatalysts with a hierarchical structure. *Langmuir*, **26**, 7174–7180.
8. The original paper on this plot is: Williamson, G.K. and Hall, W.H. (1953) X-ray line broadening from filed aluminum and wolfram. *Acta Metall.*, **1**(1), 22–31.
9. Ohtani, B. and Sano, M. to be submitted.
10. Nosaka, Y. and Nosaka, A. (2004) Nyumon Hikarishokubai (Japanese), Tokyo Tosho, Tokyo, p. 59.
11. (a) Colombo, D.P. Jr., and Bowman, R.M. (1996) Does interfacial charge transfer compete with charge carrier recombination? A femtosecond diffuse reflectance investigation of TiO_2 nanoparticles. *J. Phys. Chem.*, **100**(47), 18445–18449; (b) Colombo, D.P. Jr., and Bowman, R.M. (1995) Femtosecond Diffuse-reflectance spectroscopy of TiO_2 powders. *J. Phys. Chem.*, **99**(30), 11752–11756; (c) Colombo, D.P. Jr.,, Roussel, K.A., Saeh, J., Skinner, D.E., Cavaleri, J.J., and Bowman, R.M. (1995) Femtosecond study of the intensity dependence of electron–hole dynamics in TiO_2 nanoclusters. *Chem. Phys. Lett.*, **232**(3), 207–214.
12. Ohtani, B., Kominami, H., Bowman, R.M., Colombo, P. Jr.,, Noguchi, H., and Uosaki, K. (1998) Femtosecond diffuse reflectance spectroscopy of aqueous titanium(IV) oxide suspension: correlation of electron–hole recombination kinetics with photocatalytic activity. *Chem. Lett.*, **27**, 579–580(1998).
13. Ohtani, B., Ogawa, Y., and Nishimoto, S.-I. (1997) Photocatalytic activity of amorphous-anatase mixture of titanium(IV) oxide particles suspended in aqueous solutions. *J. Phys. Chem. B*, **101**(19), 3746–3752.
14. Ohtani, B., Azuma, Y., Li, D., Ihara, T., and Abe, R. (2007) Isolation of anatase crystallites from anatase-rutile mixed particles by dissolution with hydrogen peroxide and ammonia. *Trans. Mater. Res. Soc. Jpn.*, **32**(2), 401–404.
15. Ohtani, B., Prieto-Mahaney, O.O., Li, D., and Abe, R. (2010) What is Degussa (evonik) P25? Crystal composition analysis, reconstruction from isolated pure particles, and photocatalytic activity test. *J. Photochem. Photobiol., A*, **216**, 179–182.
16. Klug, H.P. and Alexander, L.E. (1974) *X-Ray Diffraction Procedures for Polycrystalline and Amorphous Materials*, 2nd edn, John Wiley & Sons, Inc., New York, p. 687.

17. (a) Nosaka, Y., Kishimoto, M., and Nishino, J. (1998) Factors governing the initial process of TiO_2 photocatalysis studied by means of in-situ electron spin resonance measurements. *J. Phys. Chem. B*, **102**(50), 10279–10283; (b) Nakaoka, Y. and Nosaka, Y. (1997) ESR investigation into the effects of heat treatment and crystal structure on radicals produced over irradiated TiO_2 powder. *J. Photochem. Photobiol., A*, **110**(3), 299–305.

18. Ikeda, S., Sugiyama, N., Murakami, S.-Y., Kominami, H., Kera, Y., Noguchi, H., Uosaki, K., Torimoto, T., and Ohtani, B. (2003) Quantitative analysis of defective sites in titanium(IV) oxide photocatalyst powders. *Phys. Chem. Chem. Phys.*, **5**, 778–783.

19. (a) Kominami, H., Matsuura, T., Iwai, K., Ohtani, B., Nishimoto, S.-I., and Kera, Y. (1995) Ultra-highly active titanium(IV) oxide photocatalyst prepared by hydrothermal crystallization from titanium(IV) alkoxide in organic solvents. *Chem. Lett.*, **24**(8), 693–694; (b) Kominami, H., Kato, J.-I., Murakami, S.-Y., Kera, Y., Inoue, M., Inui, T., and Ohtani, B. (1999) Synthesis of titanium(IV) oxide of ultra-high photocatalytic activity: high-temperature hydrolysis of titanium alkoxides with water liberated homogeneously from solvent alcohols. *J. Mol. Catal. A: Chem.*, **144**(1), 165–171.

20. (a) Murakami, N., Prieto-Mahaney, O.O., Abe, R., Torimoto, T., and Ohtani, B. (2007) Double-beam photoacoustic spectroscopic studies on transient absorption of titanium(IV) oxide photocatalyst powders. *J. Phys. Chem. C*, **111**(32), 11927–11935; (b) Murakami, N., Prieto-Mahaney, O.O., Torimoto, T., and Ohtani, B. (2006) Photoacoustic spectroscopic analysis of photoinduced change in absorption of titanium(IV) oxide photocatalyst powders: a novel feasible technique for measurement of defect density. *Chem. Phys. Lett.*, **426** (1–3), 204–208.

21. For example,Bard, A.J. (1982) Design of semiconductor photo-electrochemical systems for solar-energy conversion. *J. Phys. Chem.*, **86**(2), 172–177.

22. (a) Nakabayashi, S., Fujishima, A., and Honda, K. (1983) Experimental-evidence for the hydrogen evolution site in photocatalytic process on Pt/TiO_2. *Chem. Phys. Lett.*, **102**(5), 464–465; (b) Baba, R., Nakabayashi, S., Fujishima, A., and Honda, K. (1985) Investigation of the mechanism of hydrogen evolution during photocatalytic water decomposition on metal-loaded semiconductor powders. *J. Phys. Chem.*, **89**(10), 1902–1905.

23. (a)Examples of papers in which the property-activity correlation is discussed are: Enriquez, R., Agrios, A.G., and Pichat, P. (2007) Probing multiple effects of TiO_2 sintering temperature on photocatalytic activity in water by use of a series of organic pollutant molecules. *Catal. Today*, **120**(2), 196–202; (b) Ryu, J. and Choi, W. (2008) Substrate-specific photocatalytic activities of TiO_2 and multiactivity test for water treatment application. *Environ. Sci. Technol.*, **42**(1), 294–300.

24. Nishimoto, S.-I., Ohtani, B., Sakamoto, A., and Kagiya, T. (1984) Photocatalytic activities of titanium(IV) oxide prepared from Titanium(IV) Sulfate. *Nippon Kagaku Kaishi*, 246–252.

25. Nishimoto, S.-I., Ohtani, B., Kajiwara, H., and Kagiya, T. (1985) Correlation of the crystal structure of titanium dioxide prepared from titanium tetra-2-propoxide with the photocatalytic activity for redox reactions in aqueous propan-2-ol and silver salt solutions. *J. Chem. Soc., Faraday Trans. 1*, **81**, 61–68.

26. (a) Prieto-Mahaney, O.O., Murakami, N., Abe, R., and Ohtani, B. (2009) Correlation between photocatalytic activities and structural and physical properties of titanium(IV) oxide powders. *Chem. Lett.*, **38**(3), 238–239; (b) Ohtani, B., Prieto-Mahaney, O.O., Amano, F., Murakami, N., and Abe, R. (2010) What Are titania photocatalysts? – An exploratory correlation of photocatalytic activity with structural and physical properties. *J. Adv. Oxidat. Technol.*, **13**(3), 247–261.

27. Rothenberger, G., Moser, J., Grätzel, M., Serpone, N., and Sharma, D.K. (1985) Charge carrier trapping and recombination dynamics in small semiconductor particles. *J. Am. Chem. Soc.*, **107**(26), 8054–8059.

28. (a) Amano, F., Prieto-Mahaney, O.O., Terada, Y., Yasumoto, T., Shibayama, T., and Ohtani, B. (2009) Decahedral single-crystalline particles of anatase titanium(IV) oxide with high photocatalytic activity. *Chem. Mater.*, **21**(13), 2601–2603; (b) Ohtani, B., Amano, F., Yasumoto, T., Prieto-Mahaney, O.O., Uchida, S., Shibayama, T., and Terada, Y. (2010) Highly active titania photocatalyst particles of controlled crystal phase, size, and polyhedral shape. *Top. Catal.*, **53** (7–10), 455–461.
29. Sugishita, N., Kuroda, Y., and Ohtani, B. (2011) Preparation of Decahedral Anatase Titania Particles with high-Level photocatalytic Activity. *Catal. Today*, **164**(1), 391–394.
30. Amano, F., Yasumoto, T., Prieto-Mahaney, O.O., Uchida, S., Shibayama, T., and Ohtani, B. (2009) Photocatalytic activity of octahedral single-crystalline mesoparticles of anatase titanium(IV) oxide. *Chem. Commun.*, **45**, 2311–2313.
31. Ohtani, B., Amano, F., and Prieto-Mahaney, O.O. to be submitted.
32. Ohno, T., Tokieda, K., Higashida, S., and Matsumura, M. (2003) Synergism between rutile and anatase TiO_2 particles in photocatalytic oxidation of naphthalene. *Appl. Catal., A*, **244**(2), 383–391.
33. Hurum, D.C., Agrios, A.G., Gray, K.A., Rajh, T., and Thurnauer, M.C. (2003) Explaining the enhanced photocatalytic activity of Degussa P25 mixed-phase TiO_2 using EPR. *J. Phys. Chem. B*, **107**(19), 4545–4549.
34. (a) Asahi, R., Morikawa, T., Ohwaki, T., Aoki, K., and Taga, Y. (2001) Photocatalysts sensitive to visible light–response. *Science*, **295**(5555), 627–627; (b) Asahi, R., Morikawa, T., Ohwaki, T., Aoki, K., and Taga, Y. (2001) Visible-light photocatalysis in nitrogen-doped titanium oxides. *Science*, **293**(5528), 269–271.
35. (a) Mitoraj, D., Beranek, R., and Kisch, H. (2010) Mechanism of aerobic visible light formic acid oxidation catalyzed by poly(tri-s-triazine) modified titania. *Photochem. Photobiol. Sci.*, **9**(1), 31–38; (b) Mitoraj, D. and Kisch, H. (2010) On the mechanism of urea-induced titania modification. *Chem. Eur. J.*, **16**(1), 261–269.
36. Yan, X., Ohno, T., Nishijima, K., Abe, R., and Ohtani, B. (2006) Is methylene blue an appropriate substrate for a photocatalytic activity test? A study with visible-light responsive titania. *Chem. Phys. Lett.*, **429**, 606–610.
37. Ohtani, B. (2012) Revisiting the original works related to photocatalysis: a review on papers in the early stage of photocatalysis studies. *Electrochemistry*, **80** submitted.

5
Modified Photocatalysts

Nurit Shaham-Waldmann and Yaron Paz

5.1
Why Modifying?

The many virtues of titanium dioxide, among which are relatively high efficiency, high stability, low toxicity, and low price made it the current photocatalyst of choice for water treatment. Indeed, the number of scientific manuscripts on water treatment has grown over the years almost exponentially [1]. Yet, the current situation is far from being satisfactory, as evidenced by the fact that commercial photocatalytic units for water treatment are scarce even after 30 years of expectations.

Bridging over this discrepancy between expectations and reality requires analyzing various parameters controlling the effectiveness of photocatalysis and finding ways to improve the performance of the photocatalyst. Generally speaking, these controlling parameters may be divided into three categories. The first category consists of parameters directly related to the physicochemical properties of the photocatalyst (phase, size, energy levels structure, surface morphology, etc.). The second category groups together parameters related to the design of the photoreactors (falling film reactors vs stirred reactors, type of optics, supported vs nonsupported photocatalysts, etc.). The third category consisted of operational parameters that connect the physicochemical properties of the photocatalyst and the specific design of the photocatalytic reactors, thus optimizing the output of the photocatalyst. Contact time, pH, photon flux, and contaminants' concentration are among the relevant parameters here.

The gradual understanding that optimizing the operational parameters is not enough for implementation led a growing number of researchers to the conclusion that modifying the physicochemical properties of the photocatalyst is a must for obtaining products that may successfully compete with existing technologies. This chapter tries to analyze the various attitudes aimed at modifying the photocatalyst properties, that is, analyzes these approaches that alter properties related to the first class of parameters, en route for large-scale commercial application. There are several reviews on modified TiO_2 photocatalysts, for example [2–6]. These reviews either provide a number of specific model systems or describe specific ways for modification. Unlike these reviews, this chapter is written from the point of view of

Photocatalysis and Water Purification: From Fundamentals to Recent Applications, First Edition. P. Pichat.

the specific parameters or effects that are sought. This point of view is reflected by the way that the core of the chapter is organized, that is, according to the parameters (or effects) that are altered on modification. This structuring of the chapter was done deliberately in order to help invoking new attitudes for modification that may find their origins in more than one type of modification.

While this chapter focuses on approaches for modifying the physicochemical properties of the photocatalyst, hence gives less attention to the other two classes of parameters, it should be noted that there is some in-class coupling and interclass coupling between the various parameters governing the photocatalytic treatment of water.

It is noteworthy that to prevent overlapping with other chapters, this chapter on modified photocatalysts does not refer to photocatalysts other than titanium dioxide. For the same reason, its coverage of modifying the photocatalyst toward activity under visible light is deliberately very limited. In writing this chapter, the authors have tried to deduce conclusions based on large number of scientific reports and manuscripts, reflecting large number of different experimental setups. While the diversity of the experimental setups may be challenging, the fact that the model contaminants can be classified into a limited number of types assists in generalizing the accumulated data. Among these contaminants, one finds dyes (in particular, methylene blue (MB), rhodamine B, malachite green) and chlorinated aromatic compounds such as 4-chlorophenol.

5.2
Forms of Modification

Modifying the photocatalyst may take three main forms: bulk modification, surface modification, and the use of composite materials. The term *bulk modification* refers, for example, to controlling the phase of the photocatalyst or its electronic properties, usually by adding guest chemical species (ions, for most cases) into the bulk of the photocatalyst (Figure 5.1). Many dopants have been used so far including noble metal ions, metallic cations, and nonmetallic ions. The dopants may be incorporated into the matrix either during its preparation, for example, by introducing ammonia during a sol–gel preparation process [7], or post preparation, for example, by ion implantation [8].

Surface modification is manifested by altering the properties of the surface of the photocatalyst, for example, the surface area, affinity toward chemical species, and surface electronic properties. These properties may be altered with or without using chemical species that are removed at some stage of the preparation procedure (Figure 5.2a). Of specific relevance to this class are photocatalysts modified by molecular imprinting [9] (Figure 5.2b). In what follows the tailoring of the surface of a photocatalyst by an ultrathin layer, that is, a layer that is thin enough not to completely block the properties of the substrate (usually less than 5–10 molecular layers) also belongs to this category (Figure 5.2c). Here, a popular technique is the chemisorption–calcination cycle (CCC) technique, based on chemisorption of

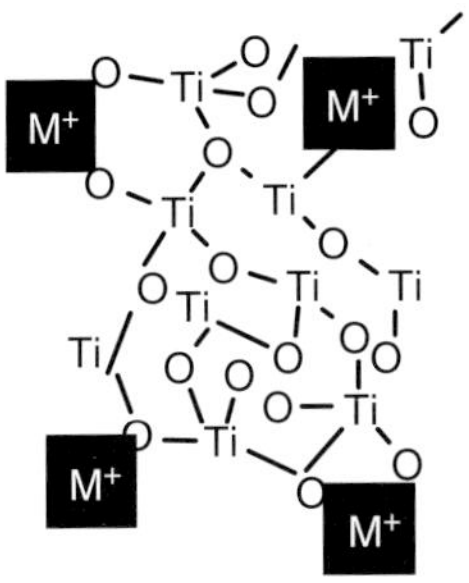

Figure 5.1 A schematic representation of bulk modification by cation-doping.

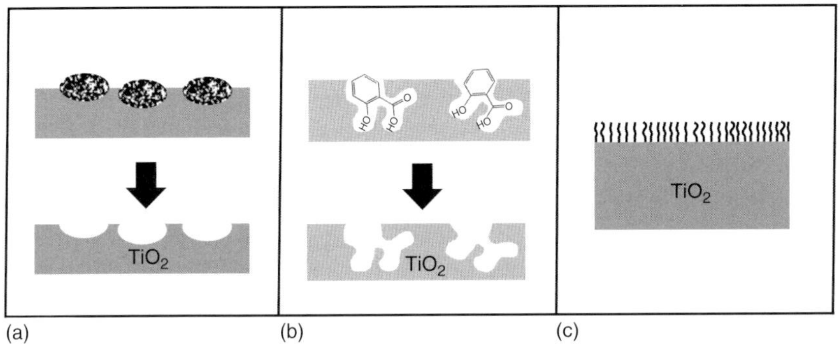

Figure 5.2 Various approaches for surface modification: (a) meso-porosity by introducing polymers or agglomerated surfactants (b) molecular imprinting (c) modification by ultrathin layers.

metal complexes, followed by oxidizing the organic (ligand) part, thus forming ultrathin films strongly adhered to the photocatalyst surface [10].

Composite photocatalysts are defined as materials comprising of active titanium dioxide domains in close proximity to domains made of inert compounds or other photoactive compounds. Here, the major difference between composite photocatalysts and bulk-modified photocatalysts is the formation of a segregated phase in the former case. Several types of composites are reported. One of these types consist of composites made of an inert skeleton onto which titanium dioxide is coated (Figure 5.3a), for example, composite membranes made of a porous alumina skeleton onto which mesoporous TiO_2 was grown by the sol–gel method using a mixture of titanium tetra iso propoxide, acetic acid, and a surfactant (Tween 80) [11]. Another type of a composite photocatalyst consists of two (or more) domains, both of which are exposed to the outer environment so that the non-TiO_2 domains serve as part of the photocatalytic process. Examples include partial coverage of activated carbon with titanium dioxide (by atomic layer deposition [12] or by physical contact [13]) and structures made from titanium dioxide and inert oxides such as zeolites [14] (Figure 5.3b). Of specific interest are composites made of titanium dioxide substrate, onto which islands of inert materials are grown (Figure 5.3c). In this context, of particular interest are nanoislands of noble metals (such as platinum, gold, and silver [15]). These nanoislands can be easily formed from a variety of metallic

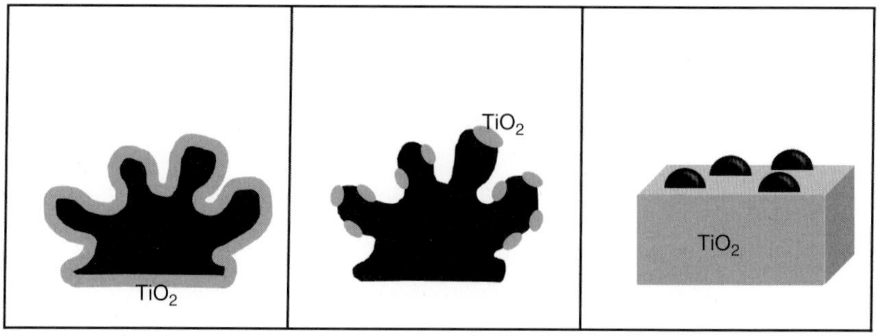

Figure 5.3 Various approaches of modification by formation of composite structures: (a) an inert skeleton completely coated with titanium dioxide (b) an inert substrate partially covered with titanium dioxide (c) a TiO_2 substrate partially coated with micro- or nano-islands of an inert material.

complexes, for example, chloroplatinic acid, hexahydroxyplatinic acid, platinum dinitrodiammine [16], in many cases by photoassisted reduction process [17].

While the above-mentioned classification relates either to the modification of pure TiO_2 or TiO_2 modified by inorganic additives, one should also consider systems where organic modifiers were used. Typical modifiers include organosilanes [18], polymers [19], and self-assembled monolayers [20]. Regardless of the purposes for which these modifiers were used, and regardless of the benevolent effects found with these systems, they all share one crucial characteristic: the organic modifier is doomed to be degraded over time, unless elaborate measures (from physical barriers to sacrificial agents) are taken to prevent it. For photocatalysis in air, this problem is aggravated considerably by the remote degradation effect [20, 21]. Luckily, this effect is less pronounced in aqueous media.

5.3
Modified Physicochemical Properties

In the following paragraphs, the intensive work performed so far on modifying titanium dioxide is analyzed and classified according to the specific effects that the modifications induced, namely, crystallinity, surface morphology, surface area and adsorption, oxygen adsorptivity, concentration of surface OH, specificity, products' control, deactivation, recombination rates and charge separation, visible light activity, charging–discharging phenomena, and mass transport effects.

5.3.1
Crystallinity and Phase Stability

TiO_2 exists in three main phases: anatase, rutile, and brookite. While rutile is thermodynamically most stable, anatase is being regarded as the more active phase for the degradation of most compounds. The phase transformation from anatase to rutile is a gradual process beginning at 600–650 °C. Such temperatures are often required to obtain good adherence between the photocatalyst and its substrate. For this reason (and for other reasons related to obtaining high surface area), there is an interest in improving the stability of the anatase phase. In this framework, it was found that modifying titanium dioxide with silica and zirconia may inhibit the anatase to rutile phase transformation [22]. It is noteworthy that a recent study reported a smaller average diameter of TiO_2 particles in a TiO_2–ZrO_2 composite (30 nm in diameter for the composite vs 65 nm for the pristine) [23]. Changes in the crystalline properties of TiO_2, such as grain size, were obtained not only in composites but also on doping with ions. Examples include doping with silver ions [24], tin ions [25], and manganese ions [26].

5.3.2
Surface Morphology, Surface Area, and Adsorption

Surface roughness may play a role in the efficacy of the photocatalyst. Roughness having a characteristic length in the order of few hundred nanometers induces high reflectivity, thus preventing the use of TiO_2 in applications that require high transparency in the visible range of the spectrum, such as self-cleaning glass [27]. This requirement is less stringent in the treatment of water, so far as the reflected photons are being utilized, and indeed, this is usually well handled in most designs of photocatalytic water treatment reactors. Moreover, high roughness may assist bacteria to attach to the surface. This can be of high importance when using photocatalysis for antifouling, for example, in photocatalytic antifouling reverse osmosis membranes [28].

At the nanometer scale, the effect of increasing roughness is often connected with enhanced surface area (see below). While many scientific manuscripts refer to high roughness as being equivalent not only to high surface area but also as reflecting small grains, such a correlation does not necessarily exist. In fact, small crystallites can be instrumental in the preparation of films having very low roughness. In this context, it is noteworthy to mention that doping TiO_2 with silver not only reduced the average size of the crystallites but also led to a smoother surface [24]. This was interpreted as indicating smaller grains in the Ag-doped TiO_2 in comparison with undoped titania. It was postulated that the smaller grain photocatalyst had larger surface area and accordingly, enhanced adsorption of malic acid (by a factor of 2) was observed.

The kinetics of degradation of contaminants in water can be almost always represented by Langmuir-Hinshelwood (LH) kinetics (Chapters 3, 4, and 12). Although it was shown that such kinetics do not necessarily mean preadsorption [29], there is no doubt that most cases involve adsorption, hence the larger the adsorption is, the higher the rate of degradation [30].

Surface charge modification may considerably affect the adsorption of charged reactants and products and, accordingly, is expected to affect the photocatalytic rates. The surface charge of bare TiO_2 is positive below pH 5 and negative above pH 7 [31]. By loading metal oxides and by adsorbing charged compounds, it is possible to alter the point of zero charge (PZC) of the surface, hence the surface charge at a given pH. Accordingly, loading silica or nafion onto the surface of TiO_2 lowered the PZC and consequently accelerated the photocatalytic degradation of cationic substances such as tetramethylammonium (TMA) [32]. It should be pointed out that there are cases in which the presence of surface charge may affect the degradation rates of neutral species also, as was demonstrated with *N*-nitrodimethylamine (NDMA) on TiO_2-loaded nafion. Here, the high concentration of protons within the nafion layer facilitated the formation of Lewis acid complex with NDMA, thus enhancing its photocatalytic degradation [33].

One of the most dominant parameters affecting adsorption is surface area. For this reason, modifying the photocatalyst in a manner that increases its surface area was, and still is, the subject of many studies. While it is possible to check to what extent the method of preparation affects the surface area, it is practically impossible to generalize quantitatively the effect of preparation procedure on surface area and photocatalytic activity, because of the large number of operational parameters, from the type and concentration of contaminant to photon flux and photocatalyst concentration. It should be noted that the adsorption capacity for a given compound depends not only on the surface area but also on the type of functional groups on the surface, the surface morphology, its local curvature, particle size, and even, the tendency of the particles to aggregate.

Controlling the grain size of the photocatalyst may alter the surface roughness and, in certain cases, also the surface area. In that sense, composites containing zirconia or silica revealed increased stability not only toward temperature-induced phase transformation but also toward loss of surface area during sintering [22]. Of interest here is also the observation that doping with cations of Sn^{4+} was shown to increase the surface area [25].

The use of surfactants and polymers as structure-directing agents that are burnt away on TiO_2 formation is quite common. Among the many compounds used for that purpose one may outline pluoronic acids [34] and polyethylene glycol (PEG) [35]. The pluoronic acids served to induce highly ordered cubic mesoporous WO_3/TiO_2 films. In contrast, addition of PEG led to the formation of nonordered mesopores, increasing the surface area of the photocatalyst by up to 70% (from 70 to 116 $m^2\ g^{-1}$), depending on the molecular weight of the PEG.

Chemical vapor deposition was used to obtain titanium dioxide films having fractal-type morphology. The basic concept in producing this nanostructured morphology was the simultaneous use of two different precursors: $TiCl_4$ and titanium tetraisopropoxide (TTIP) [36]. Hierarchically structured nanofibrous architecture was produced also by pulsed laser deposition, characterized by assembling of ultra-fine nanoparticles into a hierarchical structure having a surface area of 250 $m^2\ g^{-1}$ without prepatterning or templating [37]. High surface area is obtained also in self-organized nanoporous titanium dioxide [38] and in TiO_2 nanotubes/nanocolumns [39] prepared by anodization of titanium in the presence of HF or other salts and complexes of fluoride ions. While these structures were typically mentioned in the context of photovoltaics or gas-phase photocatalysis, there are also a few

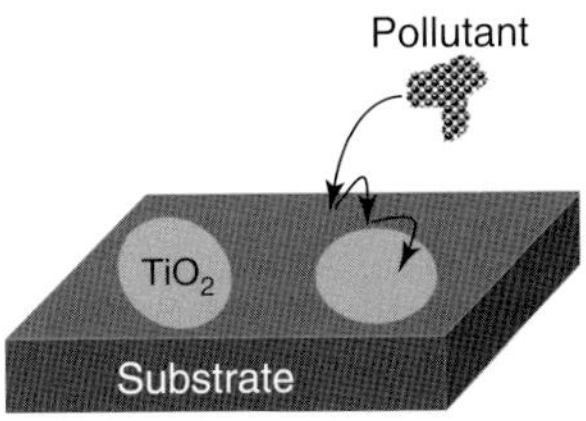

Figure 5.4 The principle of "Adsorb & Shuttle" mechanism.

manuscripts about water treatment, including inactivation of *Escherichia coli* by an immobilized film composed of a TiO_2 nanotube array [40].

Another means by which photocatalysts can be modified to enhance adsorption is by forming a high surface area porous inert skeleton onto which a thin photocatalyst layer is grown or deposited (Figure 5.3a,b). A good example of this approach was the use of porous alumina membranes onto which thin films of titania were grown by the sol–gel method, forming mesoporous photocatalytic membranes with high water permeability and controlled pore size distribution [11].

One of the most common ways to promote the adsorption of contaminants is to use composite photocatalysts that operate via the "Adsorb and Shuttle" (A&S) mechanism (Figure 5.4). The basic concept is to use inert domains located at the vicinity of the photocatalyst to adsorb the pollutants. Once adsorbed, the pollutants surface diffuse from the inert adsorptive sites to the photocatalytic sites. The net result is an increase in the concentration of the contaminants at the photocatalytic degradation domains, leading to enhanced degradation rates. It should be noted that direct contact between the two types of domains is a prerequisite. In certain cases (degradation of phenol by activated carbon mixed together with P25 particles [41]), direct contact, leading to a synergistic effect, can be obtained by simple mixing. However, for most cases, simple mixing of inert particles and photocatalytic particles may not work synergistically but, on the contrary, can be quite deleterious because of competitive adsorption between the inert particles and the photoactive particles.

One of the first studies on this effect was performed by the group of Yoneyama who compared the effect of various adsorptive substrates such as silica, alumina, activated carbon, and zeolites on the photocatalytic degradation of propyzamide [42], propionaldehyde [43], and pyridine [44]. As part of this study, the diffusion coefficient of adsorbed propionaldehyde on the adsorptive substrates was measured [45]. It was established that in cases where the adsorption constant on the adsorptive sites is low, the decomposition rate is determined by the amount of adsorbed contaminant, while if the adsorption constant is very high, plenty of adsorbed molecules can be found on the support, yet, the degradation rates are low, because of insufficient mobility.

The average diffusion distance that the adsorbed molecules have to cross in order to get from the adsorptive domains to the photocatalytic regions may affect considerably the efficacy of the A&S approach. Regardless of whether the composite material is made of an inert skeleton, partially covered with photocatalytic domains, or of a photocatalytic skeleton onto which inert domains are introduced, this distance depends on the partial loading of each component, the relative exposed area, and the shape of the domains. This issue was discussed by Avraham-Shinman and Paz [13], who studied the effect of loading on surface area, the morphology of the domains, and the photocatalytic degradation rate of rhodamine 6G (R6G).

Most pollutants may adsorb on both the inert substrate and the photocatalyst. This might cause an inherent problem in decoupling between adsorption, surface diffusion, and reactivity while trying to study composite photocatalytic systems (Chapter 12). Here, studying the effect of dark-light cycles and correlating it with both the relative amount of each component and the structural parameters related to the way by which the TiO_2 is deposited (i.e., cracks, relative area coverage, etc.) may provide the answer for this problem. Unfortunately, such analysis is rarely performed.

Carbonaceous materials are among the most popular inert components in A&S composite systems. Among these materials, one finds activated carbon (AC) [13], carbon [12], carbon nanotubes (CNT) [46, 47], and graphene [48]. Photocatalytic degradation of MB in aqueous solution was investigated using TiO_2 immobilized on activated carbon fibers (ACFs). The composite material was prepared by using epoxy to bind the two components, followed by calcination at 460 °C in N_2 atmosphere. The TiO_2/ACF composite could be used repeatedly without a decline in photocatalytic degradation ability. After six cycles, the amount of MB removal for the TiO_2/ACF composite was still slightly higher than that for fresh P25 TiO_2 in suspension. Through measurement of chemical oxygen demand in the solution and the concentration of ammonium generated during degradation of MB, it was confirmed that MB molecules are mineralized instead of just adsorbed by the ACFs [49].

In many model systems, for example, TiO_2-CNTs [47] and TiO_2-graphene [48], the measured changes in degradation rates are attributed not only to A&S effects but also to electronic effects (see below). Likewise, the significant enhancement in the photocatalytic degradation of rhodamine B with Ag-deposited TiO_2 nanoparticles under visible light was attributed to both enhanced adsorption on the Ag–TiO_2 surface and electron trapping on the Ag islands [50]. Optimal islands were 2–3 nm in diameter, well dispersed within particles of 15 nm in diameter (2% by weight). Higher concentration was found to be detrimental. Dark adsorption measurements indicated higher adsorption of dye on the Ag–TiO_2 particles. The role of the mechanism was inferred indirectly from the fact that under visible light, the rhodamine B was degraded through deethylation (considered to be a surface process) in the presence of Ag–TiO_2 particles, whereas a through-chromophore degradation (considered to be predominantly a bulk process) was found with unloaded TiO_2 particles.

The A&S phenomenon is not restricted to carbonaceous materials or to metallic islands. Oxides may operate by this mechanism as well. Such systems may comprise of silica, alumina, zeolites of various types [42], as well as tungsten trioxide [51].

It is interesting to note that the point of view regarding A&S mechanism varies from focusing on photocatalysis (i.e., enhanced photocatalytic degradation because of enhanced adsorption) to focusing on adsorption (namely, regeneration of adsorbent by light exposure). To some extent, this difference in attitude may reflect whether the authors designed the system to work in dark-light cycles or not and whether light exposure was carried out simultaneously together with adsorption. Nevertheless, regardless of the point of view, introducing inert components to take advantage of the A&S mechanism seems to be quite successful, and no wonder that it was adopted by many manufacturers of commercial, photocatalytic water and air purification systems.

5.3.3 Adsorption of Oxygen

Reduction of adsorbed oxygen is probably the dominant mechanism for preventing recombination [52]. Therefore, modifying the photocatalyst in a manner that may increase the adsorption of oxygen may assist photocatalysis, in particular under low concentration of oxygen, as might happen in the liquid phase.

To increase the local concentration of oxygen, composite film comprising of polytetrafluoroethylene (PTFE) particles, platinum nanoparticles, and TiO_2 particles was deposited on ITO-covered glass. The PTFE was chosen based on its affinity toward both O_2 and the model pollutant (1,2,4-trichlorobenzene). Indeed, increasing the local concentration of oxygen led to faster reduction of the oxygen on the platinum nanoislands, consequently enhancing the degradation rate of the model pollutant [53].

5.3.4 Concentration of Surface OH

The primary cause for the photocatalytic activity of titanium dioxide is believed to be the formation of OH radicals on oxidation of adsorbed water by the photogenerated holes [54–56] (Chapter 1). Accordingly, the photocatalytic activity increases with increase in the density of surface hydroxyl groups [57]. This gives a rationale for modifying the surface to increase the concentration of hydroxyls on the surface or at least to study what is the effect of modifying titanium dioxide on the concentration of surface OH.

There are several works showing that doping with metal ions (for example, tin ions [25]) may change the surface hydrophilicity, which is strongly related to the concentration of surface OH. In a different work, TiO_2 nanoparticles were synthesized through a well-controlled sol–gel process in the presence of NaCl that was removed following calcination by washing with DI water. XPS measurements revealed an O_1s peak comprising of one peak at a binding energy of 530.0 eV, ascribed to bulk O^{2-}, and a second peak at 532.0 eV, ascribed to surface -OH. Comparing the ratio between the two peaks led to the conclusion that the concentration of surface OH was twice as high in samples formed with NaCl. This increase was manifested in the degradation rate of MB and formaldehyde, where a faster rate by a factor of 2–3 was observed [58]. No change in the crystallinity was observed; nevertheless, the average diameter of the nanoparticles formed with NaCl was slightly smaller (12.3 vs 15.3 nm). It was proposed that the mechanism responsible for the higher concentration of surface -OH species stemmed from sodium atoms being connected to oxygen, which, upon washing, were replaced by hydrogen ions later. Exchange of sodium ions with hydrogen ions was found to be benevolent also in the case of TiO_2 on soda lime glass, although in this case replacing the sodium ions was performed before the formation of the TiO_2 film to prevent deleterious effects by the sodium [59].

Binary oxides of TiO_2–ZrO_2 nanoparticles prepared by a solid-state reaction revealed higher activity in the photocatalytic degradation of phenol in water, compared with pure TiO_2 (and in fact compared also with P25 [23]). The dependence of the activity on the relative percentage of ZrO_2 (peaking at 50% ZrO_2) was found to correlate with the amount of surface OH as measured by FTIR. This led the authors to claim that at least part of the enhanced activity of the binary mixture was due to the increased concentration of surface OH, generating more OH radicals. Other factors may also play a role in this case, as the authors mentioned based on the dependence of the activity, at a constant TiO_2/ZrO_2 ratio, on the preparation process. Here, samples prepared by the solid-state dispersion (SSD) method (first preparing nanoparticles of each component by the sol–gel technique, then mixing with ethanol and calcining) yielded faster rates than samples prepared by a one-step process.

Modification of the surface of the photocatalyst may change not only the concentration of surface hydroxyls but also their mobility, thus affecting photocatalytic rates. Surface fluorination of titanium dioxide involves a simple ligand exchange between fluoride ions and surface hydroxyls on the photocatalyst. It was shown that fluorinated surfaces may increase the photocatalytic oxidation of various aqueous-phase pollutants, such as TMA [60] and phenol [61]. On the basis of the fact that surface fluorides ($E^0(F^*/F^-) = 3.6$ V vs NHE) do not react with TiO_2 VB holes, this enhanced degradation rates were attributed to the enhanced generation of mobile free OH. This explanation is in line with the observation that F-TiO_2 was more effective than pure TiO_2 for the photooxidation of acid orange 7 and phenol but not for the degradation of dichloroacetate [62]. Another implication of this mechanism is the rate retardation in oxidation reactions governed by direct hole-transfer mechanism, as a consequence of hindered adsorption of the pollutants.

5.3.5 Specificity

For many years, the issue of treating streams containing co-existing pollutants did not receive much attention in the academia. To large extent, this stemmed from the understandable approach that for studying scientific principles, one has to work with simple model systems having small number of well-defined parameters. Nevertheless, on considering commercialization, there is no way to avoid practical problems, among which is the treatment of water containing co-existing pollutants.

Polluted water might contain highly hazardous contaminants together with organic contaminants of low toxicity. In many cases, the toxic stuff is in low concentration, whereas the less toxic compounds are the majority. In such cases, it is preferable to degrade the more toxic compounds, even at the expense of lower degradation rates for the less harmful components. Moreover, many low-toxicity contaminants can be degraded by biological means, whereas many of the highly hazardous materials are nonbiodegradable, because of their high toxicity [63]. The fact that the adsorption coefficient of many of the chlorinated xenobiotic compounds on titanium dioxide is very low because of their hydrophobic nature, which further aggravates this problem.

Therefore, means that promote the degradation of specific contaminants in a stream containing several compounds are required and can be of importance in mixed streams. Unfortunately, obtaining specificity is far from trivial because photocatalytic oxidation is governed by a free radical mechanism that is expected to operate in a manner that hardly differentiates between different contaminants. However, since photocatalysis requires adsorption or at least close proximity between the pollutant and the photocatalyst, achieving specificity is possible via controlling the mass transport between the oxidizing (or reducing) species and the molecules of the pollutant.

Specificity in removing pollutants from water can be obtained either by governing operational parameters or by modifying the photocatalyst. In the literature, there are quite a few operational parameters that may change the photocatalytic degradation rate ratio between two or more co-existing contaminants on using nonmodified TiO_2 particles or films. The reported isoelectric point (PZC) for TiO_2 varies to some extent between pH 6 [64] and 7.5 [65]. This means that pH values higher than that are favorable for the adsorption of positively charged contaminants, while pH values lower than the PZC are favorable for the adsorption of negatively charged contaminants. Hence, controlling the pH of the medium, thus controlling the surface charge, may serve, at least potentially, to obtain preferential degradation. In this context, a straightforward example is the pH effect on the photocatalytic reduction of Hg(II), in a mixture prepared by dissolving $HgCl_2$ and $K_2Cr_2O_7$ [66]. Another example is the photocatalytic degradation of a mixture of 4-hydroxybenzoic acid and benzamide [67]. Here, at pH 4, the degradation rate of 4-hydroxybenzoic acid was much higher than that of benzamide, whereas at pH 8, the opposite was observed.

Although changing the acidity primarily affects the adsorption of charged contaminants, it has also a role on the photocatalysis of those neutral molecules that tend to dissociate into charged species. It should be emphasized that the effect of pH is a complex one, as the pH not only influences the adsorption of charged contaminants but also alters the concentration of surface hydroxyls and shifts the position of the valence and the conduction bands, not to mention its effect on the dissociation of many contaminants [68]. In cases where the photocatalytic degradation of neutral molecules leads to the formation of charged species, one may expect that varying the pH will affect mainly the degradation of the intermediates [69], as was indeed found in the degradation of 4-chlorophenol [70].

The presence of inorganic anions may influence the specificity, as was found in the photocatalytic degradation of salicylic acid, aniline, and ethanol [71]. Here, the presence of chlorides, sulfates, and phosphates reduced the oxidation rates of these compounds. However, the extent by which the rates were reduced, at a predetermined pH, was different for each compound. This difference apparently suggested a way for preferential degradation. Nevertheless, the effect of the cosolutes on the selectivity was very mild; therefore, introducing ions to induce specificity could not be regarded as practical. An interesting way to separate Cr(VI) and Hg(II) ions was presented by Wang *et al.* The method is based on the complexation of Hg(II) following the addition of chloride ions. This complexation retards the photocatalytic reduction of Hg(II) without affecting that of Cr(VI), thus facilitating the separation of the two, despite their similar redox potential (0.85 V for Hg^{+2}/Hg and 0.82 V for $Cr_2O_7^{-2}/Cr^{+3}$) [66]. Other operational parameters with nonmodified titanium dioxide that may affect the specificity are the irradiation with sub-bandgap light [72] and the photocatalyst concentration [73]. Overall, although some specificity can be obtained by altering operational parameters, they are impractical in real life: first, because many compounds fall into the same category and second, because the difference in the rates of degradation of pollutants belonging to different categories is quite modest.

Choosing the right type of commercially available titanium dioxide may provide some specificity. For most cases, the exact reasons for the brand dependence are obscure (Chapter 4). While it is obvious that the brand dependence stems from differences in size, crystalline phase, oxygen vacancies, and surface morphology, a well-defined correlation between these properties and specificity is yet to be understood. One example for brand specificity is the photocatalytic degradation of 1,2-diphenylhydrazine (DPH) and benzidine. Here, the ratio between the mineralization rate of DPH to that of benzidine was 25, 15, 8, and 3 on using particles of PC500 (Millennium Inorganic Chemicals), P25 (Degussa), PC50 (Millennium Inorganic Chemicals), and UV100 (hombikat) [73], respectively. In explaining these results, it was suggested that P25 is most adequate for the degradation of molecules for which adsorption (and desorption of intermediates) is slow, whereas UV100 is adequate for molecules for which adsorption (and desorption of intermediates) is relatively fast. Another factor that may explain the effects of specificity in the oxidation/reduction of charged species (for example, Hg^{+2} [66]) is the PZC, located at pH 6.2 for Hombikat UV100 and at pH 7.1 for Degussa P25.

At present, there are various approaches for modifying titanium dioxide in a manner that induces specificity. These approaches can be grossly divided into surface overcoating, surface doping, specific A&S, and molecular imprinting.

A word of caution had to be added here. Since the data on photocatalytic degradation of mixed streams is far from being satisfactory, one has to rely, when discussing specificity, on sets of experiments, many of which were performed with single pollutants. The photocatalytic degradation rates in mixed streams often reflect a competitive adsorption situation. Hence, relative rates that are measured based on single-component experiments of various contaminants should be regarded as qualitative guidelines for each approach and not as predictive values for mixed stream situations, as indeed was found for a mixture containing oxalic acid, formic acid, and formaldehyde [74]. As a rule of thumb, the degradation rates of strongly adsorbed contaminants are not expected to vary considerably in the presence of other contaminants, whereas the degradation rates of contaminants that adsorb weakly are expected to change significantly in the presence of more strongly bound contaminants [63].

5.3.5.1 TiO_2 Surface Overcoating

The covering of the surface of the photocatalyst was shown to alter the degradation rate of pollutants. Since the change varies with the pollutant species, such modification may be utilized to obtain specificity. Sulfated TiO_2 can be prepared by reacting H_2SO_4 with $Ti(OH)_4$. It was found that the ratio of the mineralization rate constant of heptane to that of toluene was 5.75 for sulfated titanium dioxide, while for non-sulfated TiO_2, made by calcining unsulfated $Ti(OH)_4$, this ratio was only 2.7. This selective enhancement in the degradation of heptane correlated well with increased coverage and with the adsorption strength on the strong Lewis acid sites produced by sulfation [75].

There are quite a few reports on overcoating titanium dioxide with organic layers as a means to obtain specificity. Most cases involving TiO_2 surface overcoating utilized molecules having hydrophobic moieties, designed to increase the adsorption of hydrophobic contaminants. For example, the specificity for the photocatalytic degradation of the endocrine disrupter 4-nonylphenol in the presence of concentrated phenol was increased on grafting of *n*-octyltriethoxysilane [76]. Other examples include modification with methyltrimethoxysilane to increase the degradation rate of rhodamine B, which does not adsorb appreciably on pure TiO_2 [77], and the overcoating of TiO_2 with a hydrophobic organosilicone layer that enhanced the photocatalytic degradation of the nonpolar pesticide permethrin [78].

One of the first works on specific degradation by organic overcoating was performed by the group of Thurnauer who showed that surface modification of TiO_2 with cysteine resulted in enhanced photocatalytic reduction of lead because of the change in the redox properties of the semiconductor and the increased adsorption of the metal ions [79]. This effect of shifting the bands to negative potential leading to enhanced reduction properties on strong adsorption of electron-donating redox couples was later demonstrated by modifying the photocatalyst with bidentate mercapto-carboxylic acids. Here, it was found that in order to enhance the reduction rates of heavy metal ions such as Pb^{2+} or Cd^{2+}, the surface modifier must contain a carboxy group that binds to both the colloidal surface and the metal ions. The presence of a mercapto group in an α-position may contribute to the binding of both. The modifier should also have some kind of a hole trap (for example, a hydrocarbon side chain) that increases the distance between the photogenerated charge carriers [6].

Another example of an overcoating effect is the modification of titanium dioxide nanoparticles with the electron-donating chelating agent arginine that binds to TiO_2 through its COO^- functional group. This modification resulted in enhanced specific adsorption of nitrobenzene, because of the interaction of the π-system of the nitrobenzene ring with the arginine amino group, which led to faster reductive photodecomposition, compared with unmodified TiO_2 [80].

For commercial water treatment or for any long-term application, covering the surface with organic molecules is problematic because of the instability of the organic modifiers under UV light. To prevent degradation of adsorptive sites, it was proposed to overcoat the TiO_2 particles with a thin layer of inorganic carbon, produced by thermal carbonization of poly(vinyl alcohol) under N_2 environment [81]. This layer was found to be very beneficial in promoting the adsorption of MB and, as a consequence, in promoting its photocatalytic degradation.

5.3.5.2 Composites Comprised of TiO_2 and Metallic Nanoislands

It is well established that metallic nanoislands of noble metals, such as platinum or palladium ,serve as electron sinks, thus accelerating the cathodic process of oxygen reduction [82]. This property can be used to differentiate between contaminants that differ significantly in their photocatalytic degradation mechanism [83]. For example, metallic nanoislands increased the degradation rate of methanol and ethanol, while decreasing the rate of degradation of dichloropropionic acid (DCP) and trichloroethylene (TCE) [84]. Likewise, platinization of Hombikat UV100 TiO_2 resulted in doubling the photocatalytic degradation ratio of dicholoacetic acid (DCA) to 4-chlorophenol [85].

The transfer of electrons to the metallic islands implies that the locus of photocatalytic degradation of pollutants that are reductively degraded is the metal islands. In this case, the specificity can be enhanced by promoting specific adsorption on the metal islands. Indeed, the photocatalytic reduction of nitrobenzene was enhanced by its specific adsorption on silver nanoislands [86]. Likewise was the degradation of bis-(2-dipyridyl) disulfide, where adsorption was promoted by specific Ag-S interaction. It is noteworthy that this mechanism was demonstrated also with other coupled electron sinks such as carbon nanotubes, where enhanced reduction rates of CCl_4 were reported [47].

It was previously mentioned that the PZC may be used to induce specificity. In this context, Pt nanoparticles deposited on titanium dioxide may assist in obtaining specificity by this mechanism, as they may alter the surface charge of the photocatalyst [69].

5.3.5.3 Doping with Metal Ions and Oxides

Generally speaking, the effect of metal ions doping on photocatalysis is a complex function of the d-electron configuration, the oxidation states of the dopants, the energy states locations, the dopants' concentration, and whether the dopants are interstitial or substitutional. For most dopants, a correlation can be found between their effect on reductive degradation versus their effect on oxidation processes, that is, regardless whether they serve as electrons traps or holes traps. This is true also for dopants with closed shell configurations (Li^+, Mg^{2+}, Al^{3+}, etc.) that hardly influence the degradation rates. Nevertheless, two dopants seem to deviate from this rule: V^{4+} that increases reduction relative to oxidation and Ru^{3+} that increases oxidation relative to reduction. To explain this deviation, it was claimed that V^{4+} and Ru^{3+} act as irreversible hole and electron traps, respectively, thus preventing the trapped charge carriers from migrating to the solid–solution interface [87].

Relative selectivity can be obtained by hindering the degradation of one species while maintaining the degradation rate of another. This was demonstrated (albeit in the gas phase and not in water) by modifying TiO_2 with silico-tungstic acid (STA, $H_4SiW_{12}O_{40}\cdot nH_2O$), thus retarding the degradation rates of ethanol and acetone but not that of nitroglycerin [88].

5.3.5.4 Utilizing the "Adsorb and Shuttle" Mechanism to Obtain Specificity

The "A&S" mechanism described in the previous section as a means to increase overall reaction rates can be utilized to obtain specificity. The principle is to utilize variations in the affinities that the inert domains may have toward specific pollutants in order to increase the concentration of these pollutants near the photocatalyst surface and consequently to enhance their rate of degradation.

The fact that the synergy effect on combining TiO_2 and activated carbon varied from one pollutant to another (for example, synergy factor of 2.4 for 4-chlorophenol but only 1.3 for 2,4-dichlorophenoxyacetic acid) may provide a means to enhance the degradation of certain contaminants [89]. This specificity can be amplified to some extent by taking into consideration that the specific synergetic effect may vary also with the type of activated carbon that is used (i.e., hydrophobic AC made by high-temperature activation vs. hydrophilic AC made by low-temperature activation).

The acidity of a surface may affect specificity, as charged molecules may be more easily adsorbed on surfaces that are oppositely charged and repelled from surfaces that are similarly charged. Accordingly, it should be possible to take advantage of the difference in surface acidity between TiO_2 and SiO_2, ZrO_2, or Al_2O_3 to induce specificity on using composites. Care should be taken that the adsorption on the inert oxides is not too strong because strong adsorption might prevent surface diffusion..

Although the use of binary oxides may significantly increase the photocatalytic degradation of contaminants that hardly adsorb on titanium dioxide and may help in preventing the emission of intermediates, its specificity is inherently quite low (similar to the low specificity obtained by controlling pH while utilizing pure TiO_2).

A different way to utilize the "A&S" approach to get high specificity is by introducing immobile self-assembled monolayers that act as organic molecular recognition sites (MRS). The MRS are located on inert domains, in the vicinity of the photocatalyst. These MRS are predesigned to physisorb target molecules with very high specificity. Once adsorbed, the target molecules surface-diffuse from site to site toward the interface between the inert domains and the photocatalytic domains, where they are destroyed (Figure 5.5) [90].

Since the MRS are organic, care should be taken to prevent their degradation by oxidizing species that are formed on the titanium dioxide surface. This means that the MRS could not be constructed directly on the photocatalytic titanium dioxide surface. Moreover, it was necessary to assure that the MRS, located on inert sites adjacent to photocatalytic domains, were not prone to an attack by oxidizing species leaving the photocatalyst surface, where they had been formed [20]. Fortunately,

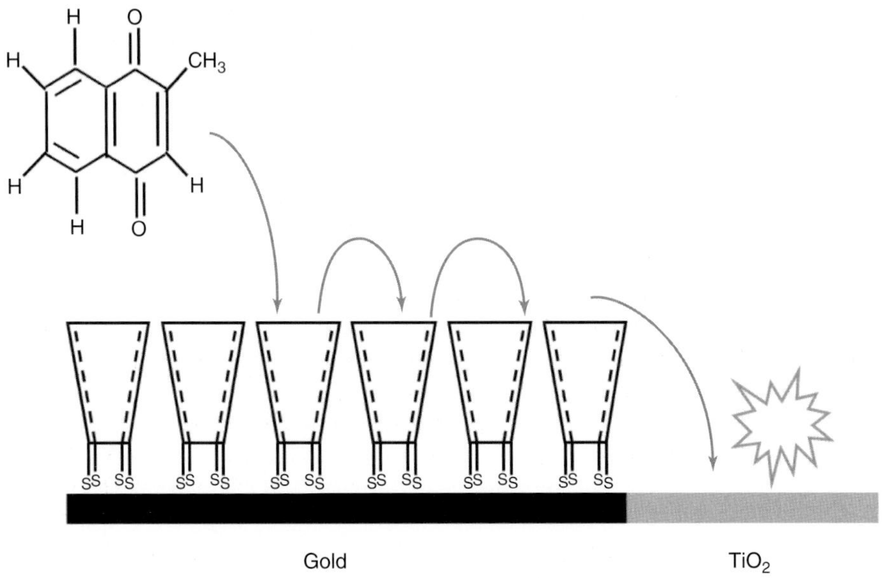

Figure 5.5 Specificity by the "Adsorb & Shuttle" mechanism, utilizing chemisorbed thiolated β-cyclodextrin as molecular recognition sites [Ref. 88].

this instability toward remote degradation was not found when the self-assembled monolayers were anchored through a thiol functional group to metallic substrates such as gold and platinum. In addition, the remote degradation effect was by far less problematic in water compared with air and under 365 nm light compared with 254 nm light [91].

The first system demonstrating this approach used thiolated β-cyclodextrin as the molecular recognition host, attached to gold microstripes, and 2-methyl-1,4-naphtoquinone (2MNQ) as the model contaminant [90, 92]. The host molecules were chemisorbed on micrometer-sized gold stripes located next to micrometer-sized titanium dioxide stripes on silicon wafers. β-Cyclodextrin is a cyclic oligosaccharide made of seven glucopyranose units forming a torus-like structure with a cavity of 0.78 nm in diameter. This molecule was chosen based on the well-known size-selective affinity between this molecule and a variety of apolar guests with sizes that match with that of the cyclodextrin cavity.

The decrease in the concentration of 2MNQ on exposure to 365 nm light for 30 min in the presence of TiO_2/gold microstripes, onto which thiolated β-cyclodextrin molecules were chemisorbed, was 60%. This decrease should be compared with 18.7% upon using TiO_2/gold microstripes having the same geometry but without thiolated β-cyclodextrin molecules. A control experiment using nonmodified TiO_2 with the same size of active area revealed a decrease that was slightly better than that of the non-MRS gold/TiO_2 samples (22.3%), evidently showing that the fast degradation observed with MRS gold/TiO_2 samples was due to the presence of the MRS and not due to the presence of the gold. Parallel experiments performed with benzene revealed that the degradation rate of benzene not only increased upon using the microstriped structure designed for 2MNQ but also, in fact, decreased. In other words, the predesigned system revealed a very large increase in the degradation rate of 2MNQ while negatively affecting the degradation rate of benzene.

This specificity effect was repeated on treating water containing a mixture of these contaminants. Here, the ratio of the reaction rate for 2MNQ to benzene was 0.49 on TiO_2, 0.71 with gold/TiO_2 stripes, and 4.23 upon using gold/TiO_2 microstripes onto which thiolated β-cyclodextrin molecules were chemisorbed. Hence, the feasibility of enhancing the preferential degradation of selected contaminants by constructing MRS at the vicinity of titanium dioxide domains was clearly demonstrated. Since both 2MNQ and benzene form inclusion complexes with β-cyclodextrin [93, 94], the opposite effects observed for these two pollutants can be rationalized by very low surface mobility of benzene because of its high complexing energy.

The degradation rate of 2MNQ on MRS-Au/TiO_2 was studied also as a function of the width of the micropatterned stripes. An inverse correlation between the domains' width (i.e., the diffusion distance) and the rate of degradation was observed, as could be expected for any "A&S" mechanism. That the enhanced degradation of 2MNQ was due to 2MNQ diffusion from the MRS to the TiO_2 domains and not due to out-diffusion of oxidizing species formed on the TiO_2 was supported by the observation that the degradation of the 2MNQ was not accompanied by the degradation of the MRS.

Another model contaminant that showed enhanced rate of photocatalytic degradation because of the presence of β-cyclodextrin close to titanium dioxide microdomains was the dye Chicago Blue Sky 6 (CB). Here, an aqueous solution containing a mixture of 0.01 mM of CB and 0.01 mM of the dye R6G was exposed to 365 nm UV light (0.09 mW cm^{-2}). The degradation of the CB dye and R6G was calculated based on the changes in their UV–VIS spectra, as reflected in their 625 and 527 nm peaks, respectively. It was found that the degradation of the CB was faster by 70% when using the MRS containing samples. This enhancement was not observed for R6G.

The approach of utilizing the "A&S" concept to obtain specificity was demonstrated not only in the liquid phase but also for the photocatalytic degradation of the semivolatile molecule diisopropyl methylphosphonate (DIMP), known as *simulant* for sarin, a nerve gas. The MRS consisted of chemisorbed 1,1-mercaptoundecanoic acid (MUA), with tail proton substituted by Cu^{+2} (MUACu) [95]. The relevance of this system to water treatment is in the way by which results were modeled, taking into account, among other factors, reversible adsorption on the reactor's walls, en route to estimating the increase in the rate constant of the first step in the degradation. This first step of degradation is of importance particularly in systems where the intermediate products are less hazardous than the initial reactants.

5.3.5.5 Mesoporous Materials

Selection by size can be achieved by developing photocatalytic materials that contain well-defined specific pores. While materials with well-defined pores (zeolites, for example) are routinely used in the field of thermal catalysis, preparation of photocatalysts having such properties is more complicated (yet possible at least to some extent).

Treating of a microporous titanosilicate photocatalyst with HF was found to increase its activity in a preferential manner. This way, in a mixture containing phenol and 2,3-dihydroxynaphthalene (DHN), a complete abatement of DHN was observed within 10 min, while the concentration of phenol hardly changed [96]. It was suggested that the internal cavities of the photocatalyst offer a protective environment against degradation for species that can easily diffuse inside. An alternative explanation claimed that the origin for specificity could be the ability to accommodate contaminants in the vicinity of the pore mouth.

Specificity was also demonstrated on intercalating titanium dioxide in sheet silicates of clay [97]. Here, the interlayer distance between adjacent layers in the clay served to enhance the photocatalytic degradation of molecules that were small enough to penetrate between the layers. Accordingly, short-chain carboxylic acids were degraded faster with the modified photocatalyst than with a nonmodified photocatalyst, whereas a long-chain carboxylic acid (capric acid) was degraded faster with the nonmodified photocatalyst.

To this class belongs also the grafting of nanoparticles of titanium dioxide onto the pore surface of mesoporous silicates. Here, the degradation rate of α-terpinol was enhanced by a factor of 4 compared with P25, while that of the dye R6G was reduced by a factor of 2 [98]. Likewise, the ratio between 4-chlorophenol removal and methanol removal was altered significantly upon intercalation of TiO_2 into montmorillonite clay. This change was attributed to a filtering-like mechanism [99].

5.3.5.6 Molecular Imprinting

Another way to obtain preferential photocatalytic degradation is to imprint cavities of the target molecules on the photocatalyst's surface during its preparation [100]. Once the preparation step is completed, the molecular template is chemically or physically removed, thus leaving a void on the surface that can repeatedly accommodate and degrade molecules that fit the molecular cavity (Figure 5.6).

While this seems to be the ultimate approach for obtaining very high specificity, it suffers from the fact that the popular sol–gel method used for preparation of TiO_2 (utilizing titanium alkoxides) yields an inactive amorphous phase that requires high-temperature treatment (>300 °C) in order to form the active anatase phase. Such temperature might be incompatible with molecular imprinting as the imprinted molecules might be burnt off before stabilization of the structure. Several strategies were developed to overcome this obstacle, among which are forming the anatase phase at low temperatures [9] and imprinting on a thin, nonanatase shell overcoating an anatase core. The shell is composed of titanium dioxide [101], silica [102], and also (subjected to the above discussed restrictions) a polymer [103].

The quality of specificity in molecularly imprinted photocatalysts can be evaluated based on the so-called *preferential (selectivity) factor*, defined as the ratio between the relative degradation rate of the target molecules and the relative degradation

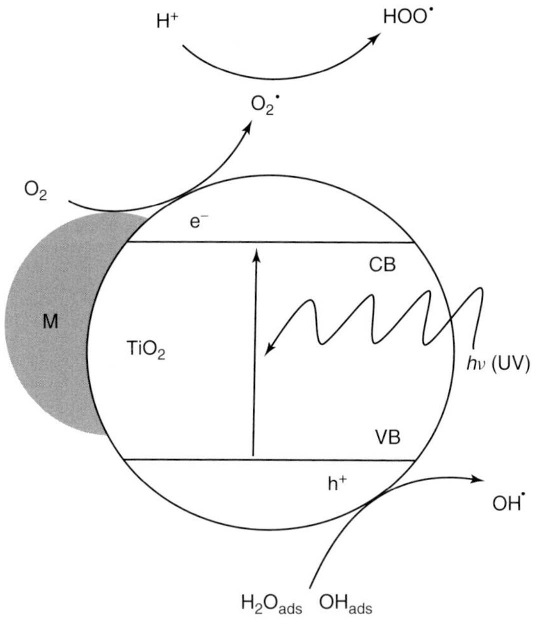

Figure 5.6 Photocatalysis in composite particles comprising titanium dioxide and noble metals.

rate of a guest molecule. Here, the word "relative" means the degradation rate with the modified photocatalyst divided by the degradation rate of the nonmodified photocatalyst. That way, the preferential factor takes into account the nonspecific rate enhancement, for example, because of an increase in the specific surface area. A comparison between various systems was performed by Nussbaum *et al.* [104]. According to their work, typical preferential factors range between 2 and 6, depending on the type of targets and guests and on the geometrical and chemical similarities between them. Generally speaking, the larger the difference is, the higher is the expected preferential factor. For modified photocatalysts made of TiO_2 core and imprinted polymer shell, the preferential factors in degrading nitrophenols increased not only as the chemical difference between the target and the guest was increased (as expected) but also as the concentration of the target was decreased [105]. It should be noted that different preferential factors can be obtained on exchanging between the template molecule and the guest molecule. For example, the preferential factor was found to be 2.7 with 2-chlorophenol as the template and 4-chlorophenol as the guest but only 1.3 with 4-chlorophenol as the template and 2-chlorophenol as the guest [106].

It was mentioned above that the degradation rate is expected to follow, under LH assumptions, the amount adsorbed on the surface and accordingly should (more or less) increase as the surface area is increased. To study whether molecular imprinting alters this relation, Nussbaum *et al.* [104] used the data published in [9, 102, 103, 105, 106]. For various template molecules and guest molecules, they defined the adsorption ratio as the ratio of the amount that was adsorbed on the imprinted photocatalyst to the amount adsorbed on nonmodified P25. Likewise, the *k* ratio was defined as the ratio between the degradation rate constants measured with imprinted photocatalysts and the degradation rate constants measured with nonmodified P25. The *k* ratio values were then divided by the respective adsorption ratios. According to this analysis, a value of "1" means that molecular imprinting did not alter the relation between surface area and photocatalytic degradation rates. Values as high as 3.5–5 were obtained for chlorophenols. As expected, the adsorption ratios for systems in which the guest molecule was different from the template molecule (0.7–0.85) were lower than for systems in which the guest molecule was identical to the template molecule (1.3–1.4). For all chlorophenols, the *k* ratio was significantly higher than 1 (2–6), leading to the high values obtained on dividing the *k* ratio by the adsorption ratio. From this analysis, it was deduced that while the effect of imprinting on adsorption is mild, the effect of imprinting on photocatalytic degradation rate is quite strong, probably because of the geometry of the adsorbed molecules that favors degradation.

In contrast, in nitrophenols, the degradation rates inclined to scale like the adsorption, that is, the values obtained on dividing the *k* ratios by the respective adsorption ratios were between 1.1 and 1.5. Here, again, and as expected, the adsorption ratios were higher whenever the template molecule and the guest molecule were identical.

One may imagine that the preferential factor depends on the actual ratio between the imprinted surface area and the nonimprinted surface area. The higher the relative coverage by imprinted sites is, the higher is the expected preferential factor. Therefore, it was recently proposed to use inert ultrathin layer in between the imprinted sites as a means to increase the preferential factor (to some extent, at the expense of overall degradation rates) [107].

Overall, the imprinting approach, although in its infancy, seems to be very promising because of its simplicity, for those contaminants that can bind directly to TiO_2 precursors.

5.3.6 Products' Control

It is generally taken for granted that photocatalysis leads to complete mineralization (i.e., production of CO_2, water, and inorganic ions). However, long-lived intermediate products or even nonmineralized end products are often observed. The problem of semimineralization may arise under restricted contact time in continuous flow reactors and, in particular, if mixed streams are to be decontaminated. In certain cases, the end products might be more toxic than the original pollutants. Here, the best example is probably the gas-phase photocatalytic degradation of tricholoroethylene that yields phosgene and dichloroacetyl chloride as intermediate products [108]. In the liquid phase, photocatalytic degradation of *s*-triazine derivatives yields the recalcitrant end product cyanuric acid [109]. Another example for liquid phase is toluene (and other benzene derivatives) that often produces end products that might deactivate the surface of the photocatalyst (see below). Semimineralization (whether partial oxidation or partial reduction) is not necessarily negative. On the contrary, it may provide a route toward green production of chemicals having a significant commercial value, provided that the selectivity is high enough and the required separation processes are inexpensive.

While recalcitrant products can be dealt by synergistic combination of several techniques (for example, sonolysis and photocatalysis for treating chlorinated and fluorinated aromatic compounds [110, 111]), it can be less expensive to prevent the problem by developing means to control the end products.

Such control may be achieved by utilizing modified photocatalysts. As shown hereby, although there is no single type of modification that is specifically adequate for this purpose, it can be claimed that the route toward products' control goes, for most cases, through modifying the adsorption of the intermediate products. Accordingly, some coupling between preferential degradation and product selectivity cannot be avoided. Indeed, the same operational parameters that govern specificity were found to affect products' selectivity. Among these are pH (affecting the products of methanol [84]) and the type of solvents that are used [14]. Similar to specificity, there are multiple approaches for modifying TiO_2 to control products' selectivity.

5.3.6.1 Surface Modification by Molecular Imprinting

Surface modification may alter the type of end products. This way, modifying the surface of titanium dioxide by an iron–porphyrin complex changed the selectivity in the photocatalytic monooxygenation of cyclohexane, in a manner that increased the alcohol-to-ketone ratio [112]. Nevertheless, modification with organic compounds is problematic for long-term operation as discussed above.

Performing photocatalytic reactions in molecularly imprinted photocatalysts may release less intermediate products, an issue of high importance when the intermediate products are of high toxicity. Thus, an aqueous solution of pentachlorophenol (PCP) was rapidly detoxified, as measured biologically by using luminescent bacteria, using P25 whose surface was modified by imprinting. At the same time, bare P25 yielded no more than a slight detoxification [103]. Another example is the photocatalytic degradation of diethyl phthalate (DEP) on particles having a molecularly imprinted inorganic shell [102], where imprinting had the benevolent effect of reducing the accumulation of toxic aromatic by-products, such as phthalic acid and diethyl 2- and 3-hydroxyphthalate.

The effect of imprinting photocatalysts on the release of intermediates was studied also with molecularly imprinted polymer overcoating a TiO_2 core (MIP) [105]. It was found that the same type of intermediates (3-nitrocatechol and 1,2,4-benzenetriol in the degradation of 2-nitrophenol and 1,2,4-benzenetriol in the degradation of 4-nitrophenol) were released with and without imprinting; however, the accumulation levels of the major aromatic intermediates produced with the imprinted photocatalyst were lower than that with neat TiO_2.

The introduction of templates that are similar (yet not identical) to the target molecules (often coined "pseudo templates") may also affect the type of released intermediates. In that manner, the degradation of PCP on dinitrophenol-imprinted MIP did not release any *p*-chloranil or tetrachlorohydroquinone (TCHQ) that are usually found during photocatalytic degradation over P25 [103]. It was suggested that the pseudotemplate affected the mechanism of PCP degradation, possibly because of some interaction between the PCP molecules and amino groups in the footprint cavities. This example suggests that pseudotemplates can be used not only to reduce the concentration of intermediates but also as a tool to direct photocatalytic reactions toward preferred products.

5.3.6.2 Composites Comprised of TiO_2 and Metallic Nanoislands

Photocatalysts containing nanoislands of noble metals were shown to produce end products that may be slightly different (by type and relative concentration) from end products that are produced in the absence of modification, as demonstrated with the degradation of oxalic acid in an aqueous solution on a Pt/TiO_2 and an Ag/TiO_2 composite photocatalysts [113]. This effect was attributed to enhanced charge separation and reduction of dioxygen. Lee *et al.* proposed that more basic alkylamines tend to produce amines that are more N-alkylated through the anoxic mechanism on Pt/TiO_2. This change is not necessarily benevolent as was observed in the degradation of alkylamine where toxic by-products were produced with composite Pt/TiO_2 but not with unloaded TiO_2 [114]. Another example, this time for a reductive process is that of selenate ions (SeO_4^{2-}) [115]. Here, in the absence of silver nanodeposits, H_2Se generation occurred only when the Se(VI) ions in the suspension were exhausted, whereas on Ag/TiO_2 particles, the H_2Se was generated simultaneously during the reduction of Se(VI). It was proposed that the photogenerated electrons were transferred from TiO_2 to the selenium through the silver nanodomains, leading to the accumulation of electrons on selenium deposits which eventually led to self-reduction from Se^0 to Se^{2-}.

The photocatalytic formation of hydrogen from ammonia and methyl amine was studied in aqueous suspensions containing metal-loaded TiO_2. Ammonia was decomposed to H_2 and nitrogen with a stoichiometric ratio under a deaerated condition, without any side reactions such as formation of nitrite and nitrate. The formation of O_2 was negligible. The rate of H_2 formation drastically changed depending on the co-catalyst loaded on the TiO_2 particles, the type of TiO_2 samples, and the pH of the suspension. For a series of co-catalysts whose overpotential followed the sequence Ir $>$ Cu $>$ Ag $>$ Au $>$ Pd $>$ Pt, the lower the hydrogen overpotential was, the higher the H_2 evolution was measured. The highest rate of H_2 evolution was obtained at pH 10.7. Other end products included NH_4^+ and HCO_3^- [116].

A drastic accelerating effect was reported also for Ag cluster loaded on TiO_2 in the reductive photocatalysis of nitrobenzene to aniline. The effect was attributed to the improvement of the charge separation efficiency [86].

5.3.6.3 Doping with Metal Ions

Surface doping with lanthanide ions was found to increase the photocatalytic degradation rate of *p*-chlorophenoxyacetic acid. It was claimed that this increase was due to the formation of a Lewis acid–base complex between the lanthanide ion and the substrate. Although specificity toward other contaminants was not analyzed, it is conceivable that this enhancement is not specific because lanthanides are known for their ability to form complexes with acids, amines, aldehydes, alcohols, and thiols. The specific interaction between the dopant and the acid affected intermediates distribution, as the only intermediate that was found was *p*-chlorophenol, instead of hydroquinone and chloroquinone that are found in the case of pure TiO_2 [117].

5.3.6.4 Nonmetallic Composite

The inert adsorptive substrate in composites comprising of TiO_2 and a nonmetallic inert compound (usually affecting photocatalysis by the A&S mechanism) has an effect on the type of emitted products. For example, the use of composite particles made of P25 and activated carbon totally prevented the appearance of any intermediates in the photocatalytic degradation of gas-phase TCE [63]. This effect can vary with the type of activated carbon that is used (hydrophobic or hydrophilic) and may provide a means to enhance the degradation of a specific pollutant.

The extent of dispersion of TiO_2 within zeolites plays a key role in the decomposition of NO. High photocatalytic efficiency and high selectivity for the formation of N_2 in the photocatalytic decomposition of NO were reported for tetrahedral coordination of titania within the zeolite as N_2 formation was predominate for a octahedral coordination [118]. Shimizu *et al.* [119] reported the use of TiO_2 pillared clays for the selective photooxidation of liquid hydrocarbons. The difference between the relative adsorption on TiO_2 and TiO_2-pillared clays (mica, montmorillonite, and saponite) manifested itself also in the distribution of species that were produced during the photocatalytic degradation of benzene. The selectivity enhancement in the partial oxidation probably originates from the hydrophobic nature of the pillared clays.

5.3.6.5 TiO_2 Morphology and Crystalline Phase

High selectivity (99%) in visible-light-driven partial photocatalytic oxidation of benzyl alcohol into benzaldehyde was achieved by using single-crystalline rutile TiO_2 nanorods. The nanorodes were composed of a core made by calcining electrospun nanofibers and a shell that was grown hydrothermally from $Ti(OBu)_4$. It was proposed that the visible-light-induced partial oxidation was due to absorption of visible light by the benzyl alcohol, followed by electron transfer to the rutile. The unidirectionality of the nanorods favored the electron transfer along the (110) plane. Hence, it can be said that the obtained partial oxidation (if only under visible light) was the outcome of modifying TiO_2 by preparing it in the form of nanorods having well-defined directionality [120].

5.3.7 Reducing Deactivation

Poisoning and deactivation of catalysts is a well-known problem in heterogeneous catalysis, in particular under elevated temperatures, where sintering of the catalytic particles (leading to lower surface area) or growth of metallic nanodomains (reducing the contour length of the metallic domains) might occur. Unlike most catalytic processes, photocatalysis takes place usually at ambient temperatures, hence phenomena such as particle sintering or metal islands growth is unlikely. Yet, deactivation of TiO_2 poses a severe limitation on commercialization of TiO_2 for water treatment. Here, the dominant mechanism for deactivation is the production of end products that strongly adsorb on the photocatalyst surface. For example, toluene and benzaldehyde are photocatalytically oxidized to benzoic acid that tends to adsorb irreversibly on the titanium dioxide surface [121]. Another example, more related to air treatment than to water treatment, is the formation of thin layers of silica on the photocatalyst surface following the photocatalytic degradation of organosilicones [122]. While many researchers have encountered deactivation phenomena, it seems to us that there is no enough awareness among researchers in the area of photocatalysis to the importance of deactivation processes. This lack of awareness is reflected in the small number of reports dedicated to long-term performance compared with the large number of manuscripts reporting on (rather successful) short-term performance.

In the context of photocatalysis, prevention of deactivation can be obtained by providing the right tools to degrade the problematic end products. One option is to utilize the thermal catalysis properties [123] of a composite made of titanium dioxide decorated with nanoislands of noble metals such as gold or platinum [124, 125].

An interesting effect of carbon doping in resisting deactivation was reported recently [126]. Here, a one-step, hydrothermal method for the synthesis of N, C, co-doped titanium dioxide was performed, having ammonia as the nitrogen source and a variety of alcohols (methanol, ethanol, isopropanol, 1-butanol, 2-butanol, tert-butanol) as carbon sources. The crystallite size of the doped TiO_2 was not affected by the nature of the alcohol. Nevertheless, the photocatalytic efficiency, and, more relevant to this section, the deactivation of the photocatalyst, depended on the type of organic molecule used as a carbon source. Specifically, utilizing 2-butanol and 1-butanol as carbon precursors yielded high resistance to deactivation during multiple catalyst reuse in the treatment of phenol-containing aqueous solutions compared to pristine TiO_2, Degussa P25, or N, C-doped TiO_2 made from other carbon sources. It was proposed that the source of carbon in the doped photocatalyst had an effect on the type of formed by-products and eventually also on the tendency for poisoning.

5.3.8
Recombination Rates and Charge Separation

A general limitation in the efficiency of any photocatalytic process is the recombination of the photogenerated electrons and holes, which takes place both in the bulk (where the charge carriers are formed) and at the surface (Chapter 4). Considerable efforts have been made to suppress recombination by enhancing charge carriers' separation. These efforts, not necessarily in the context of water treatment, are summarized in several reviews [5, 127, 128]. These modifications can be grouped into structure modification; composites containing nonphotocatalytic noble metals, oxides, or semiconductors; doping with metal/nonmetal ions; and surface modification by ultrathin films.

5.3.8.1 Structure Modification

It has been shown that primary TiO_2 particles may, under specific conditions, aggregate in solution thus forming a three dimensional network. A novel energy transfer mechanism between the primary particles ("the Antenna mechanism") was suggested [129], leading to improved photocatalytic activity. In that context, modifications that assist this type of aggregation leading to strong electronic coupling between primary particles are likely to affect charge separation and consequently photocatalysis efficiency.

The performance of the nanostructured TiO_2 with and without Fe-dopants was shown to depend strongly on the preparation and the Fe-content. Results were explained noting that the Fe(III)-doped TiO_2 nanoparticles form extended networks in aqueous suspensions, that is, by aligning the atomic planes of different particles in a parallel manner. In addition, it was claimed that deaggregation processes play a major role in the observed phenomenon [56].

5.3.8.2 Composites–Metal Islands

The most common way by which photocatalysts were modified to reduce recombination is by introducing nanometer-sized islands of noble metals such as gold, platinum [130], palladium, and silver [131] that serve as sinks for the photogenerated electrons, thus facilitating electron transfer to dioxygen (or to other electron acceptors) (Figure 5.6). The recombination is retarded by the formation of a Schotkey barrier between the metallic nanoislands and the photocatalyst, as verified for silver [132] and platinum by a variety of techniques, among which were electrochemistry, time-resolved spectroscopy [133, 134], and photocurrent measurements that used Fe^{3+}/Fe^{2+} ions as electron shuttles [135].

Factors such as particle size, lattice defects, and oxygen vacancies seem to affect the efficiency of charge transport between the photocatalyst and the metal islands. The charge separation effect depends also on the type of metal. Indeed, a comparative transient absorption spectra study of Pt/TiO_2 and Au/TiO_2 suggested that Pt is more efficient in accepting electrons than [134]. This difference may stem from a more effective trapping and pooling of photogenerated electrons and/or because of a higher reactivity of platinum in assisting the reduction of oxygen on its surface [136].

Core–shell structures made of a metal core and a TiO_2 shell may serve to reduce recombination rates in the absence of electron scavengers such as oxygen. The basic principle here is to temporarily store the photogenerated electrons, thus facilitating oxidation processes at the outer surface. Some idea of the ability to accommodate electrons can be inferred from following shifts in the plasmon resonance bands of the metal. For example, Ag/TiO_2 structure were obtained by reducing Ag^+ ions, to form Ag nanoparticles onto which TiO_2 was deposited by hydrolyzing titanium isopropoxide [137, 138]. Enhanced photo decolorization rates of aqueous solution of methyl orange were then observed in such core–shell structures. Part of the enhanced rates was attributed to the existence of multiheterojunctions between Ag/anatase, Ag/rutile, and anatase/rutile [139].

5.3.8.3 Composites Comprising Carbonaceous Materials

Carbonaceous materials such as graphene or carbon nanotubes may reveal, upon coupling with titanium dioxide, properties that are similar to those of metallic nanoislands, that is, improved charge separation following electron migration from the photocatalyst to the inert component. This effect may be coupled with adsorptive effects (Section 5.3.3) as demonstrated in the photocatalytic degradation of MB by TiO_2-graphene [48]. Here, the rate of OH generation was estimated by the electron paramagnetic resonance (EPR) spin-trap technique, using 5,5-dimethyl-1-pyrroline-*N*-oxide (DMPO) as the spin-trapping reagent. Higher concentration of OH radicals was found in the composite case. This higher concentration was attributed not only to reduced rate of recombination of holes on the TiO_2 particle but also to a mechanism involving superoxide anions formed on the graphene that reacted with adsorbed water to produce OH radicals.

A good example of coupling between charge separation phenomena and adsorptive effects was presented with mats made of poly(acrylonitrile) nanofibers containing both TiO_2 and carbon nanotubes. Here, faster photocatalytic reduction rate of CCl_4 on the CNT-containing mats was explained by the reduction of CCl_4, taking place not on the photocatalyst surface but on the carbon nanotube, where the pollutant is readily adsorbed [47]. Charge separation can be counterproductive in those rare cases where the reaction is bound to take place in one locus, whereas the relevant active species (electrons, holes, or OH radicals) are shuttled to the other component. For example, Cr(VI) is degraded to Cr(III) reductively yet tends to adsorb on TiO_2 and not on OH-functionalized CNTs. Accordingly, the shuttling of electron from the photocatalyst to the OH-functionalized CNTs does not increase the rate of reduction but in fact, can be quite deleterious [140] (Figure 5.7).

5.3.8.4 Composites Composed of TiO_2 and Nonoxide Semiconductors

Coupling between different semiconductors in photocatalytic systems may induce, under proper matching of their conduction band and valence band, vectorial transfer of photogenerated charge carriers [141]. Coupling two semiconductors that seem to be adequate based on an analysis of their energy bands as measured when they are separated does not guarantee enhanced charge separation because other parameters that govern efficiency (such as surface area, crystallinity, defect density, and surface charge at a given pH) may be altered during the coupling process [5]. Generally, the oxidation power is lower in such configurations; nevertheless, in some cases, this lower oxidation power may be sufficient, as is the case for TiO_2/CdS under visible light where the hole on the CdS surface ($Eg = 2.5$ eV) cannot oxidize water to form OH radicals and yet OH radicals are formed [142]. It was proposed that the formation of OH radicals occurs in a stepwise mechanism, involving three electrons (or actually three photons). Two electrons (together with

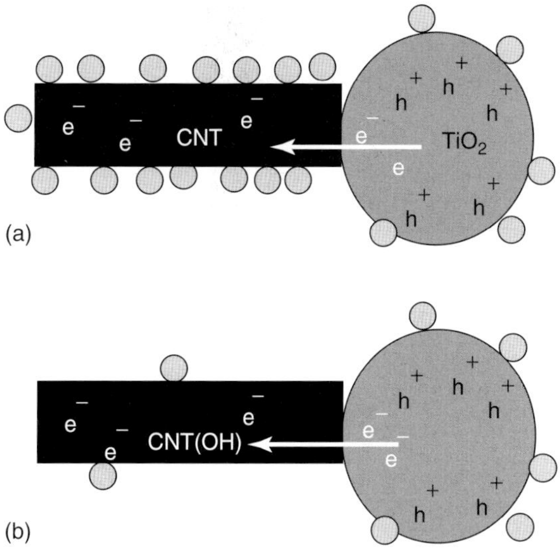

Figure 5.7 Contradicting effects in composites comprising TiO_2 and carbon nanotubes: (a) enhanced photocatalytic reduction of CCl_4 on TiO_2/CNT (b) retarded photocatalytic reduction of Cr(VI) on TiO_2/CNT(OH).

$2H^+$ and oxygen) form hydrogen peroxide and another electron forms superoxide anion on the TiO_2 surface. Then, the superoxide anion and the hydrogen peroxide react to give OH radical that may attack organic molecules. Regardless of the need for three photons in order to form one OH radical, the CdS/TiO_2 composite is unlikely to be used for water treatment because of the tendency of CdS to photocorrode, releasing highly toxic cadmium ions.

5.3.8.5 **Composites Composed of TiO_2 and Other Oxides**

Improved charge separation can be obtained not only by coupling titanium dioxide and metals or semiconductors but also in composites containing titanium dioxide with other oxides, the most common of which are WO_3 [143], SnO_2 [144], ZnO [145], and Fe_2O_3 [146].

A layered structure made of TiO_2 overlayers (70 nm in thickness) on WO_3 was studied in the context of photoinduced hydrophilicity and photocatalytic degradation of adsorbed MB [143]. Although the top surface was fully covered by TiO_2, the presence of the WO_3 layer was found to enhance photoinduced superhydrophilic conversion. When the TiO_2 and WO_3 layers were separated by an insulating layer comprising of silica (40 nm), the surface did not become highly hydrophilic under the same exposure to light. The photocatalytic oxidation of MB adsorbed onto the surface and the photocatalytic reduction of silver nitrate in an aqueous solution were investigated under UV black light. The oxidation reaction was enhanced by the presence of the WO_3 layer, however, the reduction reaction was not, suggesting that photogenerated holes accumulated in the TiO_2 layer, whereas electrons accumulated in the WO_3 layer. This explained also why TiO_2/WO_3

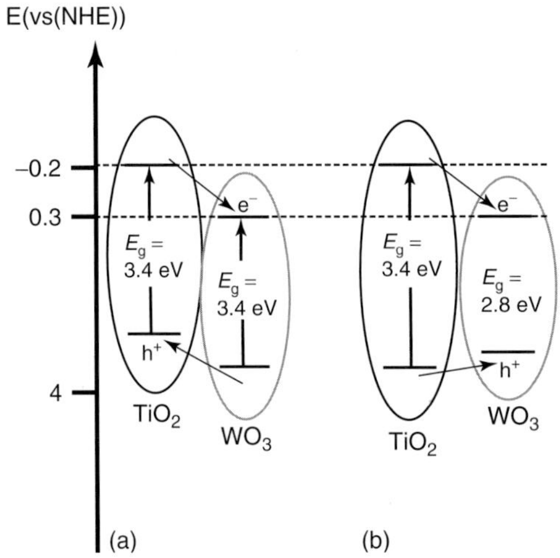

Figure 5.8 Schematic energy diagram for: (a) TiO_2/amorphous WO_3 (b) TiO_2/crystalline WO_3 [after Ref. 143].

layered film formed on a grounded electroconductive substrate displayed a higher hydrophilicizing rate than when coated onto an insulating substrate.

The conduction band of WO_3 is located at 0.3 V VCE (for both crystalline and amorphous tungsten trioxide) compared with −0.2 for TiO_2. Accordingly, a bandgap excitation of WO_3/TiO_2 composite caused effective electron transport from TiO_2 to WO_3 (possibly through formation of W(V) species) that generated high anodic photocurrent during the oxidation of oxalic acid in solution [147]. The preparation method may affect the bandgaps and band edge levels and consequently the tendency for charge separation. Indeed, a WO_3/TiO_2 electrode having an amorphous-like WO_3 phase exhibited higher photocurrent than crystalline WO_3/TiO_2, WO_3, or TiO_2. This was explained by hole transfer from the WO_3 to the TiO_2 because of the larger Eg of amorphous WO_3 in comparison to that of the crystalline (3.4 vs 2.8 eV, respectively), leading to a situation where the valence band of the crystalline WO_3 is located below that of TiO_2 [147] (Figure 5.8).

Polycrystalline WO_3/TiO_2 and MoO_3/TiO_2 samples were prepared from H_2WO_4 and ammonium paramolybdate, respectively. The surface acidity of the prepared materials was shown to increase with the loading of tungsten oxide/molybdenum oxide, in correlation with enhanced rate of photocatalytic degradation of 1,4-dichlorobenzene. It was claimed that surface acidity, namely, the protonated surface hydroxyl groups, promote the electron transfer to adsorbed oxygen molecules [148]. Although the authors did not explicitly negate the possibility of an "A&S" effect in this case, the fact that 1,4-dichlorobenzene is uncharged seem to support the proposed mechanism. Too much loading of WO_3 might have an adverse effect since at high concentration the tungsten atoms might act as recombination centers [149].

Coupling SnO_2 ($E_g = 3.8$ eV) and TiO_2 was reported to lead to very rapid photocatalytic decolorization of Acid Orange under bias, probably due to favorable location of the conduction bands of the two components (E_{cb} of 0 V vs NHE and −0.5 V vs NHE for SnO_2 and TiO_2, respectively, at pH 7 [144]). Enhanced photoelectrocatalytic rates were observed also in the degradation of naphthol blue [150]. Here, the influence of the presence of oxygen on the short-circuit photocurrent and the open-circuit photovoltage became less and less significant as the SnO_2 content was increased, suggesting that the injection of electrons from the titanium dioxide to the SnO_2 was very rapid. Coupling between SnO_2 and TiO_2 was performed not only by the preparation of composite particles but also by forming a layered structure, where a TiO_2 layer overcoated a SnO_2 underlayer. An increased rate in photocatalytic oxidation reactions and a decreased rate in photocatalytic reduction reactions was found, explained by interfacial electron transfer from TiO_2 to SnO_2 [151].

The valence band of ZnO is located slightly higher than that of TiO_2. Since the two semicondctors have similar bandgaps (3.37 eV for ZnO and 3.2 eV for TiO_2) the bottom of the conducting band of ZnO is also located higher than that of TiO_2. This means that in coupled composite particles, electrons will accumulate on the TiO_2, whereas holes will accumulate on the ZnO regardless of absorption location, thus reducing recombination [145].

5.3.8.6 **Doping with Metals**

Metal-ion dopants can alter the lifetime of the photogenerated charge carriers by trapping them, thus facilitating the counter carrier to reach the surface of the photocatalyst [152–154]. Some reports suggest a direct relationship between charge carrier lifetime and photocatalytic activity of titanium dioxide doped with metal ion dopants. On the other hand, the dopants might act as recombination centers [155], in particular at high concentrations. Two different types of traps are considered. Shallow-trapped holes are in thermally activated equilibrium with free holes and exhibit very high oxidation potential that may increase both lifetime and photocatalytic activity. Deeply trapped holes, which are identified chemically as surface-bound hydroxyl radicals, may increase the lifetime of carriers but at a partial expense of their ability to reduce/oxidize, and therefore they are often considered as recombination centers [156]. An enhanced degradation rate of MB [157] and phenol [158] in aqueous solution by Au-doped TiO_2 was observed. In this case, the improvement was attributed to the increased absorption in the visible region by intercalation of Au in the titania structure, by electron scavenging by Au(III) ions and by the decrease in the Fermi level, subsequently improving interfacial charge-transfer process at the TiO_2 interface by the gold nanoparticles. In addition, electrons from the conduction band migrated to the Au on the surface and thus improved charge separation. The presence of ionic Ag^+ on the photocatalyst surface assisted in trapping the electrons by a similar mechanism, thus enhancing photocatalytic degradation of aniline by Ag/TiO_2 [159].

The photocatalytic activity of Fe^{3+}-doped TiO_2 depends on dopant concentration [152]. Here, dopants at low concentration act as shallow trap sites for both photogenerated electrons and holes thus inhibit recombination. If the concentration is too high, the two types of sites react through quantum tunneling. This detrimental dopant concentration dependency was reported also for Ag deposition on TiO_2 [50].

The optimal Fe^{3+} dopant concentration for different particle sizes was identified in the degradation of $CHCl_3$ in the liquid phase, and this concentration was found to decrease with increasing particle size [160]. Doping of titanium dioxide with Fe^{3+} was shown to increase the rate of photocatalytic oxidation of methanol to formaldehyde [129]. Here, it was also proposed that the Fe^{3+} ions act as shallow traps for electrons, facilitating rapid transfer of electrons to molecular oxygen. In parallel, the formation of three dimensional networks of such particles was observed by transmission electron microscope (TEM). It was claimed that the formation of this networks facilitates energy transfer from particles as antenna systems that lead to improved photocatalytic activities of the colloidal particles.

5.3.8.7 **Doping with Nonmetals**

Doping with nitrogen may yield not only substitutional N 2p states but also localized states above the valence band edge arising from dopants at interstitial positions (oxygen vacancies), acting as strong hole-traps [161, 162]. These theoretical predictions were confirmed later experimentally by Nakamura *et al.* who traced N-induced mid-gap surface states by studying the photooxidation of various hole scavengers [163]. In fluorinated TiO_2 surfaces (i.e., TiO_2 whose surface hydroxyls were exchanged with fluoride anions) the surface F atoms act as electron trapping sites, due to the strong electronegativity of fluorine. Consequently, photocatalytic reduction processes, for example, that of TCA (trichloroacetate) tend to be retarded on these substrates [11].

5.3.9
Visible Light Activity

The relatively large energy gap (3.2 eV) of TiO_2 facilitates the utilization of no more than a few percent of the solar energy impinging on earth. For this reason many attempts were carried out to red-shift the absorption edge of the photocatalyst. These attempts include surface modification via organic materials (sensitization by dye molecules), coupling with semiconductors, coupling with metals (thus taking advantage of surface plasmon resonances), doping with nonmetals such as N, C, F, S, and metal doping or bandgap modification by creating oxygen vacancies.

For lack of space in this chapter, and in accordance with the extensive coverage of this topic in another chapter of this book, dedicated to visible light activity, it was decided not to include this highly important aspect of modified photocatalysts in this chapter.

5.3.10
Charging–Discharging

The possibility to use composite materials to reversibly store photoelectrons was first realized by Tatsuma and Fujishima, who showed that WO_3 in WO_3/TiO_2 hybrid structures stored photogenerated electrons under UV illumination. These electrons can be later released under dark conditions [164, 165].

The proposed charging mechanism involved the following steps:

1) Light absorption by titanium dioxide and formation of electrons and holes

$$TiO_2 + h\nu \rightarrow TiO_2{}^{*}\,(e^- + h^+) \tag{5.1}$$

2) Electron transport from the titanium dioxide to tungsten trioxide.

$$e^-\ (TiO_2) \rightarrow e^-\ (WO_3) \tag{5.2}$$

3) Oxidation of water to yield H^+

$$2H_2O + 4h^+ \rightarrow O_2 + 4H^+ \tag{5.3}$$

4) Intercalation of H^+ to form H_xWO_3.

$$WO_3 + e^- + H^+ \rightarrow HWO_3 \tag{5.4}$$

Under dark conditions a self-discharge process occurs at the intercalated tungsten trioxide particle, where the previously transported electrons are consumed:

$$HWO_3 \rightarrow e^- + H^+ + WO_3 \tag{5.5}$$

and

$$O_2 + 2HWO_3 \rightarrow H_2O_2 + 2WO_3 \tag{5.6}$$

Other composite materials such as Cu_2O/TiO_2 [166], SnO_2/TiO_2 [167], MoO_3/TiO_2 [168], and polyoxometalates/TiO_2 have also shown similar abilities; however, WO_3 is regarded as the most viable and effective. A comparison between three types of MoO_3/TiO_2 films (mixed, bilayer, and adjacent (often called *separated*)) on indium tin oxide revealed that while all types can be charged upon exposure to UV light, there were distinct differences in the self-discharging time, the longest being that of the separated composite [168]. Another parameter that affects the discharging time (apart of the material used for energy storage) is the surface ratio and weight ratio between the two components, as demonstrated with WO_3/TiO_2 [169].

The energy-storing effect of these composite materials was utilized usually in the context of corrosion prevention [164] or water splitting under dark conditions [166] and less for water decontamination.

5.3.11 Mass Transfer

One of the biggest challenges in the application of membranes for filtration is fouling, which imposes high energy costs and low flux limitations. The use of titanium dioxide, modified into hierarchically structured nanofibrous architecture with multiscale organization facilitated the use of photocatalysis for handling of fouling while maintaining high flux and low pressure drop [37]. These structures, made of either pure TiO_2 or of TiO_2 composites (for example, TiO_2 nanofibers/ZnO nanorods) were implemented on conventional polymer membranes, en route for commercialization [145].

5.3.12
Facilitating Photocatalysis in Deaerated Suspensions

Oxygen plays an important role in photocatalysis, both for trapping electrons (thus reducing recombination rate) and as an integral part of the degradation reactions. Accordingly, absence of oxygen may hinder photocatalysis. The presence of oxygen may be important even when the first step is reductive, as it happens in the case of TCA, which accepts a conduction band photoelectron thus forming dichloroacetate radical that is rapidly mineralized upon reacting with oxygen.

Modification of titanium dioxide with nanosized islands of platinum may facilitate an anoxic path for water treatment, as demonstrated with TCA. Here, the dichloroacetate radical ions react with valence band holes yielding CO_2 and dichlorocarbene (CCl_2). The latter is then hydrolyzed in a process that does not require oxygen, to give chloride ions [170].

Summary

Modification of titanium dioxide is a viable way to overcome inherent problems associated with the use of the photocatalyst for water treatment. These problems are beyond maximizing quantum efficiency and relate to a variety of aspects including specificity, deactivation, and formation of by-products. Their complexity is further aggravated by their dependency in operational parameters and in the specific model system. This chapter discusses the various ways by which the photocatalyst can be modified. Modifications may take place at the surface of the photocatalyst, at its bulk, or by forming composite structures, utilizing almost any type of material (metals, organic compounds, oxides, semiconductors, carbonaceous materials, etc.). While, at present, it is impossible to pinpoint a "silver-bullet" modification that will address most of the aspects discussed in this chapter, it is still hoped that such solution will be found. At any case, we will not be surprised if such a solution will incorporate more than one of the approaches described in this chapter.

References

1. Paz, Y. (2009) in *Advances in Chemical Engineering*, Photocatalytic Technologies, Vol. **36** (eds H.I. de-Lasa and B.S. Rosales), Elsevier, pp. 289–336.
2. Choi, W. (2006) Pure and modified TiO_2 photocatalysts and their environmental applications. *Catal. Surv. Asia*, **10**, 16–28.
3. Paz, Y. (2010) Composite titanium dioxide photocatalysts and the "adsorb & shuttle" approach: a review. *Solid-State Phenom.*, **162**, 135–162.
4. Paz, Y. (2011) Self-assembled monolayers and titanium dioxide: from surface patterning to potential applications. *Beilstein J. Nanotechnol.*, **2**, 845–861.
5. Zhang, H., Chen, G., and Bahnemann, D.W. (2009) Photoelectrocatalytic materials for environmental applications. *J. Mater. Chem.*, **19**, 5089–5121.
6. Thurnauer, M.C., Rajh, T., and Tiede, D.M. (1997) Surface modification of TiO_2: correlation between structure, charge separation and reduction

properties. *Acta Chem. Scand.*, **51**, 610–618.

7. Sakthivel, S. and Kisch, H. (2003) Photocatalytic and photoelectrochemical properties of nitrogen-doped titanium dioxide. *Chem. Phys.*, **4**, 487–490.
8. Anpo, M., Ichihashi, Y., Takeuchi, M., and Yamashita, H. (1998) D esign of unique titanium dioxide photocatalysit by an advanced metal ion-implantation method and photocatalytic reactions under visible light irradiation. *Res. Chem. Intermed.*, **24**, 143–149.
9. Sharabi, D. and Paz, Y. (2010) Preferential photodegradation of contaminants by molecular imprinting on titanium dioxide. *Appl. Catal. Environ.*, **95**, 169–178.
10. Tada, H., Jin, Q., Nishijima, H., Yamamoto, H., Fujishima, M., Okuoka, S.-I., Hattori, T., Sumida, Y., and kobayashi, H. (2011) Titanium (IV) dioxide surface- modified with iron oxide as a visible light photocatalyst. *Angew. Chem.*, **50**, 3501–3505.
11. Choi, H., Stathatos, E., and Dionysiou, D.D. (2006) Sol-gel preparation of mesoporous photocatalytic TiO_2 film and TiO_2/ Al_2O_3 composite membranes for environmental applications. *Appl. Catal. Environ.*, **63**, 60–67.
12. Dey, N.K., Kim, M.J., Kim, K.-D., Seo, H.O., Kim, D., Kim, Y.D., Lim, D.C., and Lee, K.H. (2011) Adsorption and photocatalytic degradation of methylene blue over TiO_2 films on carbon fiber prepared by atomic layer deposition. *J. Mol. Catal. A: Chem.*, **337**, 33–38.
13. Avraham-Shinman, A. and Paz, Y. (2006) Photocatalysis by composite particles containing inert domains. *Isr. J. Chem.*, **46**, 33–43.
14. Liu, B.-J., Torimoto, T., and Yoneyama, H. (1998) Photocatalytic reduction of carbon dioxide in the presence of nitrate using TiO_2 nanocrystal photocatalyst embedded in SiO_2 matrices. *J. Photochem. Photobiol., A*, **115**, 227–230.
15. Tada, H., Ishida, T., Takao, A., and Ito, S. (2004) Drastic enhancement of TiO_2-photocatalyzed reduction of nitrobenzene by loading Ag clusters. *Langmuir*, **20**, 7898–7900.
16. Herrmann, J.M., Disdier, J., and Pichat, P. (1982) Photoassisted platinum deposition on TiO_2 powder using various platinum complexes. *J. Phys. Chem.*, **90**, 6028–6034.
17. Kraeutler, B. and Bard, A.J. (1978) Heterogeneous photocatalytic preparation of supported catalysts. Photodeposition of platinum on TiO_2 powder and other substrates. *J. Am. Chem. Soc.*, **100**, 4317–4318.
18. Ramanathan, K., Avnir, D., Modestov, A., and Lev, O. (1997) Sol-Gel derived rmosil-exfoliated graphite-TiO_2 composite floating catalyst: photodeposition of copper. *Chem. Mater.*, **9**, 2533–2540.
19. Cho, S. and Choi, W. (2001) Solid-phase photocatalytic degradation of PVC–TiO_2 polymer composites. *J. Photochem. Photobiol., A*, **143**, 221–228.
20. Haick, H. and Paz, Y. (2001) Remote photocatalytic activity as probed by measuring the degradation of self-assembled monolayers anchored near micro-domains of titanium dioxide. *J. Phys. Chem. B*, **105**, 3045–3051.
21. Lee, M.C. and Choi, W. (2002) Solid phase photocatalytic reaction on the soot/TiO_2 interface: the role of migrating OH radicals. *J. Phys. Chem. B*, **106**, 11818–11822.
22. Fu, X., Clark, L.A., Yang, Q., and Anderson, M.A. (1996) Enhanced photocatalytic performance of titania-based binary metal oxides: TiO_2/SiO_2 and TiO_2/ZrO_2. *Environ. Sci. Technol.*, **30**, 647–653.
23. Kambur, A., Pozan, G.S., and Boz, I. (2012) Preparation, characterization and photocatalytic activity of TiO_2–ZrO_2 binary oxide nanoparticles. *Appl. Catal. Environ.*, **115–116**, 149–158.
24. Herrmann, J.M., Tahiri, H., Ait-Ichou, Y., Lassaletta, G., Gonzalez–Elipe, A.R., and Fernandez, A. (1997) Characterization and photocatalytic activity in aqueous medium of TiO_2 and Ag-TiO_2 coatings on quartz. *Appl. Catal. Environ.*, **13**, 219–228.
25. Arpac, E., Sayılkan, F., Asilturk, M., Tatar, P., Kiraz, N., and Sayılkan, H. (2007) Photocatalytic performance of

Sn-doped and undoped TiO_2 nanostructured thin films under UV and vis-lights. *J. Hazard. Mater.*, **140**, 69–74.

26. Gracia, F., Holgado, J.P., Caballero, A., and Gonzalez-Elipe, A.R. (2004) Structural, optical, and photoelectrochemical properties of Mn^{+}-TiO_2 model thin film photocatalysts. *J. Phys. Chem. B*, **108**, 17466–17476.
27. Paz, Y., Luo, Z., Rabenberg, L., and Heller, A. (1995) Photo-oxidative self-cleaning transparent titanium dioxide films on glass. *J. Mater. Res.*, **10**, 2842–2848.
28. Kallio, T., Alajoki, S., Pore, V., Ritala, M., Laine, J., Leskel, M., and Stenius, P. (2006) Antifouling properties of TiO_2: photocatalytic decomposition and adhesion of fatty and rosin acids, sterols and lipophilic wood extractives. *Colloids Surf., A Physicochem. Eng. Asp.*, **291**, 162–176.
29. Turchi, C.S. and Ollis, D.F. (1990) Photocatalytic degradation of organic water contaminants: mechanisms involving hydroxyl radical attack. *J. Catal.*, **122**, 178–192.
30. Kopac, T. and Bozgeyik, K. (2010) Effect of surface area enhancement on the adsorption of bovine serum albumin onto titanium dioxide. *Colloids Surf. B Biointerfaces*, **76**, 265–271.
31. Kormann, C., Bahnemann, D.W., and Hoffmann, M.R. (1991) Photolysis of chloroform and other organic molecules in aqueous TiO_2 suspensions. *Environ. Sci. Technol.*, **25**, 494–500.
32. Park, H. and Choi, W. (2005) Photocatalytic reactivities of nafion-coated TiO_2 for the degradation of charged organic compounds under UV or visible light. *J. Phys. Chem. B*, **109**, 11667–11674.
33. Lee, J., Choi, W., and Yoon, J. (2005) Photocatalytic degradation of N-nitrosodimethylamine: mechanism, product distribution and TiO_2 surface modification. *Environ. Sci. Technol.*, **39**, 6800–6807.
34. Pan, J.H. and Lee, W.I. (2006) Preparation of highly ordered cubic mesoporous WO_3/TiO_2 films and their photocatalytic properties. *Chem. Mater.*, **18**, 847–853.
35. Sun, W., Zhang, S., Liu, Z., Wang, C., and Mao, Z. (2008) Studies on the enhanced photocatalytic hydrogen evolution over Pt/PEG-modified TiO_2 photocatalysts. *Int. J. Hydrogen Energy*, **33**, 1112–1117.
36. Goossens, A. and Maloney, E.L. (1998) Gas-phase synthesis of nanostructured anatase TiO_2. *Chem. Vap. Deposition*, **4**, 109–114.
37. Fonzo, F.D., Casari, C.S., Russo, V., Brunella, M.F., Bassi, A.L., and Bottani, C.E. (2009) Hierarchically organized nanostructured TiO_2 for photocatalysis applications. *Nanotechnology*, **20**, 015604–015610.
38. Macak, J.M., Sirotna, K., and Schmuki, P. (2005) Self-organized porous titanium oxide prepared in Na_2SO_4/NaF electrolytes. *Electrochim. Acta*, **50**, 3679–3684.
39. Bauer, S., Kleber, S., and Schmuki, P. (2006) TiO_2 Nanotubes: tailoring the geometry in H_3PO_4/HF electrolytes. *Electrochem. Commun.*, **8**, 1321–1325.
40. Baram, N., Starosvetsky, D., Starosvetsky, J., Epshtein, M., Armon, R., and Ein-Eli, Y. (2009) Enhanced inactivation of E. Coli bacteria using immobilized porous TiO_2 photoelectrocatalysis. *Electrochim. Acta*, **54**, 3381–3386.
41. Matos, J., Laine, J., and Herrmann, J.-M. (1998) Synergy effect in the photocatalytic degradation of phenol on a suspended mixture of titania and activated carbon. *Appl. Catal. Environ.*, **18**, 281–291.
42. Torimoto, T., Ito, S., Kuwabata, S., and Yoneyama, H. (1996) Effects of adsorbents used as supports for titanium dioxide loading on photocatalytic degradation of propyzamide. *Environ. Sci. Technol.*, **30**, 1275–1281.
43. Takeda, N., Torimoto, T., Sampath, S., Kuwabata, S., and Yoneyama, H. (1995) Effect of inert supports for titanium dioxide loading on enhancement of photodecomposition rate of gaseous propionaldehyde. *J. Phys. Chem.*, **99**, 9986–9991.

44. Sampath, S., Uchida, H., and Yoneyama, H. (1994) Photocatalytic degradation of gaseous pyridine over zeolite-supported titanium dioxide. *J. Catal.*, **149**, 189–194.
45. Takeda, N., Ohtani, M., Torimoto, T., Kuwabata, S., and Yoneyama, H. (1997) Evaluation of diffusibility of adsorbed propionaldehyde on titanium dioxide-loaded adsorbent photocatalyst films from its photodecomposition rate. *J. Phys. Chem. B*, **101**, 2644–2649.
46. Kedem, S., Schmidt, Y., Paz, Y., and Cohen, Y. (2005) Composite polymer nanofibers with carbon nanotubes and titanium dioxide particles. *Langmuir*, **21**, 5600–5604.
47. Kedem, S., Rozen, D., Cohen, Y., and Paz, Y. (2009) Enhanced stability effect in composite polymeric nanofibers containing titanium dioxide and carbon nanotubes. *J. Phys. Chem. C*, **113**, 14893–14899.
48. Zhao, D., Sheng, G., Chen, C., and Wang, X. (2012) Enhanced photocatalytic degradation of methylene blue under visible irradiation on graphene@TiO_2 dyade structure. *Appl. Catal. Environ.*, **111–112**, 303–308.
49. Yuan, R., Guan, R., Shen, W., and Zheng, J. (2005) Photocatalytic degradation of methylene blue by a combination of TiO_2 and activated carbon fibers. *J. Colloid Interface Sci.*, **282**, 87–91.
50. Sung-Suh, H.M., Choi, J.R., Hah, H.J., Koo, S.M., and Bae, Y.C. (2004) Comparison of Ag deposition effects on the photocatalytic activity of nanoparticulate TiO_2 under visible and UV light irradiation. *J. Photochem. Photobiol., A*, **163**, 37–44.
51. Kwon, Y.T., Song, K.Y., Lee, W.I., Choi, G.J., and Do, Y.R. (2000) Photocatalytic behavior of WO_3-loaded TiO_2 in an oxidation reaction. *J. Catal.*, **191**, 192–199.
52. Gerischer, H. and Heller, A. (1991) The role of oxygen in photooxidation of organic molecule on semiconductor particles. *J. Phys. Chem.*, **95**, 5261–5267.
53. Uchida, H., Katoh, S., and Watanabe, M. (1998) Photocatalytic degradation of trichlorobenzene using immobilized TiO_2 containing poly(tetra-uoroethylene) and platinum metal catalyst. *Electrochim. Acta*, **43**, 2111–2116.
54. Enríquez, R., Agrios, A.G., and Pichat, P. (2007) Probing multiple effects of TiO_2 sintering temperature on photocatalytic activity in water by use of a series of organic pollutant molecules. *Catal. Today*, **120**, 196–202.
55. Gao, R., Stark, J., Bahnemann, D.W., and Rabani, J. (2002) Quantum yields of hydroxyl radicals in illuminated TiO_2 nanocrystallite layers. *J. Photochem. Photobiol., A*, **148**, 387–391.
56. Wang, C.Y., Pagel, R., Dohrmann, J.K., and Bahnemann, D.W. (2006) Antenna mechanism and deaggregation concept: novel mechanistic principles for photocatalysis. *C.R. Chim.*, **9**, 761–773.
57. Kobayakawa, K., Nakazawa, Y., Ikeda, M., Sato, Y., and Fujishima, A. (1990) Influence of the density of surface hydroxyl groups on titania photocatalytic activities. *Ber. Bunsen Ges.*, **94**, 1439–1443.
58. Wang, J., Liu, X., Li, R., Qiao, P., Xiao, P., and Fan, J. (2011) TiO_2 Nanoparticles with increased surface hydroxyl groups and their improved photocatalytic activity. *Catal. Commun.* doi: 10.1016/j.catcom.2011.12.028
59. Paz, Y. and Heller, A. (1997) Photo-oxidatively self-cleaning transparent titanium dioxide films on soda lime glass: the deleterious effect of sodium contamination and its prevention. *J. Mater. Res.*, **12**, 2759–2766.
60. Vohra, M.S., Kim, S., and Choi, W. (2003) Effects of surface fluorination of TiO_2 on the photocatalytic degradation of tetramethylammonium. *J. Photochem. Photobiol., A*, **160**, 55.
61. Minero, C., Mariella, G., Maurino, V., and Pellizzetti, E. (2000) Photocatalytic transformation of organic compounds in the presence of inorganic anions. 1. Hydroxyl-mediated and direct electron-transfer reactions of phenol on a titanium dioxide-fluoride system. *Langmuir*, **16**, 2632–2641.

62. Park, H. and Choi, W. (2004) Effects of TiO_2 surface fluorination on photocatalytic reactions and photoelectrochemical behaviors. *J. Phys. Chem. B*, **108**, 4086–4093.
63. Paz, Y. (2006) Preferential photodegradation–Why and How? *C.R. Chim.*, **9**, 774–787.
64. Wang, K.H., Hsieh, Y.H., and Chen, L.J. (1998) The heterogeneous photocatalytic degradation, intermediates and mineralization for the aqueous solution of cresols and nitrophenols. *J. Hazard. Mater.*, **59**, 251–260.
65. Rodenas, L.A.G., Weisz, A.D., Magaz, G.E., and Blesa, M.A. (2000) Effect of light on the electrokinetic behavior of TiO_2 particles in contact with Cr(VI) aqueous solutions. *J. Colloid Interface Sci.*, **230**, 181–185.
66. Wang, X., Pehkonen, S.O., and Ray, A.K. (2004) Photocatalytic reduction of Hg(II) on two commercial TiO_2 catalysts. *Electrochim. Acta*, **49**, 1435–1444.
67. Robert, D., Piscopo, A., and Weber, J.-V. (2004) First approach of the selective treatment of water by heterogeneous photocatalysis. *Environ. Chem. Lett.*, **2**, 5–8.
68. Duffy, J.E., Anderson, M.A., Hill, C.G. Jr.,, and Zeltner, W.A. (2000) Photocatalytic oxidation as a secondary treatment method following Wet Air oxidation. *Ind. Eng. Chem. Res.*, **39**, 3698–3706.
69. Jaffrezic-Renault, N., Pichat, P., Foissy, A., and Mercier, R. (1986) Study of the effect of deposited Pt particles on the surface charge of TiO_2 aqueous suspensions by potentiometry, electrophoresis and labelled ion adsorption. *J. Phys. Chem.*, **90**, 2733–2738.
70. Theurich, J., Lindner, M., and Bahnemann, D.W. (1996) Photocatalytic degradation of 4-chlorophenol in aerated aqueous titanium dioxide suspensions: a kinetic and mechanistic study. *Langmuir*, **12**, 6368–6376.
71. Abdullah, M., Low, G.K.-C., and Matthews, R.W. (1990) Effects of common inorganic anions on rates of photocatalytic oxidation of organic carbon over illuminated titanium dioxide. *J. Phys. Chem.*, **94**, 6820–6825.
72. Agrios, A.G., Gray, K.A., and Weitz, E. (2004) Narrow-band irradiation of a homologous series of chlorophenols on TiO_2: charge-transfer complex formation and reactivity. *Langmuir*, **20**, 5911–5917.
73. Muneer, M., Singh, H.K., and Bahnemann, D. (2002) Semiconductor-mediated photocatalysed degradation of two selected priority organic pollutants, benzidine and 1,2-diphenylhydrazine, in aqueous suspension. *Chemosphere*, **49**, 193–203.
74. Li, Y., Lu, G., and Li, S. (2003) Photocatalytic production of hydrogen in single component and mixture systems of electron donors and monitoring adsorption of donors by in situ infrared spectroscopy. *Chemosphere*, **52**, 843–850.
75. Muggli, D.S. and Ding, L. (2001) Photocatalytic performance of sulfated TiO_2 and Degussa P-25 TiO_2 during oxidation of organics. *Appl. Catal. Environ.*, **32**, 181–194.
76. Inumaru, K., Murashima, M., Kasahara, T., and Yamanaka, S. (2004) Enhanced photocatalytic decomposition of 4-nonylphenol by surface-organografted TiO_2: a combination of molecular selective adsorption and photocatalysis. *Appl. Catal. Environ.*, **52**, 275–280.
77. Matthews, L.R., Avnir, D., Modestov, A.D., Sampath, S., and Lev, O. (1997) The incorporation of titania into modified silicates for solar photodegradation of aqueous species. *J. Sol-Gel Sci. Technol.*, **8**, 619–623.
78. Hidaka, H., Nohara, K., Zhao, J., Serpone, N., and Pelizzetti, E. (1992) Photo-oxidative degradation of the pesticide permethrin catalysed by irradiated TiO_2 semiconductor slurries in aqueous media. *J. Photochem. Photobiol., A*, **64**, 247–254.
79. Rajh, T., Ostafin, A., Micic, O.I., Tiede, D.M., and Thurnauer, M.C. (1996) Surface modification of small particle TiO_2 colloids with cysteine for enhanced photochemical reduction: an EPR study. *J. Phys. Chem.*, **100**, 4538–4545.
80. Makarova, O.V., Rajh, T., Thurnauer, M.C., Martin, A., Kemme, P.A., and

Cropek, D. (2000) Surface modification of TiO_2 nanoparticles for photochemical reduction of nitrobenzene. *Environ. Sci. Technol.*, **34**, 4797–4803.

81. Tsumura, T., Kojitani, N., Izumi, I., Iwashita, N., Toyoda, M., and Inagaki, M. (2002) Carbon coating of anatase-type TiO_2 and photoactivity. *J. Mater. Chem.*, **12**, 1391–1396.
82. Wang, C.-M., Heller, A., and Gerischer, H. (1992) Palladium catalysis of O_2 reduction by electrons accumulated on TiO_2 particles during photoassisted oxidation of organic compounds. *J. Am. Chem. Soc.*, **114**, 5230–5234.
83. Fox, M.A., Ogawa, H., and Pichat, P. (1989) Regioselectivity in the semiconductor-mediated photooxidation of 1,4-pentadiol. *J. Org. Chem.*, **54**, 3847–3852.
84. Chen, J., Ollis, D.F., Rulkens, W.H., and Bruning, H. (1999) Photocatalyzed oxidation of alcohols and organochlorides in the presence of native TiO_2 and metallized TiO_2 suspensions. Part (i): photocatalytic activity and pH influence. *Water Res.*, **33**, 661–668.
85. Lindner, M., Theurich, J., and Bahnemann, D.W. (1997) Photocatalytic degradation of organic compounds: accelerating the process efficiency. *Water Sci. Technol.*, **35**, 79–86.
86. Tada, H., Teranishi, K., Inubushi, Y., and Ito, S. (2000) Ag nanocluster loading effect on TiO_2 photocatalytic reduction of Bis(2-dipyridyl)disulfide to 2-mercaptopyridine by H_2O. *Langmuir*, **16**, 3304–3309.
87. Choi, W., Termin, A., and Hoffmann, M.R. (1994) Effects of metal-Ion dopants on the photocatalytic reactivity of quantum-sized TiO_2 particles. *Angew. Chem. Int. Ed.*, **33**, 1091–1092.
88. Muradov, N.Z., Raiss, A.T., Muzzey, D., Painter, C.R., and Kemme, M.R. (1996) Selective photocatalytic destruction of airborne VOCs. *Sol. Energy*, **56**, 445–453.
89. Matos, J., Laine, J., and Herrmann, J.-M. (2001) Effect of the type of activated carbons on the photocatalytic degradation of aqueous organic pollutants by UV-irradiated titania. *J. Catal.*, **200**, 10–20.
90. Ghosh-Mukerji, S., Haick, H., Schvartzman, M., and Paz, Y. (2001) Selective photocatalysis by means of molecular recognition. *J. Am. Chem. Soc.*, **123**, 10776–10777.
91. Zemel, E., Haick, H., and Paz, Y. (2002) Photocatalytic destruction of organized organic monolayers chemisorbed at the vicinity of titanium dioxide surfaces. *J. Adv. Oxid. Technol.*, **5**, 27–32.
92. Ghosh-Mukerji, S., Haick, H., and Paz, Y. (2003) Controlled mass transport as a means for obtaining selective photocatalysis. *J. Photochem. Photobiol., A*, **160**, 77–85.
93. Pang, L. and Whitehead, M.A. (1992) Atom-atom potential analysis of the complexing characteristics of cyclodextrins (host) with benzene and p-dihalobenzenes (guest). *Supramol. Chem.*, **1**, 81–92.
94. Lee, J.Y. and Park, S. (1998) Electrochemistry of guest molecules in thiolated cyclodextrin self-assembled monolayers: an implication for size-selective sensors. *J. Phys. Chem. B*, **102**, 9940–9945.
95. Sagatelian, Y., Sharabi, D., and Paz, Y. (2005) Enhanced photodegradation of diisopropyl methyl phosphonate by the "adsorb & shuttle" approach. *J. Photochem. Photobiol., A*, **174**, 253–260.
96. Llabres, F.X., Xamena, I., Calza, P., Lamberti, C., Prestipino, C., Damin, A., Bordiga, S., Pelizzetti, E., and Zecchina, A. (2003) Enhancement of the ETS-10 titanosilicate activity in the shape-selective photocatalytic degradation of large aromatic molecules by controlled defect production. *J. Am. Chem. Soc.*, **125**, 2264–2271.
97. Yoneyama, H., Haga, S., and Yamanaka, S. (1989) Photocatalytic activities of microcrystalline TiO_2 incorporated in sheet silicates of clay. *J. Phys. Chem.*, **93**, 4833–4837.
98. Aronson, B.J., Blanford, C.F., and Stein, A. (1997) Solution-phase grafting of titanium dioxide onto the pore surface of mesoporous silicates: synthesis and structural characterization. *Chem. Mater.*, **9**, 2842–2851.

99. Pichat, P., Khalaf, H., Tabet, D., Houari, M., and Saidi, M. (2005) Ti-montmorillonite as photocatalyst to remove 4-chlorophenol in water and methanol in air. *Environ. Chem. Lett.*, **2**, 191–194.
100. Lee, S.W., Ichinose, I., and Kunitake, T. (1998) Molecular imprinting of azobenzene carboxylic acid on a TiO_2 ultrathin film by the surface sol-gel process. *Langmuir*, **14**, 2857–2863.
101. Shen, X., Zhu, L., Yu, H., Tang, H., Liu, S., and Li, W. (2009) Selective photocatalysis on molecular imprinted TiO_2 thin films prepared via an improved liquid deposition method. *New J. Chem.*, **33**, 1673–1679.
102. Shen, X., Zhu, L., Huang, C., Tang, H., Yu, Z., and Deng, F. (2009) Inorganic molecular imprinted titanium dioxide photocatalyst: synthesis, characterization and its application for efficient and selective degradation of phthalate esters. *J. Mater. Chem.*, **19**, 4843–4851.
103. Shen, X., Zhu, L., Liu, G., Tang, H., Liu, S., and Li, W. (2009) Photocatalytic removal of pentachlorophenol by means of an enzyme-like molecular imprinted photocatalyst and inhibition of the generation of highly toxic intermediates. *New J. Chem.*, **33**, 2278–2285.
104. Nussbaum M., Paz, Y. (2012) Environmental approaches by molecular imprinting on titanium dioxide in *Handbook of Molecular Imprinting: Advanced Sensor Applications* (eds S.W. Lee and T. Kunitake, Pan Stanford Publishing) ISBN: 10: 9814316652, ISBN: 13: 978-9814316651.
105. Shen, X., Zhu, L., Liu, G., Yu, H., and Tang, H. (2008) Enhanced photocatalytic degradation and selective removal of nitrophenols by using surface molecular imprinted titania. *Environ. Sci. Technol.*, **42**, 1687–1692.
106. Shen, X., Zhu, L., Li, J., and Tang, H. (2007) Synthesis of molecular imprinted polymer coated photocatalysts with high selectivity. *Chem. Commun.*, 1163–1165.
107. Nussbaum, M. and Paz, Y. (2012) Ultra-thin SiO_2 layers on TiO_2: improved photocatalysis by enhancing products' desorption. *Phys. Chem. Chem. Phys.*, **14**, 3392–3399.
108. Nimlos, M.R., Jacoby, W.A., Blake, D.M., and Milne, T.A. (1993) Direct mass spectrometric studies of the destruction of hazardous wastes. 2. Gas-phase photocatalytic oxidation of trichloroethylene over TiO_2: products and mechanisms. *Environ. Sci. Technol.*, **27**, 732–740.
109. Minero, C., Maurino, V., and Pelizzetti, E. (1997) Heterogeneous catalytic transformation of s-triazine derivatives. *Res. Chem. Intermed.*, **23**, 291–310.
110. Peller, J., Wiest, O., and Kamat, P.V. (2003) Synergy of combining sonolysis and photocatalysis in the degradation and mineralization of chlorinated aromatic compounds. *Environ. Sci. Technol.*, **37**, 1926–1932.
111. Theron, P., Pichat, P., Petrier, C., and Guillard, C. (2001) Water treatment by TiO_2 photocatalysis and/or ultrasound: degradation of phenyltrifluoromethylketone, a trifluoroacetic acid-forming pollutant, and octan-1-ol, a very hydrophobic pollutant. *Water Sci. Technol.*, **44**, 263–270.
112. Molinari, A., Amadelli, R., Antolini, L., Maldotti, A., Battioni, P., and Mansuy, D. (2000) Phororedox and photocatalytic processes on Fe(III)–porphyrin surface modified nanocrystalline TiO_2. *J. Mol. Catal. A: Chem.*, **158**, 521–531.
113. Iliev, V., Tomova, D., Bilyarska, L., Eliyas, A., and Petrov, L. (2006) Photocatalytic properties of TiO_2 modified with platinum and silver nanoparticles in the degradation of oxalic acid in aqueous solution. *Appl. Catal. Environ.*, **63**, 266–271.
114. Lee, J. and Choi, W. (2004) Effect of platinum deposits on TiO_2 on the anoxic photocatalytic degradation pathways of alkylamines in water: dealkylation and N-alkylation. *Environ. Sci. Technol*, **38**, 4026–4033.
115. Tan, T.T.Y., Yip, C.K., Beydoun, D., and Amal, R. (2003) Effects of nano-Ag particles loading on TiO_2 photocatalytic

reduction of selenate ions. *Chem. Eng. J.*, **95**, 179–186.

116. Kominami, H., Nishimune, H., Ohta, Y., Arakawa, Y., and Inaba, T. (2012) Photocatalytic hydrogen formation from ammonia and methyl amine in an aqueous suspension of metal-loaded titanium(IV) oxide particles. *Appl. Catal. Environ.*, **111–112**, 297–302.

117. Ranjit, K.T., Willner, I., Bossmann, S.H., and Braun, A.M. (2001) Lanthanide oxide-doped titanium dioxide photocatalysts: novel photocatalysts for the enhanced degradation of p-chlorophenoxyacetic acid. *Environ. Sci. Technol.*, **35**, 1544–1549.

118. Zhang, J., Hu, Y., Matsuoka, M., Yamashita, H., Minagawa, M., Hidaka, H., and Anpo, M. (2001) Relationship between the local structures of titanium oxide photocatalysts and their reactivities in the decomposition of NO. *J. Phys. Chem. B*, **105**, 8395–8398.

119. Shimizu, K.-I., Kaneko, T., Fujishima, T., Kodama, T., Yoshida, H., and Kitayama, Y. (2002) Selective oxidation of liquid hydrocarbons over photoirradiated TiO_2 pillared clays. *Appl. Catal. Gen.*, **225**, 185–191.

120. Li, C.-J., Xua, G.-R., Zhanga, B., and Gonga, J.R. (2012) High selectivity in visible-light-driven partial photocatalytic oxidation of benzyl alcohol into benzaldehyde over single-crystalline rutile TiO_2 nanorods. *Appl. Catal. Environ.*, **115–116**, 201–208.

121. Mendez-Roman, R. and Cardona-Martinez, N. (1998) Relationship between the formation of surface species and catalyst deactivation during the gas-phase photocatalytic oxidation of toluene. *Catal. Today*, **40**, 353–365.

122. Hay, S.O., Obee, T.N., and Thibaud-Erkey, C. (2010) The deactivation of photocatalytic based air purifiers by ambient siloxanes. *Appl. Catal. Environ.*, **99**, 435–441.

123. Bollinger, M.A. and Vannice, M.A. (1996) A kinetic and DRIFTS study of low-temperature carbon monoxide oxidation over Au-TiO_2, catalysts. *Appl. Catal. Environ.*, **8**, 417–443.

124. Lee, S.L., Scott, J., Chiang, K., and Amal, R. (2009) Nanosized metal deposits on titanium dioxide for augmenting gas-phase toluene photooxidation. *J. Nanopart. Res.*, **11**, 209–219.

125. Uner, D., Tapan, N.A., Ozen, I., and Uner, M. (2003) Oxygen adsorption on Pt/TiO_2 catalysts. *Appl. Catal. Gen.*, **251**, 225–234.

126. Dolat, D., Quici, N., Kusiak-Nejman, E., Morawski, A.W., and Li, P.G. (2012) One-step, hydrothermal synthesis of nitrogen, carbon co-doped titanium dioxide (N,C TiO_2) photocatalysts. Effect of alcohol degree and chain length as carbon dopant precursors on photocatalytic activity and catalyst deactivation. *Appl. Catal. Environ.*, **115–116**, 81–89.

127. Thompson, T.L. and Yates, J.T. Jr., (2006) Surface science studies of the photoactivation of TiO_2s new photochemical processes. *Chem. Rev.*, **106**, 4428–4453.

128. Kumar, S.G. and Devi, L.G. (2011) Review on modified TiO_2 photocatalysis under UV/visible light: selected results and related mechanisms on interfacial charge carrier transfer dynamics. *J. Phys. Chem. A*, **115**, 13211–13241.

129. Wang, C.-Y., Bottcher, C., Bahnemann, D.W., and Dohrmann, J.K. (2003) A comparative study of nanometer sized Fe(III)-doped TiO_2 photocatalysts: synthesis, characterization and activity. *J. Mater. Chem.*, **13**, 2322–2329.

130. Pichat, P., Herrmann, J.M., Disdier, J., Mozzanega, M.N., and Courbon, H. (1984) Modification of the anatase electron density by ion doping or metal deposit and consequences for photoassisted reactions. *Stud. Surf. Sci. Catal.*, **19**, 319–326.

131. Wold, A. (1993) Photocatalytic properties of TiO_2. *Chem. Mater.*, **5**, 280–283.

132. Sclafani, A. and Herrmann, J.-M. (1998) Influence of metallic silver and of platinum-silver bimetallic deposits on the photocatalytic activity of titania (anatase and rutile) in organic and aqueous media. *J. Photochem. Photobiol., A*, **113**, 181–188.

133. Yamakata, A., Ishibashi, T., and Onishi, H.G. (2001) Water and oxygen-induced kinetics of photogenerated electrons in TiO_2 and Pt/TiO_2: a time-resolved infrared absorption study. *J. Phys. Chem. B*, **105**, 7258–7262.
134. Furube, A., Asahi, T., Masuhara, H., Yamashita, H., and Anpo, M. (2001) Direct observation of a picosecond charge separation process in photoexcited platinum-loaded TiO_2 particles by femtosecond diffuse reflectance spectroscopy. *Chem. Phys. Lett.*, **336**, 424–430.
135. Park, H. and Choi, W. (2003) Photoelectrochemical ivestigation on electron transfer mediating behaviors of polyoxometalate in UV-illuminated suspensions of TiO_2 and Pt/TiO_2. *J. Phys. Chem. B*, **107**, 3885–3890.
136. Bamwenda, G.R., Tsubota, S., Nakamura, T., and Haruta, M. (1995) Photoassisted hydrogen production from a water-ethanol solution: a comparison of activities of Au-TiO_2 and Pt-TiO_2. *J. Photochem. Photobiol., A*, **89**, 177–189.
137. Hirakawa, T. and Kamat, P.V. (2004) Photoinduced electron storage and surface Plasmon modulation in Ag@TiO_2 clusters. *Langmuir*, **20**(14), 5645–5647.
138. Hirakawa, T. and Kamat, P.V. (2005) Charge separation and catalytic activity of Ag@TiO_2 core-shell composite clusters under UV-irradiation. *J. Am. Chem. Soc.*, **127**, 3928–3934.
139. Yu, J., Xiong, J., Cheng, B., and Liu, S. (2005) Fabrication and characterization of Ag–TiO_2 multiphase nanocomposite thin films with enhanced photocatalytic activity. *Appl. Catal. Environ.*, **60**, 211–221.
140. Shaham Waldmann, N. and Paz, Y. (2010) Photocatalytic reduction of Cr(VI) by titanium dioxide coupled to functionalized CNTs: an example of counter – productive charge separation. *J. Phys. Chem. C*, **114**, 18946–18952.
141. Serpone, N., Maruthamuthu, P., Pichat, P., Pelizzetti, E., and Hidaka, H. (1995) Exploiting the interparticle electron transfer process in the photocatalysed oxidation of phenol, 2-chlorophenol and pentachlorophenol: chemical evidence for electron and hole transfer between coupled semiconductors. *J. Photochem. Photobiol., A*, **85**, 247–255.
142. Wu, L., Yu, J.C., and Fu, X. (2006) Characterization and photocatalytic mechanism of nanosized CdS coupled TiO_2 nanocrystals under visible light irradiation. *J. Mol. Catal. A: Chem.*, **244**, 25–32.
143. Miyauchi, M., Nakajima, A., Watanabe, T., and Hashimoto, K. (2002) Photoinduced hydrophilic conversion of TiO_2/WO_3 layered thin films. *Chem. Mater.*, **14**, 4714–4720.
144. Vinodgopal, K. and Kamat, P.V. (1995) Enhanced rates of photocatalytic degradation of an azo dye using SnO_2/TiO_2 coupled semiconductor thin films. *Environ. Sci. Technol.*, **29**, 841–845.
145. Bai, H., Liu, Z., and Sun, D.D. (2012) A hierarchically structured and multifunctional membrane for water treatment. *Appl. Catal. Environ.*, **111–112**, 571–577.
146. Nolan, M. (2011) Electronic coupling in iron oxide-modified TiO_2 leads to a reduced band gap and charge separation for visible light active photocatalysis. *Phys. Chem. Chem. Phys.*, **13**, 18194–18199.
147. Higashimoto, S., Sakiyama, M., and Azuma, M. (2006) Photoelectrochemical properties of hybrid WO_3/TiO_2 electrode. Effect of structures of WO_3 on charge separation behavior. *Thin Solid Films*, **503**, 201–206.
148. Papp, J., Soled, S., Dwight, K., and Wold, A. (1994) Surface acidify and photocatalytic activity of TiO_2, WO_3/TiO_2, and MoO_3/TiO_2 photocatalysts. *Chem. Mater.*, **6**, 496–500.
149. Keller, V., Bernhardt, P., and Garin, F. (2003) Photocatalytic oxidation of butyl acetate in vapor phase on TiO_2, Pt/TiO_2 and WO_3/TiO_2 catalysts. *J. Catal.*, **15**, 129–138.
150. Vinodgopal, K., Bedja, I., and Kamat, P.V. (1996) Nanostructured semiconductor films for photocatalysis. Photoelectrochemical behavior of SnO_2/TiO_2 composite systems and its role in photocatalytic degradation

of a textile azo dye. *Chem. Mater.*, **8**, 2180–2187.
151. Hattori, A., Tokihisa, Y., Tada, H., and Ito, S. (2000) Acceleration of oxidations and retardation of reductions in photocatalysis of a TiO_2/SnO_2 bilayer-type catalyst. *J. Electrochem. Soc.*, **147**, 2279–2283.
152. Yu, J., Yu, H., Ao, C.H., Lee, S.C., Yu, J.C., and Ho, W. (2006) Preparation, characterization and photocatalytic activity of in situ Fe-doped TiO_2 thin films. *Thin Solid Films*, **496**, 273–280.
153. Sakthivel, S., Janczarek, M., and Kisch, H. (2004) Visible light activity and photoelectrochemical properties of nitrogen-doped TiO_2. *J. Phys. Chem. B*, **108**, 19384–19387.
154. Pichat, P. (1987) Surface properties, activity and selectivity of bifunctional powder photocatalysts, **11**, 135–140.
155. Dvoranova, D., Brezová, V., Mazúr, M., and Malati, M.A. (2002) Investigations of metal-doped titanium dioxide photocatalysts. *Appl. Catal. Environ.*, **37**, 91–105.
156. Bahnemann, D.W., Hilgendorff, M., and Memming, R. (1997) Charge carrier dynamics at TiO_2 particles: reactivity of free and trapped holes. *J. Phys. Chem. B*, **101**, 4265–4275.
157. Li, X.Z. and Li, F.B. (2001) Study of Au/Au^{3+}-TiO_2 photocatalysts toward visible photooxidation for water and wastewater treatment. *Environ. Sci. Technol.*, **35**, 2381–2387.
158. Sonawane, R.S. and Dongare, M.K. (2006) Sol–gel synthesis of Au/TiO_2 thin films for photocatalytic degradation of phenol in sunlight. *J. Mol. Catal. A: Chem.*, **243**, 68–76.
159. Kumar, A. and Mathur, N. (2004) Photocatalytic oxidation of aniline using Ag+-loaded TiO_2 suspensions. *Appl. Catal. Gen.*, **275**, 189–197.
160. Zhang, Z.B., Wang, C.C., Zakaria, R., and Ying, J.Y. (1998) Role of particle size in nanocrystalline TiO_2-based photocatalysts. *J. Phys. Chem. B*, **102**, 10871–10878.
161. Valentin, C.D., Pacchioni, G., Selloni, A., Livraghi, S., and Giamello, E. (2005) Characterization of paramagnetic species in N-doped TiO_2 powders by EPR spectroscopy and DFT calculations. *J. Phys. Chem. B*, **109**, 11414–11419.
162. Mrowetz, M., Balcerski, W., Colussi, A.J., and Hoffmann, M.R. (2004) Oxidative power of nitrogen-doped TiO_2 photocatalysts under visible illumination. *J. Phys. Chem. B*, **108**, 17269–17273.
163. Nakamura, R., Tanaka, T., and Nakato, Y. (2004) Mechanism for visible light responses in anodic photocurrents at N-doped TiO_2 film electrodes. *J. Phys. Chem. B*, **108**, 10617–10620.
164. Tatsuma, T., Saitoh, S., Ohko, Y., and Fujishima, A. (2001) TiO_2–WO_3 Photoelectrochemical anticorrosion system with an energy storage ability. *Chem. Mater.*, **13**, 2838–2842.
165. Tatsuma, T., Saitoh, S., Ohko, Y., and Fujishima, A. (2002) Energy storage of TiO_2–WO_3 photocatalysis systems in the gas phase. *Langmuir*, **18**, 7777–7779.
166. Yasomanee, J.P. and Bandara, J. (2008) Multi-electron storage of photoenergy using Cu_2O–TiO_2 thin film photocatalyst. *Sol. Energy Mater. Sol. Cells*, **92**, 348–352.
167. Subasri, R. and Shinohara, T. (2003) Investigations on SnO_2–TiO_2 composite photoelectrodes for corrosion protection. *Electrochem. Commun.*, **5**, 897–902.
168. Takahashi, Y., Ngaotrakanwiwat, P., and Tatsuma, T. (2004) Energy storage TiO_2–MoO_3 photocatalysts. *Electrochim. Acta*, **49**, 2025–2029.
169. Park, H., Baka, A., Jeonb, T.H., Kimc, S., and Choi, W. (2012) Photochargeable and dischargeable TiO_2 and WO_3 heterojunction electrodes. *Appl. Catal. Environ.*, **115–116**, 74–80.
170. Kim, S. and Choi, W. (2002) Dual photocatalytic pathways of trichloroacetate degradation on TiO_2: effects of nanosized platinum deposits on kinetics and mechanism. *J. Phys. Chem. B*, **106**, 13311–13317.

6
Immobilization of a Semiconductor Photocatalyst on Solid Supports: Methods, Materials, and Applications

Didier Robert, Valérie Keller, and Nicolas Keller

6.1
Introduction

Achieving a successful commercial implementation of the photocatalysis technology will require being able to operate the photocatalyst in a manner consistent with the criteria and constraints of industrial implementation, whether it be applications in liquid phase or gas phase. Practically, powdery photocatalyst can be used suspended or fixed on a solid support. TiO_2 suspensions have better efficiency than immobilized catalysts [1–4], as illustrated in Figure 6.1, for example, in the case of TiO_2 fixed on macroscopic β-SiC cellular foam by sol–gel method for the treatment of water-containing pesticide molecules [4].

However, in the case of liquid-phase applications, industrial-scale production of suspended, finely powdered TiO_2 for wastewater treatment would raise problems, such as separating the catalyst from treated water and recycling or regenerating it. Because of the small size of the TiO_2 particles (less than 0.5 μm for agglomerates and generally 10–100 nm for elementary particles), complete separation is very difficult and expensive, and the process is therefore not viable on large scales. Indeed, in the case of solar photocatalysis, the benefit obtained from free solar energy does not compensate for the expenses caused by the need to filter out the oxide particles. Decantation could be a solution but it needs huge tanks to stock the contaminated and treated water. On the other hand, this process is very long because the TiO_2 particles decant very slowly. The other advantage is that a supported catalyst is easier to recycle.

In the case of photocatalytic applications in gas phase, since fluidized-bed remains are difficult to operate and implement on a fully controlled way, it is essential to use supported photocatalyst for fixing the catalyst and thus enabling processing the system onstream, without the catalyst being driven out of the reactor with the gas flow [5, 6]. The problem can be solved by fixing the semiconductor photocatalyst on an inert support, because it could help to eliminate the costly phase separation process. Over the recent years, much work has been done in this area. This has led to the use of a variety of supports [2–4, 7–12] (silica gel, quartz optical fibers, glass fiber, glass beads, SiC foam, ceramics,

Photocatalysis and Water Purification: From Fundamentals to Recent Applications, First Edition. P. Pichat.

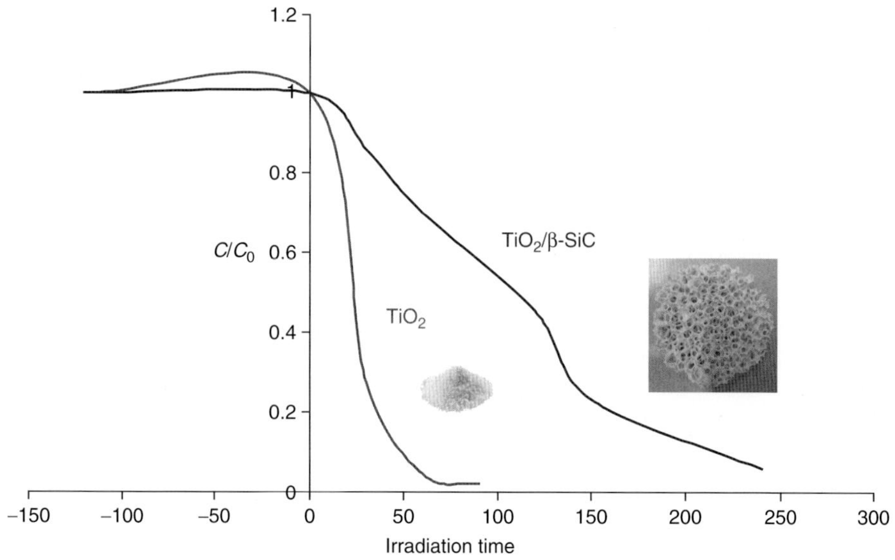

Figure 6.1 Comparison of TiO_2 sol–gel powder suspensions and TiO_2 supported on β-SiC foam as photocatalytic systems (with same loading in TiO_2) for the elimination of Diuron in water.

cellulose membranes, polymer films, etc.). The other area of development has been related to the methods for depositing catalyst on the support, on which Shan *et al.* [13] have recently published a short review on immobilization of TiO_2 onto supporting materials. According to these authors, a good support must be generally transparent to UV light, but in practice, this is rarely the case; the physicochemical bonds between TiO_2 and the surface must be strong without any negative effect on the degrading properties; the support must have a high specific surface; the support must have a good capacity to adsorb the organic compounds to be degraded and a physical configuration favoring the liquid–solid separation; and the support must be chemically inert. The first essential quality for a good support is that the support–catalyst junction should resist strain derived from particle to particle and particle–fluid mechanical interactions in the reactor environment in order to avoid detachment of catalyst particles from the support.

The design of photocatalytic supported materials is a key aspect in the development of highly efficient photocatalytic processes and reactors, which should take into account and optimize the interaction between the light, the active catalyst, and the reactants, for maximizing both flow/exposed area contact and utilization of radiated energy. Therefore, the photocatalyst form and the photocatalytic reactor engineering are linked.

By supports, we should consider here materials that

1) in liquid-phase applications solve the strongly restrictive problem of the costly and time-consuming powdery photocatalyst filtration

2) in gas-phase applications allow photocatalytic reactors to operate differently than that in a wall-coated mode by increasing the active photocatalyst content per reactor volume unit.

In some cases, the support not only provides its macroscopic structure to the photocatalytic materials but also plays an active role within the reactor, for example, in terms of flow distribution by static mixer phenomena. Therefore, this has ruled out the thematic scope, different kinds of powdery support materials that are often associated to the concept of "supported photocatalysts." Thus, various support materials such as activated charcoal, zeolithe, and mesoporous silica, as well as one-dimensional supports at the nanometric scale, such as carbon nanotubes/nanofibers, or at the micrometric scale, such as silica or quartz fibers, are not considered here. They are used in photocatalysis by analogy to the thermal catalysis field, for dispersing supported active phases and increasing their exposed reactive surface, as well as for providing enhanced mass transfer or adsorption properties. However, unless they can be macronized into macroscopic bodies, their usual use in a powdery form does not allow them in any way to solve the above-reported operation restrictions for liquid- and gas-phase processes.

In some laboratory studies, the photocatalyst is immobilized by direct deposition on the walls of the photoreactor, which are often made of Pyrex or quartz, but this way of immobilizing the photocatalyst is out of the scope of this review that is devoted to solid supports. Finally, one of the aspects not discussed in this chapter concerns self-cleaning materials including self-cleaning glass, which are commercialized by companies such as Pilkington, PPG industries, and Saint-Gobain. In the case of self-cleaning materials, superhydrophilic properties intervene in addition to photocatalytic properties.

In this chapter, we first present the main techniques used for immobilizing photocatalyst semiconductors on different solid supports. The use of a previously prepared catalyst (for example, in the case of commercial photocatalysts such as P25 from Degussa [14]) is the simplest method to provide support coating. Generally, it is necessary to carry out a thermal treatment for fixing the catalyst. Other chemical or physical methods used on a laboratory scale, such as sol–gel process [4], chemical vapor deposition (CVD) [10], and electrophoretic deposition [15], help to generate the catalyst *in situ* on the support. The main support materials used for immobilizing photocatalysts (packed-bed materials, cellular monoliths, and optical fibers) and providing its macroscopic structure to the photocatalysts is described, and finally, laboratory and industrial applications of supported photocatalysts are reported, as well as the main commercially available photocatalytic products.

6.2
Immobilization Techniques

In photocatalysis, the way of immobilizing TiO_2 particles onto substrates remained a research field less investigated at the laboratory scale than the development of new photocatalytic materials or the determination of kinetic rate laws toward specific chemical targets. However, this aspect is of great interest when targeting the implementation of the photocatalysis technology and searching for economically viable processes. One could refer to a recent patents' overview published by Paz [16], and reporting on numbers of patents applied to protect specific TiO_2/substrate systems as well as the corresponding methods of producing them. Several strategic approaches could be categorized, for example, using directly crystalline TiO_2 powders or TiO_2 precursors, binderless or binder-through anchorage, depending on the substrate chemical nature. Besides the usual loss of activity due to the physical immobilization of the photocatalyst, the immobilized photocatalyst should be able to intrinsically maintain as much as possible the activity of the powdery materials.

The binder-through approach addresses two aspects of the immobilization of TiO_2: (i) the binder directly fixes the photocatalyst onto the substrate and (ii) in case of a fragile organic support, it can prevent the photocatalytic attack of the support by physically isolating it from the active TiO_2. Therefore, the use of fully inorganic binders should be preferred to prevent any possible long- or even middle-term destroying of the photocatalytic materials, such as in the case of the commercial photocatalytic paper developed by the Ahlstrom Company, using a colloidal SiO_2 binder (see the adequate Section 6.4). In addition, polymers and organosilane polymers containing organic functional groups [17] can also be used. The interest in using oxygen-containing functional groups is due to the direct formation of TiO_2–polymer bindings during the preparation steps of TiO_2 particles on or within the films [18]. Josset *et al.* [19] proposed the use of a mixed system, made of a monodispersed silica sol modified with bifunctional *epoxy* $Y{-}(CH_2)_n{-}Si(OH)_3$ organosilanetriol molecules. The high concentration of active silanol functions allowed chemical bonding with the TiO_2 surface and a high degree of cross-linking by formation of 2D and 3D siloxane networks. During curing, functionalized SiO_2 sol particles align into a dense packed structure and are covalently incorporated into the siloxane network for providing hardness and scratch resistance to the composite. The formation of the polysiloxane protective film occurred through condensation steps between $-Si(OH)_3$ functional groups and by reacting further with hydroxyl surface groups of TiO_2 particles. The organic function of this bifunctional organosilanetriol molecule, that is, the Si-bonded *Y epoxy* groups, is responsible for the grafting of the film onto the organic surface of the substrate. The main drawback of the binder-through approach remains linked to the strong reduction in terms of TiO_2 surface available for adsorption and reaction, due to partial or even total TiO_2 encapsulation in the binder coating. In addition, one needs to keep in mind that noninorganic centers could be possibly subject to an attack by photogenerated oxidative species, participating to the long-term instability of the materials.

The binderless approach has been investigated using many routes and methods, including thermal treatments, sol–gel methods, CVD methods, direct air hydrolysis, high-temperature impregnation techniques, spray pyrolysis, electrophoretic deposition, the so-called layer-by-layer (LbL) – (multilayer)-approach.

The simplest immobilizing method remains to perform a *thermal treatment* after the prior deposition of the photocatalyst. By contrast to many other methods, it directly uses crystalline TiO_2 powder and is thus very convenient to implement. The photocatalyst is mainly deposited first on the support surface by impregnation of the photocatalyst suspension at an optimized concentration. Two key aspects should be kept in mind, which are also the limits of the method. First, the temperature of the treatment should be suitable in regard to the chemical and thermal stability of the support. This rules out fragile supports such as polymers, suffering from a low-temperature upper limit of resistance, or limits the use of carbon materials that, even if they can sustain higher temperatures, are however usually oxidized under air at 500–600 °C. Second, the temperature should not affect the activity – unless with a positive impact – of the immobilized photocatalyst by modifying bulk and/or surface physicochemical properties. For instance, the photoactivity of TiO_2 P25 has been reported to increase with the increase in the thermal treatment temperature under vacuum until 400 °C [14]. The increase in photoactivity was attributed to the increase in the adsorbed oxygen uptake, resulting in the generation of more superoxide anion radicals under UV irradiation and thus resulting in better separation of photoexcited electrons and holes. Above this temperature, the increase in the rutile content was detrimental to the intrinsic activity of the TiO_2 particles. Too high temperatures are also detrimental to the maintenance of a reasonable specific surface area for adsorbing reactant molecules. In general, in the case of support materials of low specific surface area, adhesion can be improved to some extent by stabilizing an interlayer between the support and the TiO_2 active phase. For instance, without any exclusive, this interlayer can be formed by thermal oxidation in the case metal foams, by formation of SiO_2 or mixed Al_2O_3–SiO_2 sol–gel films or by electrochemical formation of alumina at the surface of aluminum honeycomb-shaped monolith [20].

Sol–gel methods are among the most widely used immobilization methods, due to the relatively low processing costs as well as the flexible and tunable implementation onto a wide range of substrates in terms of shape, size, and chemical nature. Bulk sol–gel processing is very well documented, and titanium alkoxides, titanium tetrachloride, or titanium halogenide are commonly used as TiO_2 precursors [21]. Dip-coating and spread-coating are the two main processing methods for immobilizing sol–gel TiO_2 onto substrates, but only dip-coating is really applicable for coating substrates with complex geometry. This process requires the full control of the pull out rate for immersing the substrate into colloidal suspensions of particles as well as that of withdrawn from the solution. If the many parameters affecting the process can be controlled – such as the physicochemical properties of the precursor solution, the immersion/withdrawn atmosphere, or the number of coating cycle – sol–gel remains the most versatile and tunable preparation method for immobilizing TiO_2 onto supports. The versatility of the method is very powerful for often tune *on-demand* bulk and surface physicochemical properties of the coated TiO_2 photocatalysts. The chemical modification of the sol–gel coating process can be performed for increasing the intrinsic photocatalytic activity of the TiO_2-based photocatalyst or enlarging its light absorption range, as well as for enhancing the photocatalyst adherence, limiting the closure of the film porosity, or lowering its crystallization temperature. Indeed, obviously, here again, the chemical and thermal stability of the support determines the upper-limit values in terms of temperatures at which it can be subjected during the final calcination step, required for crystallizing TiO_2 and creating adequate chemical binding with the substrate.

Both the chemical nature and high purity of the substrate were reported to be important during the implementation of the sol–gel method and notably during the final thermal calcination step, for enabling chemical binding with TiO_2 and avoiding reduction of the intrinsic activity of the TiO_2 material, respectively. For instance, the sodium content of glass supports (or that of other alkali metal impurities) should not be overlooked. Indeed, fused silica does not contain any sodium, whereas borosilicate and soda-lime glasses contain 4 and 14 wt% of sodium, respectively. The heat treatment used to get immobilized anatase form or to strengthen the TiO_2 anchorage onto glass substrates requires that the content of sodium or other alkali metal impurities should be maintained at a minimum level for avoiding their diffusion from the glass substrate to the TiO_2 film, which is strongly detrimental in terms of photocatalytic activity [22]. Indeed, Tada and Tanaka [23] have reported a marked difference in the photocatalytic activity of TiO_2 films coated on quartz and glass substrates, which was interpreted by a shortening of the photocarrier diffusion length induced by sodium ions. More recently, Ghazzal *et al.* established a direct inverse correlation between the photocatalytic bleaching of aqueous solution of Orange II dye over sol–gel TiO_2 films immobilized on different glass substrates varying in terms of sodium content (fused silica as well as borosilicate and soda-lime glasses) and the amount of sodium observed inside the TiO_2 films, resulting from its diffusion during the final calcination step of the sol–gel procedure [24]. Therefore, the choice of the adequate glass beads as support points out the necessity to compromise between different parameters such as the impurity diffusion rate, the thermal stability, the surface Si–OH density, and the light transmission, without ruling out crucial cost aspects, because replacing glass or even borosilicate glass beads by their corresponding quartz analogs can be drastically unacceptable depending on the targeted applications.

To solve the problem of the quality of the glass substrate, Robert and coworkers [25] have clearly demonstrated that TiO_2 pollution by sodium resulting from the thermally activated diffusion from soda-lime glass substrates during crystallization could be efficiently prevented by the deposition of a thin and transparent SiN_x barrier. Taking the photocatalytic bleaching of Orange II as the model reaction, they evidenced that only a 100 nm thick SiN_x interfacial layer was sufficient for avoiding the sodium poisoning of the TiO_2 coating, with no impact of the transparency property of the photoactive glass.

Similarly, the influence of both chemical nature and purity of the substrate was also evidenced by Fernandez *et al.* during their investigation on the influence of commonly used substrates such as glass, quartz, and stainless steel. They have prepared TiO_2 films by dip-coating on glass and quartz and by TiO_2 P25 electrophoresis on stainless steel [26]. They have reported that the photocatalytic activity was strongly influenced by the support's purity, TiO_2 coated on silica being much more active than when anchored on stainless steel or glass, with Na^+, Si^{4+}, Fe^{3+}, and Cr^{3+} cationic impurities diffusing from the glass and stainless steel substrates to the TiO_2 layer and acting as recombination sites for charge carriers. They also seemed to provide a less crystallized film. The weak activity of the TiO_2 deposit on steel was attributed to the thermal treatment at 600 °C performed, leading to the appearance of the rutile phase and to the diffusion of Fe^{3+} and Cr^{3+} ions.

The impact of the substrate nature on the microstructural properties of sol–gel TiO_2 has been investigated by Ma *et al.*, using glass, ITO/glass, and p-Si(111) substrates [27]. The photocatalytic efficiency, evaluated in terms of kinetic rate constants, was ranked in the, ITO/glass > glass > p-Si(111) substrate order. The authors could not exclude that the differences resulted from variations in TiO_2 particle sizes or from the presence of both anatase and rutile phases, leading to mixed systems. They claimed that the superiority of the ITO/glass could result from a good photogenerated charge separation, favored by the semiconductor coupling phenomena and evidenced by surface photovoltage spectrometry (SPS) and electric-field-induced SPS analyses.

CVD is another interesting method for immobilizing highly pure TiO_2 onto supports. In a typical CVD process, substrates are subjected to a gaseous flow of a single- or multicomponent volatile precursor in an inert atmosphere at controlled pressure and controlled temperature, and decomposition of the volatile precursors takes place at the substrate surface, resulting in the formation of thin films. The operating conditions, and the temperature especially, should be chosen for allowing crystallization of the TiO_2 particles while both chemical and thermal stability of the support have to be taken into account. CVD processes can be implemented in different ways, governed by the type of precursors used, the type of substrates, the operating conditions, and the desired degree of purity, crystallinity, or thin film uniformity. Apart from the classic atmospheric pressure chemical vapor deposition (APCVD) [28, 29], one can mainly cite plasma-enhanced CVD, using plasma for increasing the reaction rates of precursors [30], or metal-organic chemical vapor deposition (MOCVD). If the CVD does not meet the restrictive requirements imposed by the chemical nature of the substrate, MOCVD is an interesting variant that takes advantage of the use of organometallic precursors for significantly decreasing the deposition temperature. In general, the different CVD methods should benefit from investigation in chemical reaction engineering for allowing the implementation of the deposition methods over nonplanar supports or even supports with complex geometry, as needed in the field of supported photocatalysis. In certain cases, and especially for porous or microfibrous substrates, chemical vapor infiltration (CVI) can be preferred to CVD or MOCVD, as it allows a better deposition of TiO_2 within the porosity of the support [31].

Electrophoretic deposition is an interesting method allowing the formation – on substrates with complex geometry – of films with a controlled thickness. On the basis of the application of a potential between two electrodes (the substrate to be coated acting as cathode, whereas anode being usually platinum), this method is restricted to metallic substrates, or at least conductive ones, so that it could be suitable for coating metallic foams [13]. Attachment of TiO_2 in the case of stainless substrates was reported to be very strong, with no particle release during the photocatalytic tests. One should not forget that this method often requires a postcoating annealing at temperatures about 500 °C, so that the support stability should be kept in mind.

Finally, the *electrostatic multilayer self-assembly deposition*, also named as *layer-by-layer*, is an emerging method for immobilizing photocatalysts on substrates of almost every environmental shape and every size. Developed and further rationalized by Decher starting from the early 1990s, it has, over the past 20 years, become a powerful tool for building polyelectrolyte multilayer films and more complex multicomponent systems with nanoscale precision [32]. It is based on the concept of multiple weak interactions across the interface between adjacent layers, which are mostly electrostatic in nature, and consists thus in the building of multilayer thin films by the "dipping" or "spraying" alternate deposition of polyanions and polycations on the surface of a substrate, previously either negatively or positively charged. This method differs from many others by its simplicity of implementation and by benefiting from a fully controlled and homogeneous deposition, from the very good adhesion of the obtained films, and from the ease and versatility of implementation of the technology, whatever the complexity of the substrate geometry [33]. The LbL assembly method has been recently adapted to the alternate deposition of TiO_2 and adequate polyelectrolytes, so that it recently emerged for immobilizing TiO_2 nanostructures for photocatalytic applications, for example, polyelectrolyte multilayers containing commercial Aeroxide TiO_2 P25 were used to degrade rhodamine B [34], whereas Krogman *et al.* [35] have investigated the performance of LbL films containing TiO_2 nanoparticles for the decomposition of chloroethyl ethyl sulfide, a model chemical warfare agent. In the field of air pollutant removal, Shibata *et al.* [36] and Nakajima *et al.* [37] studied the photocatalytic oxidation of gaseous 2-propanol using LbL films containing TiO_2 nanosheets, whereas Dontsova *et al.* [38] targeted the removal of H_2S under UV-A. Maintaining the film porosity during the LbL building up, for preventing TiO_2 nanoparticle encapsulation by two sandwiching and neighbored polyelectrolyte layers and thus for preventing TiO_2 from a negative unavailability toward the reactant flow, remains one of the key aspects that needs to be improved.

6.3 Supports

The design of supported photocatalytic materials is driven by the need to overcome – or at least to minimize – inherent drawbacks resulting directly from the photocatalyst immobilization: (i) low exposed surface area-to-volume ratios, (ii) possible mass transfer limitations, (iii) increase in pressure drops within the catalytic reactor, (iv) complexity in providing the light to the photocatalyst and to the core of the reactor especially, and (v) possible catalyst washout or fouling related to the photocatalyst/support adherence. Some of the drawbacks may be more pronounced or even specific to some reactor configurations and to some designs of supported photocatalytic materials. Depending on the case, these drawbacks can be minimized or even totally overcome. Taken this into account, we highlight here on the use of packed-bed photocatalytic materials, cellular monolithic solids with both honeycomb-shaped and alveolar materials, and optical fiber-based photocatalysts.

6.3.1 Packed-Bed Photocatalytic Materials

The use of immobilized supported photocatalysts based on an inert body coated with a photocatalytic phase led to the concept of packed-bed reactors. Such materials have the merit of increasing the contact surface between the pollutant and the catalyst, that is, the exposed surface area per reactor volume unit. As reported by many authors, the only limitation to the range of materials used as catalyst supports is related to the possibility for physical or chemical binding of the active phase to the support and to the endless imagination of the researcher, which enable taking advantage of the large variety of artificial as well as natural materials, from simple to more complex and sophisticated. However, first, setting aside photocatalytic beads made from pure TiO_2-immobilized supported photocatalysts for packed-bed reactors consist mainly of photocatalytic beads composed of inert beads as support coated with the photocatalytically active phase, usually TiO_2. Although they are among the most used in industrial reactors, packed-bed reactors remain less developed in photocatalysis, mainly as a substantial complexity arises first and foremost from the need to introduce the light into the reactive bed [39]. Inhomogeneous flow distribution and possible pressure drops (depending on the mean object size) may also restrict the implementation of such materials to some extent. Second, setting aside uncommon materials – put forward for valuably trying to tap some natural (often local) and thus low-cost resources, the main need for supports transparent to irradiation led mainly to investigate glass beads as support material, in the form of hollow or bulk microbeads, although glass tubes [40] or glass Raschig rings [41] can also be used.

Different grades of glass substrates can be used, containing variable amounts of sodium or some other alkali metal ions in lower amounts, and differing also in terms of UV light transmission, thermal stability, and surface Si–OH density, for example, basic glass, fused silica, borosilicate glass, soda-lime glass, Pyrex glass, or Quartz. Only focusing on the Si–OH density at the surface of glass beads, some authors mentioned that soda-lime glass beads can be preferred to other commercial glasses, since they combine a higher Si–OH density to a good UV light transmission and chemical inertness [42, 43]. However, other studies highlighted that the sodium content of the glass support should not be overlooked if the preparation of the TiO_2 photocatalyst immobilized on the glass substrate requires heat treatment to achieve the anatase form or to strengthen the TiO_2 anchorage onto the substrate. For example, the choice of the glass grade should be taken into account to maintain the sodium content at a minimum level, for avoiding its diffusion from the glass substrate to the TiO_2 film (see the above-detailed "sol-gel" paragraph within the Section 6.2).

The adherence stability of such immobilized TiO_2 photocatalysts remains a critical point that is not often extensively investigated in comparison to the activity evaluation, although it is of high importance for long-term use in realistic operation conditions [44, 45]. The adherence stability is directly related to the preparation procedure for obtaining the supported photocatalyst. Besides the deposition of crystallized TiO_2 powder followed by a high-temperature treatment, sol–gel coating methods are initiated by the hydrolysis of titanium precursors with final thermal bonding at moderate temperatures, usually in the 350–550 °C range. They are usually preferred to relatively costly methods such as CVD and physical vapor deposition and benefit from an easy-to-process and flexible nature, which can lead to TiO_2-based photocatalysts exhibiting highly tunable bulk and surface physicochemical properties. Furthermore, surface –OH groups – and reactive Si–OH groups in the case of silica-based supports especially – play an important role by being involved in the formation of stable Ti–O–Si bonds during the calcination step. Therefore, pretreating the support (e.g., by NaOH-etching of glass beads) is a way for increasing the surface density of Si–OH anchorage sites. It can also increase the support surface roughness and thus positively acting for immobilizing TiO_2, in the sense that a rougher surface texture, such as in quartz sands, could provide a larger surface area for anchoring TiO_2.

Even less investigated than glass substrates, various polymeric materials have been used for immobilizing TiO_2. Besides the use in wall-coated reactors of polythene or polyethylene polymer film as host for anchoring TiO_2, a real use of polymeric materials in packed-bed reactors remained scarce. One can mention studies dealing with polystyrene beads [46]. Polymer materials are often reported as benefiting from chemical inertness, mechanical stability, innocuity, low costs, and abundance. Surprisingly, the thermal stability of polymers such as polystyrene (expanded or not) is sometimes put forward, but it remains largely weaker than that of other transparent support materials such as glass. Furthermore, the viable use of polymers is highly questionable – if no additional protective coating is added – due to the long-term stability limitation caused by low resistance to the photocatalytic oxidative attack toward polymeric carboned chains.

In some liquid-phase applications, immobilized TiO_2 led also to the concept of floating photocatalysts, for taking advantage of the proximity with the air/water interface for maximizing both the light utilization (especially in solar-light-driven processes) and the oxygenation of the photocatalyst (especially for nonstirred reactions), as well as for facilitating their postuse recovery (e.g., [47]). Apart from TiO_2 hollow microspheres, floating photocatalysts include TiO_2 coated on various supports such as – without any exclusive – SiO_2 hollow glass microbeads ($\phi = 2$ mm) [48, 49], porous lava ($\phi = 2$ mm) [50], natural porous silica pumice ($\phi = 10$ mm) [51, 52], exfoliated vermiculite ($\phi = 0.2$–0.5 mm) [51], exfoliated graphite ($\phi = 0.2$ mm) [53], or expanded polystyrene beads [47]. Even some of them are at the limit of being considered as a powder rather than as a macroscopic body in terms of grain or bead mean diameters; they take advantage of a very low volume density for being more easily recovered from water after use.

6.3.2 Monolithic Photocatalytic Materials

The need for an increase in the exposed surface area per volume unit and for a good contact between the reactant fluid and the photocatalyst without any increase in the operating pressure drop has also put forward the use of cellular monoliths as photocatalyst supports. Cellular monolithic solids are made up of connected arrays of struts or plates and are characterized by the geometric structure of the cells, that is, both the shape and size of the cells and the way the cells are distributed. Like crystal lattice, they are described by a typical cell with certain symmetry elements, and therefore are also called *lattice monolithic solids*. However, they differ strongly from crystal lattice in terms of scale, since the unit cell of cellular materials is on the order of millimeter to micrometer magnitude, so that they can be seen both as structures and materials. The two main cellular solid structures are honeycombs and foams, first existing as natural materials such as wood and cork for honeycomb-like materials and like trabecular bone, plant parenchyma, coral, and sponge for foams. Furthermore, they can now be engineered from polymers, metals, ceramics, and composites.

Whatever their internal structure, the implementation of cellular monolith materials greatly simplifies both building and maintenance of photocatalytic systems when compared to packed-bed configurations by simplifying both loading and offloading operations without any dismantling of the entire system.

Honeycomb-shaped or square-channel materials are the most commonly targeted cellular monolith solids used as photocatalyst supports. Belonging to the family of multichannel materials, they have been intensely studied from both modeling and experimental approaches [54–58]. Strictly speaking, honeycomb-shaped materials should exhibit hexagonal channels, but this terminology is improperly used for other channel geometry such as square-channel (Figure 6.2). In general, they enable operating like car-exhaust systems and to process high flows at low pressure drops, as well as increasing the surface area-to-reactor volume ratio and therefore the exposed surface. Indeed, the resulting monolithic photoreactor is usually reported to provide a surface-to-volume ratio of about 10–100 times greater than plates or beads substrates with the same outer dimensions [60].

The main drawback of the multichannel monolith is low light efficiency, with a poor irradiation distribution in the monolith. The light flux within the monolith has been modeled in different configurations, for example, by Hossain *et al.* [57] or Taranto *et al.* [54], for predicting the irradiation of the coated channels. Hossain *et al.* notably predicted that the light flux already decreases by 50% at the entrance of square channels because of the shadowing of the incoming diffuse light at the channel entrance by the wall [57]. In addition, only about 10% of the initial light flux is maintained at a distance of about 1–2 aspect ratio (i.e., length-to-width ratio for channels), and this decreases down to 1% at a distance of 3–4 aspect ratio.

Regarding the chemical nature of the multichannel monoliths used as a support in photocatalysis, they mainly consist of metallic structures (e.g., aluminum [61]) as

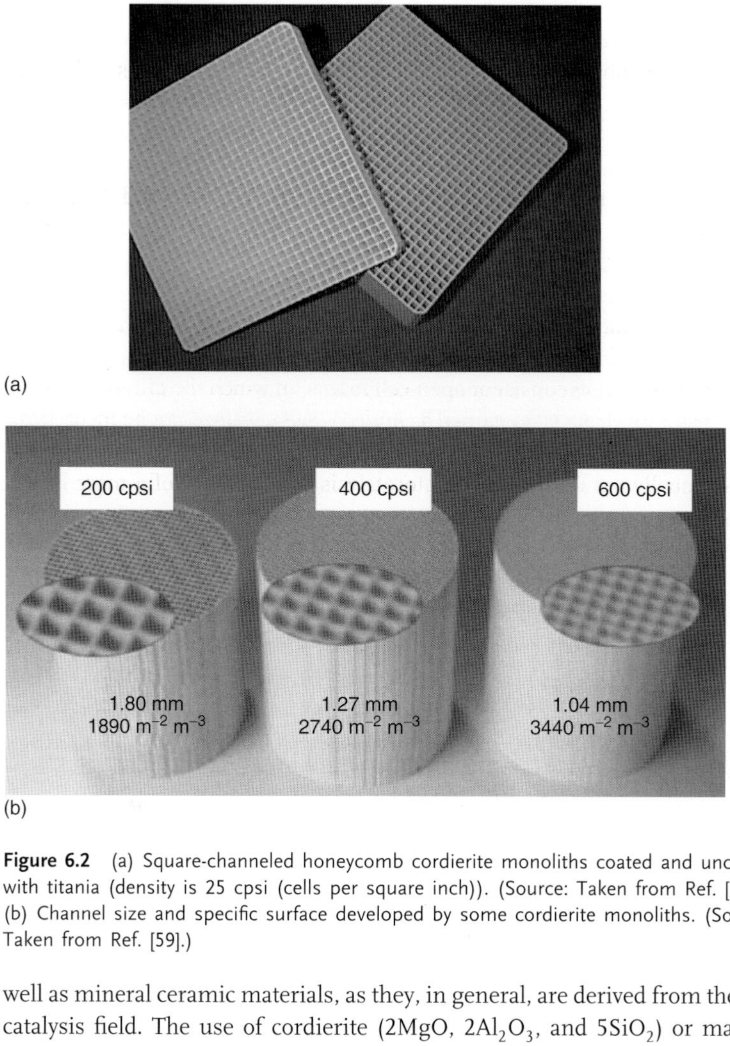

Figure 6.2 (a) Square-channeled honeycomb cordierite monoliths coated and uncoated with titania (density is 25 cpsi (cells per square inch)). (Source: Taken from Ref. [55].) (b) Channel size and specific surface developed by some cordierite monoliths. (Source: Taken from Ref. [59].)

well as mineral ceramic materials, as they, in general, are derived from the thermal catalysis field. The use of cordierite (2MgO, $2Al_2O_3$, and $5SiO_2$) or magnesium silicate monoliths is noteworthy. The mineral monoliths being, in most of the cases, only used for supporting the TiO_2 photocatalyst, their chemical nature was not often deeply discussed.

Scarce works have reported on the direct incorporation of the TiO_2 photocatalyst into the ceramic monolith walls, by including a TiO_2 precursor (hydroxylated titanium dioxide gel) into the paste before extrusion, together with natural silicates used as inorganic binders, in such a way that the TiO_2 was incorporated in the monolithic matrix [62].

An interesting strategy for improving the effective length of nontransparent multichannel monoliths deals with the combination of both monolith and optical fiber approaches for providing the light deeper into the core of the monolith. This approach is described in more detail later.

A totally different alternative to nontransparent ceramic monoliths has been proposed by Sanchez *et al.*, with the use of polyethylene terephthalate (PET) polymer monoliths as the photocatalytic support [63]. Such PET monoliths (WaveCore PET150-9/S, Wacotech GmbH & Co. KG from Bielefeld, Germany), primarily used as thermal insulator in passive solar systems, have 2.23 $m^2\ m^{-3}$ geometric surface area, 9 mm × 9 mm pitch cross section, 0.15 mm wall thickness, and 45 $kg\ m^{-3}$ density. They have the advantages of being UV transparent and lightweight to implement them in low-cost photocatalytic reactors. Taking the degradation of trichloroethylene as the model reaction, the authors evidenced that the PET polymer monoliths can be efficiently used under UV-A after application of a protective SiO_2 layer to prevent oxidation of the fragile polymer, and simultaneously maintain the light transparency of the material (Figure 6.3).

Alveolar foam materials consist of open-cell foams, in which the different cells are connected through struts (also named as *bridges*), even *distingo* can be theoretically made with close-cell foams, in which neighboring cells are isolated from each other, and usually not used in the photocatalysis field because of a more limited light penetration depth within the structure.

Recently, in the field of thermal catalysis, cellular foam structures received increasingly more scientific and industrial interest as catalytic support in order to overcome some of the drawbacks of "conventional" packed beds [65–67]. This new medium has a highly permeable porous structure with high porosity (0.60–0.97),

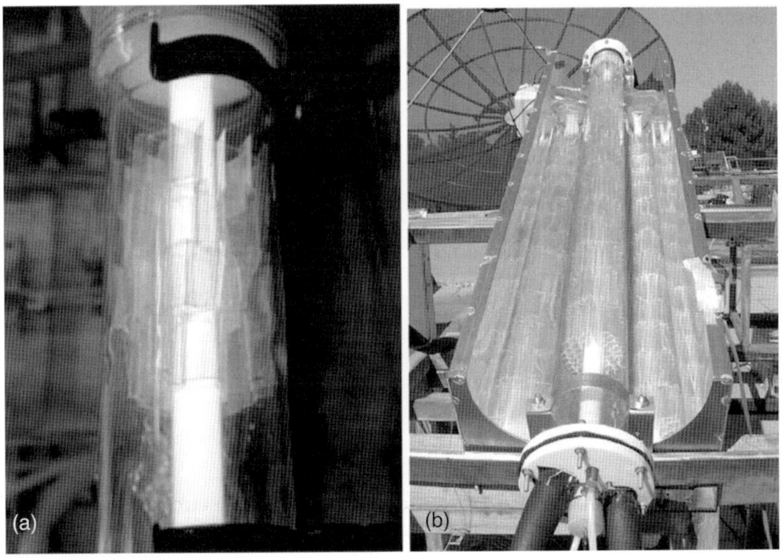

Figure 6.3 (a) Lab-scale photoreactor showing the arrangement of the PET monoliths around a UV-A fluorescent lamp. (Source: Taken from Ref. [63].) (b) Tubular photoreactor from CIEMAT (Madrid, Spain), operating with 14 photocatalytic PET monoliths and using a nonconcentrating compound parabolic concentrator (CPC) for reflecting sunlight complemented by an inner artificial UV-A radiation source. (Source: Taken from Ref. [64].)

which enables a considerable reduction of the pressure drop along the catalyst bed even at high gaseous space velocity, together with suitable mass transfer properties and an easy control of the external porosity. Cellular foams are also known to act as efficient static mixer and to allow improving the reactant-to-surface contact probability ratio. They take also advantage of a strong gain of exposed surface inside reactors, resulting in an increased surface-to-reactor volume ratio, for strongly increasing the amount of TiO_2 immobilized inside photoreactors. More specific to photocatalysis applications, they provide a better light transmission within the core of the reactor in comparison to channeled cellular monoliths, at least, obviously, as the cell size is not too tiny. Therefore, designing cellular foams for use as photocatalyst support should achieve a compromise between large cell sizes, allowing the maximum penetration of light inside the reactor, and large exposed surface area for anchoring TiO_2 and increasing the amount of immobilized TiO_2 per reactor volume unit. Within this respect, in contrary to the thermal catalysis field, large open cells of about 4500 μm are usually reported to provide the higher photocatalytic efficiencies for air as well as for water treatment [19, 68, 69].

By contrast, a priori, photocatalysis does not take direct advantage of the continuous aspect of the foams, as reported in thermal catalysis, which improves heat transfer properties through the whole foam. Indeed, the continuous connection of the different catalyst domains provided by the solid ligaments (or struts) in the foam material increases the effective thermal conductivity on the entire system without thermal breaking points as encountered with packed beds. This property could at best be beneficial in the implementation of photocatalysis, if deactivated photocatalysts require to be thermally regenerated, for example, by treatments in air at temperatures greater than 500 °C for cleaning TiO_2 deactivated by surface poisoning molecules, such as sulfates (produced from sulfur-containing reactants), nitrates (produced from nitrogen-containing reactants), or partially oxidized aromatic intermediates.

Taking polyurethane polymeric foam as example, Figure 6.4 shows the most characteristic parameters of cellular foams, namely, the cell size (Φ), the window

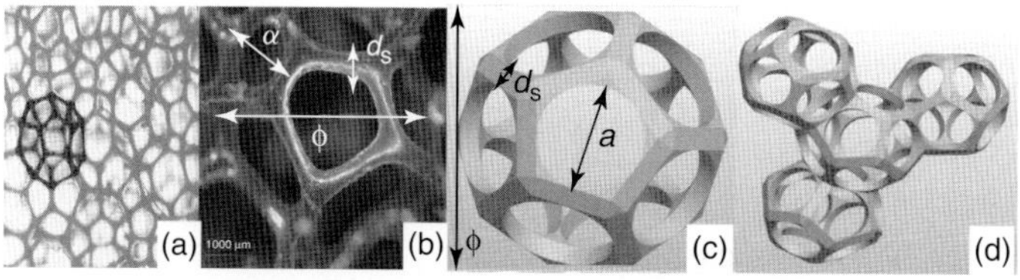

Figure 6.4 (a) High degree of interconnectivity through the entire matrix of open-cell cellular foams evidenced by taking polyurethane polymeric foam as example. (b) The most characteristic parameters are the cell size (Φ), the strut diameter (d_s), and the window size or pore diameter (a), which can be measured by several techniques and correlates with the pore density (i.e., the number of pores per linear inch, ppi). (c,d) Schematic drawing of the model developed by Edouard, with packing of regular pentagonal dodecahedron. (Source: After Refs. [19, 70].)

size (a) (or pore diameter, which correlates with the number of pores per linear inch, ppi), and the strut diameter (d_s). According to Edouard *et al.* [71] in a recent state-of-the-art review of the main theoretical geometrical models, geometric modeling of cellular foams is significantly more difficult than that of conventional packed beds or even of honeycomb-like materials because of the more intricate geometry of foam structures. Commercial solid foams consist of a complex three-dimensional array of interconnected struts with an irregular lump of solid at the strut intersection. In the literature, various foam cell geometric idealizations have been proposed and the recent model developed by Edouard and coworkers [70], based on the packing of regular pentagonal dodecahedron, revealed to be quite close to the real structure of a unit cell of solid foam. Taking into account simple relationships that only involve the porosity and either the cell diameter (or the pore or the strut), it enables to describe the specific surface area, pressure drop, and external solid–fluid mass transfer.

The alveolar cellular foams used as support in photocatalysis are polymers, metals, carbon, and ceramic foams.

Till now, *polymer foams* are scarcely used in photocatalysis, although they are very tunable in terms of geometry for being adapted to various reactor designs and benefit from low processing costs that could globally strongly lower the costs associated with the implementation of photocatalysis applications. Indeed, first, polymer foams could be irreversibly damaged if used in nonadequate operating conditions: they suffer from a low thermal and chemical stability, so that temperature treatments or non-neutral fluid/support should not be implied, neither during the deposition of the photocatalyst nor for regenerating possibly deactivated photocatalysts (see above, e.g., thermal or soda-based washing regenerations). Then, such polymer foams would also be restricted to the treatment of nonaggressive fluids. In addition, the sensitivity of polymer foams requires protecting them by a passivating coating to prevent them from the photocatalytic oxidative degradation. Josset *et al.* [19] recently showed the interest in using reticulated polyurethane foams – passivated by a protecting polysiloxane coating – as support for active TiO_2 particles for designing flow-through 3D-structured photocatalytic reactors for purifying bioaerosols contaminated with *Legionella pneumophila* bacteria (Figure 6.5). This photocatalytic material could thus work at ultralow pressure drops as a result of a large cell size, with a 4500–4800 μm average cell size and a 97% open porosity. It also took advantage of a strong increase in the exposed surface area per volume unit, that is, close to 500%, compared to the corresponding wall-coated photoreactor with similar geometry. Very efficient decontamination levels toward airborne *L. pneumophila* were reached in a single-pass test mode.

Targeting a possible use as low-cost drinking water filter where bacterial contamination of the surface water is a health risk, such as in developing countries, similar polyurethane foams were also used by Jain and Pradeep [72] for uniformly supporting bactericidal silver nanoparticles and implementing a continuous disinfecting process, with a logarithmic abatement of 5 over *Escherichia coli* in contaminated water (water flow rate of 0.5 l min^{-1} and residence time close to 1 s). The nitrogen atoms of the polyurethane foam surface seemed to be beneficial to the

Figure 6.5 SEM images of (a) the polyurethane (PU) foam surface passivated by a smooth polysiloxane protective coating, (b) the homogeneous TiO_2 coating, and (c) a scratched zone evidencing the thickness of both polysiloxane and TiO_2 coatings on the PU surface. (Source: Taken from Ref. [19].)

binding of silver nanoparticles with the foam, through direct nitrogen–silver interactions.

The most used cellular foams for immobilizing photocatalysts are metallic foams (mainly made of nickel or aluminum) and ceramic foams (alumina or cordierite).

Usually, *metallic foams* suffer from a low specific surface area that needs to be increased by different methods, such as chemical and/or thermal pretreatments, or by coating the foam with an oxide interlayer. Whatever the method, the need is to provide a better anchorage for the supported TiO_2 and to increase its density per foam unit volume. Taking the 2,4-dichlorophenol degradation in water as the model reaction, Plantard *et al.* [73–75] pointed out the interest of using aluminum foams for immobilizing TiO_2 P25 compared to a commercial Ahlstrom photocatalytic paper, although the slurry reactor still remained the more efficient configuration. They reported that TiO_2 anchorage first requires a fluorhydric acid surface treatment followed by the deposition of several silica layers and an intermediate thermal treatment at 200 °C, before TiO_2 could be impregnated and finally stabilized onto the support by a treatment at 400 °C. In the case of TiO_2-supported nickel cellular foam photocatalytic materials, in aqueous phase, there is photocatalytic inactivation of *Chlorella* [76] and degradation of quinoline and industrial wastewater [77]. In addition, TiO_2/nickel foams were used for degrading gaseous acetaldehyde [78] and integrated into a 15 channel prototype for designing an indoor air purifier working at a 1 m s^{-1} velocity with low pressure drops [79]. A preheating treatment, resulting in the formation of an NiO interlayer, was of advantage by increasing the low specific surface area of nickel foams, with positive impact on the photocatalytic efficiency, and also because the NiO coating prevents the injection of photogenerated electrons from TiO_2 to the metal foam substrate [80]. However, still low, the surface area was further increased by Yuan *et al.* by coating the nickel foams with an Al_2O_3–SiO_2 sol–gel film as the interlayer for promoting the activity.

Alumina and cordierite cellular foams are strongly more resistant than polymer or carbon foams and benefit from a reactive oxygen-containing surface, prone to bind with TiO_2 or TiO_2 precursors, even they have a relatively low specific surface area due to the high temperatures used for the ceramic shaping (Figure 6.6) [56]. For instance, the method developed by Schwartzwalder and Sommer and used by Plesch *et al.* for shaping alumina foams operates with a final sintering at 1200 °C to preserve a high porosity. After annealing at 600 °C, TiO_2 P25 films prepared by a wash coating process, displayed an adequate adhesion at the alumina foam surface and a thickness of 5–10 μm. The authors showed that increasing the pore size from 25 to 10–15 ppi was beneficial to the degradation rate in liquid phase, taking phenol degradation as the model reaction, due to a better flow of solution through the macroporous foam structure and the better access of light to the active TiO_2 surface, even if they observed a 25% loss of activity after immobilization compared to a standard slurry configuration [81, 82]. It was worth noting that no measurable release of deposited TiO_2 was observed after successive photocatalytic cycles, and the prepared films exhibited sufficiently good adhesion, even in the case of strongly cracked TiO_2 thick films. Alumina foam was also used in the photocatalytic degradation of H_2S and CH_3SH by Kato *et al.* [83] for immobilizing TiO_2 powder coated by photodeposited nanosized Ag particles. Taking the photooxidation of 1,8-diazabicyclo[5.4.0]undec-7-ene in water as the target reaction, the recent work of Ochuma *et al.* also evidenced the interest in using a TiO_2-coated alumina reticulated foam monolith (at 15 ppi) inserted inside an annular photoreactor compared to a conventional slurry reactor [84].

Targeting the development of air cleaners incorporating three-dimensional through-pore structure filters, the use of cordierite foams for supporting TiO_2 for photocatalysis purpose was reported by Yao *et al.*, TiO_2 nanoparticles being immobilized firmly onto the cordierite surface through a simple impregnation procedure followed by calcination at 500 °C [85]. TiO_2/cordierite foam exhibited a high bactericidal rate toward five types of airborne or droplet-based infectious pathogens (*E. coli, Pseudomonas aeruginosa, L. pneumophila, Klebsiella pneumoniae,* and methicillin-resistant *Staphylococcus aureus*), as well as a high photocatalytic degradation ability on gaseous acetaldehyde. So, it could display not only high bactericidal performance to remove pathogens from the air and from water droplets but also strong decontaminating/deodorizing functionality.

Works on the use of alveolar *self-bonded β-SiC foams* in photocatalysis remain very scarce. The more interesting and flexible synthesis method for preparing alveolar open-cell β-SiC foam with medium specific surface area is the shape memory synthesis (SMS) replica method, in which a preshaped polyurethane foam is transformed into its corresponding carbide (Figure 6.7). The SMS process has been first developed at the laboratory scale for transforming an sp^2 or sp^3 carbon perform into the corresponding β-SiC structure, from the nano- to the

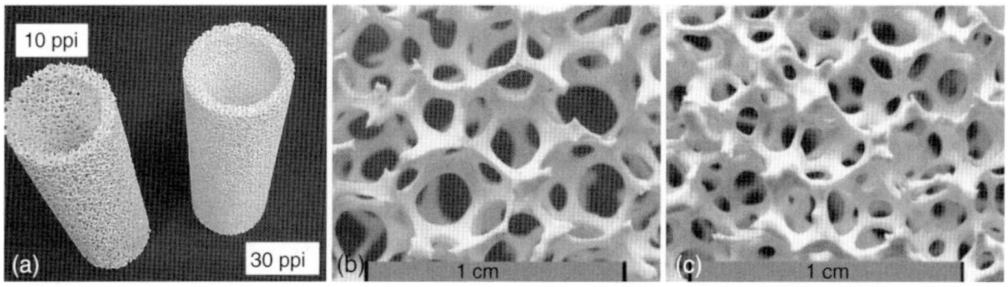

Figure 6.6 Cellular alumina foams with different pore sizes: (a) 10 and 30 ppi, (b) 10 ppi, and (c) 15 ppi. (Source: Taken from Refs. [56] and [82].)

milliscale range, before being upscaled and now industrialized at the SICAT Company (Willstätt, Germany) [86]. The material can be shaped in various forms including extrudates, spheres, rings, trilobes, and structured material such as open-cell foams. In the first step, a preshaped polyurethane foam is infiltrated by a silicon-containing resin, before drying and polymerization at 250 °C. The reactive step is then performed at 1360 °C under argon, with the attack of carbon by SiO vapor formed by the reaction between silicon and the resin oxygenated compounds, to give β-SiC. Finally, residual carbon is removed by combustion at 700 °C. Then, the SMS process reflects the maintenance of the original macrostructural features of a solid, that is, its macroscopic shape, during the synthesis. Thus, β-SiC foam can be manufactured with tunable shapes for being adjusted to the photoreactor geometry and size.

In 2009, Rodriguez *et al.* [87] studied the aqueous-phase photocatalytic degradation of ammonia over different geometries of supports dip-coated with TiO_2 P25 (cordierite monoliths, stainless steel, and 1000 μm mean window size alveolar β-SiC foams), focusing on physicochemical properties of the TiO_2 coating in terms of stability, adherence, structure, and morphology. Whatever the support, increasing the TiO_2 concentration of the impregnation suspension caused a decrease in the coating adherence. During the deposition of TiO_2 onto the β-SiC foam, cracks were formed, but did not affect the photocatalyst grip. With no TiO_2 amount being mentioned, they concluded that TiO_2/β-SiC foam is a promising material for internally structuring classical reactors.

Kouamé *et al.* [4] published a more detailed study, in which stable TiO_2/β-SiC open-cell foams with a strong anchorage onto the SiC foam surface of TiO_2 nanosized particles were obtained through sol–gel process in acidic conditions (Figure 6.7). The efficiency of TiO_2/β-SiC foam-based photocatalytic reactors was proportional to the size of the β-SiC foam cells until an optimum size of about 4500 μm. Taking the aqueous phase degradation of 3-(3,4-dichlorophenyl)-1,1-dimethyl-urea), known as the *Diuron® phenylurea herbicide,* under UV irradiation as the test reaction, and even if the performance remained lower compared to that of a slurry reactor, TiO_2/β-SiC foam appeared as a promising material for internally structuring classical flow-through reactors for targeting the photocatalytic treatment of wastewater in continuous flow mode [68]. In particular, investigation on the photocatalytic material aging and reuse was promising for large-scale liquid-phase applications. Indeed, no decrease in the apparent kinetic rate constants was observed as a function of successive runs, meaning that (i) no loss of TiO_2 particles occurred by washing out and (ii) there was no poisoning of the photocatalyst surface by adsorption of the Diuron intermediate by-products (no fouling of the catalyst surface).

One can note that the SMS method can be modified for synthesizing chemically modified β-SiC foams by introducing various transition metal doping elements during the synthesis for tuning bulk and surface properties of the resulting β-SiC foams, without altering the SMS concept. Modifications can affect the specific surface area, porosity, surface acidity, and mechanical strength, affecting the

Figure 6.7 Optical images of (a) a pre-shaped disk of polyurethane foam used as β-SiC foam precursor, (b) the resulting disk of self-bonded β-SiC foam obtained after SMS, and (c) the TiO_2/β-SiC foam, with an example of TiO_2 nanosized particle coating through acidic sol–gel process. SEM images of (d) the β-SiC foam and (e,f) the TiO_2/β-SiC foams [4]. Unpublished results from the PhD Thesis of R. Masson, 2012, University of Strasbourg.

efficiency of the photocatalytic materials and the implementation of photocatalytic processes. For instance, TiO_2-β-SiC foams can be directly prepared by titanium incorporation during the SMS process, resulting in the synthesis of TiC-β-SiC composite foams, in which the TiC material can be subsequently selectively oxidized into TiO_2, thanks to the β-SiC oxidation resistance [88, 89]. Actually, this TiO_2-β-SiC composite foams still strongly lack efficiency compared to TiO_2-supported β-SiC foam photocatalysts but would be a way to simplify the preparation of the photocatalytic materials by avoiding the TiO_2 deposition step, which as a result would completely rule out one of the main restrictive problems encountered in supported photocatalysis that is related to the stability of the TiO_2 anchorage on the support material.

The SMS method can be modified for synthesizing open-cell carbon foams rather than β-SiC foams by performing the process without incorporating micronized Si powder and by subsequent pyrolysis under Ar at a temperature greater than 700 °C. Cellular carbon foams of 4500 μm cell size were successfully used for supporting TiO_2 P25 and increasing by a factor 4 the photocatalytic conversion of gas-phase methanol compared to the corresponding wall-coated tubular reactor [69]. They take advantage of being easier to process than the β-SiC foams and of displaying a better temperature and chemical resistance compared to polymer foams, so that no additional protective coating should be implied, even if they remained largely more chemically fragile than self-bonded β-SiC foams. Compared to β-SiC foams, they suffer unfortunately from their very low specific surface area and from a low density of oxygenated surface groups that could act as grafting groups for anchoring TiO_2, so that the TiO_2 amount per unit of reactor volume remained closer to that with polymer foams. This drawback could be partially overcome by activating the carbon foam surface through chemical or temperature treatments, even if great

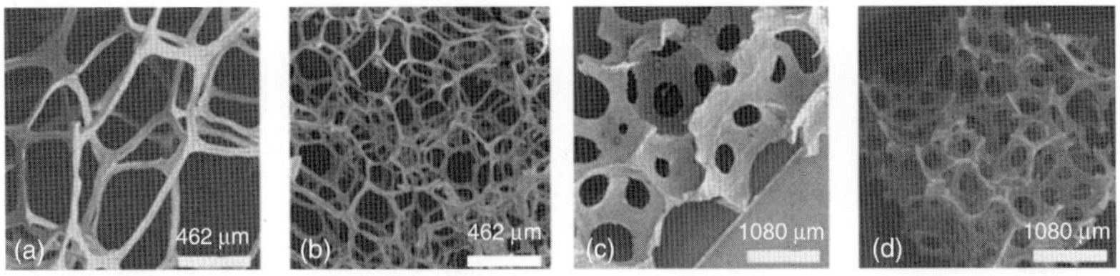

Figure 6.8 (a–d) Macromolecular TiO_2 foams with tunable pore and bridge size, synthesized according to the new concept of "integrative chemistry." (Source: Taken from Ref. [90].)

care should be taken for not fragilizing too strongly the carbon foam and thus not reducing too much its mechanical stability.

Finally, we would like to put forward the synthesis of macromolecular TiO_2 cellular foams with tunable pore and bridge size, according to the new concept of "integrative chemistry," combining *soft chemistry* and *soft matter* [90]. This would be a support-free way of immobilizing the TiO_2 photocatalyst inside reactors, which could solve adherence problems, but evaluation of their photocatalytic efficiency for air or water treatment has not been reported so far (Figure 6.8).

Noncellular monolith materials have been recently developed by Zahraa and coworkers [91] for investigating the impact of the geometry of a monolith support on the efficiency of photocatalysts for air cleaning, using the gas-phase degradation of methanol as the model reaction. They used Nd-YAG-laser-assisted stereolithography for shaping epoxy resins as *on-demand* three-dimensional supports for TiO_2 layers and manufactured various configurations labeled as static mixer, double spiral, quadruple spiral, alveolar crossed channels, and star geometry (Figure 6.9).

They developed a model that discriminates the basic phenomena involved, that is, light absorption, hydrodynamic and transfer processes, as well as reaction kinetics. They reported that the geometry has practically no influence on the external mass transfer rate, which depends essentially on the flow rates. By contrast, the monolithic support geometry has a strong influence on the kinetics of photocatalysis. Indeed, the static mixer monolithic geometry absorbs more photons per unit area of support, leading thus to a higher degradation rate per unit of surface. However, they concluded that among the investigated configurations, the crossed channels support has the most appropriate geometry if the aim is to design a photoreactor that is as compact as possible. The drawback of weaker absorption of photons per unit area of support, resulting in a lower rate per surface unit compared to other structures, was overcome by a higher surface-to-volume ratio.

6.3.3
Optical Fibers

The concept of optical fibers for immobilizing TiO_2 is based on the double role that optical fibers can play, by simultaneously acting as TiO_2 photocatalyst support and as light distributing guide for remote light transmission. The concept was

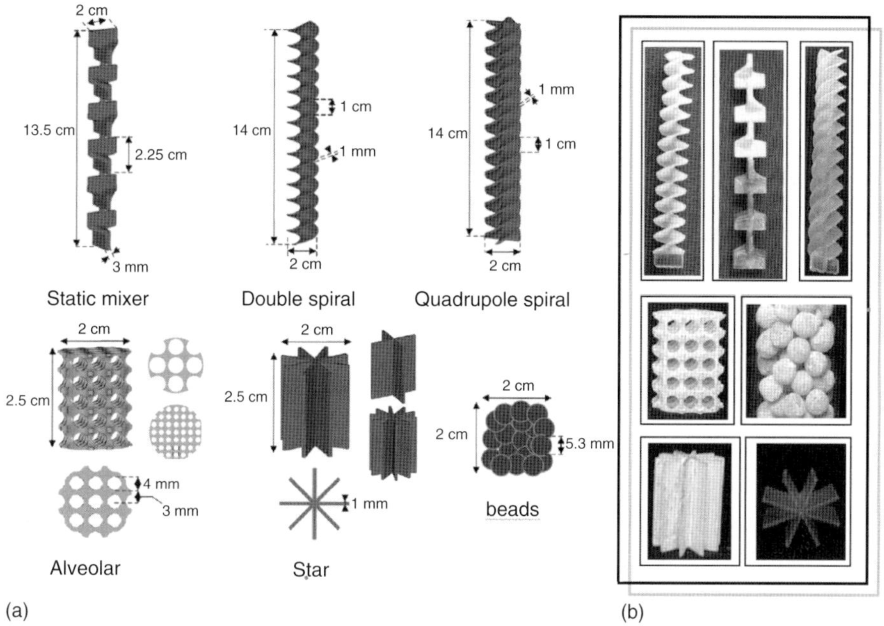

Figure 6.9 Monolithic supports for TiO_2 layers in static mixer, double and quadruple spiral, alveolar crossed channels, and star geometry configurations, as well as usual spherical beads, all being three-dimensionally shaped by Nd-YAG-laser-assisted stereolithography of epoxy resin. The crossed channel and star configurations can be tuned in terms of channel and star plates, respectively. (Source: Adapted from Ref. [91].) (a) Schematic view and (b) optical images of the stereolithography-shaped monolithic supports in the different configurations and beads.

first published by Marinangeli and Ollis [92, 93], even if they calculated that the catalyst could deactivate due to the heat buildup in the bundle array, before being experimentally evidenced in a TiO_2-coated quartz fiber reactor with the liquid-phase photodegradation of 4-chlorophenol [94]. Both characterization and modeling foundations for the concept of optical fiber reactors (OFRs) for liquid- and gas-phase photomineralization of organic pollutants were finally laid by Peill and Hoffmann [95], with applications in both liquid- and gas-phase degradation of organic pollutants [96–98]. Taking the gaseous benzene degradation as the model reaction, Wang and Ku [98] reported that 2 orders of magnitude could be gained under similar operating conditions by replacing an annular fixed-film photoreactor by the OFR.

Here, the support body clearly plays an active role inside the photoreactor, with the enhancement of both uniformity and distribution of the UV light within a given reaction volume and the ability of the remote delivery of light, so that such system can thus be used for *in situ* treatment of contaminated sites. The coating of a thin TiO_2 film on the outer surface of the optical fibers is required for preventing the UV light from total reflection inside the fiber core and thus allowing the UV light to radially refract out of the fibers and penetrate through the TiO_2 layer. By contrast, the use of optical fibers for immobilizing TiO_2 has intrinsic drawbacks inherent to the operating concept, foremost among which is the exponential decay of the light intensity along the axial direction of the coated fiber, even it might be partly overcome by changing the refractive index of the fiber downward the fiber length. Wang and Ku [99] reported 99% refraction after 10 cm (i.e., 1% transmission), whereas an uncoated stripped fiber still transmits 40% after 20 cm. Second, the opposite direction between light transmission and reactant diffusion results in the lowering of the overall reaction rates, due to the internal mass transfer resistance within the TiO_2 film and the increased charge recombination rates caused by the generation of charge carriers far from the photocatalyst–fluid interface [97]. Also, the geometry of thin optical fibers, taking up usually about 20–30% of the reactor volume but providing a relatively low surface area, strongly decreases the active volume ratio within the reactor.

Taking that drawback into account, both optical fiber and monolithic structure approaches were combined with the design of an optical fiber monolith reactor (OFMR), consisting of the distribution of optical fibers coated with TiO_2 inside a ceramic multichannel monolith, with the insertion of a single optical fiber into each of the parallel channels [60, 100]. In this configuration, both the stripped optical fibers and the inner surface of monolith channels were coated with TiO_2, as thin and thick films, respectively, so that the TiO_2 film coating the monolith walls was thick enough to fully absorb the incident UV light refracted out of the fibers. The OFMR configuration increases by more than an order of magnitude the irradiated catalyst surface area per unit volume of liquid treated in the reactor compared to a continuous annular reactor or the OFR. Using *o*-dichlorobenzene and phenanthrene as model compounds in wastewater treatment, experimental results of Lin and Valsaraj [60] showed about 2 orders of magnitude higher apparent quantum efficiency compared with continuous annular and slurry reactors, whereas the apparent rate constants were twice higher in the case of dichlorobenzene but twice lower toward phenanthrene.

Experimental and modeling results highlighted that the geometric parameters are of high importance and significantly affected the overall degradation efficiency [100]. The fiber diameter (i) affects the fiber-to-channel diameter ratio and thus the extent of mass transfer, (ii) affects the photocatalyst coating surface area and thus the light propagation length, (iii) determines the monolith channel number, and (iv) further affects the quantum efficiency with a given incident light intensity as well as the throughput in the monolith. Higher degradation efficiency was observed for multichannel monoliths filled with large fibers and with a smaller fiber-to-channel diameter ratio at the expense of a lower liquid throughput. The authors evidenced an optimum thickness of the TiO_2 film on the optical fibers at 0.4 μm, and a short light propagating length of about 5–6 cm, after which the UV light is almost extinct. As a result, the configuration remains limited by the effective length of the monolith support, since the light distribution profile inside each monolith channel showed an exponential

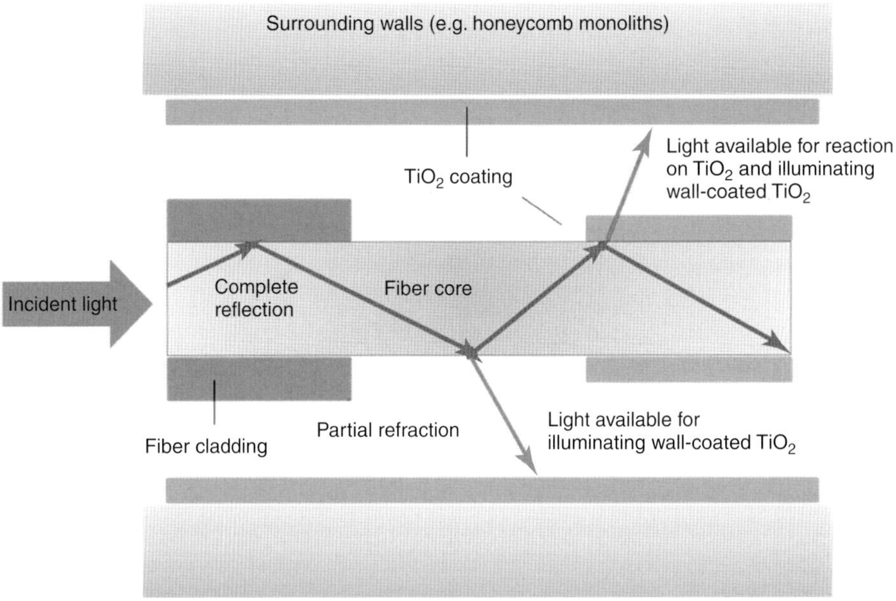

Figure 6.10 Schematic drawing of light propagation within a TiO_2-coated optical fiber. In addition, it shows the irradiation of the surrounding wall-coated TiO_2 by partial refraction out of the fiber, leading to the concept of internally irradiated monolith reactors (IIMRs). (Source: Adapted from Refs. [99, 101].)

decay in propagation along the quartz fiber core and penetration into the TiO_2 film.

More recently, for extending the effective length of the monolith, this approach was modified by Moulijn and coworkers [102], leading to the concept of internally irradiated monolith reactors (IIMRs), in which two side-light fibers were located in diagonally opposite corners inside each channel of the honeycomb-shaped ceramic monolith with inner walls coated with TiO_2, but – by contrast – were not covered by a TiO_2 film (Figure 6.10). Compared with OFR and OFMR, the emitted light of IIMR can reach the photocatalyst–reactant interface without being strongly attenuated by the photocatalytic layer. This configuration provides an extra design flexibility since the light propagation process from the source to the photocatalyst–reactant interface is decoupled from the physical properties of the catalyst. It takes also advantage of an easier TiO_2 anchorage process onto ceramic walls versus quartz fibers, as well as of a longer-lasting lifetime in terms of coating stability and fragility. The IIMR photonic efficiency achieved in the selective cyclohexane photooxidation is 0.062, which is lower than the efficiency of 0.151 obtained with a top irradiation slurry reactor but higher than that obtained in an annular slurry reactor and a reactor configuration with side-light fibers immersed in TiO_2 slurry, reaching a photonic efficiency of 0.008 and 0.002, respectively.

6.4 Laboratory and Industrial Applications of Supported Photocatalysts

Currently, industrial applications of photocatalysis for water purification are not numerous around the world because the individual treatment of water is not easy and little developed in industrialized countries. However, this technology can be effective than and competitive with other technologies of water treatment (adsorption, ozonation, or biological treatment) for certain types of aqueous effluents. This is the case for some agricultural effluents or industrial wastewater that may contain pollutants recalcitrant to traditional treatment methods. Even if they are present in low concentrations, the regular rejection of these pollutants known as *persistent in water* can cause a risk to the environment over the long term. Organic pollutants that can be effectively removed by photocatalysis (Table 6.1) include pharmaceuticals (antibiotics, analgesics, and hormones); aromatic compounds present in industrial waste such as tanneries, poultry (phenols, benzoic acid, aniline, and indole), or plasticizers; flame retardant; and endocrine disruptors (polychlorinated biphenyls (PCBs) and bisphenol A). One of the many advantages of photocatalysis is to be unselective; however, degradation rates differ markedly.

Among the applied research with supported photocatalyst, Shifu and Gengyu [103] showed that the floating TiO_2–SiO_2 photocatalyst beads can be used for the photocatalytic degradation of common organophosphorus pesticides (dichlorvos, monocrotophos, parathion, and phorate). The filtration and resuspension of photocatalyst powders can be avoided because the density of TiO_2–SiO_2 beads is less than 1.0 g cm^{-3}. Therefore, they can float on the water surface and so they are directly irradiated by sunlight. Figure 6.11 shows the efficiency and the unselectivity of the process, as 65×10^{-3} mmol·l of four organophosphorus pesticides of three

Table 6.1 Some examples of photocatalysis works with supported catalyst.

Product family	Molecules	Supported photocatalysts	References
Pesticides	Dichlorvos, monocrotophos, parathion, and phorate	TiO_2-SiO_2/beads	[103]
	Dicarzol	TiO_2-nonwoven paper	[104]
	Diuron	TiO_2-β-SiC foam	[4, 68]
	Carbofuran	TiO_2/glass plate	[105]
	Alachlor	TiO_2/glass tube	[40]
Dyes	Drimarene	TiO_2/glass fibers	[106]
	Indigo carmine and Congo red	TiO_2-nonwoven paper	[104]
	Methylene blue	TiO_2-activated carbon fiber	[107]
Aromatic and organochloride compounds	Aniline	TiO_2/Ni plate	[108]
Endocrine disruptor	17-β-Oestradiol	TiO_2/Ti-6Al-4V alloy plate	[112]

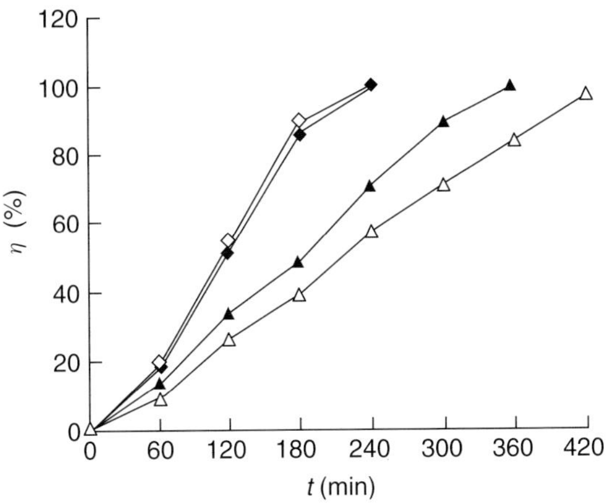

Figure 6.11 Relationship between the photodegradation efficiency and the solar irradiation time with floating $TiO_2{-}SiO_2$ photocatalyst beads: ◇, dichlorvos; ♦, monocrotophos; ▲, parathion; and △, phorate [103].

different structures can be completely photocatalytically degraded into phosphate ions after 420 min of irradiation by sunlight.

Two main *commercially available photocatalytic products* have been developed by the Finnish Ahlström and French Saint-Gobain Quartz companies. Photocatalytic papers have been engineered by the Ahlstrom Company (Pont-Evêque, France). They consist of nonwoven textiles made of synthetic cellulosic fibers, on which TiO_2 Millennium PC 500 ($18\,g\,m^{-2}$) is fixed onto the fibers by coating/compression through a size press process, using Snowtex 50 colloidal SiO_2 at $20\,g\,m^{-2}$ (Nissan Chemical) as the inorganic binder for anchoring TiO_2 onto the cellulosic textile fibers [109]. In addition, the binder acts for protecting the cellulosic textile fibers from any possible long-term photocatalytic degradation, whereas the photocatalytic material takes advantage of the use of a carbon-free binder for preventing any possible long- or even middle-term destruction of the photocatalytic materials. Different grades are proposed, including or not extra UOP 2000 zeolite ($2\,g\,m^{-2}$) as adsorbent particles within the coated inorganic paste, or sandwiching an adsorptive activated charcoal inner layer between two layers of nonwoven textile fibers (Figure 6.12). Both cases aimed at enhancing the adsorption of the pollutants and of the reaction intermediate by-products on the textile fibers for improving mineralization efficiencies and globally smoothing possible pollution peaks.

These photocatalytic materials are easy to be integrated in a photocatalytic reactor. They have been used by many laboratories for the treatment of volatile organic compounds (VOCs) and also for water treatment [104, 106, 110]. This photocatalytic material is used in industrial systems: Phytocat® and Phytomax® developed, respectively, by the French companies RESOLUTION (*www.resolutionlesite.com*) and AGRO-ENVIRONMENT (*www.agroenvironnement.com*) for the treatment of water

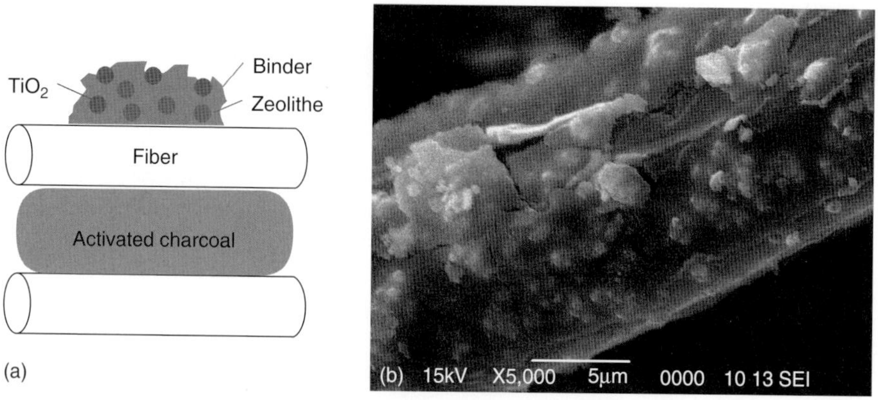

Figure 6.12 Conceptual drawing (a) and SEM image (b) of the photocatalytic paper developed by the Ahlstrom Company: 1048 Grade: TiO_2 + zeolite (75 g m^{-2}).

effluents containing pesticides. These two processes have an annual treatment capacity of about 12 m^3.

Saint-Gobain Quartz has developed the Quartzel® photocatalytic substrate, made from entangled needle-punched quartz fibers supporting sol–gel TiO_2 nanoparticles (Figure 6.13). With long curly fibers of 8–12 μm in diameter, it has an areal weight of 120 g m^{-2}, corresponding to a density of 80 g m^{-2} for bare fibers on which 40 g m^{-2} of TiO_2 sol–gel has been coated and exposes an overall specific surface area of 40 m^2 g^{-1}. In addition, Saint-Gobain Quartz has also internally developed other photocatalytic products remaining noncommercialized up to now [111]. They have designed new photocatalytic felts based on needle-punched fibers, differing from the commercial Quartzel® substrate by the amount of TiO_2, lowered down to 7, 10, or 15 wt% compared to 20 wt%. They have also developed a photocatalytic textile made of woven fibers coated with TiO_2 (450 g m^{-2} global areal weight but the coated TiO_2 density in grams per meter squared was

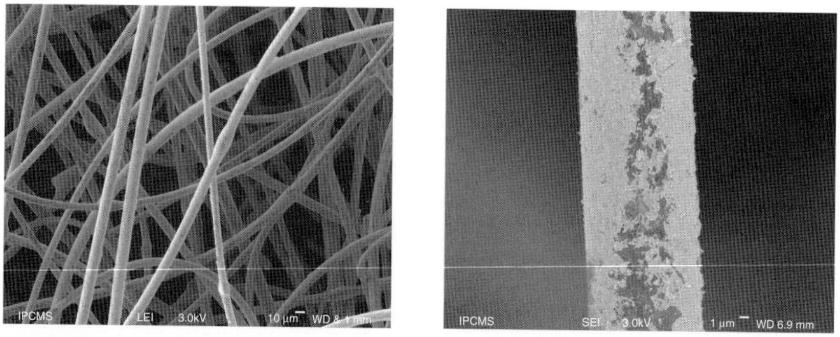

Figure 6.13 SEM images of the Quartzel® photocatalytic felt substrate developed by Saint-Gobain Quartz.

maintained confidential), reported to benefit from a greater mechanical resistance compared to that shown by both Ahlstrom photocatalytic papers or Quartzel® felts.

Usually implemented inside reactors with a thickness of 20 mm, the Quartzel® photocatalytic felt has several advantages.The product exhibits a good UV transmission, resulting from the excellent UV transmission of the quartz fiber support and from the moderate volume density of the entangled fibers, for providing light into the core of the reactor rather than only at its external surface. It has also a very high exposed surface area compared, for instance, to honeycomb-shaped materials, and it can operate at low pressure drops, even about three times higher than on three-dimensional cellular carbon foams [71]. Finally, the complete mineral nature of the material, with no organic binder, together with the high purity of quartz fibers avoids any sensitivity to oxidation and any photocatalyst poisoning by sodium, respectively.

Actually, *industrial* applications of photocatalysis for water purification are used very little in the world. In the case of the treatment of air, photocatalysis with *supported catalyst* can be applied to many fields, whether it is indoors or outdoors. For example, some companies commercialize autonomous reactors for air purification (VOC treatment or microorganisms) and photocatalytic modules integrated into air-conditioning systems. In these devices, the titanium dioxide used as a photocatalyst is usually fixed on a suitable support.

6.5 Conclusion

According to the recent market survey and analysis carried out by the *BBC Research* high-tech specialist (*http://www.bbcresearch.com/*) under the auspices of the *European Photocatalysis Federation* (*www.photocatalysis-federation.eu*), the global market for photocatalyst-based products has increased from €500 millions in 2005 to €750 millions in 2010 and is expected to increase further up to €1500 millions in 2015. Among the six market sectors identified (construction, consumers, automotive, environment, energy, and health), the largest part accounting for about 75% of the global market and acting in the coming years as the driving force for the implementation of the photocatalysis technology, with still worldwide concern, is the construction sector with both outdoor and indoor products (self-cleaning glass, cement, concrete, and painting). However, the other application domains (that are of concern to both air and water treatment) are also expected to increase by 2015, from €150 to 300 millions. This study also indicated that, even if Asia is actually – and in the coming years will remain – the major market with 60% of the global revenues, the European market will comfort its second ranking, with a yearly 20% progression, well ahead of the American market. The photocatalysis market still suffers from the involvement of a limited number of industrial actors, as a result of the persistence of some technological bottlenecks. In the forthcoming decade, some of them should be overcome, at least partially, in order to reduce investment risks and thus to allow more development and growth of the market.

References

1. Matthews, R. and Mc Evoy, S. (1992) Photocatalytic degradation of phenol in the presence of near-UV irradiated titanium dioxide. *J. Photochem. Photobiol. A Chem.*, **64**, 231–246.
2. Sabate, J., Anderson, M., Aguado, M., Gimenez, J., and Cervera-March, S. (1992) Comparison of TiO_2 powder suspensions and TiO_2 ceramic membranes supported on glass as photocatalytic systems in the reduction of chromium(VI) S. *J. Mol. Catal.*, **71**, 57–68.
3. Chester, G., Anderson, M., and Read, H. (1993) A jacketed annular membrane photocatalytic reactor for wastewater treatment: degradation of formic acid and atrazine. *J. Photchem. Photobiol. A: Chem.*, **71**, 291–297.
4. Kouamé, N.A., Robert, D., Keller, V., Keller, N., Pham, C., and Nguyen, P. (2011) Preliminary study of the use of β-SiC foams as a photocatalytic support for water treatment. *Catal. Today*, **161**, 3–7.
5. Ao, C.H., Lee, S.C., and Yu, J.C. (2003) Photocatalyst TiO_2 supported on glass fiber for indoor air purification: effect of NO on the photodegradation of CO and NO_2. *J. Photochem. Photobio A: Chem.*, **156**, 171–177.
6. Iguchi, Y., Ichiura, H., Kitaoka, T., and Tanaka, H. (2003) Preparation and characteristics of high performance paper containing titanium dioxide photocatalyst supported on inorganic fiber matrix. *Chemosphere*, **53**, 1193–1199.
7. Nogueira, R.F.P. and Jardim, W.F. (1996) TiO_2-fixed-bed reactor for water decontamination using solar light. *Sol. Energy*, **56**(5), 471–477.
8. Lu, M.C., Roam, G.D., Chen, J.N., and Huang, C.P. (1993) Factors affecting the photocatalytic degradation of dichlorvos over titanium dioxide supported on glass. *J. Photochem. Photobiol. A: Chem.*, **76**, 103–110.
9. Tennakone, K., Tilakaratne, C.T.K., and Kottegoda, I.R.M. (1995) Photocatalytic degradation of organic contaminants in water with TiO_2 supported on polythene films. *J. Photchem. Photobiol. A: Chem.*, **87**, 177–179.
10. Ding, Z., HU, Y., Yu, P.L., Lu, G.Q., and Greenfield, P.F. (2001) Synthesis of anatase TiO_2 supported on porous solids by chemical vapor deposition. *Catal. Today*, **68**, 173–182.
11. Pozzo, R.L., Giombi, J.L., Baltanas, M.A., and Cassano, A.E. (2000) The performance in a fluidized bed reactor of photocatalysts immobilized onto inert supports. *Catal. Today*, **62**, 175–187.
12. Pozzo, R.L., Baltanas, M.A., and Cassano, A.E. (1997) Supported titanium oxide as photocatalyst in water decontamination: state of the art. *Catal. Today*, **39**, 219–231.
13. Shan, A.Y., Mohd Ghazi, T.I., and Rashid, S.A. (2010) Immobilisation of titanium dioxide onto supporting materials in heterogeneous photocatalysis: a review. *Appl. Catal. Gen.*, **389**, 1–8.
14. Yu, J.C., Lin, J., Lo, D., and Lam, S.K. (2000) Influence of thermal treatment on the adsorption of oxygen and photocatalytic activity of TiO_2. *Langmuir*, **16**(18), 7304–7308.
15. Djosic, M.S., Miskovic-Stankovic, V.B., Janackovic, T., Kacarevic-Popovic, Z.M., and Petrovi, R.D. (2006) Electrophoretic deposition and characterization of boehmite coatings on titanium substrate. *Colloids Surf. A*, **274**, 185–191.
16. Paz, Y. (2010) Application of TiO_2 photocatalysis for air treatment: Patents' overview. *Appl. Catal. B: Environ.*, **99**(3–4), 448–460.
17. (2004) Thermoset resin laminated wood. US Patent Publication No. US2004/0081834 A1.
18. (2006) A novel TiO_2 material and the coating methods thereof. International Patent Publication No. WO 2006/008434.
19. Josset, S., Hajiesmaili, S., Begin, D., Edouard, D., Pham-Huu, C., Lett, M.-C., Keller, N., and Keller, V. (2010) UV-A photocatalytic treatment of *Legionella pneumophila* bacteria

contaminated airflows through three-dimensional solid foam structured photocatalytic reactors. *J. Hazard. Mater.*, **175**(1–3), 372–381.

20. Ishikawa, Y. and Matsumoto, Y. (2001) Electrodeposition of TiO_2 photocatalyst into nano-pores of hard alumite. *Electrochim. Acta*, **46**, 2819–2824.
21. Jolivet, J.P., Henry, M., and Livage, J. (1994) *De la Solution à L'oxyde*, EDP Sciences et CNRS Édition, Paris.
22. Paz, Y. and Heller, A. (1997) Photo-oxidatively self-cleaning transparent titanium dioxide films on soda lime glass: the deleterious effect of sodium contamination and its prevention. *J. Mater. Res.*, **12**, 2759–2766.
23. Tada, H. and Tanaka, M. (1997) Dependence of TiO_2 photocatalytic activity upon its film thickness. *Langmuir*, **13**, 360–364.
24. Ghazzal, M.M., Chaoui, N., Aubry, E., Koch, A., and Robert, D. (2010) A simple procedure to quantitatively assess the photoactivity of titanium dioxide films. *J. Photochem. Photobiol. A Chem.*, **215**, 11–16.
25. Aubry, E., Ghazzal, M.M., Demange, V., Chaoui, N., Robert, D., and Billard, A. (2007) Poisoning prevention of TiO_2 photocatalyst coatings sputtered on soda-lime glass by intercalation of SiNx diffusion barriers. *Surf. Coat. Technol.*, **201**, 7706–7712.
26. Fernandez, A., Lassaletta, G., Jimenez, V.M., Justo, A., Gonzalez-Elipe, A.R., Herrmann, J.-M., Tahiri, H., and Ait-Ichou, Y. (1995) Preparation and characterization of TiO_2 photocatalysts supported on various rigid supports (glass, quartz and stainless steel). Comparative studies of photocatalytic activity in water purification. *Appl. Catal. B: Environ.*, **7**, 49–63.
27. Ma, Y., Qiu, J.-B., Cao, Y.-A., Guan, Z.-S., and Yao, J.-N. (2001) Photocatalytic activity of TiO_2 films grown on different substrates. *Chemosphere*, **44**, 1087–1092.
28. Nolan, M.G., Pemble, M.E., Sheel, D.W., and Yates, H.M. (2006) One step process for chemical vapour deposition of titanium dioxide thin films incorporating controlled structure nanoparticles. *Thin Solid Films*, **515**, 1956–1962.
29. Dunnill, C.W.H., Aiken, Z.A., Pratten, J., Wilson, M., Morgan, D.J., and Parkin, I.P. (2009) Enhanced photocatalytic activity under visible light in N-doped TiO_2 thin films produced by APCVD preparations using t-butylamine as a nitrogen source and their potential for antibacterial films. *J. Photochem. Photobiol. A*, **207**, 244–253.
30. Nizard, H., Kosinova, M.L., Fainer, N.I., Rumyantsev, Y.M., Ayupov, B.M., and Shubin, Y.V. (2008) Deposition of titanium dioxide from TTIP by plasma enhanced and remote plasma enhanced chemical vapor deposition. *Surf. Coat.Technol.*, **202**, 4076–4085.
31. Sarantopoulos, C. (2007) Photocatalyseurs à base de TiO_2 préparés par infiltration chimique en phase vapeur (CVI) sur supports microfibreux. PhD dissertation. Institut National Polytechnique of Toulouse, France, October 19th, 2007.
32. Decher, G., Hong, J.-D., and Schmitt, J. (1992) Buildup of ultrathin multilayer films by a self-assembly process: III. Consecutively alternating adsorption of anionic and cationic polyelectrolytes on charged surfaces. *Thin Solid Films*, **210/211**, 831–835.
33. Decher, G. (1997) Fuzzy nanoassemblies: toward layered polymeric multicomposites. *Science*, **277**, 1232–1237.
34. Priya, D.N., Modak, J.M., and Raichur, A.M. (2009) LbL fabricated poly(styrene sulfonate)/TiO_2 multilayer thin films for environmental applications. *ACS Appl. Mater. Interfaces*, **1**(11), 2684–2693.
35. Krogman, K.C., Zacharia, N.S., Grillo, D.M., and Hammond, P.T. (2008) Photocatalytic layer-by-layer coatings for degradation of acutely toxic agents. *Chem. Mater.*, **20**(5), 1924–1930.
36. Shibata, T., Sakai, N., Fukuda, K., Ebina, Y., and Sasaki, T. (2007) Photocatalytic properties of titania nanostructured films fabricated from titania nanosheets. *Phys. Chem. Chem. Phys.*, **9**(19), 2413–2420.
37. Nakajima, A., Akiyama, Y., Yanagida, S., Koike, T., Isobe, T., Kameshima, Y.,

and Okada, K. (2009) Preparation and properties of Cu-grafted transparent TiO_2-nanosheet thin films. *Mater. Lett.*, **63**(20), 1699–1701.

38. Dontsova, D., Steffanut, P., Felix, O., Keller, N., Keller, V., and Decher, G. (2011) Photocatalytically active polyelectrolyte/nanoparticle films for the elimination of a model odorous gas. *Macromol. Rapid Commun.*, **32**, 1145–1149.
39. Paz Y., (2009) in *Advances in Chemical Engineering Photocatalytic Technologies*, Vol. 36 (Eds. H.I. de Lasa and B. Serrano Rosales), Academic Press, pp. 289–336, 1–365.
40. Ryu, C.S., Kim, M.-S., and Kim, B.-W. (2003) Photodegradation of alachlor with the TiO_2 film immobilised on the glass tube in aqueous solution. *Chemosphere*, **53**, 765–771.
41. Mansilla, H.D., Bravo, C., Ferreyra, R., Litter, M.I., Jardim, W.F., Lizama, C., Freer, J., and Fernandez, J. (2006) Photocatalytic EDTA degradation on suspended and immobilized TiO_2. *J. Photochem. Photobiol. A Chem.*, **181**, 188–194.
42. Qiu, W. and Zheng, Y. (2007) A comprehensive assessment of supported titania photocatalysts in a fluidized bed photoreactor: Photocatalytic activity and adherence stability. *Appl. Catal. B Environ.*, **71**, 151–162.
43. Takeda, S., Yamamoto, K., Hayashi, Y., and Matsumoto, K. (1999) Surface OH group governing wettability of commercial glasses. *J. Non-Cryst. Solids*, **249**, 41–46.
44. Kanki, T., Hamasaki, S., Shinpei, S., Noriaki, T., Toyoda, A., and Hirano, K. (2005) Water purification in a fluidized bed photocatalytic reactor using TiO_2-coated ceramic particles. *Chem. Eng. J.*, **108**, 155–160.
45. Keshmiri, M., Mohseni, M., and Troczynski, T. (2004) Development of novel TiO_2 sol–gel-derived composite and its photocatalytic activities for trichloroethylene oxidation. *Appl. Catal. B Environ.*, **53**, 209–219.
46. Fabiyi, M.E. and Skelton, R.L. (2000) Photocatalytic mineralisation of methylene blue using buoyant TiO_2-coated polystyrene beads. *J. Photochem. Photobiol. A: Chem.*, **132**, 121–128.
47. Magalhaes, F. and Lago, R.M. (2009) Floating photocatalysts based on TiO2 grafted on expanded polystyrene beads for the solar degradation of dyes. *Sol. Energy*, **83**, 1521.
48. Denny, F., Scott, J., Pareek, V., Peng, G.D., and Amal, R. (2009) CFD modelling for a TiO_2-coated glass-bead photoreactor irradiated by optical fibres: photocatalytic degradation of oxalic acid. *Chem. Eng. Sci.*, **64**, 1695–1706.
49. Chiou, C.-S., Shie, J.-L., Chuang, C.-Y., Liu, C.-C., and Chang, C.-T. (2006) Di-n-butyl phthalate using photoreactor packed with TiO_2 immobilized on glass beads. *J. Hazard. Mater. B*, **137**, 1123–1129.
50. Ma, M., Li, Y., Chen, W., and Li, L. (2010) Preparation of cost-effective TiO_2-outerloaded porous lava composites using supercritical CO_2 and their photocatalytic activity for methylene blue degradation. *Chin. J. Catal.*, **31**(10), 1221–1226.
51. Machado, L.C.R., Torchia, C.B., and Lago, R.M. (2006) Floating photocatalysts based on TiO_2 supported on high surface area exfoliated vermiculite for water decontamination. *Catal. Commun.*, **7**, 538–541.
52. Chuan, X.-Y., Hirano, M., and Inagaki, M. (2004) Preparation and photocatalytic performance of anatase-mounted natural porous silica, pumice, by hydrolysis under hydrothermal conditions. *Appl. Catal. B Environ.*, **51**, 255–260.
53. Ramanathan, K., Avnir, D., Modestov, A., and Lev, O. (1997) Sol-gel derived ormosil-exfoliated graphite-TiO_2 composite floating catalyst: photodeposition of copper. *Chem. Mater.*, **9**, 2533–2540.
54. Taranto, J., Frochot, D., and Pichat, P. (2007) Modeling and optimizing irradiance on planar, folded, and honeycomb shapes to maximize photocatalytic air purification. *Catal. Today*, **122**, 66–77.
55. Hall, R.J., Bendfeldt, P., Obee, T.N., and Sangiovanni, J.J. (1998) Computational and experimental studies of UV/titania photocatalytic oxidation of

VOCs in honeycomb monoliths. *J. Adv. Oxid. Technol.*, **3**, 243–252.
56. Raupp, G.B., Alexiadis, A., Hossain, M.M., and Changrani, R. (2001) First-principles modeling, scaling laws and design of structured photocatalytic oxidation in reactors for air purification. *Catal. Today*, **69**, 41–49.
57. Hossain, M.M., Raupp, G.B., Hay, S.O., and Obee, T.N. (1999) Three-dimensional developing flow model for photocatalytic monolith reactors. *AIChE J.*, **45**, 1309–1321.
58. Obee, T.N. and Brown, R. (1995) TiO_2 Photocatalysis for indoor Air applications: effects of humidity and trace contaminant levels on the oxidation rates of formaldehyde, toluene, and 1,3-butadiene. *Environ. Sci. Technol.*, **29**(5), 1223–1231.
59. Moulijn, J.A., Kreutzer, M.T., Nijhuis, T.A., and Kapteijn, F. (2011) Monolithic catalysts and reactors: high precision with Low energy consumption. *Adv. Catal.*, **54**, 249–327.
60. Lin, H.F. and Valsaraj, K.T. (2005) Development of an optical fiber monolith reactor for photocatalytic wastewater treatment. *J. Appl. Electrochem.*, **35**, 699–708.
61. Taranto, J., Frochot, D., and Pichat, P. (2009) Photocatalytic air purification: comparative efficacy and pressure drop of a TiO_2-coated thin mesh and a honeycomb monolith at high air velocities using a 0.4 m^3 close-loop reactor. *Sep. Purif. Technol.*, **67**(2), 187–193.
62. Ávila, P., Sánchez, B., Cardona, A.I., Rebollar, M., and Candal, R. (2002) Influence of the methods of TiO_2 incorporation in monolithic catalysts for the photocatalytic destruction of chlorinated hydrocarbons in gas phase. *Catal. Today*, **76**(2–4), 271–278.
63. Sanchez, B., Coronado, J.M., Candal, R., Portela, R., Tejedor, I., Anderson, M.A., Tompkins, D., and Lee, T. (2006) Preparation of TiO_2 coatings on PET monoliths for the photocatalytic elimination of trichloroethylene in the gas phase. *Appl. Catal. B: Environ.*, **66**, 295.
64. Portela, R., Suárez, S., Tessinari, R.F., Hernández-Alonso, M.D., Canela, M.C., and Sánchez, B. (2011) Solar/lamp-irradiated tubular photoreactor for air treatment with transparent supported photocatalysts. *Appl. Catal. B: Environ.*, **105**(1–2), 95–102.
65. Richardson, J.T., Peng, Y., and Remue, D. (2000) Properties of ceramic foam catalyst supports: pressure drop. *Appl. Catal. Gen.*, **204**, 19–34.
66. Groppi, G. and Tronconi, E. (2000) Design of novel monolith catalyst supports for gas/solid reactions with heat exchange. *Chem. Eng. Sci.*, **55**(12), 2161–2171.
67. Lacroix, M., Nguyen, P., Schweich, D., Pham-Huu, C., Savin-Poncet, S., and Edouard, D. (2007) Pressure drop measurements and modeling on SiC foams. *Chem. Eng. Sci.*, **62**(12), 3259–3267.
68. Kouamé, N.A., Robert, D., Keller, V., Keller, N., Pham, C. and Nguyen, P. (2012) TiO_2/SiC foam-structured photoreactor for continuous waste water treatment. *Environ. Sci. Pollut. Res.*, **19**, pp. 3727–3734.
69. Hajiesmaili, S., Josset, S., Bégin, D., Pham-Huu, C., Keller, N., and Keller, V. (2010) 3D solid carbon foam-based photocatalytic materials for vapor phase flow-through structured photoreactors. *Appl. Catal. A: Gen.*, **382**(1), 122–130.
70. Huu, T.T., Lacroix, M., Pham-Huu, C., Schweich, D., and Edouard, D. (2009) Towards a more realistic modeling of solid foam: use of the pentagonal dodecahedron geometry. *Chem. Eng. Sci.*, **64**(24), 5131–5142.
71. Edouard, D., Lacroix, M., Pham, C., Mbodjic, M., and Pham-Huu, C. (2008) Experimental measurements and multiphase flow models in solid SiC foam beds. *AIChE J.*, **54**(11), 2823–2832.
72. Jain, P. and Pradeep, T. (2005) Potential of silver nanoparticle-coated polyurethane foam as an antibacterial water filter. *Biotechnol. Bioeng.*, **90**(1), 59–63.
73. Plantard, G., Correia, F., and Goetz, V. (2011) Kinetic and efficiency of TiO_2-coated on foam or tissue and TiO_2-suspension in a photocatalytic reactor applied to the degradation of

the 2,4-dichlorophenol. *J. Photochem. Photobiol. A: Chem.*, **222**, 111–116.
74. Plantard, G., Goetz, V., Correia, F., and Cambon, J.P. (2011) Importance of a medium's structure on photocatalysis: using TiO_2-coated foams. *Sol. Energy Mater. Sol. Cells*, **95**, 2437–2442.
75. Plantard, G., Goetz, V., and Sacco, D. (2011) TiO_2-coated foams as a medium for solar catalysis. *Mater. Res. Bull.*, **46**, 231–234.
76. Xiong, Z., Zhang, G., Xiong, L., Fu, L., and Pan, H. (2011) Photocatalytic inactivation of chlorella by TiO_2/foam nickel. *Appl. Mech. Mater.*, **71–78**, 2944–2947.
77. Zhu, S., Yang, X., Wang, G.-N., Zhang, L.-L., and Zhu, H.-F. (2011) Facile preparation of P-25 films Dip-coated nickel foam and high photocatalytic activity for the degradation of quinoline and industrial wastewater. *Int. J. Chem. Reactor Eng.*, **9**(1), 2605–2611.
78. Yuan, J., Hui, H., Chen, M., Shi, J., and Shangguan, W. (2008) Promotion effect of Al_2O_3-SiO_2 interlayer and Pt loading on TiO_2/nickel-foam photocatalyst for degrading gaseous acetaldehyde. *Catal. Today*, **139**, 140–145.
79. Yang, L., Liu, Z., Shi, J., Hu, H., and Shangguan, W. (2007) Design consideration of photocatalytic oxidation reactors using TiO_2-coated foam nickels for degrading indoor gaseous formaldehyde. *Catal. Today*, **126**(3–4), 359–368.
80. Hu, H., Xiao, W.J., Yuan, J., Shi, J.W., Chen, M.X., and Shangguan, W.F. (2007) Preparations of TiO_2 film coated on foam nickel substrate by sol–gel processes and its photocatalytic activity for degradation of acetaldehyde. *J. Environ. Sci.*, **19**, 80–85.
81. Plesch, G., Gorbar, M., Vogt, U.F., Jesenak, K., and Vargova, M. (2009) Reticuled macroporous ceramic foam supported TiO_2 for photocatalytic applications. *Mater. Lett.*, **63**, 461–463.
82. Vargova, M., Plesch, G., Vogt, U.F., Zahoran, M., Gorbar, M., and Jesenak, K. (2011) TiO_2 thick films supported on reticulated macroporous Al_2O_3 foams and their photoactivity in phenol mineralization. *Appl. Surf. Sci.*, **257**, 4678–4684.
83. Kato, S., Hirano, Y., Iwata, M., Sano, T., Takeuchi, K., and Matsuzawa, S. (2005) Photocatalytic degradation of gaseous sulfur compounds by silver-deposited titanium dioxide. *Appl. Catal. B: Environ.*, **57**, 109–115.
84. Ochuma, I.J., Osibo, O.O., Fishwick, R.P., Pollington, S., Wagland, A., Wood, J., and Winterbottom, J.M. (2007) *Catal. Today*, **128**, 100.
85. Yao, Y., Ochiai, T., Ishiguro, H., Nakano, R., and Kubota, Y. (2011) Antibacterial performance of a novel photocatalytic-coated cordierite foam for use in air cleaners. *Appl. Catal. B: Environ.*, **106**(3–4), 592–599.
86. (a) Keller, N., Pham-Huu, C., Roy, S., Estournès, C., Guille, J., and Ledoux, M.J. (1999) Influence of the reaction conditions on the formation of a high surface area silicon carbide for use as heterogeneous support material. *J. Mater. Sci.*, **34**(13), 3189–3202; (b) Nhut, J.M., Vieira, R., Pesant, L., Tessonnier, J.P., Ehret, G., Keller, N., Pham-Huu, C., and Ledoux, M.J. (2002) Synthesis and catalytic uses of carbon and silicon carbide nanostructures. *Catal. Today*, **76**, 11–32; (c) The SICAT company, located at Industriepark-B310 Industriestraße 1 D-77731 Willstätt, Germany, is provding customized beat silicon carbide based catalyst supports, *www.sicatcatalyst.com.*
87. Rodriguez, P., Meille, V., Pallier, S., and Sawah, M.A.A. (2009) Deposition and characterisation of TiO_2 coatings on various supports for structured (photo)catalytic reactors. *Appl. Catal. A: Gen.*, **360**, 154–162.
88. Nguyen, P. and Pham, C. (2011) Innovative porous SiC-based materials: From nanoscopic understandings to tunable carriers serving catalytic needs. *Appl. Catal. A: Gen.*, **391**, 443–454.
89. Keller, N., Keller, V., Barraud, E., Garin, F., and Ledoux, M.J. (2004) Synthesis and characterisation of a new medium surface area TiO_2-β-SiC material for use as photocatalyst. *J. Mater. Chem.*, **14**(12), 1887–1892.

90. Carn, F., Colin, A., Achard, M.-F., Deleuze, H., Sanchez, C., and Backov, R. (2005) Anatase and rutile $Ti0_2$ macrocellular foams: Air-liquid foaming Sol–gel process towards controlling cell sizes, morphologies, and topologies. *Adv. Mater.*, **17**(1), 62–66.
91. Furman, M., Corbel, S., Le Gall, H., Zahraa, O., and Bouchy, M. (2007) Influence of the geometry of a monolithic support on the efficiency of photocatalyst for air cleaning. *Chem. Eng. Sci.*, **62**, 5312–5316.
92. Marinangeli, R.E. and Ollis, D.F. (1977) Photoassisted heterogeneous catalysis with optical fibers. I. Isolated single fiber. *AIChE J.*, **23**, 415–426.
93. Marinangeli, R.E. and Ollis, D.F. (1982) Photo-assisted heterogeneous catalysis with optical fibers. Part III: photoelectrodes. *AIChE J.*, **28**, 945–955.
94. Hofstadler, K., Bauer, R., Novalic, S., and Heisler, G. (1994) New reactor design for photocatalytic wastewater treatment with TiO_2 immobilized on fused-silica glass fibers: photomineralization of 4-chlorophenol. *Environ. Sci. Technol.*, **28**, 670–674.
95. Peill, N.J. and Hoffmann, M.R. (1995) Development and optimization of a TiO_2-coated fiber-optic cable reactor: photocatalytic degradation of 4-chlorophenol. *Environ. Sci. Technol.*, **29**, 2974–2981.
96. Danion, A., Disdier, J., Guillard, C., Abdelmalek, F., and Jaffrezic-Renault, N. (2004) A novel photocatalytic monolith reactor for multiphase heterogeneous photocatalysis. *Appl. Catal. B: Environ.*, **52**, 213–223.
97. Choi, W., Ko, J.Y., Park, H., and Chung, J.S. (2001) Investigation on TiO_2-coated optical fibers for Gas-phase photocatalytic oxidation of acetone. *Appl. Catal. B Environ.*, **31**, 209–220.
98. Wang, W. and Ku, Y. (2003) Photocatalytic degradation of gaseous benzene in air streams by using an optical fiber photoreactor. *J. Photochem. Photobiol. A*, **159**, 47–59.
99. Wang, W. and Ku, Y. (2003) The light transmission and distribution in an optical fiber coated with TiO_2-particles. *Chemosphere*, **50**, 999–1006.
100. Lin, H.F. and Valsaraj, K.T. (2006) Optical fiber monolith reactor for photocatalytic wastewater treatment: mathematical model and experimental validation. *AIChE J.*, **52**(6), 2271–2280.
101. Van Gerven, T., Mul, G., Moulijn, J., and Stankiewicz, A. (2007) A review of intensification of photocatalytic processes. *Chem. Eng. Process: Proc. Int.*, **46**(9), 781–789.
102. Du, P., Carneiro, J.T., Moulijn, J.A., and Mul, G. (2008) A novel photocatalytic monolith reactor for multiphase heterogeneous photocatalysis. *Appl. Catal. A: Gen.*, **334**, 119–128.
103. Shifu, C. and Gengyu, C. (2005) Photocatalytic degradation of organophosphorus pesticides using floating photocatalyst TiO_2-SiO_2/beads by sunlight. *Sol. Energy*, **79**, 1–9.
104. Guillard, C., Disdier, J., Monnet, C., Dussaud, J., Malato, S., Blanco, J., Maldonado, M.I., and Herrmann, J.M. (2003) Solar efficiency of a new deposited titania photocatalyst: chlorophenol, pesticide and dye removal applications. *Appl. Catal. Environ.*, **46**, 319–332.
105. Tennakone, K., Tilikaratne, C.T.K., and Kottegoda, I.R.M. (1997) Photomineralization of carbofuran by TiO_2-supported catalyst. *Water Res.*, **31**, 1909–1912.
106. Alinsafi, A., Evenou, F., Abdulkarim, E.M., Pons, M.N., Zahraa, O., Benhammou, A., Yaacoubi, A., and Nejmeddine, A. (2007) Treatment of textile industry wastewater by supported photocatalysis. *Dyes Pigm.*, **74**, 439–445.
107. Fu, P., Luan, Y., and Dai, X.,.P. (2004) Preparation of activated carbon fibers supported TiO_2 photocatalyst and evaluation of its photocatalytic reactivity. *J. Mol. Catal. A: Chem.*, **221**, 81–88.
108. Wenhua, L., Hong, L., Sao'an, C., Jianqing, Z., and Chunan, C. (2000) Kinetics of photocatalytic degradation of aniline in water over TiO_2 supported on porous nickel. *J. Photochem. Photobiol., A Chem.*, **131**, 125–132.
109. (2009) European Patent EP1069950 B1, Ahlstrom Research and Competence Center.

110. Atheba, P., Robert, D., Trokourey, A., Bamba, D., and Weber, J.V. (2009) Design and study of a cost-effective solar photoreactor for pesticide removal from water. *Water Sci. Technol.*, **60**, 2187–2201.

111. Huchon, R. (2006) Activité photocatalytique de catalyseurs déposés sur différents supports (<< medias >>). Application à la conception d'un photoréacteur pilote PhD dissertation. Claude Bernard University, Lyon.

112. Coleman, M.H., Eggins, B.R., Byrne, J.A., Palmer, F.L., and King, E. (2000) Photocatalytic degradation of 17-b-oestradiol on immobilised TiO_2. *Appl. Catal. B: Environ.*, **24**, L1–L5.

7
Wastewater Treatment Using Highly Functional Immobilized TiO_2 Thin-Film Photocatalysts

Masaya Matsuoka, Takashi Toyao, Yu Horiuchi, Masato Takeuchi, and Masakazu Anpo

7.1
Introduction

Environmental pollution as well as the lack of natural energy resources has drawn much attention to the vital need for ecologically clean materials and chemical technologies. Photocatalysis is one of the most promising solutions not only to address the environmental contaminants but also as an effective way to harness solar energy as an alternative natural resource because of the exhaustion of conventional energy sources. Among the various semiconducting metal oxides investigated, it has been reported that titanium oxide (TiO_2) or such titanate compounds can act as efficient photocatalysts for the decomposition of toxic organic compounds in air or water (change in Gibbs free energy: $\Delta G < 0$) [1–11] as well as for energy storage reactions ($\Delta G > 0$) such as water splitting [12–16] or CO_2 reduction with water [17–19] under irradiation of the 3–4% of the UV light from sunlight, the most clean and unlimited energy resource. Thus, since the pioneering discovery of the photosensitization effect of TiO_2 electrodes on the electrolysis of water into H_2 and O_2 by Honda and Fujishima in 1972 [12], photocatalysis by TiO_2 semiconductors has been widely studied for the efficient conversion of solar light energy into useful chemical energy. In this reaction, OH^- ions are oxidized into O_2 by the photoformed holes at the TiO_2 photoanode and, at the same time, H^+ ions are reduced into H_2 by the photoformed electrons at the counter electrode, enabling a stoichiometric water splitting reaction. Since the late 1970s when Bard *et al.* reported the photocatalytic decomposition of cyanide in aqueous TiO_2 suspensions [1, 2], TiO_2 photocatalysts began to be applied for the detoxification of various harmful compounds in both water and air because of their photoinduced oxidation ability [3–11]. Reiche and Bard [2] also reported that TiO_2 particles loaded with Pt can be considered "a short-circuited photoelectrochemical cell" providing both the oxidizing and reducing sites for the reaction. Various active oxygen species having oxidation ability such as O_2^- or $OH^{\bullet}$ and produced by the following processes have been reported to be responsible for detoxification reactions on Pt-loaded TiO_2 photocatalysts (Chapter 1) [7–10].

Photocatalysis and Water Purification: From Fundamentals to Recent Applications, First Edition. P. Pichat.

$$e^- + O_2 \rightarrow O_2^- \text{ (at Pt surface)}$$

$$h^+ + OH^- \rightarrow OH^{\bullet} \text{ (at } TiO_2 \text{ surface)}$$

Heterogeneous photocatalysis utilizing powdered TiO_2 is, thus, considered as one of the most efficient and advanced oxidation technologies for water and air purification involving harmful compounds. However, these purification systems have a drawback, especially in water purification systems, in that the final filtration process of the TiO_2 powder from water is costly. Therefore, the fixation or immobilization of TiO_2 photocatalysts on various substrates to avoid the filtration process is strongly desired. To address these concerns, a number of techniques have been applied to immobilize TiO_2 photocatalysts onto various substrates (see the chapter 6). These techniques involve sol–gel methods, sputtering methods, as well as the deposition treatment of TiO_2 powder using inorganic binders. The present chapter concentrates on the application of immobilized TiO_2 photocatalysts for various water treatment processes. Moreover, recent advances in the preparation of visible-light-responsive TiO_2 thin films are also introduced (Chapter 8).

7.2 Application of a Cascade Falling-Film Photoreactor (CFFP) for the Remediation of Polluted Water and Air under Solar Light Irradiation

The construction of a cascade falling-film photoreactor (CFFP) has been reported, and its efficiency has been compared with that of a compound parabolic collector (CPC) photoreactor using a TiO_2 slurry under solar light irradiation [20–22] (Chapter 15). The schematic diagram of the CFFP and slurry CPC photoreactor is shown in Figure 7.1. The CPC pilot plant was made up of a 1 m^2 CPC photoreactor

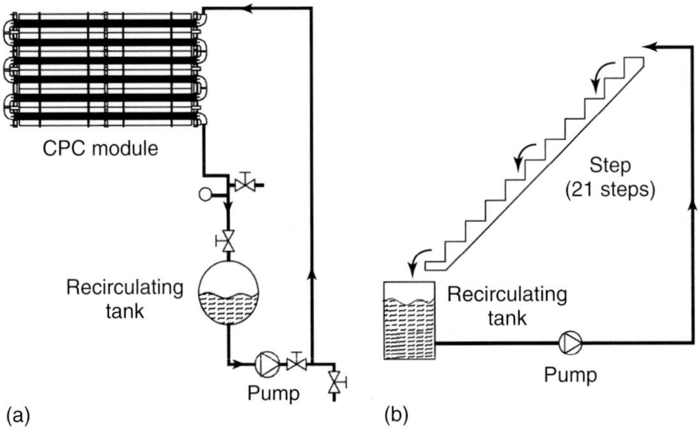

Figure 7.1 The two pilot photoreactors used in the experiment performed in the presence of (a) TiO_2 powder (CPC photoreactor) and (b) coated titania (CFFP photoreactor).

with one tank and one pump. The CPC photoreactor consisted of eight Pyrex tubes connected in a series and mounted on a fixed platform tilted at an angle of 37°. The TiO_2 (Millennium PC500) slurry flowed within the CPC pilot plant at a rate of 20 L min^{-1} and the total volume of the CPC pilot plant (35 L) was divided into two parts: the total irradiated volume in Pyrex tubes (22 L) and dead volume (tank + connecting tubing: 13 L). Each Pyrex tube was surrounded by a CPC reflector made of a highly reflective anodized aluminum sheet held by a galvanized frame. The reflection simulation results of the CPC reflector showed that sunlight was effectively reflected to the Pyrex tube with negligible optical loss, independent of the incident angle of sunlight [21]. Figure 7.2 shows the rates of 4-chlorophenol (4-CP) and total organic carbon (TOC) degradation over the accumulated sunlight energy irradiated on the CPC pilot plant as a function of TiO_2 slurry concentration. No increase in efficiency could be observed in the region above 0.5 g L^{-1}. This value was, thus, selected for the TiO_2 slurry concentration of the CPC photoreactor because too high a concentration of TiO_2 can induce a detrimental screening effect. Figure 7.3 shows the

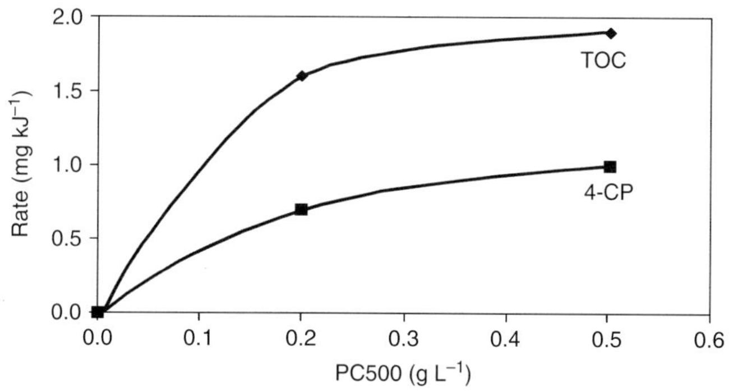

Figure 7.2 Rates of 4-CP and TOC degradation over accumulated light energy (in milligram removed per kilojoule collected) as a function of the TiO_2 Millennium PC500 concentration.

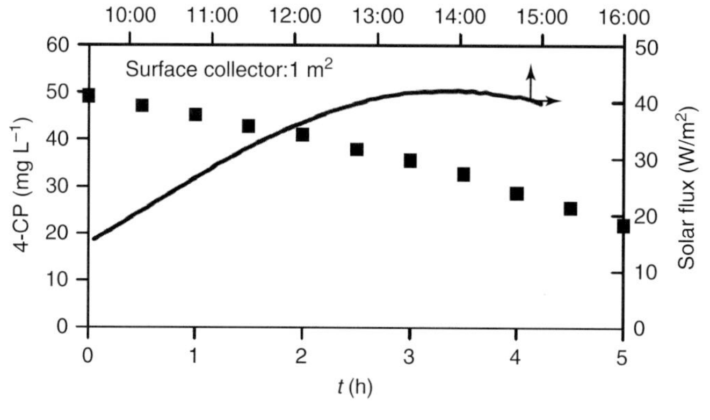

Figure 7.3 Variation of 4-CP concentration in the CPC photoreactor and UV-radiant flux as a function of local time during photocatalytic degradation with Millennium PC500 performed on 1 m^2 of irradiated photoreactor surface.

changes in the 4-CP concentration for the CPC photoreactor and solar UV-radiant flux as a function of the irradiation time. It was well demonstrated that 4-CP was efficiently decomposed under sunlight irradiation by the use of a CPC photoreactor.

In contrast to the CPC photoreactor, which utilizes the TiO_2 slurry as a photocatalyst, the CFFP utilized nonwoven paper (natural and synthetic fibers, 2 mm thick) coated with PC500 using an inorganic SiO_2 binder (TiO_2 content: 27 g m^{-2}). As shown in Figure 7.1, the CFFP solar reactor consisted of a rectangular (2 m × 0.5 m = 1 m^2), 70 mm high, stainless-steel staircase vessel having 21 steps. A 1.36 m^2 sheet of the nonwoven paper deposited with the PC500 photocatalyst was placed on the 21 steps of the CFFP photoreactor with a solar-radiation-collecting surface of 1 m^2. At the beginning of the experiment, 25 L of distilled water containing the organic compounds was circulated in the dark at 7.5 L min^{-1} in order to reach the same constant concentration as in the CPC photoreactor. Figure 7.4 shows the comparison of the degradation rate of organic compounds (4-CP and TOC) when using suspended Millennium PC500 or Degussa P25 for CPC and Millennium PC500 for nonwoven paper in the CFFP photoreactor. Independent of the kind of photocatalyst and photoreactor used, the degradation rate of 4-CP was found to be around 1.8 mg kJ^{-1} and for TOC, 1.2 mg kJ^{-1}, indicating that there was neither mass transfer limitation nor decrease in efficiency induced by the binder in the CFFP photoreactor. In some cases, lower mass transfer rates can

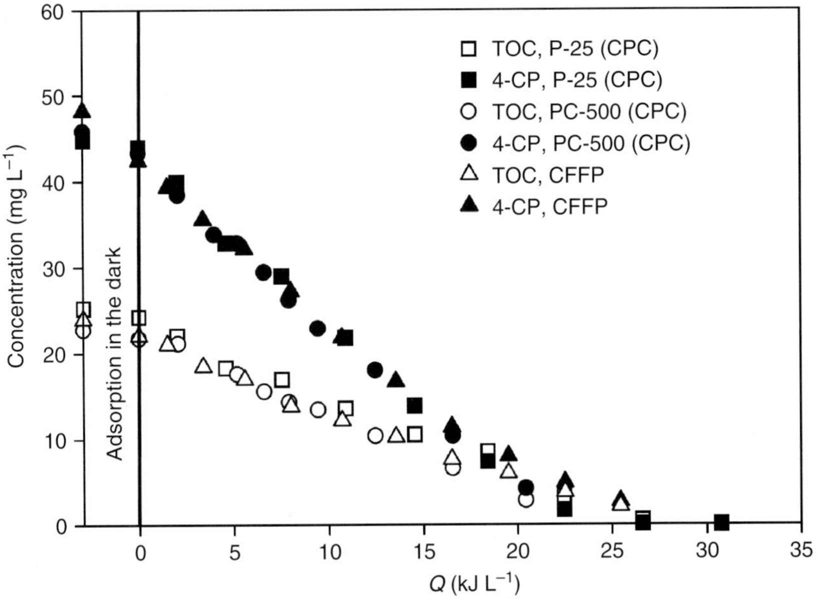

Figure 7.4 Comparison of the degradation of 4-CP and TOC when using suspended Millennium PC500 and Degussa P25 in CPC and Millennium PC500 coated on nonwoven paper in the CFFP photoreactor. Q (kJ L^{-1}) corresponds to the accumulated energy incident on the photoreactor per unit of volume of aqueous sample.

be a possible drawback for systems using the immobilized TiO_2 photocatalysts. However, these results clearly demonstrate that solar water purification using the immobilized TiO_2 photocatalyst at reaction rates similar to those obtained with slurries or the powdered TiO_2 photocatalyst was possible and comparative. The most significant advantage of applying the CFFP photoreactor is that unnecessary processes such as the final tedious filtration can be avoided, making it a good alternative to photoreactors using TiO_2 slurries or powders.

In a recent work, Suarez *et al.* have applied TiO_2–SiMgOx ceramic plates as a photocatalyst to a continuous flow Pyrex glass reactor system in which the photocatalyst plates were fixed in positions of focused sunlight using a CPC for the decomposition of trichloroethylene (TCE) in gas phase [23]. Magnesium silicate was used to prepare the ceramic plates and heat treated at 800 °C for 4 h. The resulting SiMgOx plates were immersed in a suspension of TiO_2 (G-5 Millenium: 30 g L^{-1}) by a slip-casting technique and heat treated at 500 °C for 4 h to produce the TiO_2–SiMgOx ceramic plate. As shown in Figure 7.5, TCE was

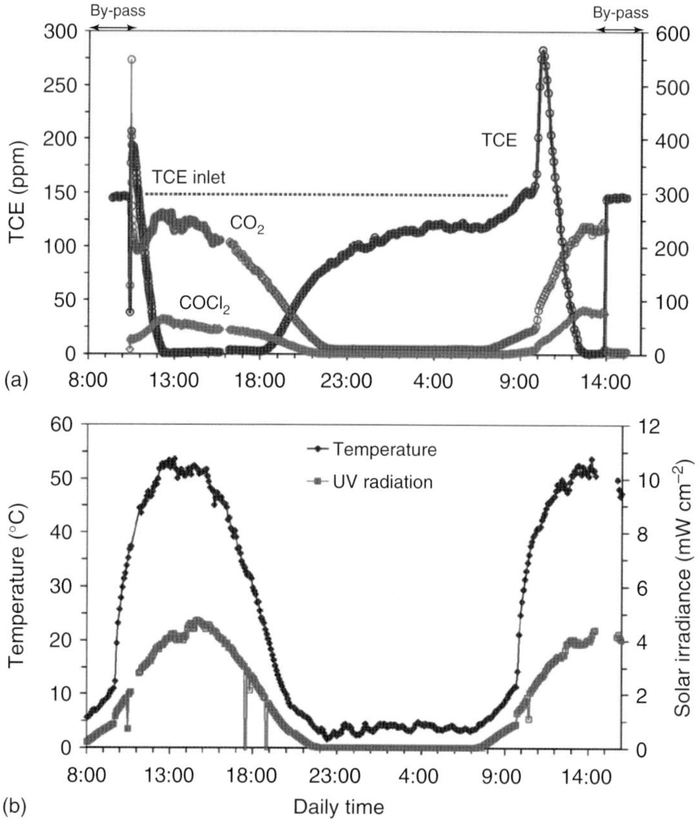

Figure 7.5 (a) Variation of the TCE concentration and CO_2, $COCl_2$ formation and (b) temperature and UV irradiation during a day–night cycle for the TiO_2–SiMgOx hybrid composite. Total gas flow = 1 L min^{-1}, (TCE) = 150 ppm, and air–gas balance.

effectively converted into CO_2 as a major product and $COCl_2$ as a minor product, their conversion corresponding well to the intensity of sunlight irradiation. These results demonstrated that a TiO_2 thin film catalyst deposited on a ceramic substrate can be applied for the remediation of polluted air under solar light irradiation.

7.3
Application of TiO_2 Thin-Film-Coated Fibers for the Remediation of Polluted Water

An effective method to immobilize TiO_2 photocatalysts on solid materials is the coating of inorganic fibers with TiO_2 thin films. In fact, Matsunaga *et al.* [24] have reported the preparation of Pd-deposited mesoporous TiO_2/SiO_2 fibers and their application for the decomposition of organic compounds in water. TiO_2/SiO_2 fibers with a gradient surface structure were fabricated by a precursor technique using polycarbosilane [25]. Polycarbosilane (–SiH(CH_3)–CH_2–)*n* containing $Ti(OC_4H_9)_4$ (50 wt%) was shaped into a fiber by melt spinning and then matured in air at 70 °C for 100 h. During this process, the titanium compounds oozed from the preceramic polymer by the so-called bleed-out phenomenon and moved to the surface of the fiber. After maturation, the fiber material was cured in air at 200 °C and then calcined in air at 1200 °C. The prepared fiber with a diameter of around 10 μm had an anatase TiO_2 surface layer (crystalline size: 8 nm) in the range 5–500 nm on the amorphous SiO_2 fiber. This TiO_2-sintered surface did not drop off after heat cycling, washing, or rubbing. TiO_2-coated SiO_2 fibers (TiO_2/SiO_2 fibers) were applied for photocatalytic reactions. The quartz tubular reactor (inner volume: 7 cm^3) with an UV lamp was filled with 2 g of fibers, and their photocatalytic activity was determined using air containing 50 ppm of acetaldehyde at a flow rate of 0.2 l min^{-1} under UV light irradiation at an intensity of 3 mW cm^{-2} (wavelength: 352 nm) [25]. Under these conditions (single pass), acetaldehyde was completely decomposed accompanied by the generation of CO_2. The fibers were also applied for photocatalytic coliform sterilization. The fibers (0.2 g) were placed in 20 ml of wastewater containing coliform at a concentration of 2×10^6 mL^{-1}. Under irradiation of UV light (wavelength: 352 nm, 2 mW cm^{-2}), all the coliform in the wastewater was found to be completely sterilized within 3 h.

In order to increase the photocatalytic activity of the TiO_2/SiO_2 fibers, Matsunaga *et al.* [24] treated the TiO_2/SiO_2 fibers with diluted HF aqueous solution. The concentration of the HF solution was 1.5 wt%. After etching, mesopores were created on the surface of the TiO_2/SiO_2 fibers because silica was easily dissolved by HF solution, while TiO_2 was stable against HF solution and not dissolved. Figure 7.6 shows the surface and cross section of the TiO_2/SiO_2 fiber after HF etching for 75 min. Many coherently bound TiO_2 particles (average particle size: 8 nm) were found on the surface, and mesopores were also found in the surface between the TiO_2 particles. It was clearly observed that mesoporous structures had formed in the TiO_2 surface layer. The photocatalytic activity of the mesoporous TiO_2/SiO_2 fiber was evaluated by methanesulfonate (MSA) production from dimethylsulfoxide (DMSO) aqueous solution (1000 ppm), as DMSO reacts with OH radicals to

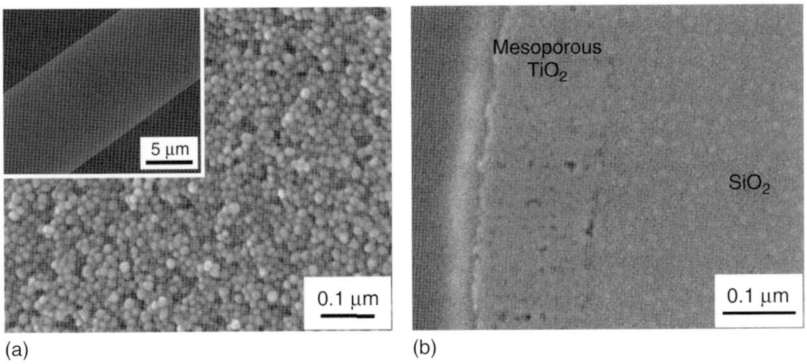

Figure 7.6 Scanning electron microscope (SEM) images of (a) surface and (b) cross section of the obtained mesoporous TiO_2/SiO_2 fibers.

produce MSA and the reaction rate can be a good index to estimate the activity of the photocatalysts. Here, the mesoporous TiO_2/SiO_2 fiber was deposited with the Pd cocatalyst to increase the photocatalytic activity. As shown in Figure 7.7, DMSO was efficiently converted into MSA on Pd-deposited mesoporous TiO_2/SiO_2 fibers. Figure 7.7 also shows the results of durability test. The MSA concentration was almost always constant regardless of the number of repetitions (up to four). These results suggested that the mesopore structure of the TiO_2 layer stably existed on the SiO_2 fiber without showing any peeling of the SiO_2 fiber surface. These

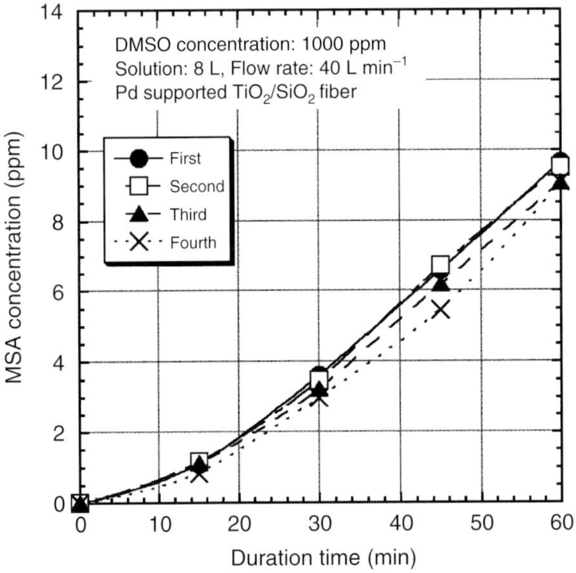

Figure 7.7 Effect of the duration time and repetition of reactions on the MSA concentration. Catalyst: Pd-supported mesoporous TiO_2/SiO_2 fiber.

observations confirmed that Pd-deposited mesoporous TiO_2/SiO_2 fibers can be applied as stable and effective photocatalysts for water purification.

7.4
Application of TiO_2 Thin Film for Photofuel Cells (PFC)

In recent years, photofuel cells (PFCs) have been constructed by using TiO_2 thin film as a photoanode and applied for electric energy production employing various kinds of organic compounds as fuel [26, 27]. PFC is a photoelectrochemical system that decomposes biomass and related compounds with a nanoporous TiO_2 film photoanode and an O_2-reducing cathode generating photocurrents. Kaneko *et al.* [26, 27] have reported that as long as the compound is either liquid or soluble (or solubilized) in water, it can be photodecomposed including polymeric compounds to produce electricity. Figure 7.8 shows the energy diagram of PFC (at pH 12) using a nanoporous TiO_2 film photoanode and a Pt cathode soaked in an aqueous solution of ammonia. Here, NH_3 was decomposed into N_2 and H^+ by photoformed holes at the TiO_2 photoanode, and O_2 was reduced to H_2O through its reaction with photoformed electrons and H^+, generating a photocurrent. The oxidation reaction proceeded as follows:

$$2NH_3 + O_2 + \text{UV light} \rightarrow \text{Photocurrents} + N_2 + 3H_2O$$

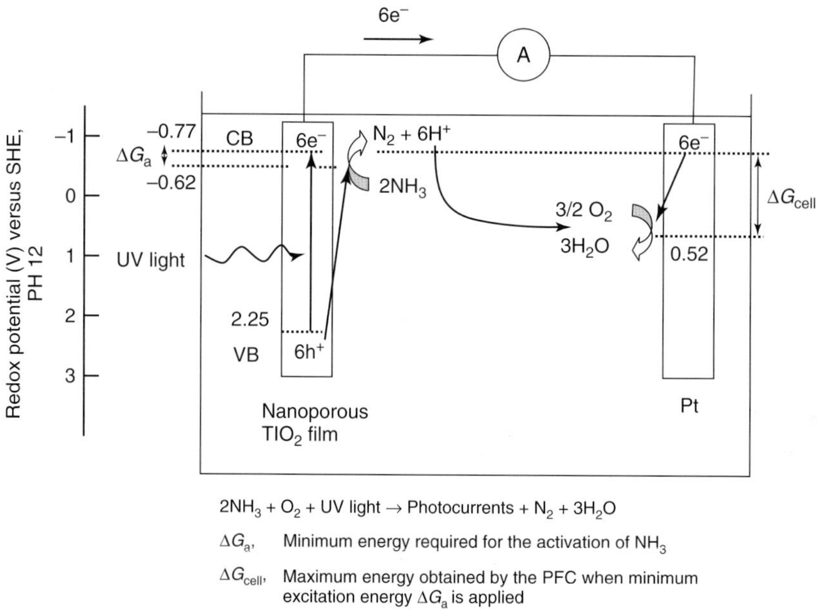

Figure 7.8 Energy diagram of PFC (at pH 12) using a nanoporous TiO_2 film photoanode and Pt cathode soaked in an aqueous solution of ammonia.

Table 7.1 Photofuel cell (PFC) characteristics of various compounds and their aqueous solutions containing 0.1 M Na_2SO_4 electrolyte by using a nanoporous TiO_2 photoanode (1 cm × 1 cm) and Pt/Pt black foil cathode (1 cm × 1 cm) under O_2 atmosphere with irradiation from a 500 W Xe lamp (light intensity: 503 mW cm^{-2}) operated at 25 °C.

Fuel (concentration) (M)	Solvent (pH[a])	V_{oc}/V	J_{sc} (mA cm^{-2})	FF
Methanol[b]	None	0.54	0.80	0.23
Methanol (50 vol%)	Water (not controlled)	0.44	0.76	0.28
Glucose (0.5)	Water (5)	0.64	0.50	0.32
Ammonia (10)	Water (12)	0.84	0.53	0.63
Glysine (0.5)	Water (5)	0.76	0.45	0.45
Glutamic acid (0.5)	Water (1)	0.90	0.64	0.42
Phenylalanine (0.5)	Water (13)	0.90	0.61	0.53
Gelatin (2 wt%)	Water (1)	0.64	0.23	0.32
Collagen (3 mg/ml)	Water (1)	0.62	0.16	0.34
Lignosulfonic acid (0.5 wt%)	Water (not controlled)	0.57	0.02	0.51
Polyethylene glycol (2 wt%)	Water (5)	0.60	0.28	0.27

[a] pH was adjusted by either H_2SO_4 or NaOH when necessary.
[b] Tetrabutylammonium perchlorate (TBAP) (0.1 M) electrolyte was used instead of Na_2SO_4.

In this system, the open circuit photovoltage (V_{oc}) was determined by the potential difference between the energy level of the conduction band of the TiO_2 photoanode (E_{CB}) and the redox potential of O_2 reduction at the cathode. PFC fabricated with a TiO_2/FTO (fluorine-doped tin oxide) photoanode and Pt-black/Pt cathode under 1 atm O_2 atmosphere using 10 M NH_3 as fuel showed an open circuit photovoltage (V_{oc}) of 0.84 V, short circuit photocurrent density (J_{sc}) of 0.53 mA cm^{-2} and fill factor (FF) of 0.63 under UV irradiation. The incident photon-to-current conversion efficiency (IPCE) was 19% based on the incident monochromatic light at 340 nm (intensity: 1.47 mW cm^{-2}). It was also found that PFC can generate a photocurrent by using various organic compounds as fuel (Table 7.1) [26, 27], demonstrating that PFC can be utilized for photocatalytic water purification systems for both the decomposition of various organic substances in water and the production of electric power under solar light irradiation.

7.5
Preparation of Visible-Light-Responsive TiO_2 Thin Films and Their Application to the Remediation of Polluted Water

Titanium dioxide is an attractive candidate as a photocatalyst for the reduction and purification of toxic compounds in polluted water and air because of its high reactivity as well as chemical stability. However, TiO_2 semiconductors have a relatively large bandgap of 3.2 eV, corresponding to wavelengths shorter than 388 nm so that the TiO_2 by itself can make use of only the 3–4% of solar energy reaching the earth, necessitating a UV light source for its application as a photocatalyst. Thus, TiO_2 photocatalysts that can operate efficiently and effectively under both UV and visible light irradiation would be the ideal for practical and widespread use. Here, we deal with recent advances in the preparation methods of visible-light-responsive TiO_2 thin films with special attention focused on their application to the remediation of polluted water.

7.5.1
Visible-Light-Responsive TiO_2 Thin Films Prepared by Cation or Anion Doping

Metal cation doping into TiO_2 photocatalysts has been carried out in various ways such as wet chemical method, high-temperature treatment, or ion implantation method [28]. Choi *et al.* [29] have performed a systematic study of TiO_2 nanoparticles doped with 21 metal ions by the sol–gel method and found that the presence of metal ion dopants significantly influenced the photoreactivity, charge carrier recombination rates, and interfacial electron transfer rates. Anpo *et al.* [30] have reported that ion implantation of various transition metal ions such as V, Cr, Mn, Fe, and Ni accelerated by high voltage into TiO_2 makes a large shift in the absorption band of TiO_2 toward the visible light region to be possible. In fact, Cr ion implantation on a TiO_2 thin film deposited on transparent porous Vycor glass (PVG) by an ionized cluster beam (ICB) method led to a smooth shift in the light absorption edge of TiO_2 to the visible light region, enabling the photocatalytic decomposition of NO into N_2 and O_2 under visible light ($\lambda > 450$ nm) irradiation at 275 K [30, 31]. Maeda and Yamada [32] have investigated the visible light photocatalytic activity of metal-doped TiO_2 films prepared by dip coating method. The absorption edge of the Cu- and Fe-doped TiO_2 films shifted toward longer wavelengths compared with that of the nondoped TiO_2 film, showing the occurrence of effective bandgap narrowing in these films. The photoinduced decomposition reaction was investigated by measuring the degradation of stearic acid using Fourier transform infrared spectroscopy (FT-IR). Figure 7.9 shows the photoinduced degradation of stearic acid as a function of the visible light irradiation time. The vertical axis expresses the absorption intensity normalized to that before irradiation. Visible-light-induced photocatalytic degradation could clearly be observed for the Cu-doped TiO_2 film, while the Fe- or Al-doped films hardly showed any photocatalytic activity. However, it was also found that the photocatalytic activity under UV light irradiation was reduced for the Cu- and Fe-doped TiO_2 films as compared with TiO_2 before doping. In these films, the impurity levels formed by the doped metal oxides acted not only as absorption centers for visible light but also as the recombination centers of the electron/hole pairs. Thus, it was shown that recombination of the photogenerated electron/hole pairs takes place efficiently in the Fe-doped TiO_2 film, while the charge separation of the photogenerated electron/hole pairs takes place effectively in the Cu-doped TiO_2 film. These results showed that cation doping by the dip coating method may not increase the photocatalytic activity of the TiO_2 thin film under UV light

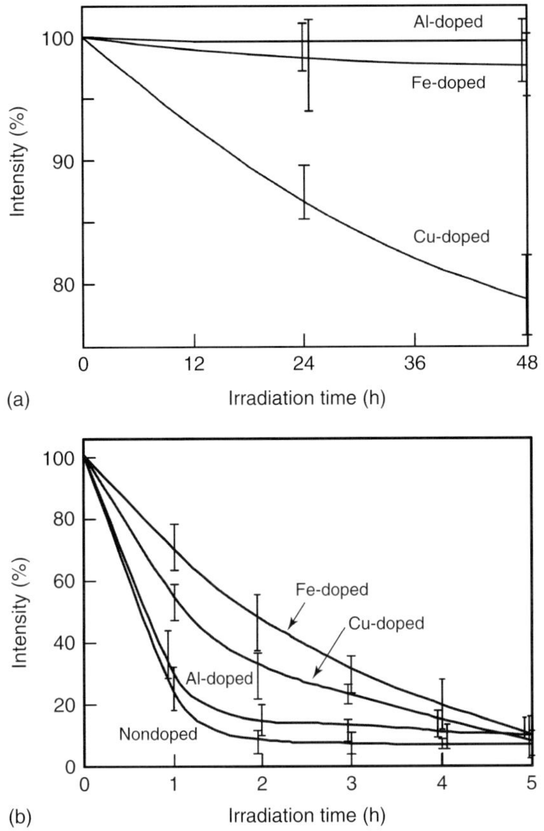

Figure 7.9 Normalized absorbance of stearic acid as a function of (a) visible light irradiation time and (b) UV light irradiation time.

irradiation; however, it may be considered effective in applications such as the purification of water and indoor air contaminated by harmful organic compounds.

In addition to cation-doped TiO_2 photocatalysts, anion (N, S)-doped TiO_2 have recently been the focus of much interest [28]. Han *et al.* [33] have reported the preparation of visible-light-responsive sulfur-doped TiO_2 nanocrystalline films by a sol–gel method based on the self-assembly technique using a nonionic surfactant to control the nanostructure and inorganic sulfur source (i.e., H_2SO_4). Various spectroscopic results involving X-ray photoelectron spectroscopy (XPS), FT-IR, and Energy dispersive X-ray spectrometry (EDX) mapping suggested that the sulfur mainly exists as anionic (S^{2-}) substituents in the TiO_2 lattice, while the S^{6+}/S^{4+} cations exist as surface sulfate groups rather than cationic substituents. The optical absorption edge of the sulfur-doped TiO_2 film (S-doped TiO_2) was red shifted toward the visible light region. The S-doped TiO_2 showed significant photocatalytic activity for the degradation of hepatotoxin microcystin-LR (MC-LR) in water under visible light irradiation. On the other hand, no significant MC-LR degradation was observed for the reference TiO_2 film, while the remaining concentrations of MC-LR were also similar under both dark and visible light conditions. Electron paramagnetic (spin) resonance (EPR) investigations also revealed that the sulfur-doped TiO_2 films exhibited a sharp signal at g of 2.004, while its intensity correlated to the sulfur content was markedly enhanced under visible light irradiation, implying the formation of localized energy states in the TiO_2 bandgap because of anion doping and/or oxygen vacancies. These results indicated that S doping into the TiO_2 thin film is also an effective method to prepare visible-light-responsive TiO_2 films, which can be applied for the remediation of polluted water.

7.5.2
Visible-Light-Responsive TiO_2 Thin Films Prepared by the Magnetron Sputtering Deposition Method

Various studies have recently been devoted to the modification of TiO_2 photocatalysts by substitutional doping of metals or nonmetals in order to extend their absorption edge into the visible light region and improve their photocatalytic activity [34–37]. However, most of these modified TiO_2 photocatalysts were in powder form and were not suitable for practical or widespread use, particularly in industrial applications. The development of stable TiO_2 thin films that can operate not only under UV but also under visible light irradiation would, therefore, be of great significance. It has been reported that visible-light-responsive TiO_2 thin films could be successfully prepared by a combination of the ionized-cluster beam (ICB) and ion-implantation techniques for the decomposition of NO into N_2 and O_2 even under visible light ($\lambda \geq 450$ nm) [30, 31]. However, this process requires two preparation steps, that is, the deposition of the TiO_2 thin film followed by ion implantation of metal ions such as V and Cr into the thin film. In fact, Kitano *et al.* [38–41] have reported the preparation of visible-light-responsive TiO_2 thin film photocatalysts by applying a radio frequency magnetron sputtering deposition method (RF-MS) and their application for water purification.

A schematic diagram of the magnetron sputtering deposition method is shown in Figure 7.10. The TiO_2 thin films were prepared by using a calcined TiO_2 plate and pure Ar gas as the sputtering gas. The system was equipped with a substrate center positioned parallel just above the source material, with a target-to-substrate distance set at 80 mm. The thin films were then prepared on a quartz or indium tin oxide (ITO) plate with the substrate temperature (Ts) held at a fixed value between a range of 473–873 K. The thickness of all the films was fixed at around 1.2 μm. Deposition of Pt (0.13 wt%) on the TiO_2 film surface was performed by immersing the TiO_2 thin film on a quartz substrate in a 10% methanol aqueous solution containing H_2PtCl_6 under light irradiation from a 500 W high-pressure Hg lamp for 10 min.

The UV–vis absorption spectra of the TiO_2 thin films prepared on a quartz substrate under various substrate temperatures (T_s) are shown in Figure 7.11. The thin film prepared at 473 K (UV-TiO_2) had almost no absorption in wavelength regions longer than 400 nm, while the film prepared at above 673 K was yellow

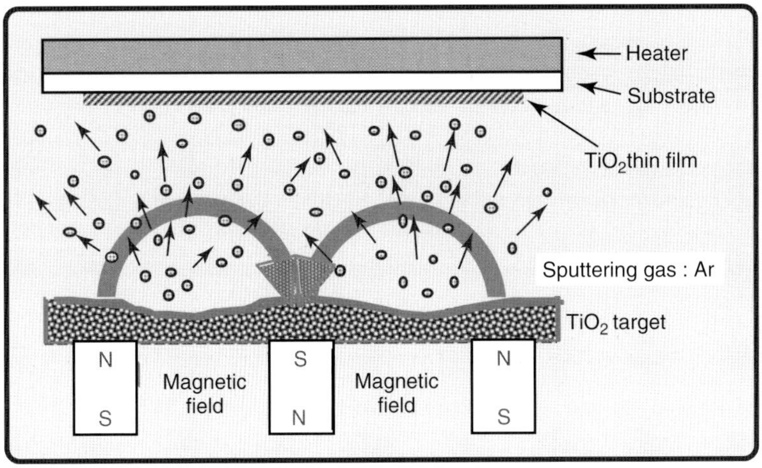

Figure 7.10 Schematic diagram of the magnetron sputtering deposition method.

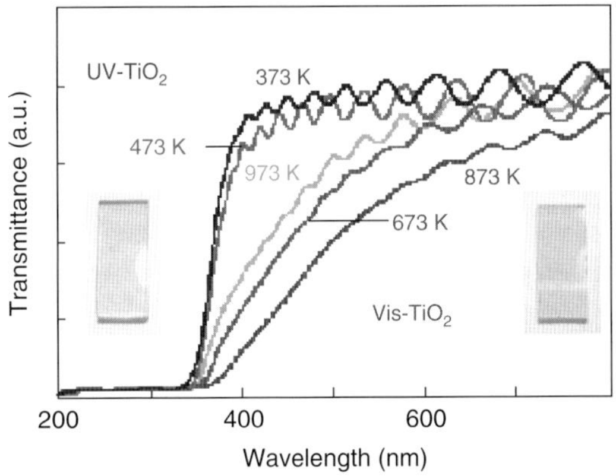

Figure 7.11 UV–vis transmission spectra of TiO_2 thin films prepared on a quartz substrate by a magnetron sputtering deposition method under various substrate temperatures. Substrate temperatures (K): (i) 373, (ii) 473, (iii) 673, (iv) 873, and (v) 973.

colored and exhibited considerable absorption in wavelength regions longer than 400 nm, thus enabling the absorption of visible light (vis-TiO_2-T_s of T_s = 673, 873). It could clearly be seen that absorption in visible light regions increased with an increase in the T_s. It was, thus, found that TiO_2 thin films that absorb visible light can be prepared by controlling the substrate temperature during a simple one-step deposition process. Visible light absorption of the vis-TiO_2 thin film can be attributed to the declined composition of the O/Ti ratio from the top surface (O/Ti ratio of 2.00) to the inside bulk (1.93), as suggested by results of Secondary ion mass spectrometry (SIMS) investigations shown in Figure 7.12 [38–41]. The declined composition plays an important role in the formation of new energy levels in the bandgap of TiO_2. Furthermore, transmission electron microscope (TEM)

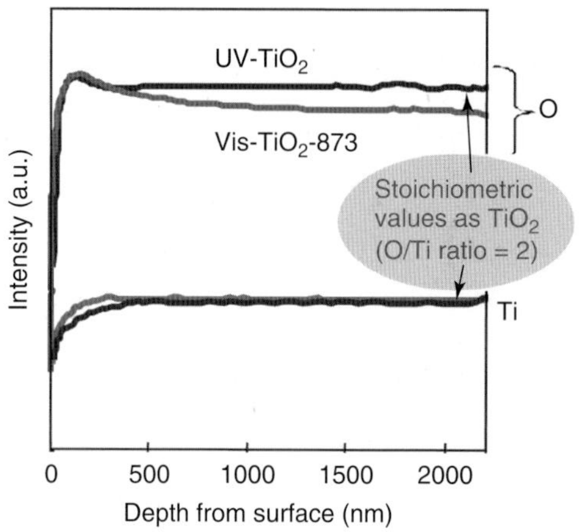

Figure 7.12 SIMS depth profiles of the composition of Ti^{4+} and O^{2-} ions of both the UV-TiO_2 and the vis-TiO_2-873 thin-film photocatalysts.

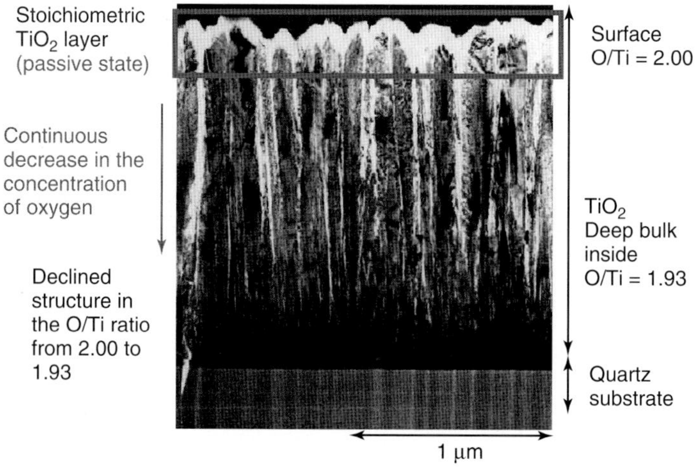

Figure 7.13 Cross-sectional TEM image of the vis-TiO_2-873 K thin film photocatalyst prepared on a quartz substrate.

investigations revealed that vis-TiO_2-873 consisted of columnar TiO_2 crystallites growing perpendicular to the substrates (Figure 7.13) [38–41].

The photocatalytic performance of these TiO_2 thin films was investigated through the oxidation of 2-propanol diluted in water under UV and visible light irradiation. As shown in Figure 7.14a, all three TiO_2 thin films (UV-TiO_2, vis-TiO_2-673, vis-TiO_2-873) exhibited photocatalytic activity for the degradation of 2-propanol under UV light irradiation, although vis-TiO_2-673 and vis-TiO_2-873 showed higher activity than UV-TiO_2. Moreover, the photocatalytic activities of the films were enhanced by the deposition of Pt on the surface. The Pt-loaded TiO_2 thin films

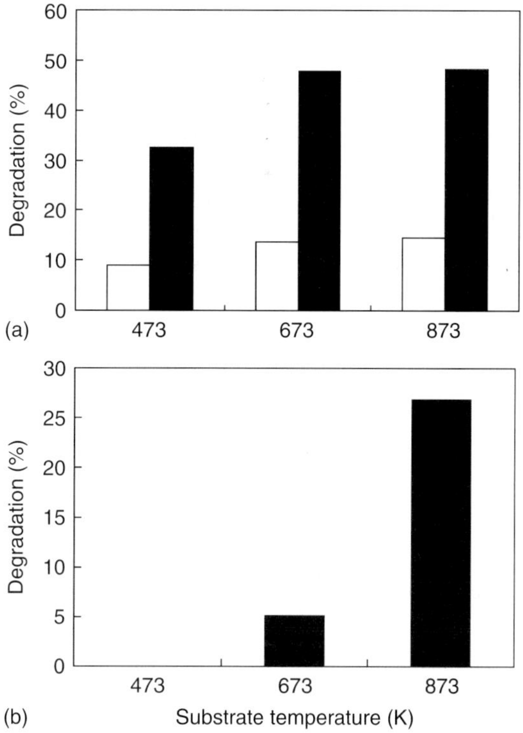

Figure 7.14 Photocatalytic degradation of 2-propanol on (white bar) TiO_2 and (black bar) Pt-loaded TiO_2 thin films after (a) UV light irradiation (full arc) for 1 h and (b) visible light irradiation ($\lambda \geq 450$ nm) for 24 h.

were referred to as Pt-UV-TiO_2, Pt-vis-TiO_2-673, and Pt-vis-TiO_2-873 in accordance with wavelengths of the absorption light and their substrate temperatures at the sputtering process. Improvement of the photocatalytic reactivity was caused by an increase in the charge-separation efficiency resulting from the electron transfer of the photoformed electrons from the TiO_2 moiety to the Pt particles [42, 43]. The photocatalytic activities of the films under visible light irradiation were also investigated (Figure 7.14b). Pt-vis-TiO_2 thin films could decompose 2-propanol in water even under visible light of wavelengths longer than 450 nm. Moreover, Pt-vis-TiO_2-873 exhibited higher activity than Pt-vis-TiO_2-673, while no reaction proceeded on Pt-UV-TiO_2 under visible light irradiation, showing a good coincidence with the results of the UV–vis absorption and photocurrent measurements. The reaction time profiles of the photocatalytic decomposition of 2-propanol on Pt-vis-TiO_2-873 were investigated and it was found that 2-propanol was decomposed into acetone and CO_2 on Pt-vis-TiO_2-873 even under visible light above 450 nm. It was, thus, confirmed that the Pt-vis-TiO_2 thin film functioned as a very stable and efficient photocatalyst for the oxidation of organic compounds into CO_2 and water even under visible light.

It was also found that vis-TiO_2-873 acted as a photocatalyst for the separate evolution of H_2 and O_2 from water as well as H_2 from water containing an organic compound. First, the visible-light-responsive TiO_2 thin film, vis-TiO_2-873, was deposited on a Ti metal foil substrate to obtain vis-TiO_2-873/Ti; then, the opposite side was deposited with Pt to prepare a vis-TiO_2-873/Ti/Pt photocatalyst by RF-MS method at a substrate temperature, T_s, of 298 K [41–43]. The prepared vis-TiO_2-873/Ti/Pt was then mounted at the center of an H-type glass container, separating the two aqueous solutions. A Nafion film was also mounted on the H-type glass container to allow proton exchange between the two aqueous solutions to compensate the H^+ consumption at the Pt side. The TiO_2 side of the photocatalyst was immersed in 1.0 M NaOH aqueous solution and the Pt side in 0.5 M H_2SO_4 aqueous solution in order to add a small chemical bias (0.826 V). As shown in Figure 7.15, H_2 evolution reactions were performed under sunlight irradiation from a UV-cut type sunlight gathering system. Before the addition of 10% methanol on the TiO_2 side, O_2 evolution was observed on the TiO_2 side accompanying the stoichiometric evolution of H_2 at the Pt side. On the other hand, the addition of 10% methanol into the aqueous solution on the TiO_2 side enhanced the H_2 evolution rate on the Pt side, while accompanying the evolution of CO_2 on the TiO_2 side as a result of methanol oxidation where methanol effectively worked as a sacrificial agent [41–43]. It was also found that H_2 evolution reactions did not proceed on UV-TiO_2/Ti/Pt thin films under solar light irradiation. These results clearly indicated that vis-TiO_2 can be applied for a H_2 evolution system from the water containing organic compounds under solar light irradiation.

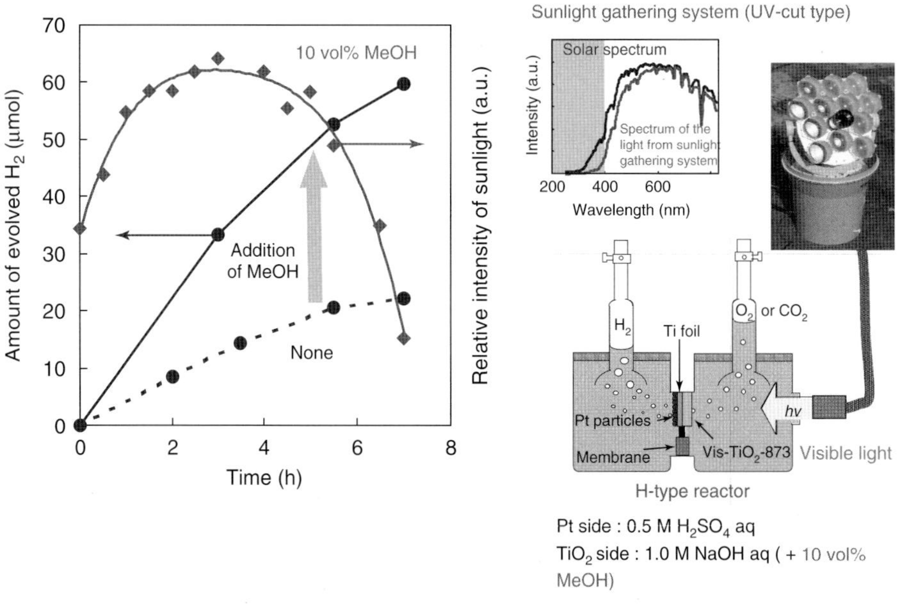

Figure 7.15 Separate evolution of H_2 on the vis-TiO_2-873 thin film before and after adding 10 vol% methanol under irradiation by light beams from a sunlight gathering system.

7.6 Conclusions

In this chapter, recent advances in the preparation and application of TiO_2 thin film photocatalysts were introduced. It was found that one of the attractive advantages of TiO_2 thin film photocatalysts against powdered or slurried TiO_2 photocatalysts was that the tedious and costly final filtration process could be avoided, especially when the photocatalysts were applied for the purification of water. The CFFP equipped with nonwoven paper coated with PC500 photocatalysts as well as Pd-deposited mesoporous TiO_2/SiO_2 fibers were found to be suitable candidates for the remediation of polluted water under continuous circulating conditions. PFCs that enable both the decomposition of various organic substances in water and the production of electric power were successfully constructed by using TiO_2 thin films as photoanodes. Furthermore, various techniques such as cation doping, anion doping, as well as the utilization of sputtering methods were shown to enable the preparation of visible-light-responsive TiO_2 thin films, which could effectively oxidize organic compounds in water. It can be expected that these methodologies for the preparation of highly functional TiO_2 thin films will open the way to their application in large-scale environmental remediation as well as solar energy conversion technologies.

References

1. Frank, S.N. and Bard, A.J. (1977) Heterogeneous photocatalytic oxidation of cyanide ion in aqueous solution at TiO_2 powder. *J. Am. Chem. Soc.*, **99**, 303–304.
2. Reiche, H. and Bard, A.J. (1979) Heterogeneous photosynthetic production of amino-acids from methane-ammonia-water at Pt-TiO_2. *J. Am. Chem. Soc.*, **101**, 3127–3128.
3. Serpone, N. and Pelizzetti, E. (eds) (1989) *Photocatalysis: Fundamentals and Applications*, John Wiley & Sons, Inc., New York.
4. Ollis, D.F. and Al-Ekabi, H. (1993) *Photocatalytic Purification and Treatment of Water and Air*, Elsevier, Amsterdam.
5. Agrios, A.G. and Pichat, P. (2005) State of the art and perspectives on materials and applications of photocatalysis over TiO_2. *J. Appl. Electrochem.*, **35**, 655–663.
6. Anderson, C. and Bard, A.J. (1995) An improved photocatalyst of TiO_2/SiO_2 prepared by a sol–gel synthesis. *J. Phys. Chem.*, **99**, 9882–9885.
7. Zhang, H., Chen, G., and Bahnemann, D.W. (2009) Photoelectrocatalytic materials for environmental applications. *J. Mater. Chem.*, **19**, 5089–5121.
8. Chen, C.C., Ma, W.H., and Zhao, J.C. (2010) Semiconductor-mediated photodegradation of pollutants under visible-light irradiation. *Chem. Soc. Rev.*, **39**, 4206–4219.
9. Anpo, M. and Kamat, P.V. (eds) (2010) *Environmentally Benign Photocatalysts*, Springer, New York.
10. Hashimoto, K., Irie, H., and Fujishima, A. (2005) TiO_2 photocatalysis: a historical overview and future prospects. *Jpn. J. Appl. Phys.*, **44**, 8269–8285.
11. Ohtani, B., Bowman, R.M., Colombo, D.P., Kominami, H., Noguchi, H., and Uosaki, K. (1998) Femtosecond diffuse reflectance spectroscopy of aqueous titanium(IV) oxide suspension: correlation of electron–hole recombination kinetics with photocatalytic activity. *Chem. Lett.*, **7**, 579–580.
12. Fujishima, A. and Honda, K. (1972) Electrochemical photolysis of water at

a semiconductor electrode. *Nature*, **238**, 37–38.
13. Domen, K., Kudo, A., and Onishi, T. (1986) Mechanism of photocatalytic decomposition of water into H_2 and O_2 over NiO - $SrTiO_3$. *J. Catal.*, **102**, 92–98.
14. Higashi, M., Abe, R., Sugihara, H., and Domen, K. (2008) Photocatalytic water splitting into H_2 and O_2 over titanate pyrochlores $Ln_2Ti_2O_7$ (Ln = Lanthanoid: Eu-Lu). *Bull. Chem. Soc. Jpn.*, **81**, 1315–1321.
15. Kitano, M., Iyatani, K., Tsujimaru, K., Matsuoka, M., Takeuchi, M., Ueshima, M., Thomas, J.M., and Anpo, M. (2008) The effect of chemical etching by HF solution on the photocatalytic activity of visible light-responsive TiO_2 thin films for solar water splitting. *Top. Catal.*, **59**, 24–31.
16. Kitano, M., Matsuoka, M., Ueshima, M., and Anpo, M. (2007) Recent developments in titanium oxide-based photocatalysts. *Appl. Catal. A: Gen.*, **325**, 1–14.
17. Iizuka, K., Wato, T., Miseki, Y., Saito, K., and Kudo, A. (2011) Photocatalytic reduction of carbon dioxide over Ag cocatalyst-loaded ALa(4)Ti(4)O(15) (A = Ca, Sr, and Ba) using water as a reducing reagent. *J. Am. Chem. Soc.*, **133**, 20863–20868.
18. Yoneyama, H. (1997) Photoreduction of carbon dioxide on quantized semiconductor nanoparticles in solution. *Catal. Today*, **39**, 169–175.
19. Ikeue, K., Yamashita, H., Anpo, M., and Takewaki, T. (2001) Photocatalytic reduction of CO_2 with H_2O on Ti-beta zeolite photocatalysts. *J. Phys. Chem. B*, **105**, 8350–8355.
20. Guillard, C., Disdier, J., Monnet, C., Dussaud, J., Malato, S., Blanco, J., Maldonado, M.I., and Herrmann, J.M. (2003) Solar efficiency of a new deposited titania photocatalyst: chlorophenol, pesticide and dye removal applications. *Appl. Catal. B: Environ.*, **46**, 319–332.
21. Blanco, J., Malato, S., Fernandez, P., Vidal, A., Morales, A., Trincado, P., Oliveira, J.C., Minero, C., Musci, M., Casalle, C., Brunotte, M., Tratzky, S., Dischinger, N., Funken, K.H., Sattler, C., Vincent, M., Collares-Pereira, M., Mendes, J.F., and Rangel, C.M. (1999) Compound parabolic concentrator technology development to commercial solar detoxification applications. *Sol. Energy*, **67**, 317–330.
22. Pichat, P., Vannier, S., Dussaud, J., and Rubis, J.P. (2004) Field solar photocatalytic purification of pesticides-containing rinse waters from tractor cisterns used for grapevine treatment. *Sol. Energy*, **77**, 533–542.
23. Suárez, S., Hewer, T.L.R., Portela, R., Hernández-Alonso, M.D., Freire, R.S., and Sánchez, B. (2011) Behaviour of TiO_2–SiMgOx hybrid composites on the solar photocatalytic degradation of polluted air. *Appl. Catal. B: Environ.*, **101**, 176–182.
24. Matsunaga, T., Yamaoka, H., Ohtani, S., Harada, Y., Fujii, T., and Ishikawa, T. (2008) High photocatalytic activity of palladium-deposited mesoporous TiO_2/SiO_2 fibers. *Appl. Catal. A: Gen.*, **351**, 231–238.
25. Ishikawa, T., Yamaoka, H., Harada, Y., Fujii, T., and Nagasawa, T. (2002) A general process for in situ formation of functional surface layers on ceramics. *Nature*, **416**, 64–67.
26. Kaneko, M., Gokan, N., Katakura, N., Takei, Y., and Hoshio, M. (2005) Artificial photochemical nitrogen cycle to produce nitrogen and hydrogen from ammonia by platinized TiO_2 and its application to a photofuel cell. *Chem. Commun.*, **12**, 1625–1627.
27. Kaneko, M., Nemoto, J., Ueno, H., Gokan, N., Ohnuki, K., Horikawa, M., Saito, R., and Shibata, T. (2006) Photoelectrochemical reaction of biomass and bio-related compounds with nanoporous TiO_2 film photoanode and O_2-reducing cathode. *Electrochem. Commun.*, **8**, 336–340.
28. Chen, X. and Mao, S.S. (2007) Titanium dioxide nanomaterials: synthesis, properties, modifications, and applications. *Chem. Rev.*, **107**, 2891–2959.
29. Choi, W., Termin, A., and Hoffmann, M.R. (1994) The role of metal ion dopants in quantum-sized TiO_2: correlation between photoreactivity and

charge carrier recombination dynamics. *J. Phys. Chem.*, **98**, 13669–13679.
30. Anpo, M. and Takeuchi, M. (2003) The design and development of highly reactive titanium oxide photocatalysts operating under visible light irradiation. *J. Catal.*, **216**, 505–516.
31. Takeuchi, M., Yamashita, H., Matsuoka, M., Anpo, M., Hirao, T., Itoh, N., and Iwamoto, N. (2000) Photocatalytic decomposition of NO under visible light irradiation on the Cr-ion-implanted TiO_2 thin film photocatalyst. *Catal. Lett.*, **67**, 135–137.
32. Maeda, M. and Yamada, T. (2007) Photocatalytic activity of metal-doped titanium oxide films prepared by sol–gel process. *J. Phys. Conf. Series*, **61**, 755–759.
33. Han, C., Pelaez, M., Likodimos, V., Kontos, A.G., Falaras, P., O'Shea, K., and Dionysiou, D.D. (2011) Innovative visible light-activated sulfur doped TiO_2 films for water treatment. *Appl. Catal. B: Environ.*, **107**, 77–87.
34. Asahi, R., Morikawa, T., Ohwaki, T., Aoki, A., and Taga, Y. (2001) Visible-light photocatalysis in nitrogen-doped titanium oxides. *Science*, **293**, 269–271.
35. Umebayashi, T., Yamaki, T., Ito, H., and Asai, K. (2002) Band gap narrowing of titanium dioxide by sulfur doping. *Appl. Phys. Lett.*, **81**, 454–456.
36. Irie, H., Watanabe, Y., and Hashimoto, K. (2003) Carbon-doped anatase TiO_2 powders as a visible-light sensitive photocatalyst. *Chem. Lett.*, **32**, 772–773.
37. Ohno, T., Akiyoshi, M., Umebayashi, T., Asai, K., Mitsui, T., and Matsumura, M. (2004) Preparation of S-doped TiO_2 photocatalysts and their photocatalytic activities under visible light. *Appl. Catal. A:Gen.*, **265**, 115–121.
38. Kitano, M., Iyatani, K., Afsin, E., Horiuchi, Y., Takeuchi, M., Cho, S.H., Matsuoka, M., and Anpo, M. (2012) Photocatalytic oxidation of 2-propanol under visible light irradiation on TiO_2 thin films prepared by an RF magnetron sputtering deposition method. *Res. Chem. Intermed.*, **38**, 1249–1259.
39. Kitano, M., Iyatani, K., Tsujimaru, K., Matsuoka, M., Takeuchi, M., Ueshima, M., Thomas, J.M., and Anpo, M. (2008) The effect of chemical etching by HF solution on the photocatalytic activity of visible light-responsive TiO_2 thin films for solar water splitting. *Top. Catal.*, **49**, 24–31.
40. Fukumoto, S., Kitano, M., Takeuchi, M., Matsuoka, M., and Anpo, M. (2009) Photocatalytic hydrogen production from aqueous solutions of alcohol as model compounds of biomass using visible light-responsive TiO_2 thin films. *Catal. Lett.*, **127**, 39–43.
41. Matsuoka, M., Kitano, M., Takeuchi, M., Anpo, M., and Thomas, J.M. (2005) Photocatalytic water splitting on visible light-responsive TiO_2 thin films prepared by a RF magnetron sputtering deposition method. *Top. Catal.*, **35**, 3–4.
42. Linsebigler, A.L., Lu, G., and Yates, J.T. (1995) Photocatalysis on TiO_2 surfaces - principles, mechanisms, and selected results. *Chem. Rev.*, **95**, 735–758.
43. Pichat, P. (1987) Surface properties, activity and selectivity of bifunctional powder photocatalysts. *New J. Chem.*, **11**, 135–140.

8
Sensitization of Titania Semiconductor: A Promising Strategy to Utilize Visible Light

Zhaohui Wang, Chuncheng Chen, Wanhong Ma, and Jincai Zhao

8.1
Introduction

TiO_2-based photocatalysis has received increasing interest in water splitting since the notable discovery of Fujishima and Honda in 1972 [1]. The specific advantages of TiO_2 are its high chemical stability and high photocatalytic activity in comparison to most other semiconductors. Also, its cost (about $2 per kg for pigment TiO_2) may not be an obstacle for many applications. However, the large band gap of TiO_2 (~3.2 eV for anatase and brookite and ~3.0 eV for rutile) limits photoabsorption to only the UV region of the solar spectrum. Given that only ~5% of the solar flux incident at the earth's surface lies in this spectral region, TiO_2 is inefficient when sunlight is used to drive the reactions unless means to extend their light response in the visible spectrum are found. Therefore, designing, fabricating, and tailoring the physicochemical and optical properties of TiO_2 is of great importance to utilize a large fraction of the solar spectrum [2–4]. For this purpose, TiO_2 is modified by various strategies such as surface sensitization by organic dyes [5] or metal complexes [6], coupling of a narrow-band-gap semiconductor [7–9], nonmetal doping [10] or codoping [11], and noble metal deposition [12].

To provide a contextual backdrop for this chapter, the authors begin with a brief overview of the principle of photosensitization. This is followed by a discussion of several prevailing sensitization methods by the use of organic chromophores (e.g., dyes and polymers), light-sensitive surface complexes, and semiconductors (or plasmonic metals). The state-of-the-art sensitization mechanisms involved in the aforementioned processes are included. In addition, some promising strategies to assist titania to work under visible light irradiation are briefly introduced, although they are beyond the scheme of sensitization in their precise significance. Throughout this chapter, the authors attempt to summarize major advancements made so far and give some perspectives to move the spectral region of sensitization of titania to the visible.

Photocatalysis and Water Purification: From Fundamentals to Recent Applications, First Edition. P. Pichat.

8.2
Principle of Photosensitization

Photosensitization is a process of initiating a reaction by the use of a substance capable of absorbing light and transferring the energy to the desired reactants. This technique is commonly used in photochemistry, physics, catalysis, and photodynamic therapy, particularly for reactions requiring light sources of certain wavelengths that are not readily available for the target reactants [13].

The molecule absorbing the photon is called as *photosensitizer (or sensitizer)* and the altered molecule/material is the *acceptor or substrate*. Both the photosensitizer and suitable light of certain wavelengths available for photosensitizer absorption are necessary for photosensitization. Similar to catalysts, in general, the ideal photosensitizers are not consumed during photosensitization reactions. Once the photosensitization reaction is complete, they should return to their original state [14]. In fact, photosensitizers usually suffer from the photobleaching reactions in which different molecules that do not absorb light are formed after the absorption of light, which typically competes with the photosensitization reaction.

There are two major reaction manners for photosensitization reaction (Figure 8.1). In the type I reaction, the excited sensitizer reacts directly with the substrate via one-electron transfer reaction to generate a radical or radical ion in both the sensitizer and the substrate. This electron transfer can proceed in either direction, governed by the redox potential of both excited sensitizer and its counterpart substrate. For example, the excited sensitizer can donate an electron to the substrate, resulting in a sensitizer radical cation ($P^{\bullet+}$) and a substrate radical anion ($S^{\bullet-}$). In the presence of oxygen, both these radicals can further react to produce oxygenated products. This reaction is undesirable because the sensitizer will be irreversibly converted to an oxidized molecule. Another possible reaction is the direct electron transfer from the coexisting electron donors, regenerating the original sensitizer (P).

In the type II reaction, the excited sensitizer transfers its excess energy to the ground-state molecular oxygen ($^3O_2$), producing the excited-state singlet oxygen ($^1O_2$) and reviving the ground-state sensitizer. $^1O_2$ is the key species involved in

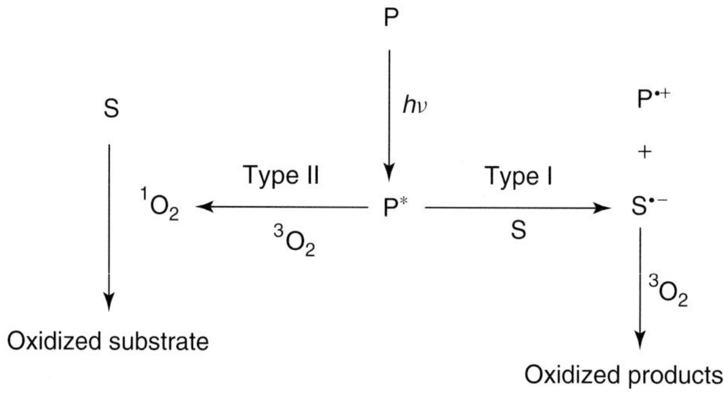

Figure 8.1 The schematic diagram of photosensitization. P, photosensitizer; S, substrate.

the sensitized oxidation of organic compounds in the absence of a heterogeneous photocatalyst. The photosensitizer is not rapidly consumed during this type of photosensitized reaction.

8.3 Dye Sensitization

A simple and efficient approach to extend catalyst absorption toward visible light is photosensitization using an appropriate dye. This photochemical reaction first found application in silver halide photography. In 1977, Watanabe and coworkers reported for the first time that the electron injection from the adsorbed rhodamine B (RhB) in its singlet excited state to the conduction band (CB) of CdS particles could result in an efficient photochemical N-deethylation of this dye [15]. In this strategy, TiO_2 itself does not absorb light directly, while the adsorbed dyes act as antennae to absorb the light energy. The visible light excites the dye sensitizer adsorbed on TiO_2 and subsequently electrons of the excited dyes are injected to the TiO_2 CB, while the valence band (VB) of TiO_2 remains unaffected in a typical photosensitization (Figure 8.2). The underlying utility of TiO_2 in dye sensitization is its ability to accept the injected electrons with high efficiency and transport these electrons to substrate electron acceptors on the TiO_2 surface.

In the absence of a redox couple, the dye sensitizer/semiconductor system can also be used in oxidative degradation of the dye sensitizer molecule itself after charge transfer. The trapped electron forms superoxide radical anion ($O_2^{\bullet-}$) by the reaction with dissolved oxygen, which results in the formation of *reactive oxygen species* (ROS) such as hydroxyl radical [2, 5, 16, 17]. These ROS are responsible for the oxidation of the organic contaminants including the dye itself or coexisting organics.

As one of the crucial parts in dye-sensitized photocatalysis, the photosensitizer should fulfill some essential characteristics [18].

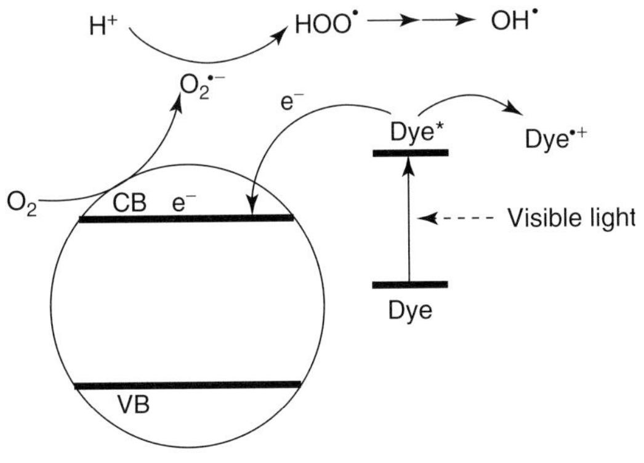

Figure 8.2 The schematic diagram of the mechanism of dye sensitization of TiO_2 under visible light illumination [5].

1) The absorption spectrum of the dyes should cover the wide visible region and even the part of the near-infrared (NIR) region.
2) The photosensitizer should be photostable (unless the self-sensitized degradation is required), and unfavorable dye aggregation on TiO_2 surface should be avoided.
3) The photosensitizer should have anchoring groups ($-SO_3H$, $-COOH$, $-H_2PO_3$, etc.) to facilitate the strong binding of dye molecules onto TiO_2 surface.
4) The excited state of the photosensitizer should be higher in energy than the CB edge of TiO_2. As a result, an efficient electron transfer between the excited dye and TiO_2 CB will be thermodynamically favorable.

The frequently used dyes include synthetic organic dyes (such as thiazines and xanthenes dyes) and transition-metal complexes such as polypyridine complexes (e.g., $[Ru(bpy)_2(4,4'-(PO_3H_2)_2)bpy)]^{2+}$ [19]), phthalocyanines (e.g., CuPcTs [20], TNCuPc [21], ZnPc [22], and AlPcTC [23]), and metalloporphyrins [24]. Dye sensitization is widely used to utilize visible light for environmental detoxification and energy conversion. This chapter exclusively discusses the examples of dyes that were used to channel the energy of visible light to drive photocatalytic chemical reactions and avoids discussing those that were applied to harvest solar irradiance for dye-sensitized solar cells (DSSCs) [18].

8.3.1 Fundamentals of Dye Sensitization

8.3.1.1 Geometry and Electronic Structure of Interface

The geometric and electronic structure of the chromophore–TiO_2 interface governs its electronic properties and the ensuing electron transfer dynamics. In general, the molecular adsorption modes on the host surface can be classified into six different modes [18]: (i) covalent attachment by direct linking groups or linking agents, (ii) electrostatic interactions, (iii) hydrogen bonding, (iv) hydrophobic interactions, (v) van der Waals' forces, and (vi) physical entrapment inside hosts such as cyclodextrins and micelles. The first kind of adsorption provides an approved strategy to accomplish a strong interlinkage between TiO_2 and the dyes of interest. The dye should have an anchoring group, which forms chemical bonds through reaction with the surface hydroxyl groups of TiO_2. Most of the photosensitizers use carboxylic acid as the anchoring group, which can coordinate to the TiO_2 surface in three ways: unidentate, chelating, and bridging bidentate mode. The bidentate mode is superior to unidentate in the stability of the anchored dye and interfacial quantum yields of electron injection due to the intimate contact with the TiO_2 surface. The chromophore geometry usually is not changed by the binding, but in some cases, especially those involving bidentate binding, the chromophore may be slightly distorted so as to provide a better match with the surface geometry.

The extent of dye adsorption depends on the nature of dye, initial dye concentration, surface properties of TiO_2, and pH. Zhao *et al.* [25] reported that the binding mode of RhB on TiO_2 was mainly the unidentate esterlike linkage by infrared vibrational spectroscopy. The electrostatic interaction between dyes and TiO_2 surface is expected to highly depend on the pH of the media, which is critical for influencing the surface charge of TiO_2 and the ionization of dye molecules. Surface modification will switch the adsorption modes of dyes on TiO_2 surface. Wang *et al.* [26] observed that RhB preferentially anchored on pure TiO_2 through the carboxylic group, while the adsorption of this dye on fluorinated TiO_2 was via the cationic moiety ($-NEt_2$ group) (Figure 8.3).

Pan *et al.* [27] recently found that water bonded to surface-bridging hydroxyls (HO_{br}) plays a crucial role in dye sensitization of TiO_2 (Figure 8.4). Water bonding is capable of increasing the polarization and acidity, forming $H_3O^+ \cdots O_{br}^-$ structures where hydronium is anchored by electrostatic interaction. This process is very slow in atmosphere but obviously accelerated in bulk water. As dye cations were added to TiO_2 suspension, they displace H_3O^+ and then form $R_n = N^+Et_2 \cdots O_{br}^-$ pair. The authors proposed that this kind of electrostatic adsorption will facilitate electron transfer and consequently promote sensitization under visible light irradiation. This opens a door toward facile improvement in the efficiency of dye sensitization and photodegradation.

The electronic properties include the alignment of the chromophore and TiO_2 energy levels, the electronic coupling between the chromophore and TiO_2 states and the spatial localization of the states. Electronic structure calculations indicate that the TiO_2 CB is created by the d-orbitals of the titanium atoms. The VB of bulk TiO_2 is mainly composed of O-2p states, with some Ti-3d and Ti-4sp character acquired through hybridization with the empty Ti-3d/-4sp CB states [28]. The dangling bonds of an unsaturated surface introduce gap states that can lower the gap energy to a few tenths of an electronvolt. Saturation of the surface by chemical bonding of water and other species makes the gap closer to the bulk value.

8.3.1.2 **Excited-State Redox Properties of Dyes**

The dyes selected to use in the dye-sensitized photocatalysis fall into two broad categories: purely organic conjugated molecules and metal–ligand complexes. The

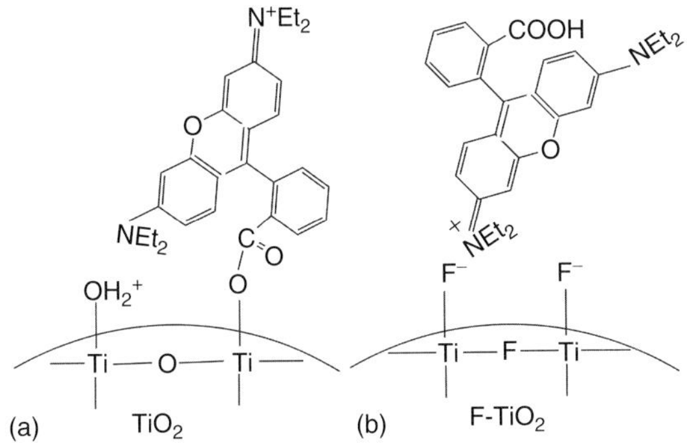

Figure 8.3 Proposed adsorption modes of RhB in aqueous pure TiO_2 (a) and fluorinated TiO_2 (b) suspensions [26].

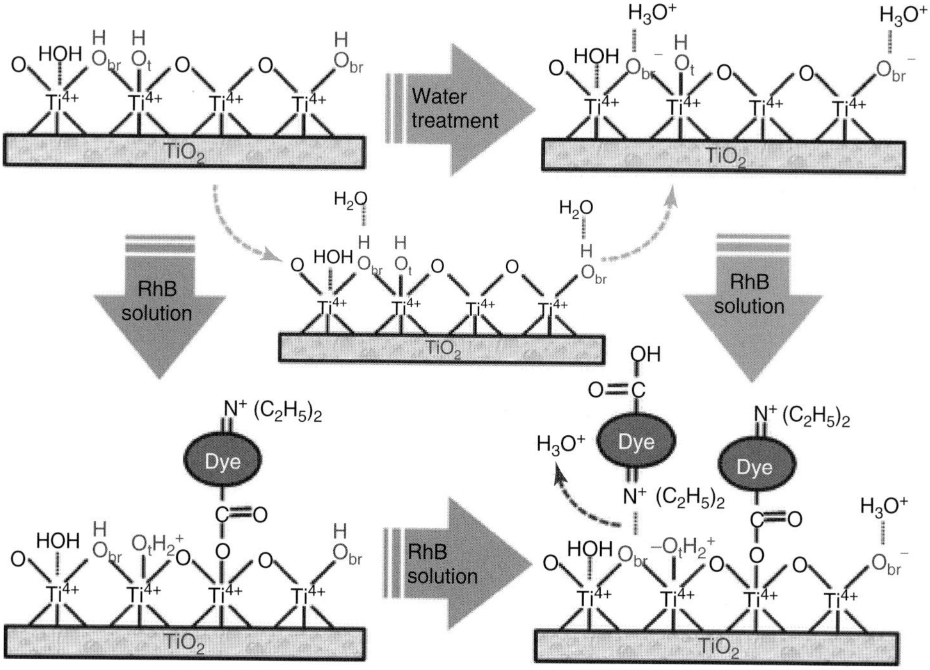

Figure 8.4 Water-mediated modulation of the surface of TiO_2 and switching of the adsorption mode of RhB [27].

photoexcited states formed by the π^* orbitals of the conjugated systems are similar in both chromophore types. The ground state of the purely organic dyes is a π-state too, but the ground state of the transition-metal chromophores is localized on the n-orbitals occupied by the metal's lone electron pairs [29]. The photoexcitation of a transition-metal complex results in an intramolecular charge transfer of the n–π^* type. The minor differences between the photoexcited states of the organic and transition-metal chromophores result from the larger excited-state delocalization in the purely organic chromophores, as well as from the positive charge on the transition metal that decelerates the transfer of the electron into the TiO_2. If there exists a spatial separation of highest occupied molecular orbital (HOMO) and lowest unoccupied molecular orbital (LUMO) of the dye molecule, excitation leads to increased electron density in the molecules close to the attachment group and TiO_2.

In general, the excited state of the dye has lower reduction potential than the corresponding ground state (Table 8.1). If its reduction potential is lower than the TiO_2 CB ($E_{CB} \approx -0.5$ V vs normal hydrogen electrode (NHE)), it is thermodynamically favorable that an electron may be injected from the excited state to the CB [5]. Consequently, CB electron and the cationic dye radicals are formed. This process has been intensively investigated in photoelectrochemical cells. Dye radical cations have been already *in situ* monitored by electron spin resonance (ESR) and time-resolved laser flash photolysis technologies [35–38].

Table 8.1 The reduction potential of the dye in an excited state and its absorption band maxima (λ_{max}).

Dyes	**Structure**	**λ_{max} (nm)**	**$E_{(D^*/D^{\bullet+})}$ vs NHE (V)**	**References**
Eosin	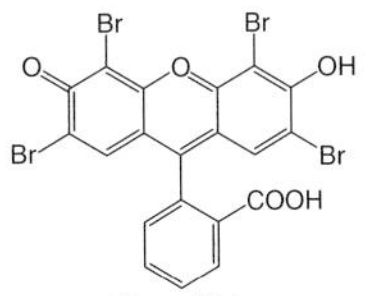	514	−1.11	[30]
Alizarin red	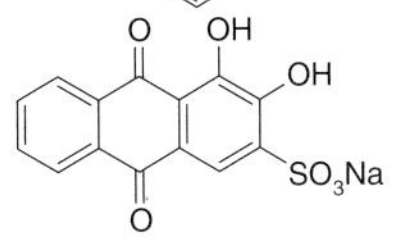	423	−1.57	[31]
Rhodamine B	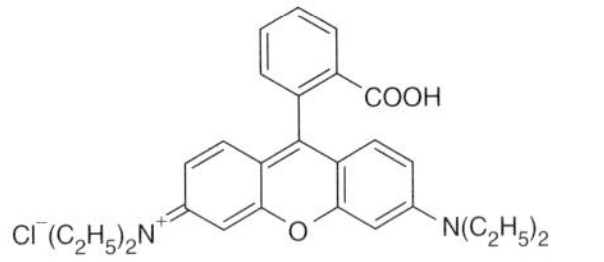	551	−1.09	[32]
Sulforhodamine B (SRB)	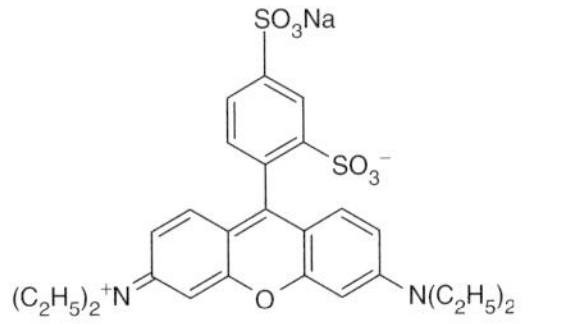	564	−1.30	[33]
Ethyl ester of fluorescein	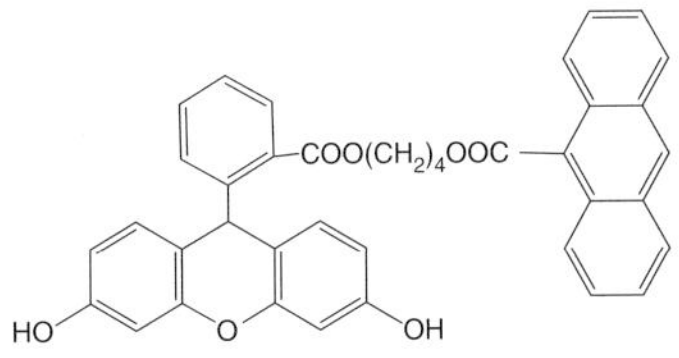	508	−1.33	[34]

Some dye radical cation with appropriate reduction potential even can be used in selective oxidation of alcohols [39].

8.3.1.3 **Electron Transfer from Dyes to TiO_2**

In the context of heterogeneous photocatalysis, electron transfer is an interfacial phenomenon between a surface and a chemisorbed (or physisorbed) species. The dynamics of electron injection from an electronically excited dye molecule into the CB of TiO_2 has been a popular topic in the field of heterogeneous photochemistry.

Coupling between the excited dye's electronic structure and the TiO_2 CB states is so strong that injection yields approaching 100% have been measured for many different dyes. Electron injection typically occurs on the sub-picosecond timescale and faster than 100 fs in some cases. Dye's electron is transferred to the CB of the TiO_2 or trapped by sites (perhaps shallow traps) within the band gap either through the CB or directly from the dye to the trap sites. It remains a challenge to experimentally and theoretically elucidate the fundamental properties of these exceptional fast injection processes. So far, the adiabatic and nonadiabatic (NA) transfer mechanisms can account for the ultrafast injection [29]. The former mechanism relies on strong chromophore coupling to one or a few TiO_2 states, but the latter one is workable on the occasion of the weak dye–semiconductor coupling. Taking the alizarin–TiO_2 system as an example, Shoute and Loppnow [40] and Ramakrishna *et al.* [41] proposed that a charge transfer complex (CTC) between alizarin and the TiO_2 surface formed on adsorption and excitation of this complex was responsible for the subsequent electron injection. In contrast, Huber *et al.* [42] characterized this transfer as adiabatic and not resulting from excitation of a charge transfer state but from a localized dye excitation event, using femtosecond transient spectroscopy. Applying the same technology, however, Kaniyankandy *et al.* [43] found the evidence of multiple electron injection in alizarin-sensitized ultrasmall TiO_2 particles. The electron injection event was followed by the formation of the dye cation at 550 nm and also the CB electron at 900 nm, which revealed a multiexponential dynamics with time constants of 100 fs, 17 ps, and 50 ps. The multiexponential injection indicated that the photoinjection event should be NA contrary to the debated literature.

The adsorption state of dye played a major role in the injection dynamics. Dyes with chromophoric centers far from the surface or isolated by nonconducting linkages exhibit slower injection time. The distance between the dye and TiO_2 surface can affect the efficiency of photoinduced electron transfer from dye to the TiO_2 particles [44]. Nilsing *et al.* [45] found that injection rates were faster for phosphate-anchored dyes than for carboxylate-anchored dyes and were deceased by aliphatic spacer (e.g., $-CH_2-$, $-CH_2-CH_2-$ and $-CH{=}CH-$) between the anchor group and dye. Ramakrishna *et al.* [46] also pointed out that the coupling of electronic states between an excited dye and its cationic state can be as important in affecting electron transfer as the coupling between the dye and TiO_2 CB states.

He *et al.* [34] reported an interesting electron transfer process in which a diester formed between fluorescein and anthracene-9-carboxylic acid (FL-AN) was used to sensitize colloidal TiO_2. The phenolic function of the fluorescein moiety was anchored to TiO_2, while the anthracene component of FL-AN served as the energy donor component without contact with the titania surface. Visible light irradiation (~475 nm) excited the fluorescein moiety and induced the electron transfer from its singlet excited state onto TiO_2 CB ($k_{et} = 2.30 \times 10^8\ s^{-1}$) (Eqs. (8.5) and (8.6)). However, exciting the anthracene side with UV light (~355 nm) first resulted in energy transfer from its singlet excited state to the lowest excited singlet state of fluorescein ($k_{enT} = 2.23 \times 10^{10}\ s^{-1}$), following which electron injection occurred form the fluorescein singlet to the TiO_2 CB (Eqs. (8.7) and (8.8)).

$$TiO_2 \cdots FL\text{-}AN \xrightarrow[475\,nm]{hv} TiO_2 \cdots {}^1FL^*\text{-}AN \tag{8.1}$$

$$TiO_2 \cdots {}^1FL^*\text{-}AN \longrightarrow TiO_2\,(e) \cdots FL^{\bullet+}\text{-}AN \tag{8.2}$$

$$TiO_2 \cdots \text{FL-AN} \xrightarrow[355\,\text{nm}]{h\nu} TiO_2 \cdots \text{FL-}^1\text{AN}^* \tag{8.3}$$

$$TiO_2 \cdots \text{FL-}^1\text{AN}^* \xrightarrow[\text{Energy transfer}]{k_{\text{enT}}} TiO_2 \cdots {}^1\text{FL}^*\text{-AN} \tag{8.4}$$

The fundamental studies on the electron transfer process may be interesting in the dye-sensitized mixed-phase TiO_2 nanocomposites. Li *et al.* [47] observed a synergistic effect between anatase and rutile TiO_2 nanoparticles when N719 dye ($(Bu_4N)_2$-[Ru(dcbpyH)$_2$(NCS)$_2$] complex (dcbpy, 4,4′-dicarboxy-2,2′-bipyridine)) was used as a sensitizer. On photoexcitation, different numbers of electrons were injected from the surface coordinated dye molecules to both anatase and rutile particles. The accumulation of electrons on TiO_2 nanoparticles could change the relative position of CB edges at the anatase–rutile boundary. If a Boltzmann distribution was used to describe the energy distribution of electrons in the rutile CB, 0.05% of the photogenerated electrons should be able to overcome the 0.2 eV barrier to the anatase phase at room temperature. Additionally, electrons injected into the rutile nanorods are probably "hot" since the energy of the dye excited state is greatly higher than the rutile CB. The electron transfer from rutile to anatase in the mixed-phase TiO_2 probably happens within the first picosecond after photoexcitation. The back reaction from anatase to rutile will be hindered by the low mobility and higher spatial confinement of free electrons in rutile. Li *et al.* [48] developed a dye/$Zn_xCd_{1-x}S$/TiO_2 nanocomposite in which the adsorbed dye in its singlet excited state can inject its electrons to the CB of $Zn_xCd_{1-x}S$ and TiO_2.

Kim *et al.* [49] proposed a novel electron transfer process in which a tin porphyrin (SnP) was used to sensitize TiO_2 in a wide pH range (pH 3–11). Initially, π-radical anion ($SnP^{\bullet -}$) was generated from the reaction (Eq. 8.5) between the excited triplet state of SnP (a strong oxidizing agent) and electron donor (EDTA). Then the subsequent electron transfer from $SnP^{\bullet -}$ to TiO_2 (Eq. 8.6) or direct electron transfer from $SnP^{\bullet -}$ to Pt occurs (Eq. 8.7). This SnP complex has markedly different photochemical behaviors from the Ru complexes, which should be ascribed to the different properties of their own excited states. For Ru complexes, their excited orbitals should be strongly coupled with Ti-3d orbitals through the chemical bond. The electron injection from the excited dye to TiO_2 CB is so fast (<10 ps) that the fluorescent decay is completely inhibited. Then a slow regeneration of dye (microseconds to milliseconds timescale) is followed through the electron transfer between the dye cation and electron donor. Therefore, the strong chemical bonding between TiO_2 and sensitizer is an essential requirement for visible light sensitization with Ru complexes. In the case of the SnP complex, the adsorption of the tin porphyrin on TiO_2 is not necessarily needed. The free excited SnP ($^3SnP^*$) is first reduced by EDTA instead of transferring electron to TiO_2 CB, due to its strong oxidizing capacity. $^3SnP^*$ favors the formation of the relatively stable and long-lived SnP π-radical anion ($SnP^{\bullet -}$), owing to the high electrophilicity of the SnP ring.

$$^3\text{SnP}^* + \text{EDTA} \longrightarrow \text{SnP}^{\bullet -} + \text{EDTA}^{\bullet +} \tag{8.5}$$

$$SnP^{\bullet-} + TiO_2 \longrightarrow SnP + TiO_2(e^-) \tag{8.6}$$

$$SnP^{\bullet-} + Pt \longrightarrow SnP + Pt(e^-) \tag{8.7}$$

8.3.2 Application of Dye Sensitization

8.3.2.1 Nonregenerative Dye Sensitization

Photosensitized degradation of organic dyes has been carried out on TiO_2 in which the organic dye serves as both a sensitizer and a substrate to be degraded. Once the dye solution is completely bleached, organic intermediates formed were unable to drive the photosensitized degradation further. The nonregenerative organic dye/TiO_2/visible light system in which the sensitizer itself degrades can be a variable option for treating dye pollutants in wastewater effluents. In this aspect, Zhao *et al.* [2, 5] and other research groups [50–52] have extensively investigated photosensitized decomposition of a wide range of dye pollutants under visible light irradiation (Table 8.2).

In general, the degradation process of the selected dye can be characterized by the following procedure: (**P1**) decolorization kinetics, (**P2**) extent of mineralization, and (**P3**) intermediates identification. **P1** can be simply realized by periodically monitoring the change of absorption band maxima (λ_{max}) of the dye pollutants using a UV–vis spectrophotometer. It is worthy to note that λ_{max} changes in different pH conditions [31] or during the photobleaching processes. For example, a hypsochromic shift of the band of some dyes containing *N*-alkylamine groups was observed due to the N-dealkylation pathway [26]. Complete decolorization does not guarantee the complete mineralization, since the cleavage of the chromophore structure of dyes will lead to decolorization while the total mineralization implies that all the organic carbon, hydrogen, and other elements such as N, P, S, and halogen should be fully converted into the corresponding inorganic species, namely, CO_2, H_2O, NO_3^-, SO_4^{2-}, and halides. Therefore, **P2** includes the data of total organic carbon (TOC), chemical oxygen demand (COD), and concentration profiles of various inorganic ions. If complete mineralization is not achieved, it is necessary to identify the residual TOC and provide further evidence for evaluating the biodegradability and toxicity of the as-formed organic intermediates (**P3**). In this step, liquid chromatography–mass spectrometry (LC–MS) and/or gas chromatography–mass spectrometry (GC–MS) is required to accurately identify their molecular structures, depending on the nature (polarity, volatility, molecular weight, etc.) of each intermediate to be measured. The ROS responsible for dye degradation can be verified by ESR spectroscopy or spin-trapping ESR [37]. On the basis of the results of **P1**, **P2**, and **P3**, a mechanism will be proposed finally. In principle, the dye photodegradation follows two competitive pathways: (i) N-dealkylation of the chromophore framework in a stepwise manner and (ii) cleavage of the whole conjugated chromophore structure.

Table 8.2 Examples of nonregenerative photosensitized degradation of dyes pollutants.

No.	Dye	Light wavelength	Technology	Main intermediates	Reaction pathways	References
1	Rhodamine B	>420 nm	UV–vis, COD, ESR, and NMR	—	N-Deethylation and chromophore destruction	[53]
2	Alizarin red	>420 nm	UV–vis, COD, NMR, IR, GC–MS, and ESR	Phthalic acid	Cleavage of C–C bonds near the C=O group of alizarin red	[31, 54]
3	Squarylium cyanine	≥430 nm	UV–vis, NMR, GC–MS, and ESR	1-Sulfopropyl-3,3-dimethyl-5-bromoindolenium-2-one	Cleavage of cyanine C=C bond	[55]
4	Sulforhodamine B	>420 nm	UV–vis, COD, ESR, NMR, and GC–MS	Diethylamine[a] and acetaldehyde[b]	Cleavage of chromophore structure[a] and N-deethylation[b]	[33]
5	Sulforhodamine B	>420 nm	UV–vis, COD, TOC, IR, NMR, and GC–MS	Diethylamine, *N*-ethylacetamide, *N*-ethylformamide, *N*,*N*-diethylacetamide, *N*,*N*-diethylformamide, HCOOH, and CH_3CHO	—	[56]
6	Sulforhodamine B	>420 nm	UV–vis, ESR, HPLC, SG-TLC, and MALDI-TOF	Sulforhodamine, *N*-ethyl-sulforhodamine, *N*-ethyl-*N'*-ethylsulforhodamine, *N*,*N*-diethylsulforhodamine, and *N*,*N*-diethyl-*N'*-ethylsulforhodamine	N-Deethylation	[57]
7	Methylene blue	Solar light	UV–vis and TOC	Azure A, Azure B, Azure C, and thionine	N-Demethylation	[58]

[a] TiO_2 was positively charged.
[b] Anionic DBS surfactant was added.

Depending on the selected dye pollutants, the photodegradation efficiency can be controlled by different influencing factors. The common strategies (although not exclusive) to enhance the rate of sensitized photodegradation are discussed in the following.

1) **Increasing the dye adsorption onto the TiO_2 surface**
   The quantity of the dye preadsorbed on the catalyst surface is of paramount importance because only these adsorbed molecules directly participate in the photoinduced electron injection and subsequent photodegradation process. To enhance the dyes' adsorption, an anionic surfactant dodecylbenzene sulfonate (DBS) was added to the TiO_2 dispersion and the change in surface charge density of titania from positive to negative occurred, thereby rendering stronger adsorption of the cationic dye RhB even at acidic pH [44]. As a result, the complete decomposition of RhB was achieved in the presence of a surfactant, while only ~45% degradation was observed in the absence of DBS. Hence, preadsorption on the surface of TiO_2 particles is a prerequisite and a valid strategy for efficient photodegradation of dyes under visible light irradiation.
2) **Enhancing the charge separation efficiency**
   Once the electron injection efficiency is settled, the overall photodegradation rate is dependent on the charge separation efficiency of photogenerated charge carriers (e.g., $dye^{\bullet+}$ and e^-_{CB}), which is controlled by two competitive processes, namely, rapid depletion via expected reactions or recombination via backward electron transfer. Zhao *et al.* [59] reported that the chemisorbed hexachloroplatinate ($PtCl_6{}^{2-}$) on TiO_2 can accelerate the self-sensitized transformation dynamics of the azo dye-ethyl orange (EO) under visible light illumination. The adsorbed Pt surface complex can facilely trap the injected electrons of TiO_2 CB to form a transient Pt(III) species. The Pt(III) intermediate subsequently reacted with preadsorbed oxygen to generate superoxide radical anions. However, the addition of Pt species exhibited no apparent effect on the extent of mineralization. In contrast, trace quantities of transition-metal ions such as Cu^{2+} and Fe^{3+} with suitable redox potentials may suppress the activation of dioxygen by the CB electrons and display unfavorable effect on visible-light-induced degradation of dyes [60] (Figure 8.5). Recently, it was found that depositing a thin Al_2O_3 overlayer on TiO_2 enhanced the visible light activities for the dechlorination of organohalogen [61]. Thin insulating Al_2O_3 layer can act as a physical barrier between electron donor and acceptor and can thus inhibit back electron transfer from TiO_2 to the dye's radical cation [25].
3) **Increasing ROS concentrations**
   A number of studies have proved that ROS such as $OH^\bullet$ and H_2O_2 were involved in the photodegradation of dye pollutants. Yang *et al.* [17] have proposed that both superoxide radical anion and the dye cationic radical were essential to the mineralization of the dyes in dye-sensitized photocatalytic reactions. Therefore, activation of dioxygen adsorbed on TiO_2 surface becomes an important theme since nearly all ROS are derived from the reduction of molecular oxygen. Since metallic platinum is an excellent catalyst for the reduction of O_2, it can serve

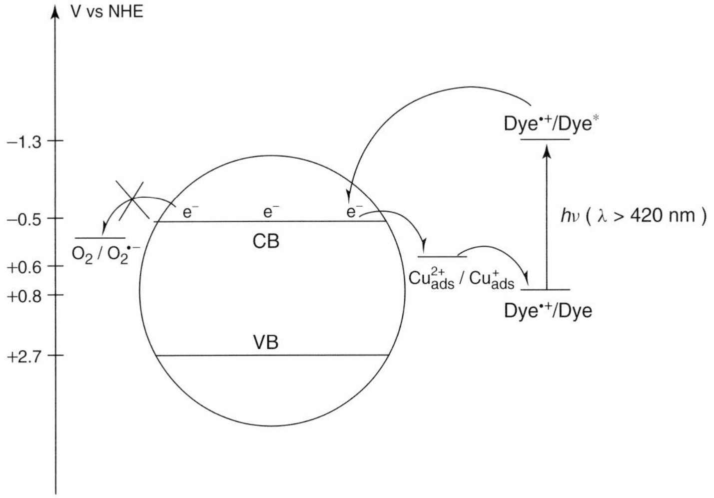

Figure 8.5 Electron transfer pathway for the addition of Cu^{2+} to TiO_2 aqueous dispersions under visible irradiation [60].

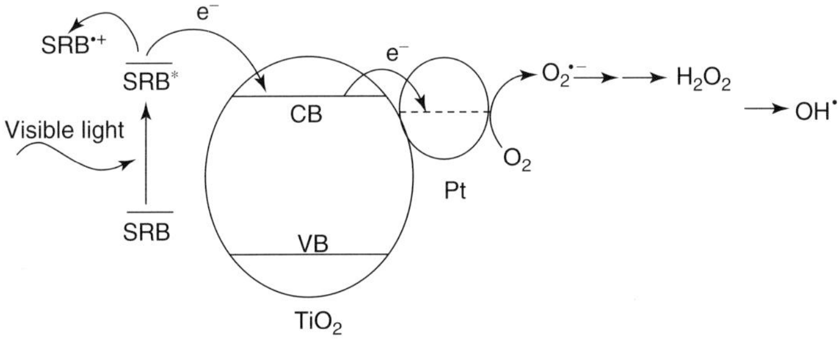

Figure 8.6 The proposed mechanism of photosensitized degradation of SRB on TiO_2/Pt particles under visible light irradiation [62].

as a good candidate for enhancing the production rate of ROS in TiO_2-based photosensitized system. Zhao *et al.* [62] observed that Pt-deposited TiO_2 was nearly fourfold more active than bare TiO_2. On the basis of ESR detection and H_2O_2 measurement, this enhancement effect was attributed to the rapid formation of relatively large quantities of ROS produced in the SRB-Pt/TiO_2 system (Figure 8.6).

8.3.2.2 **Regenerative Dye Sensitization**

In the nonregenerative mode of dye sensitization, dye pollutants can be fully decolored but it is difficult to realize their complete mineralization under visible light irradiation. Therefore, it is desirable to develop a stable photosensitization system based on photostable organic/inorganic metal complexes in which the dye sensitizer can be regenerated and the pollutants of interest, even colorless compounds, can be completely mineralized. In this regard, ruthenium(II) complexes and metal porphyrins have been widely studied.

Cho *et al.* [63] investigated the photoreductive decomposition of CCl_4 on TiO_2 sensitized by tris(4,4′-dicarboxy-2,2′-dipyridyl)ruthenium(II) complexes ($Ru(II)L_3$) under visible light irradiation. When the $Ru(II)L_3$ sensitizer was excited in the solution without TiO_2, it rapidly relaxed into the ground state through fluorescence, and hence, there was little chance of electron transfer to the substrate. However, in the presence of TiO_2, the electron injection from the excited sensitizer to the CB of TiO_2 (quantum yield, ~0.6) was faster than the direct relaxation to the ground state. The estimated driving force for the electron injection ($E(Ru(II)L_3^*/Ru(III)L_3)$-$E_{CB}$) was reported to be 0.45 V in the $Ru(II)L_3/TiO_2$ system (pH = 3), while the driving force sufficient for electron injection is as small as about 0.1 V. Since the rate of electron injection ($k_{inj} = 3.2 \times 10^7\ s^{-1}$) in the $Ru(II)L_3/TiO_2$ system is 80 times larger than that of the back electron transfer ($k_b = 4 \times 10^5\ s^{-1}$), the electrons in the CB can be immediately transferred to the adsorbed CCl_4. As a result, the active sensitizer was irreversibly converted to its oxidized form $Ru(III)L_3$. The presence of suitable electron donors is a prerequisite for regeneration of the sensitizer. Propan-2-ol, instead of methanol and acetate, seemed to be an efficient electron donor to revive the light-sensitive sensitizer $Ru(II)L_3$. The recent work by Zhao *et al.* [64] identified a highly stable photocatalyst – Pt(dcbpy)Cl_2/TiO_2 without the need for sacrificial electron donors, which promoted the visible light (λ >420 nm) sensitized degradation of 4-chlorophenol (4-CP) in the presence of dioxygen. Sacrificial additives were not required as the oxidized form of Pt(dcbpy)Cl_2 (E > 0.735 V vs Fc^+/Fc) was able to irreversibly oxidize 4-CP (E = 0.525 V vs Fc^+/Fc). Complete mineralization of 4-CP was realized via two operable reaction pathways using the Pt(II)-dye-sensitized titania nanomaterials.

Bae and Choi [19] systematically investigated the effects of the anchoring groups (carboxylate or phosphonate) in ruthenium bipyridyl complexes on the sensitized production of H_2 in aqueous dispersion containing TiO_2 that was used as a support for the sensitizers. The effects of the anchoring group on the sensitization of TiO_2 are primarily related to the formation of surface chemical bonds, but are indirectly influenced by other factors such as the nature of solvent and coexisting adsorbates. By changing the type and number of the anchoring group, not only the intrinsic properties of the surface chemical bonds but also the surface charge and the visible-light-absorbing capability are affected. However, such chemical anchoring method is not sufficiently stable in an aqueous solution and can be prepared only in the acidic pH region. In this context, Choi and Park [65] developed an alternative approach to attaching unfunctionalized Ru complex sensitizers through electrostatic attraction at the TiO_2/H_2O interface by coating with perfluorosulfonate polymer resin-Nafion. The roles of the Nafion layer on TiO_2 in the sensitized hydrogen production were proposed to be twofold: (i) to provide binding sites for cationic sensitizers ($Ru(bpy)_3^{2+}$) and (ii) to act as a H^+ reservoir and thus enhance the local activity of protons on the surface region.

In recent years, metal porphyrins have been used as photosensitizers for polycrystalline TiO_2 so as to enhance the visible light sensitivity of the TiO_2 matrix

[24, 49]. The central coordinated metal in porphyrins plays an important role in the photocatalytic reactions of the sensitized systems. It was shown that Cu(II) porphyrin–TiO_2 composites were more effective than other metal porphyrins (Mn, Fe, Zn, Co, etc.) in the photodegradation of 4-nitrophenol [24]. The peripherally substituted group within the framework of porphyrins affects the efficiency of the porphyrin–TiO_2 systems, depending on the strength of the polar group involved (e.g., –OH, RCOO–, $-SO_3$), position of the substituent and spacer length, and the electronegativity of the substituent atoms (e.g., O or Cl). However, the recent work by Wang *et al.* [24] indicated that photocatalytic activity of the Cu(II) porphyrin–TiO_2 composites was mainly determined by the amount of sensitizer adsorbed on the TiO_2 surface rather than by the nature of the substituted porphyrins themselves.

Besides organic metal complexes, inorganic metal complexes also can serve as an efficient photosensitizer due to their stability and low cost. A novel version of sensitization was reported by Kisch and coworkers [66] for titania containing up to 3% of Pt(IV), Rh(III), and Au(III) chlorides in the bulk. In these systems, the surface complex of metal chloride acted as visible-light-absorbing sensitizers and charge separation centers, while the titania matrix served as a charge trap. Upon visible light irradiation, the excited platinum complex underwent hemolytic metal-chloride bond cleavage to afford Pt(III) transient species. Subsequent electron transfer from the former to titania and from 4-CP to the chlorine atom reformed the sensitizer.

Lanthanide possesses diverse electronic band structure and versatile coordinated approaches and thus different optical properties. Fu and colleagues [67] prepared nanocrystalline TiO_2 incorporated with praseodymium(III) nitrate ($Pr(NO_3)_3$) by an ultrasound-assisted sol–gel method. The prepared sample has visible light absorptions at 444, 469, 482, and 590 nm, which are attributed to the 4f transitions $^3H_4 \rightarrow {}^3P_2$, $^3H_4 \rightarrow {}^3P_1$, $^3H_4 \rightarrow {}^3P_0$, and $^3H_4 \rightarrow {}^1D_2$ of praseodymium(III) ions, respectively. This novel $Pr(NO_3)_3$–TiO_2 visible light photocatalyst showed high activity and stability in the degradation of dye and 4-CP. The local excitation of the dispersed $Pr(NO_3)_3$ chromophore led to an interfacial charge transfer (IFCE) to TiO_2.

In conclusion, such reported organic/inorganic metal complexes work well in the regenerative mode at the laboratory scale. However, it should be emphasized that there are still some difficulties in practical application of these sensitizers due to their costs, potential metal toxicity, and long-term stability and adaptability in real wastewaters.

8.4 Polymer Sensitization

8.4.1 Carbon Nitride Polymer

Mitoraj and Kisch's [68] recent findings showed that higher melamine condensation products can act as visible light sensitizers and should be responsible for the visible light photocatalytic activity of N-doped or N-modified titania prepared from urea, instead of nitridic or NO_x species or defect states. Zhou *et al.* [69] prepared the carbon nitride (CN)-polymer-sensitized TiO_2 nanotube (TNT) arrays using dicyandiamide as the C and N sources by a modified electrodeposition method and found that photoelectrochemical and photocatalytic properties of the CN-polymer-sensitized TNT arrays significantly improved when compared with those of the TNT arrays alone. They hypothesized that the achieved high efficiency may be attributed to two major improvements: (i) the polymer affords multiple excitations derived from the adsorption of a single photon (although no direct experimental evidence was provided) and (ii) the crystalline nature of TNT and film topography allow a rapid and efficient charge transfer between the excited polymer and TNT.

8.4.2
Conducting Polymers

Conducting polymers have been extensively investigated in the field of electronics and photonics due to their good semiconducting, electronic, and optical properties. The most important characteristics of conducting polymers include sufficient stability, high absorption coefficient, and wide absorption wavelength in the visible region. Thus such conducting polymers such as polyaniline, polythiophene, and their derivatives are promising candidates for application as visible light photosensitizers (Table 8.3). The photocatalytic activity of conducting-polymer-modified TiO_2 under visible light irradiation originated from the visible light adsorption of polymers and enhanced charge separation of photogenerated charge carriers owing to the heterojunction built between polymers and TiO_2. Zhang *et al.* [79] proved that singlet oxygen was the dominant ROS in anhydrous solution in poly(fluorine-*co*-thiophene)-sensitized TiO_2 reaction systems using the NaN_3 quenching method. They tentatively proposed that the selected polymer first transformed to the singlet excited state and then its triplet excited state after absorbing photon energy, finally reacting with oxygen to generate singlet oxygen by energy transfer.

8.5
Surface-Complex-Mediated Sensitization

CTC-mediated visible light sensitization operates in a mechanism that is quite different from the well-known dye sensitization and classical band gap-excited photocatalysis. On irradiation there is an electronic transition from the ground state of the CTC system mostly consisting of adsorbate orbitals to the ligand to metal charge transfer (LMCT) state. Then the electron is transferred to TiO_2 CB through the LMCT mechanism, whereas the dye sensitization is mediated through the excited dye state [2].

Table 8.3 Comparison of the visible light efficiency of polymer/TiO_2 photosensitization systems.

Polymer	Substrate	Visible light efficiency	References
Poly(fluorine-*co*-thiophene)	Phenol	74.3% degraded (10 h)	[70]
Polypyrrole	Methyl orange	1.47 times[a]	[71]
Polythiophene	Methyl orange	85.6% decolored (120 min)	[72]
Polythiophene and its derivatives	2,3-Dichlorophenol	Poly-3-methylthiophene ≈ polythiophene > polythiophenecarboxylic acid > P3HT	[73]
Polyaniline	Methylene blue and rhodamine B	88% MB decolored (5 h); 97% RhB decolored (100 min)	[74]
Polyaniline	Methylene blue	1.57 times[a]	[75]
Polyvinyl alcohol	Methyl orange	28.6% degraded (6 h)	[76]
P3HT	Methylene blue	2.8 times[b]	[77]
Poly(fluorine-*co*-bithiophene)	Phenol	$0.0199\ min^{-1}$	[78]
CN polymer (using urea as C and N source)	Formic acid	92% degraded within 3 h	[68]
CN polymer (using dicyandiamide as C and N source)	Acid orange II	100% decolored at optimal conditions	[69]

P3HT, poly(3-hexylthiophene).
[a] Decoloration rate constant compared to that of bare TiO_2.
[b] Decoloration rate was increased by 2.8 times in bilayer TiO_2/P3HT than that in nano-TiO_2/P3HT.

8.5.1 Organic Ligand

Many organic compounds with phenolic or carboxylic groups, such as hydroxylated biphenyls, phenols, and chlorophenols [6], are able to form CTC with the TiO_2 surface for visible light absorption. Catechol–TiO_2 complex is a typical example of CTC and is also one of the smallest chromophores studied in a dye/TiO_2 system. Electron transfer processes between catechol and TiO_2 were attributed to excitation of a CTC formed on TiO_2 surface. Figure 8.7 [80] shows the optical absorption spectra of catechol alone (a), catechol chelated to a Ti^{4+} complex (b), suspended TiO_2 nanoparticles (c), and catechol adsorbed onto TiO_2 (d). Wang *et al.* assigned the new broadband centered at 400 nm to excitation of a catechol–TiO_2 charge transfer state because this absorption was absent in the pure catechol or TiO_2 cases. A theoretical calculation provided evidence that the visible light absorption is attributed to LMCT and the excited state of catechol is not significantly involved in the electron injection process.

The CTC-mediated photocatalytic degradation over TiO_2 under visible light irradiation is important for the removal of colorless pollutants. Agrios *et al.* [81]

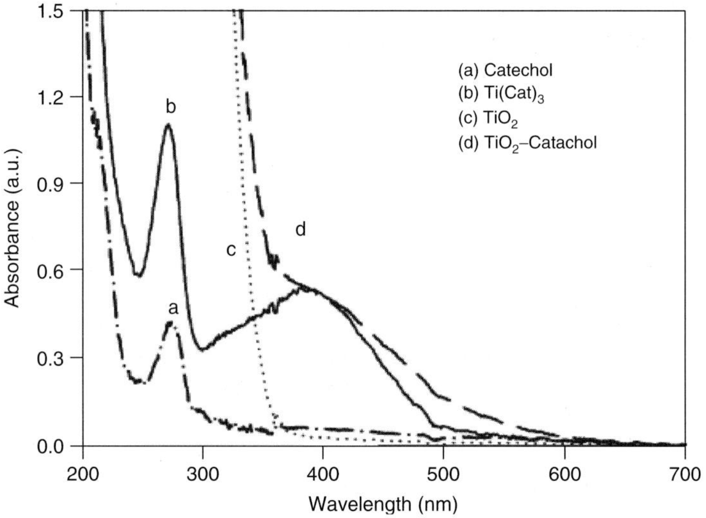

Figure 8.7 UV–vis spectra for aqueous solutions of (a) catechol, (b) Ti (catechol)$_3^{2-}$, (c) suspended TiO_2 nanoparticles, and (d) catechol adsorbed on suspended TiO_2 nanoparticles [80].

observed that 3,4,5-trichlorophenol was transformed to hexachlorodibenzo-*p*-dioxin over Degussa P25 under visible light, and they ascribed the visible light reactivity to a CTC mechanism, in which neither the photocatalyst nor the organic compounds absorb visible light by themselves. Similar results by Paul and coworkers [82] were observed in the visible-light-induced TiO_2 photocatalysis of fluoroquinolone antibacterial agents. Li *et al.* [83] reported that anatase nanoparticles showed higher visible light photocatalytic activity for phenolic compounds than Degussa P25, possibly due to the formation of surface complexes between TiO_2 and the organic substrates, and between catalyst and the degradation intermediates.

The CTC-mediated degradation process does not occur in case of benzoic and terephthalic acids but it does in the other tested aromatic organics containing hydroxyl or amino groups [84]. The spectra for TiO_2 remained unchanged after benzoic acid adsorption. The ground state of benzoic acid is approximately 1 eV below the VB edge, while its excited state is about 1 eV above the CB edge. Therefore, benzoic acid does not insert any states within the band gap of TiO_2. As a consequence, it does not create new optical bands that were lower in energy than the band gap absorption of TiO_2. Additionally, the coupling through carboxylic acid group of benzoic acid is weaker than the hydroxyl group seen with catechol. Catechol can form a chelate surface complex with Ti^{4+} with significant stability, but other dihydroxybenzenes do not form such surface complexes. Both salicylic acid and *p*-hydroxybenzoic acid have the same functional groups, but the photodegradation of the former is much faster than that of the latter. These indicated that the nature of substituent groups and their positions play an important role in the proposed CTC-mediated process. Wang *et al.* [85] reported that an electron-donating substituent in the benzene ring favored the CTC-mediated photodegradation of a pollutant, but an electron-withdrawing substituent showed an opposite effect as expected from the electron density on the ring. The order of degradation for various organic compounds followed the sequence 8-hydroxyquinoline > salicylic acid > phenol > *p*-aminobenzoic acid > *p*-hydroxybenzoic acid > aniline > benzoic acid ≈ terephthalic acid. 8-Hydroxyquinoline is more favorable to form a five-membered ring with the titanium atom, and the electron donors (both N and O atoms) can supply their nonbonded electron pair to the pyridine and benzene rings, thereby stabilizing CTC formation.

Recently, a new CTC-type visible light photocatalyst was developed by anchoring water-soluble fullerol on the surface of TiO_2 [86]. Surface complexes that absorb visible light (400–500 nm) were formed through covalent linkage between the surface titanol group (≡Ti–OH) and fullerol (Eq. (8.8)). LMCT may occur in the fullerol/TiO_2 system under visible light irradiation (Eq. (8.9)). The visible photon absorption should directly excite an electron from the fullerol π HOMO to the TiO_2 CB, forming a fullerol radical cation. The injected electron can subsequently flow to another fullerol (ground state) to produce the fullerol radical anion (Eq. (8.10)). Finally, water molecule was oxidized to evolve dioxygen (Eq. (8.11)), which was detected in a deaerated dispersion of fullerol/TiO_2 under visible light irradiation.

$$\equiv \text{Ti–OH} + \text{HO–C}_{60}(\text{OH})_{x-1} \longrightarrow \equiv \text{Ti–O–C}_{60}(\text{OH})_{x-1} + \text{H}_2\text{O} \tag{8.8}$$

$$\text{Fullerol/TiO}_2 + h\nu\,(> 420\text{nm}) \longrightarrow (\text{Fullerol}^{\bullet +})/\text{TiO}_2\,(\text{e}^-_{\text{CB}}) \tag{8.9}$$

$$\text{Fullerol/TiO}_2(\text{e}^-_{\text{CB}}) \longrightarrow (\text{Fullerol}^{\bullet -})/\text{TiO}_2 \tag{8.10}$$

$$(\text{Fullerol}^{\bullet +})/\text{TiO}_2 + 0.5\text{H}_2\text{O} \longrightarrow \text{Fullerol/TiO}_2 + \text{H}^+ + 0.25\text{O}_2 \tag{8.11}$$

8.5.2
Inorganic Ligand

In the presence of H_2O_2, the –OOH groups of H_2O_2 substitute for the surface basic –OH groups of titania (≡Ti–OH), forming a yellow surface complex of the type ≡Ti(IV)–OOH (Figure 8.8) [87]. The degradation of salicylic acid in the presence of both TiO_2 and H_2O_2 under visible light was attributed to the surface complexation of H_2O_2 with the TiO_2 surface that shows visible light response, although neither TiO_2 nor H_2O_2 absorb visible light. Upon visible light irradiation, electrons were injected from the excited surface complex $(\equiv\text{Ti(IV)–OOH})^*$ to the TiO_2 CB and generated ≡Ti(IV)–OOH$^\bullet$, which may further give rise to ≡Ti(IV)–OH and O_2. The CB electron further reacted with the adsorbed H_2O_2 on the TiO_2 surface to generate the OH radical. The magnitude of visible light response increased with an increase in the number of H_2O_2 molecules adsorbed on the TiO_2 surface.

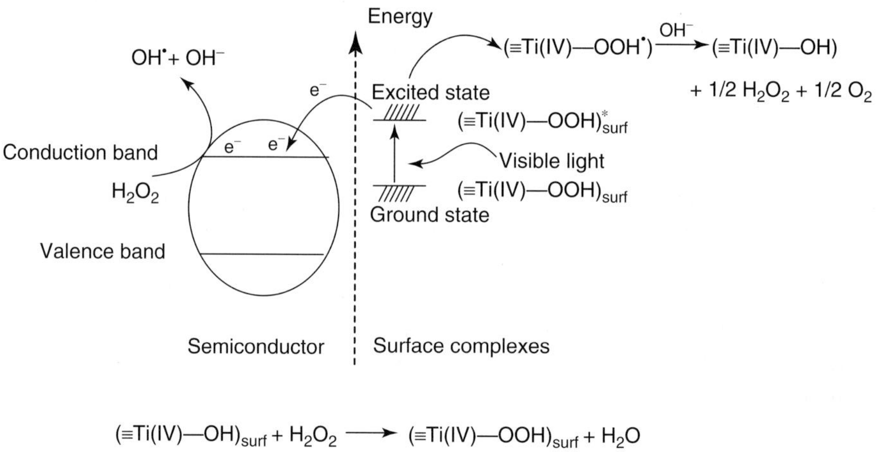

Figure 8.8 A possible mechanism of photoinduced electron transfer and interface photoreaction of the surface complex of H_2O_2 on the TiO_2 surface [87].

Recently Yu *et al.* [88] reported that TiO_2 powders grafted with metal ions (Cu(II), Cr(III), Ce(III), and Fe(III)) were capable of serving as photocatalysts sensitive to visible light. For example, the grafting of Fe(III) to form Fe(III)/TiO_2 resulted in a rather high quantum efficiency (22%) for gaseous propan-2-ol decomposition under visible light (400–530 nm). These kinds of materials were prepared by a simple impregnation technique without thermal calcinations, which suggested that these metal ions are just on the surface of titania and unlikely to be doped into the lattices. The authors ascribed the high performance of metal-ion-grafted TiO_2 to IFCE, that is, electrons in the VB of TiO_2 directly transfer to Fe(III), not via the excited state, to generate Fe(II) while holes remaining in the VB decompose organic compounds. The adsorbed oxygen could be catalytically reduced (presumably via multielectron reduction) to ROS by photoproduced Fe(II) species.

8.6 Solid Semiconductor/Metal Sensitization

In 1984, Serpone *et al.* [89] first proposed an interparticle electron transfer (IPET) process with substantial enhancement of reductive processes on TiO_2. They demonstrated that two semiconductors coupling (e.g., TiO_2 and CdS) led to an improvement in water splitting in the presence of H_2S as the sacrificial agent. In the past decades, a certain numbers of studies have described photocatalytic activity of TiO_2 coupled with metal oxides such as CdS, CdSe, Bi_2S_3, Bi_2O_3, SnO_2, WO_3, Cu_2O, Fe_2O_3, ZrO_2, and In_2O_3 [90]. The main focus of this section is to review the recent progress of IPET involving TiO_2 and other semiconductors or metal particles, comparing their properties with that of TiO_2 alone especially under visible light irradiation.

8.6.1
Small-Band-Gap Semiconductor

8.6.1.1 Basic Concepts

The formation of heterojunction structure between a narrow-band-gap semiconductor and TiO_2 with matching band potentials provides an effective way to extend the photosensitivity of TiO_2 to the visible region of the solar spectrum. There are three prerequisite conditions for such a heterogeneous semiconductor to function efficiently [91] (as shown in Figure 8.9): (i) the sensitizer with a narrow band gap (**SEM I**) should have a strong absorption in the visible region; (ii) **SEM I** should have a higher CB minimum and VB maximum compared to the wide-band-gap semiconductor (**SEM II**, usually TiO_2), for smooth injection of electrons downhill from the CB of **SEM I**; and (iii) electron injection should be fast and efficient.

Suitable energy levels are not the only factor influencing IPET, there are also coupling aspects between acceptor and donor electronic states of both the semiconductors, morphology of the particles, the surface texture, and extent of surface contact between the particles play a crucial role. Intimate contact between the coupled semiconductors is critical for achieving maximum charge separation. Such an IPET through the heterojunction is found to be dependent on the adsorption of pollutants to the sensitizer. The ability of a small-band-gap semiconductor to photosensitize a large-band-gap TiO_2 with no absorption in the visible region has many important applications from the viewpoint of solar energy conversion. Not only activity upon visible light irradiation but also an efficient charge separation and increased yield of the catalytic reaction are the main advantages of such systems (Table 8.4).

8.6.1.2 Category in Terms of Charge Transfer Process

Type I: CB Electron Transfer to TiO_2 Earliest studies concerning the coupling effect between CdS and TiO_2 via IPET can be traced back to the early 1980s [89, 101],

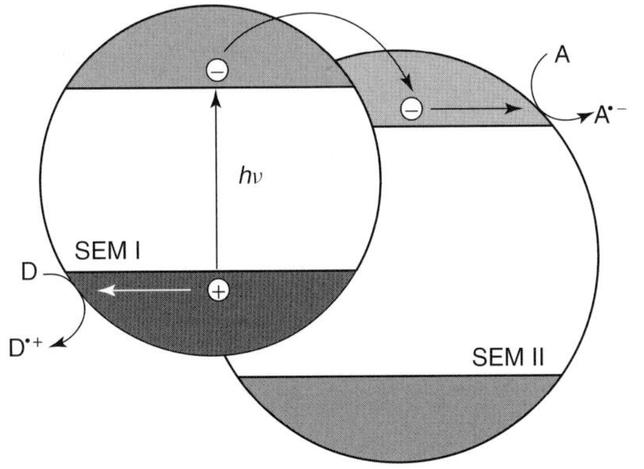

Figure 8.9 Electron injection in composite semiconductors. A, acceptor; D, donor.

Table 8.4 Comparison of the visible light efficiency of selected composite semiconductors prepared by various methods.

Photocatalyst	Preparation route	Substrate	Visible light efficiency	References
CdS/TiO_2	Microemulsion-mediated solvothermal synthesis	MB	80% MB decolored (1 h)	[7]
CdS/TiO_2	Ion exchange after preplanting CdO into ordered mesoporous TiO_2	NO gas and MB, 4-CP	30% NO gas degraded (10 min); 95% MB decolored (3 h); 70% 4-CP degraded (4 h)	[92]
CdS/TiO_2 nanotube	Solution-based method	MB	~10 times compared to Degussa P25	[93]
$CdSe/TiO_2$ nanotube	Direct current electrotechnique	ACA	~100% ACA degraded	[94]
$InVO_4/TiO_2$	Sol–gel method	Benzene	6.2 mol benzene degraded per mol catalyst	[95]
Cu_2O/TiO_2	Sol–gel method	H_2O	$R_{H2} = 6\ ml\ h^{-1}$ (pH = 9)	[96]
NiO/TiO_2	Calcination of layered metal hydroxides	MB	70% MB decolored (4 h)	[97]
α-Fe_2O_3/TiO_2	Chemical bath deposition method	MB	Similar to the pure TiO_2 nanosheets	[98]
MoS_2/TiO_2	*In situ* photoreduction deposition method	MB, 4-CP	32% MB decolored (4 h); 50% 4-CP degraded (4 h)	[99]
WS_2/TiO_2	*In situ* photoreduction deposition method	MB, 4-CP	32% MB decolored (4 h); 60% 4-CP degraded (4 h)	[99]
$CdSe/TiO_2$	Solvothermal method	MB	40% MB decolored (4 h)	[100]

MB, methylene blue; 4-CP, 4-chlorophenol; (ACA), anthracene-9-carboxylic acid.

when photocatalytic production of H_2 by CdS was found to be enhanced once TiO_2 particles were added. The smaller band gap of CdS (~2.5 eV) enables it to absorb visible light to generate holes and electrons (Table 8.5).

Since both the VB and CB of CdS photocatalysts are higher than their counterparts in the TiO_2, the electrons generated from CdS will subsequently flow from the CB of CdS to the adjacent TiO_2 CB [102–105]. Evans *et al.* [106] reported that the CB electron transfer from CdS to TiO_2 could happen on a timescale of picoseconds, leading to a significantly slower recombination for the charge carriers generated on light absorption in CdS. The enhanced charge separation under visible light irradiation has made this CdS/TiO_2 couple widely used in photocatalytic degradation of organic pollutants and water splitting.

However, the previous research on CdS/TiO_2 couples was mostly limited to colloidal systems in which CdS was often used as a dominant component. The following investigations started to use a small amount and/or size of CdS as a "sensitizer" to make the heterostructure sensitive to visible light. Wang *et al.* [107]

Table 8.5 Band positions of some common semiconductor photocatalysts in aqueous solution [90].

Semiconductor	VB (V vs NHE)	CB (V vs NHE)	Band gap (eV)	Band gap wavelength (nm)
TiO_2	+3.1	−0.1	3.2	387
SnO_2	+4.1	+0.3	3.9	318
ZnO	+3.0	−0.2	3.2	387
ZnS	+1.4	−2.3	3.7	335
WO_3	+3.0	+0.4	2.8	443
CdS	+2.1	−0.4	2.5	496
CdSe	+1.6	−0.1	2.5	729
GaAs	+1.0	−0.4	1.7	886
GaP	+1.3	−1.0	1.4	539

prepared a CdS/TiO_2 hydrogel with a CdS loading of only about 0.9% where −2 nm uniform crystalline CdS particles dispersed in TiO_2. At this size, the quantum size effect of CdS could be expected, that is, the band gap of nanosized CdS would widen and thus the energy differences between the CBs of CdS and TiO_2 should increase. Yu *et al.* [99] also reported the deposition of quantum-sized semiconductors such as WS_2, MoS_2, and CdSe on the surface of TiO_2 for the first time. Although the bulk WS_2 has a CB edge potential lower than that of TiO_2, the strong quantum dot confinement effects for nanosized WS_2 make efficient electron transfer from WS_2 to TiO_2 possible. The uses of quantum-sized particles as sensitizers principally implies several advantages as compared to organic dyes: the band gap and thereby the absorption range are easily adjustable by the size of the particles, the band edge type of absorption behavior is most favorable for effective light harvesting, and the surface properties of the particles can be modified in order to increase the photostability of the composite photocatalyst.

One of main disadvantages of CdS/TiO_2-coupled material is its susceptibility to photocorrosion. The hole generated on the VB of the sensitizer cannot oxidize the surface hydroxyls to hydroxyl radicals and hence can corrode the sensitizer itself. Fuji *et al.* [108] have proposed a solution for the photocorrosion problem by embedding CdS fine particles on a TiO_2 transparent gel matrix. CdS(bulk)–TiO_2[1] showed higher activity for H_2 production compared to CdS–TiO_2(bulk)[2] and nano-CdS/nano-TiO_2 under visible light irradiation. It is remarkable that the fabrication method of the composite had such a dramatic effect despite that the composites were almost the same.

Type II: VB Hole Transfer to TiO_2 In Type I mode, the injected electrons from **SEM I** are utilized for the subsequent photocatalytic reactions, while the residual

1) CdS(bulk)–TiO_2 refers to the preparation of TiO_2 on the bulk CdS surface.
2) CdS–TiO_2(bulk) refers to the preparation of CdS on the bulk TiO_2 surface.

holes in VB of **SEM I** usually cannot be used because of their low oxidizing power. Therefore, a new strategy utilizing the photogenerated hole is of equal importance, which was achieved by designing the electronic band structure of the sensitizer in the TiO_2-based composite. The concept of such a heterojunction-type photocatalyst, utilizing the hole transfer from the sensitizer to TiO_2, was realized by Lee and coworkers [109] in the $FeTiO_3$–TiO_2 system that showed superior activity for the degradation of propan-2-ol and 4-CP compared to Fe_2O_3–TiO_2 and TiO_2 under visible light irradiation. According to the proposed mechanism, hole transfers from the VB of $FeTiO_3$ to the TiO_2 VB leave behind electrons in the $FeTiO_3$ CB under visible light excitation. By this intersemiconductor hole transfer mechanism, the photogenerated charge carriers can be separated efficiently so that the composite semiconductors can utilize visible light to degrade the pollutants.

Very recently, Xu *et al.* [110] prepared an efficient and stable composite photocatalyst Bi_2WO_6/TiO_2 that exhibited enhanced visible-light-induced activity in photocatalytic degradation of pollutants in aqueous/gaseous phases compared with Bi_2WO_6 or TiO_2 alone. Both PL (photoluminescence) spectra and electrochemical impedance spectroscopy (EIS) analysis revealed that Bi_2WO_6 coupled with TiO_2 promoted the mobility and separation of photogenerated carriers effectively. According to the estimated E_g values of the Bi_2WO_6 and TiO_2 semiconductors, the VB level of Bi_2WO_6 is lower by 0.353 V than that of TiO_2 while the CB position does not allow the reduction of O_2. It suggests that photogenerated holes in the VB of Bi_2WO_6 after its excitation by visible light can be transferred to that of TiO_2. The large flat sheet of Bi_2WO_6 facilitates the migration of charge carriers while the high specific surface area of P25 is beneficial to the acceptance of large quantities of holes. Accordingly, the electrons on the surface of Bi_2WO_6 sheets and holes on the TiO_2 particles can separately participate in photocatalytic reactions.

8.6.2 Plasmonic Metal

8.6.2.1 Basic Concepts

Nanosized plasmonic metals are characterized by their strong interaction with resonant photons through excitation of localized surface plasmon resonance (LSPR) derived from the collective oscillation of valence electrons [12]. The LSPR-enhanced strong local electric fields in the neighborhood of the nanostructure induce surface-enhanced Raman scattering, or the formation of energetic charge carriers that can be transferred to the surroundings or relax by locally heating the nanostructure in the process of surface plasmon decay (Figure 8.10). The resonant wavelength is dependent on the nature of the metal. Gold, silver, and copper nanostructures are of particular interest because their LSPR wavelengths are within the UV–vis spectral range. The LSPR intensity and wavelength also depend on the size and shape of metallic nanoparticles. For example, small Au nanoparticles (<5 nm) did not exhibit any resonant behavior, whereas a sharp absorption band was observed when the size of Au particles was up to 50 nm [111] (Chapter 4). As for the Ag nanomaterial, a shift in the plasmon absorption band from 400 to 670 nm

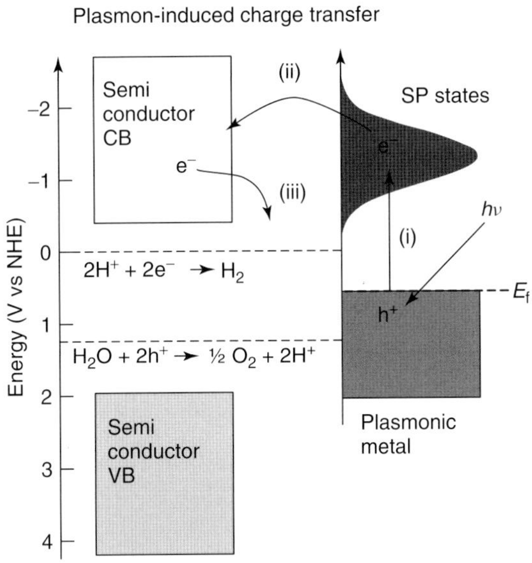

Figure 8.10 Mechanism of SPR-induced charge transfer with approximate energy levels on the NHE scale [12]. Dashed lines on the left of this figure refer to the water-splitting redox potentials. (i) Electrons near the metal Fermi level, E_f, are excited to surface plasmon (SP) states; (ii) electrons are transferred to a nearby semiconductor particle and (iii) this activates electron-driven processes such as the hydrogen-evolution half-reaction.

occurred as the Ag particle shape changed from spherical to triangular prisms [112, 113]. Therefore, by manipulating the composition, shape, and size of plasmonic nanoparticles, the LSPR intensity and wavelength can be modulated.

The unique capacity of plasmonic metal nanoparticles to absorb visible light makes them suitable for serving as a photosensitizer and hence shows significant promise in heterogeneous photocatalysis. In fact, noble metals such as Pt, Au, and Ag deposited directly on a semiconductor, serving as the cocatalyst, can assist in charge carrier separation and electron trapping [62, 114, 115]. These cocatalysts also can facilitate catalytic reactions by providing chemically active sites where relevant chemical transformations occur with lower activation barriers than on bare TiO_2 surfaces. Although this cocatalyst/TiO_2 enhances the rates of electron–hole separation and thus improves performance compared with TiO_2 itself, this composite photocatalyst only works under UV irradiation. Now this problem can be partially alleviated since it has been recognized recently that this kind of plasmonic metallic nanoparticle/semiconductor photocatalyst behaves more efficiently in various photocatalytic reactions as compared to its pure semiconductor counterpart. For example, Au/TiO_2 composite showed high efficiency for photooxidation of acetic acid and propan-2-ol and for selective oxidation of propan-2-ol to acetone, thiol to disulfide, and benzene to phenol, as well as exhibited a better photovoltaic efficacy [111, 116, 117].

8.6.2.2 **Proposed Mechanisms**

It has been evident that LSPR played a critical role in increasing the rate of photocatalytic transformations on nearby semiconductors, because many investigations experimentally demonstrated a positive correlation between LSPR intensity and the rate increment. Therefore, it has been hypothesized that LSPR of metallic nanostructure can concentrate UV–vis irradiation and then transfer it to the nearby semiconductor, thereby enhancing the energetic charge carrier available for photocatalytic reactions. There are two types of non-mutually-exclusive energy transfer mechanisms involved in enhancing the rate of photocatalytic transformations [12].

Direct Charge Transfer Before light illumination, the electron occupancy of plasmonic metals obeys the Fermi–Dirac distribution. Zhu and his colleagues [118, 119] reported that incident photons of blue light can be absorbed by Au or Ag nanoparticles through their LSPR excitation. As a consequence, for example, Ag electrons are excited from within the outermost sp band to higher energy states (also called *intraband excitation*). The excited electrons are directly injected from plasmonic metal into the CB of titania and then reduce molecular oxygen adsorbed on the surface of titania, while the electron-deficient metal has a greater affinity for capturing electrons from the nearby organic matters and thus return to its metallic state. This process was found to operate in composite photocatalysts in which the titania and nanosized metal particles are in direct contact with each other, facilitating a rapid electron transfer. Recently, Furube *et al.* [120] reported that the electron transfer from Au nanoparticles to titanium oxide takes less than 240 fs, using femtosecond transient absorption spectroscopy with an infrared probe. Sakai *et al.* [121] also observed generation of a cathodic photocurrent on an (Indium tin oxide) ITO/Au/TiO_2 electrode, proving the existence of electron injection process from plasmon-excited gold nanoparticles to titania.

This direct charge transfer mechanism is analogous to dye sensitization [2]. The plasmonic nanoparticles indeed act as a dye sensitizer, absorbing resonant photons, and transferring excited electrons to the nearby semiconductor. Because of their excellent mobility of charge carriers and high absorption cross sections, nanoscale noble metals are likely to become even more ideal sensitizers than a typical dye sensitizer. Also, this plasmonic metal sensitizer is very promising for utilizing the full solar spectrum by tuning their size or shape of nanostructures.

Radiative Energy Transfer The LSPR-mediated enhancements have been previously observed in the photocatalytic activity of semiconductors, although the two building blocks were physically separated by nonconductive spacers. In these cases, radiative energy transfer mechanism works through near-field electromagnetic and scattering processes.

Photoexcited plasmonic nanoparticles are characterized by strong electric fields that are several orders of magnitude higher than the field of incident photons. In other words, these metallic nanostructures essentially serve as nanoscale concentrators that can amplify the local light intensity. When a semiconductor comes into the proximity of these excited plasmonic nanoparticles, it will be affected by these intense fields. Therefore, the so-called near-field electromagnetic mechanism is based on the interaction of semiconductor with the strong LSPR-induced electric fields around the metallic nanostructure. Since the rate of electron–hole formation in a semiconductor is proportional to the local intensity of the electric field, the rate of LSPR-induced electron–hole pair formation increases by a few orders of magnitude, being highest in the regions of the semiconductor that are closest to the metallic nanostructure.

In addition to the near-field electromagnetic effects, which play a role in the spatially heterogeneous formation of electron–hole pairs in proximal semiconductors, an efficient scattering of resonant photons is functional. Here, the plasmonic metallic nanoparticles work as a nanomirror, because the scattering of photons will increase the possibility of photons absorption by the semiconductor by increasing the average photon path length, thereby leading to an enhanced rate of electron–hole pair formation.

8.6.2.3 **Critical Parameters**

The relative importance of the two mechanisms is governed by the optical characteristics of the building blocks and their geometric arrangements. For instance, direct electron transfer is the exclusive mechanism for composite photocatalysts where metal LSPR can be excited with lower energy than the semiconductor band gap. It means that metal nanoparticles can absorb low-energy photons (e.g., visible light) to which titania is not sensitive. The radiative energy transfer mechanism is significant for building blocks where metal LSPR and semiconductor absorbance overlap the spectra of irradiation sources. Essentially, metal LSPR excited by source photons should have sufficient energy to overcome the semiconductor band gap, but cannot extend absorption bands of this composite material compared to the semiconductor counterpart. The magnitude of LSPR-induced rate enhancements of photocatalytic reaction depends strongly on the properties of titania, including particle size, surface chemistry, and crystallinity (anatase or rutile), and on the size, shape, and coverage of noble metal deposits and their spacing between semiconductor and plasmonic metal building blocks [12]. Kowalska *et al.* [111] compared the photocatalytic activities of 15 commercial titania modified with gold under visible light irradiation. They observed that enhancement ratios (ERs) of activities of gold-loaded titania to the bare one, for rutile-rich titania (ER = 0.9–4.9), were generally larger than those for anatase-rich titania (ER = 1.6–3.0). Large rutile particles loaded with gold particles of a wide range of sizes and samples with larger gold particle size exhibited the highest level of photocatalytic activity, possibly due to greater light absorption ability for larger gold particles and longer diffusion length of electrons within the bulk of titania for larger TiO_2 particles, respectively.

In conclusion, sensitizing titania semiconductor with noble plasmonic metals has some obvious advantages over the aforementioned approaches. There is no recombination center yielded in the crystalline lattice of bulk semiconductor as doping method does. Compared to organic dyes and polymers, noble metal nanoparticles are more stable during the photocatalytic reactions. Despite the limited study using plasmonic metallic nanostructure as a visible light sensitizer for TiO_2, the encouraging results obtained recently show the promise of this approach for developing novel visible-light-response photocatalysts [122].

8.7 Other Strategies to Make Titania Visible Light Active

8.7.1 Band Gap Engineering

The main objective of band gap engineering is to narrow the band gap of nanoscale semiconductor either by elevating the VB maximum or lowering the CB minimum, in order to adequately utilize visible light. Doping as an important band-gap narrowing strategy involves inclusion or substitution of an alien atom into the TiO_2 lattice so as to induce new electronic states and optical transitions not seen in pure TiO_2. The ideal doping needs a homogeneous distribution of dopant at the atomic level in the entire TiO_2 matrix without altering the crystal structure of the nanomaterial [91]. However, in practice, some critical issues such as the origin of visible light absorption and the dopants' location in TiO_2 lattice are often not well determined and the effects of the nature of dopants are not well understood. Hence, it is difficult to specifically define the sensitizer in a certain doped TiO_2, although Szaciłowski *et al.* [123] classified it as one kind of photosensitization. Furthermore, modification of TiO_2 at the bulk or surface level is discussed in chapter 5. In this context, doping toward visible-light photocatalytic activity is briefly introduced.

8.7.1.1 Metal Doping

Doping of transition-metal ions into a wide-band-gap photocatalyst is one of the ways to develop a new visible-light-responsive photocatalyst. The first systematic study on metal doping in quantum-sized TiO_2 by Choi *et al.* [124] has been carried out by examining the photoreactivities of doped TiO_2 and their transient charge carrier recombination dynamics. However, the as-prepared metal-ion-doped titania was not visible light responsive. Many endeavors were made by many groups to examine how cation doping shifts the TiO_2 absorption properties into the visible. A variety of metals on TiO_2 have been explored, including Fe, Cr, Ni, Nb, V, Mn, Cu, Al, Co, and W [125]. Taking Cr^{3+} as a representative example of metal doping, Anpo and Takeuchi [126] showed that Cr doping of TiO_2 gives rise to visible light absorptivity. The optical band edge of mesoporous $Cr{-}TiO_2$ was assigned to the charge transfer band $Cr^{3+} \rightarrow Ti^{4+}$ or $4A_{2g} \rightarrow 4T_{1g}$ of Cr^{3+} in an octahedral configuration. The $4A_{2g} \rightarrow 4T_{2g}$ d–d transitions of Cr^{3+} contribute to a broad adsorption band above 620 nm [127, 128]. However, the main visible light absorption band may be different when different synthesis routes are used. Cr-doped TiO_2 by ion implantation has a parallel absorption band edge up to about 550 nm, but it has only a visible light absorption shoulder when impregnation method is adopted. This suggests the chemical and/or structural environment of the Cr^{3+} dopant will affect its utility as a promoter of visible light photocatalytic activity in TiO_2. It is worthy to note that visible light absorptivity in metal-doped

TiO_2 does not necessarily translate into photochemistry. Scientists [129, 130] have found redshifts in the absorption thresholds with Cr, V, Fe, and Co doping, but did not observe any enhancement in photocatalytic activity for these doped samples over pure TiO_2. According to Choi and coworkers [124, 131], visible excitations in metal-doped TiO_2 promote surface photochemistry, depending on the dopant concentration and its atomic structure. The optimum dopant content balances the role of these centers to serve as traps for effective charge separation with the adverse tendency of these sites to increase the possibility of charge recombination. Also, they concluded that most dopants with open shells were active for redshifting TiO_2 photochemistry into visible, whereas those cations with closed shell cannot trigger any photochemical reaction.

8.7.1.2 **Nonmetal Doping**

For most transition-metal oxide semiconductors, the VBs and CBs consist of dominant O-2p and metal-3d states. In such cases, the general idea for nonmetal doping is to substitute a heteratom for O and then have some dopant states within the band gap. In principle, the strict requirements for nonmetal dopant are as follows [89]: (i) the electronegativity of the selected nonmetal must be lower than that of oxygen and (ii) the nonmetal dopant, as a substitutional lattice atom, should have a radius commensurate with that of the lattice O atoms so that the dopant atoms can realize a uniform distribution within the whole host matrix.

Since the Japanese scientists' reports of visible light activity in nitrogen-doped TiO_2 in 2001 [8, 132], nonmetal doping has been intensively investigated. By far, nearly all nonmetals have been screened for suitable dopants to improve their visible light photocatalytic activity by modulating the electronic structure of titania. In particular, N, C, S, and B have been extensively examined, both experimentally and theoretically. Taking N-doped TiO_2 as an example, at least five mechanisms were proposed to explain the origin of visible light absorption, but the intrinsic structure factors controlling these electronic mechanisms, positions, and behaviors of these dopants in titania lattice matrix remain unresolved. For more information about these issues, the reader is referred to Refs. [91, 133–135].

8.7.1.3 **Codoping**

Another strategy to extend the visible response of TiO_2 is to modify the photocatalyst with two dopants (metal/nonmetal or nonmetal/nonmetal). Codoping with two or more suitable heteroatoms can achieve substantial synergistic effects. Zhao and coworkers [11] developed a novel titania-based photocatalyst synthesized by doping with both boron and Ni_2O_3 (B, Ni-doped TiO_2), which can efficiently photodecompose colorless organic pollutants under visible light irradiation (Figure 8.11). Codoping of B and Ni can play a vital role in activating visible-light absorption and thus improve photocatalytic activity through two effects: (i) incorporation of B into TiO_2 lattice extends the spectral response to the visible region and acts as an antenna and (ii) loading of Ni_2O_3 into the matrix can effectively inhibit the charge recombination and greatly enhance the photocatalytic activity.

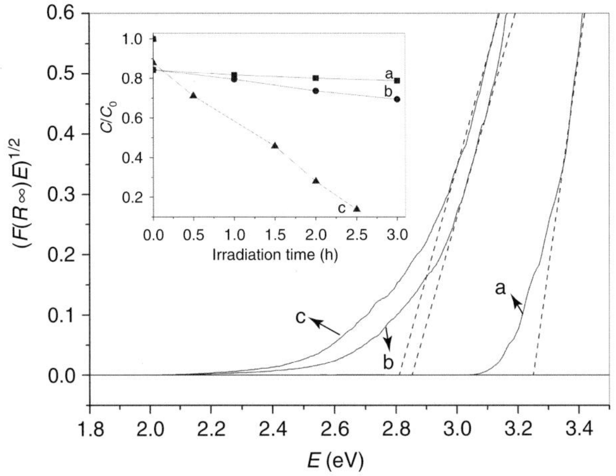

Figure 8.11 Diffuse reflectance absorption spectra of pure TiO_2 (a), boron-doped TiO_2 (b), and boron- and nickel-doped TiO_2 (c) [11]. Data are plotted as transformed Kubelka–Munk function versus the energy of light. Inset: temporal course of the photodegradation of 2,4,6-trichlorophenol (1.0×10^{-4} M; 50 ml) in aqueous dispersions containing 50 mg of catalysts under visible light irradiation.

N-F-codoped TiO_2 is an interesting example of nonmetal–nonmetal codoping [136]. The yellow photocatalyst was prepared from a mixed solution containing titanium(IV) *n*-butoxide and NH_4F. UV–vis absorption spectra showed that a new absorption band appeared in the visible range of 350–550 nm that was attributed to the doped N atoms (Figure 8.12). Wang *et al.* [137] proved that doping of F atoms can greatly improve the amount of implanted nitrogen and enhance the oxidation ability of hole due to the shift of N-2p levels.

With respect to monodoping, the codoping is a more flexible strategy to extend the photoresponse to the visible region. The synergistic effects have been often observed in the TiO_2 codoped with donor–acceptor impurities. The charge compensation between the dopants can increase the amount of the photoactive dopants and depress the bulk defects, enhancing the oxidation and reduction powers and mobility of the photogenerated carriers. However, since most dopants are located in the surface/subsurface of TiO_2, codoping can only exert its influence on the modification of electronic structure within a limited depth from the surface. Therefore, the challenge is to select suitable dopant pairs and then introduce them throughout the particle, thus utilizing their synergistic effects utmost [91]. A more detailed discussion on synergistic effects of codopants in TiO_2-based photocatalysis can be found in the authors' recent review article in *Current Organic Chemistry* [138].

8.7.2
Structure/Surface Engineering

In general, a surface photoreaction can be initiated by the photoexcitation of both the intrinsic and extrinsic absorption bands of a photocatalyst. The former is related

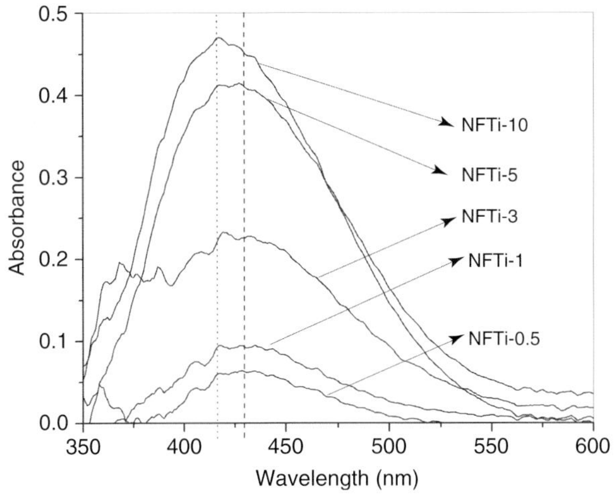

Figure 8.12 Difference in the spectra between the F-N TiO_2 (NFTi-*x*, where *x* denotes the molar ratio between precursor of NH_4F and Ti) obtained using TiO_2 as the blank [137].

to electron transition from the VB to the CB of a semiconductor and therefore requires larger energy than the band gap energy. The latter concerns the light absorption of impurity, the photoionization of original or newly formed defects, and the excitation of surface states, which needs less energy than the former and can even be sometimes realized with visible light [3]. Recently, TiO_2 was reported to absorb visible light via artificially created intrinsic point defects (i.e., oxygen vacancies, titanium vacancies, and interstitial titanium) in its crystal structure [28, 139–141]. Both oxygen vacancies and Ti interstitials result in donor levels at 0.75–1.18 and 1.23–1.56 eV below the CB, respectively, whereas Ti vacancies give rise to acceptor levels above the VB [142–144]. These different defects can contribute to an additional shoulder absorption band within the range of visible light and/or a tail absorption band in the NIR and infrared ranges, depending on the preparation techniques. Kuznetsov and Serpone [145] reported that NIR and infrared absorptivity is derived from Ti^{n+}-related ($n = 2, 3$) color centers, while the visible light absorption bands are associated with oxygen vacancies. Justicia *et al.* [146–148] found that substoichiometric anatase films exhibited visible photocatalytic activity for methylene blue (MB) degradation. They proposed that the overlap of defect states formed by oxygen substoichiometry with the CB states of titania narrowed its band gap.

Previous photochemical studies have mainly focused on the powder TiO_2 material with inhomogeneous surfaces. However, in that case, surface states are not easily controllable. Batzill's group [149] reported a new TiO_2 phase with a narrow band gap (~2.1 eV) on rutile (011) formed by the annealing of slightly reduced TiO_2, which was close to perfect for visible light absorption. However, the stability of this phase in the presence of oxygen and information on the photocatalytic efficiency of this material were not reported. Ariga *et al.* [140] made the first attempt to investigate the effect of the unique "lattice-work structure" of a TiO_2 (001) on visible light responsible photooxidation by using formic acid as a probe molecule. They found through scanning tunneling microscopy (STM) that the nanostructured surface was photochemically active toward visible light (~2.3 eV). Combined electronic and theoretical studies revealed that band-gap narrowing is caused by the low coordinated atoms on the topmost surface. However, Serpone *et al.* argued that one should be cautious with theoretical studies of band-gap narrowing especially if done by density functional theory (DFT) methods (see more examples in [150, 151]).

Recently, several groups have observed visible light photocatalytic activity for slightly reduced TiO_2 materials. For example, Martyanov *et al.* [152] observed photocatalytic oxidation of acetaldehyde with visible light by a near-stoichiometric TiO_2 prepared by oxidation of TiO or Ti_2O_3. These authors attributed the observed photocatalytic activity directly to the excitation of Ti^{3+} sites. The blue color in reduced TiO_2 originates from optical absorption in the NIR region of the visible spectrum resulting from excitation events associated with Ti^{3+}-related bulk defects. Komaguchi *et al.* [153] proposed that visible light excitation of Ti^{3+} was related to transitions of localized electrons at Ti^{3+} sites into the CB. Zuo *et al.* [139] fabricated a visible-light-responsive TiO_2 by introducing Ti^{3+} through one-step combustion method, which accounts for the H_2 generation from water under visible light irradiation.

Very recently, Mao *et al.* [154] have developed a high-temperature (e.g., 200 °C) and high-pressure (e.g., 20.0 bar) hydrogenation method to prepare hydrogenated TiO_2 that is an active and stable photocatalyst for visible-light-induced reaction. They proposed that hydrogenation of TiO_2 gives rise to TiO_2 reduction, leaving the self-doped Ti^{3+} species and the hydrogen atoms on its surface. Theoretical calculations indicated that high concentration of Ti^{3+} results in a donor level just below the CB and the residual hydrogen atoms on the surface may lead to the upper shift of the VB edge. Inspired by this scenario, Li *et al.* [141] reconstructed the surface structure of mesoporous TiO_2 with benzyl alcohol molecules (Eq. 8.12), which substantially extends its visible light absorption and photocatalytic activities.

$$2Ti^{(IV)}O_2+C_7H_8O \xrightarrow[\text{Visible light}]{N_2} 2Ti^{(III)}O(OH)+C_7H_6O \quad (8.12)$$

8.8 Conclusions

Over the past decades, considerable efforts have been expended on the exploration for visible-light-activated titania to extend the usable solar energy spectrum. Some strategies including photosensitization by organic dyes or metal complexes, coupling with narrow-band-gap semiconductor, metal ion/nonmetal ion doping, codoping, and noble metal deposition have proven to be effective for enabling TiO_2 workable upon visible light illumination and enhancing the separation of photogenerated charge carriers. The tuning of bulk/surface structure properties of TiO_2 in a facile way has also shown great promise. All these strategies to make TiO_2 visible light active provide a flexible line of action for further design, fabrication, and modification of semiconductor nanomaterial for environmental and energy applications.

To push forward the field of photosensitization of titania, however, some misconceptions and inaccuracy on visible light photocatalytic activity should be clarified here. First, photocatalytic activity obviously depends strongly on the physicochemical and optical properties of the semiconductors, such as the surface area, porosity, absorption capacity, and distribution of the active sites. For semiconductor composite materials, the original properties of each component might be modified upon coupling, which actually complicates the comparison of photocatalytic activity before and after coupling. The total photocatalytic activity of selected materials, in some cases, is the collective performance of many factors [133].

Researchers routinely characterize their photocatalysts using UV–vis absorption spectrometer. It should be emphasized that UV–vis absorption spectra only reflect the intrinsic optical property for the bulk of a solid. The actual absorption spectrum of a photocatalyst is an overlapping result of intrinsic and extrinsic absorption bands [3]. The absence of absorption bands in UV–vis absorption spectra does not necessarily mean that this examined material lacks photocatalytic activity. In contrast, the absorption of light is a necessary but not a sufficient condition to promote photocatalytic activity. Photoabsorption cannot necessarily ensure its utility as an efficient photocatalyst. The other frequent misconception is to use an organic dye molecule to characterize the visible light photocatalytic activity of TiO_2-based materials. Caution should be exercised while using dyes as the model pollutant because self-sensitized photodegradation of dyes, as discussed in Section 8.3, is likely to occur simultaneously with the band gap excitation of visible-light-responsive photocatalysts under visible light irradiation. In this case, it is difficult to make a conclusion that the photocatalytic activity of the as-prepared TiO_2-based catalyst is superior to the pristine TiO_2 under visible light illumination. Although MB is usually considered refractory to sensitized photodegradation under visible irradiation, it is still not recommended to be used as a model contaminant to evaluate the visible light activity of a TiO_2-based photocatalyst, because photosensitizated degradation of MB is very likely to occur if the energy level of TiO_2 CB is altered under some appropriate experimental conditions [155].

Acknowledgment

Parts of work described in Sections 8.3, 8.4, and 8.7 were financially supported by the 973 project (No. 2010CB933503), NSFC (Nos. 21077110, 20920102034, and 20877076), and CAS.

References

1. Fujishima, A. and Honda, D. (1972) Electrochemical photolysis of water at a semiconductor electrode. *Nature*, **238**, 37–38.
2. Chen, C.C., Ma, W.H., and Zhao, J.C. (2010) Semiconductor-mediated photodegradation of pollutants under visible-light irradiation. *Chem. Soc. Rev.*, **39**, 4206–4219.
3. Zhou, J. and Zhao, X.S. (2010) in *Environmentally Benign Photocatalysts: Applications of Titanium Oxide-Based*

Materials (eds M. Anpo and P.V. Kamat), Springer, New York, Dordrecht and Heidelberg, London, pp. 235–252.
4. Henderson, M.A. (2011) A surface science perspective on TiO_2 photocatalysis. *Surf. Sci. Rep.*, **66**, 185–297.
5. Zhao, J.C., Chen, C.C., and Ma, W.H. (2005) Photocatalytic degradation of organic pollutants under visible light irradiation. *Top. Catal.*, **35**, 269–278.
6. Kim, G. and Choi, W. (2010) Charge-transfer surface complex of EDTA-TiO_2 and its effect on photocatalysis under visible light. *Appl. Catal., B*, **100**, 77–83.
7. Yu, J.C., Wu, L., Lin, J., Li, P., and Li, Q. (2003) Microemulsion-mediated solvothermal synthesis of nanosized CdS-sensitized TiO_2 crystalline photocatalyst. *Chem. Commun.*, 1552–1553.
8. Pichat, P. (1991) in *Photochemical Conversion and Storage of Solar Energy* (eds E. Pelizzetti and M. Schiavello), Kluwer Academics Publication, pp. 277–293.
9. Gopidas, K.R., Bohorquez, M., and Kamat, P.V. (1990) Photoelectrochemistry in semiconductor particulate systems. 16. Photophysical and photochemical aspects of coupled semiconductors. Charge-transfer processes in colloidal CdS-TiO_2 and CdS-AgI systems. *J. Phys. Chem.*, **94**, 6435–6440.
10. Asahi, R., Morikawa, T., Ohwaki, T., Aoki, K., and Taga, Y. (2001) Visible-light photocatalysis in nitrogen-doped titanium oxides. *Science*, **293**, 269–271.
11. Zhao, W., Ma, W.H., Chen, C.C., Zhao, J.C., and Shuai, Z.G. (2004) Efficient degradation of toxic organic pollutants with $Ni_2O_3/TiO_{2-x}B_x$ under visible irradiation. *J. Am. Chem. Soc.*, **126**, 4782–4783.
12. Linic, S., Christopher, P., and Ingram, D.B. (2011) Plasmonic-metal nanostructures for efficient conversion of solar to chemical energy. *Nat. Mater.*, **10**, 911–921.
13. Spikes, J.D. (1989) in *The Science of Photobiology* (ed C.S. Kendric), Plenum Press, New York and London, pp. 79–110.
14. Oleinick, N.L. (2010) Basic Photosensitization, *http://www.photobiology.info/Oleinick.html* (accessed 14 January 2012).
15. Watanabe, T., Takizawa, T., and Honda, K. (1977) Photocatalysis through excitation of adsorbates. 1. Highly efficient N-deethylation of rhodamine B adsorbed to cadmium sulfide. *J. Phys. Chem.*, **81**, 1845–1851.
16. Wu, T.X., Liu, G.M., Zhao, J.C., Hidaka, H., and Serpone, N. (1999) Evidence for H_2O_2 generation during the TiO_2-assisted photodegradation of dyes in aqueous dispersions under visible light illumination. *J. Phys. Chem. B*, **103**, 4862–4867.
17. Yang, J., Chen, C.C., Ji, H.W., Ma, W.H., and Zhao, J.C. (2005) Mechanism of TiO_2-assisted photocatalytic degradation of dyes under visible irradiation: photoelectrocatalytic study by TiO_2-film electrodes. *J. Phys. Chem. B*, **109**, 21900–21907.
18. Hagfeldt, A., Boschloo, G., Sun, L., Kloo, L., and Pettersson, H. (2010) Dye-sensitized solar cells. *Chem. Rev.*, **110**, 6595–6663.
19. Bae, E. and Choi, W. (2006) Effect of the anchoring group (carboxylate vs phosphonate) in Ru-complex-sensitized TiO_2 on hydrogen production under visible light. *J. Phys. Chem. B*, **110**, 14792–14799.
20. Wang, Z., Mao, W., Chen, H., Zhang, F., Fan, X., and Qian, G. (2006) Copper(II) phthalocyanine tetrasulfonate sensitized nanocrystalline titania photocatalyst: synthesis in situ and photocatalysis under visible light. *Catal. Commun.*, 7518–7522.
21. Zhang, M., Shao, C., Guo, Z., Zhang, Z., Mu, J., Cao, T., and Liu, Y. (2011) Hierarchical nanostructures of copper(II) phthalocyanine on electrospun TiO_2 nanofibers: controllable solvothermal-fabrication and enhanced visible photocatalytic properties. *ACS Appl. Mater. Interfaces*, **3**, 369–377.
22. Li, L. and Xin, B.F. (2010) Photogenerated carrier transfer mechanism and photocatalysis properties of TiO_2 sensitized by Zn(II) phthalocyanine. *J. Cent. South Univ. Technol.*, **17**, 218–222.

23. Sun, Q. and Xu, Y. (2009) Sensitization of TiO_2 with aluminum phthalocyanine: factors influencing the efficiency for chlorophenol degradation in water under visible light. *J. Phys. Chem. C*, **113**, 12387–12394.
24. Wang, C., Li, J., Mele, G., Duan, M., Lu, X., Palmisano, L., Vasapollo, G., and Zhang, F. (2010) The photocatalytic activity of novel, substituted porphyrin/TiO_2-based composites. *Dyes Pigm.*, **84**, 183–189.
25. Zhao, D., Chen, C., Wang, Y., Ma, W., Zhao, J., Rajh, T., and Zang, L. (2008) Enhanced photocatalytic degradation of Dye pollutants under visible irradiation on Al(III)-modified TiO_2: structure, interaction, and interfacial electron transfer. *Environ. Sci. Technol.*, **42**, 308–314.
26. Wang, Q., Chen, C., Zhao, D., Ma, W., and Zhao, J. (2008) Change of adsorption modes of dyes on fluorinated TiO_2 and its effect on photocatalytic degradation of dyes under visible irradiation. *Langmuir*, **24**, 7338–7345.
27. Pan, L., Zou, J., Zhang, X., and Wang, L. (2011) Water-mediated promotion of dye sensitization of TiO_2 under visible light. *J. Am. Chem. Soc.*, **133**, 10000–10002.
28. Batzill, M. (2011) Fundamental aspects of surface engineering of transition metal oxide photocatalysts. *Energy Environ. Sci.*, **4**, 3275–3286.
29. Prezhdo, O.V., Duncan, W.R., and Prezhdo, V.V. (2008) Dynamics of the photoexcited electron at the chromophore–semiconductor interface. *Acc. Chem. Res.*, **41**, 339–348.
30. Zhang, F.L., Zhao, J.C., Shen, T., Hidaka, H., Pelizzetti, E., and Serpone, N. (1998) TiO_2-Assisted photodegradation of dye pollutants - II. Adsorption and degradation kinetics of eosin in TiO_2 dispersions under visible light irradiation. *Appl. Catal., B*, **15**, 147–156.
31. Liu, G.M., Wu, T.X., Zhao, J.C., Hidaka, H., and Serpone, N. (1999) Photoassisted degradation of dye pollutants. 8. Irreversible degradation of alizarin red under visible light radiation in air-equilibrated aqueous TiO_2 dispersions. *Environ. Sci. Technol.*, **33**, 2081–2087.
32. Zhao, J.C., Wu, T.X., Wu, K.Q., Oikawa, K., Hidaka, H., and Serpone, N. (1998) Photoassisted degradation of dye pollutants. 3. Degradation of the cationic dye rhodamine B in aqueous anionic surfactant/TiO_2 dispersions under visible light irradiation: evidence for the need of substrate adsorption on TiO_2 particles. *Environ. Sci. Technol.*, **32**, 2394–2400.
33. Liu, G.M., Li, X.Z., Zhao, J.C., Hidaka, H., and Serpone, N. (2000) Photooxidation pathway of sulforhodamine-B. Dependence on the adsorption mode on TiO_2 exposed to visible light radiation. *Environ. Sci. Technol.*, **34**, 3982–3990.
34. He, J.J., Zhao, J.C., Shen, T., Hidaka, H., and Serpone, N. (1997) Photosensitization of colloidal titania particles by electron injection from an excited organic dye-antennae function. *J. Phys. Chem. B*, **101**, 9027–9034.
35. He, J.J., Zhao, J.C., Hidaka, H., and Serpone, N. (1998) EPR characteristics of a dye/colloidal TiO_2 system under visible light irradiation. *J. Chem. Soc., Faraday Trans.*, **94**, 2375–2378.
36. He, J.J., Chen, F., Zhao, J.C., and Hidaka, H. (1998) Adsorption model of ethyl ester of fluorescein on colloidal TiO_2 and the mechanism of the interfacial electron transfer. *Colloids Surf., A*, **142**, 49–57.
37. Wang, Z.H., Ma, W.H., Chen, C.C., Ji, H.W., and Zhao, J.C. (2011) Probing paramagnetic species in titania-based heterogeneous photocatalysis by electron spin resonance (ESR) spectroscopy–a mini review. *Chem. Eng. J.*, **170**, 353–362.
38. Qu, P., Zhao, J.C., Shen, T., and Hidaka, H. (1998) TiO_2-Assisted photodegradation of dyes: a study of two competitive primary processes in the degradation of RB in an aqueous TiO_2 colloidal solution. *J. Mol. Catal. A: Chem.*, **129**, 257–268.
39. Zhang, M., Chen, C.C., Ma, W., and Zhao, J. (2008) Visible-light-induced aerobic oxidation of alcohols in a

coupled photocatalytic system Of dye-sensitized TiO_2 and TEMPO. *Angew. Chem. Int. Ed.*, **47**, 9730–9733.

40. Shoute, L.C.T. and Loppnow, G.R. (2002) Excited-state dynamics of alizarin-sensitized TiO_2 nanoparticles from resonance Raman spectroscopy. *J. Chem. Phys.*, **117**, 842–850.
41. Ramakrishna, G., Singh, A.K., Palit, D.K., and Ghosh, H.N. (2004) Slow back electron transfer in surface-modified TiO_2 nanoparticles sensitized by alizarin. *J. Phys. Chem. B*, **108**, 1701–1707.
42. Huber, R., Moser, J.E., Grätzel, M., and Wachtveitl, J. (2002) Real-time observation of photoinduced adiabatic electron transfer in strongly coupled dye/semiconductor colloidal systems with a 6 fs time constant. *J. Phys. Chem. B*, **106**, 6494–6499.
43. Kaniyankandy, S., Verma, S., Mondal, J.A., Palit, D.K., and Ghosh, H.N. (2009) Evidence of multiple electron injection and slow back electron transfer in alizarin-sensitized ultrasmall TiO_2 particles. *J. Phys. Chem. C*, **113**, 3593–3599.
44. Qu, P., Zhao, J., Zang, L., Shen, T., and Hidaka, H. (1998) Enhancement of the photoinduced electron transfer from cationic dyes to colloidal TiO_2 particles by addition of an anionic surfactant in acidic media. *Colloids Surf., A*, **138**, 39–50.
45. Nilsing, M., Persson, P., Lunell, S., and Ojamäe, L. (2007) Dye-sensitization of the TiO_2 rutile (110) surface by perylene dyes: quantum-chemical periodic B3LYP computations. *J. Phys. Chem. C*, **111**, 12116–12123.
46. Ramakrishna, S., Willig, F., and May, V. (2001) Theory of ultrafast photoinduced heterogeneous electron transfer: decay of vibrational coherence into a finite electronic-vibrational quasicontinuum. *J. Chem. Phys.*, **115**, 2743–2755.
47. Li, G., Richter, C.P., Milot, R.L., Cai, L., Schmuttenmae, C.A., Crabtree, R.H., Brudvig, G.W., and Batista, V.S. (2009) Synergistic effect between anatase and rutile TiO_2 nanoparticles in dye-sensitized solar cells. *Dalton Trans.*, 10078–10085.
48. Li, W., Li, D., Meng, S., Chen, W., Fu, X., and Shao, Y. (2011) Novel approach to enhance photosensitized degradation of rhodamine B under visible light irradiation by the $Zn_xCd_{1-x}S/TiO_2$ nanocomposites. *Environ. Sci. Technol.*, **45**, 2987–2993.
49. Kim, W., Tachikawa, T., Majima, T., Li, C., Kim, H.-J., and Choi, W. (2010) Tin-porphyrin sensitized TiO_2 for the production of H_2 under visible light. *Energy Environ. Sci.*, **3**, 1789–1795.
50. Rauf, M.A. and Ashraf, S.S. (2009) Fundamental principles and application of heterogeneous photocatalytic degradation of dyes in solution. *Chem. Eng. J.*, **151**, 10–18.
51. Khataee, A.R. and Kasiri, M.B. (2010) Photocatalytic degradation of organic dyes in the presence of nanostructured titanium dioxide: influence of the chemical structure of dyes. *J. Mol. Catal. A: Chem.*, **328**, 8–26.
52. Teh, C.M. and Mohamed, A.R. (2011) Roles of titanium dioxide and ion-doped titanium dioxide on photocatalytic degradation of organic pollutants (Phenolic compounds and dyes) in aqueous solutions: a review. *J. Alloys Compd.*, **509**, 1648–1660.
53. Wu, T.X., Liu, G.M., Zhao, J.C., Hidaka, H., and Serpone, N. (1998) Photoassisted degradation of dye pollutants. V. Self-photosensitized oxidative transformation of rhodamine B under visible light irradiation in aqueous TiO_2 dispersions. *J. Phys. Chem. B*, **102**, 5845–5851.
54. Liu, G.M., Li, X.Z., Zhao, J.C., Horikoshi, S., and Hidaka, H. (2000) Photooxidation mechanism of dye alizarin red in TiO_2 dispersions under visible illumination: an experimental and theoretical examination. *J. Mol. Catal. A: Chem.*, **153**, 221–229.
55. Wu, T.X., Lin, T., Zhao, J.C., Hidaka, H., and Serpone, N. (1999) TiO_2-Assisted photodegradation of dyes. 9. Photooxidation of a squarylium cyanine dye in aqueous dispersions under visible light irradiation. *Environ. Sci. Technol.*, **33**, 1379–1387.

56. Liu, G.M. and Zhao, J.C. (2000) Photocatalytic degradation of dye sulforhodamine B: a comparative study of photocatalysis with photosensitization. *New J. Chem.*, **24**, 411–417.
57. Chen, C.C., Zhao, W., Li, J.Y., and Zhao, J.C. (2002) Formation and identification of intermediates visible-light-assisted photodegradation sulforhodamine-B dye in aqueous TiO_2 dispersion. *Environ. Sci. Technol.*, **36**, 3604–3611.
58. Zhang, T.Y., Oyama, T., Horikoshi, S., Hidaka, H., Zhao, J.C., and Serpone, N. (2002) Photocatalyzed N-demethylation and degradation of methylene blue in titania dispersions exposed to concentrated sunlight. *Sol. Energy Mater. Sol. Cells*, **73**, 287–303.
59. Zhao, W., Chen, C.C., Ma, W.H., Zhao, J.C., Wang, D.X., Hidaka, H., and Serpone, N. (2003) Efficient photoinduced conversion of an azo dye on hexachloroplatinate(IV)-modified TiO_2 surfaces under visible light irradiation - a photosensitization pathway. *Chem. Eur. J.*, **9**, 3292–3299.
60. Chen, C.C., Li, X.Z., Ma, W.H., Zhao, J.C., Hidaka, H., and Serpone, N. (2002) Effect of transition metal ions on the TiO_2-assisted photodegradation of dyes under visible irradiation: a probe for the interfacial electron transfer process and reaction mechanism. *J. Phys. Chem. B*, **106**, 318–324.
61. Kim, W., Tachikawa, T., Majima, T., and Choi, W. (2009) Photocatalysis of dye-sensitized TiO_2 nanoparticles with thin overcoat of Al_2O_3: enhanced activity for H_2 production and dechlorination of CCl_4. *J. Phys. Chem. C*, **113**, 10603–10609.
62. Zhao, W., Chen, C.C., Li, X.Z., Zhao, J.C., Hidaka, H., and Serpone, N. (2002) Photodegradation of sulforhodamine-B dye in platinized titania dispersions under visible light irradiation: influence of platinum as a functional co-catalyst. *J. Phys. Chem. B*, **106**, 5022–5028.
63. Cho, Y., Choi, W., Lee, C.-H., Hyeon, T., and Lee, H.-I. (2001) Visible light-induced degradation of carbon tetrachloride on dye-sensitized TiO_2. *Environ. Sci. Technol.*, **35**, 966–970.
64. Zhao, W., Sun, Y., and Castellano, F.N. (2008) Visible-light induced water detoxification catalyzed by Pt^{II} dye sensitized titania. *J. Am. Chem. Soc.*, **130**, 12566–12567.
65. Park, J. and Choi, W. (2009) TiO_2-Nafion photoelectrode hybridized with carbon nanotubes for sensitized photochemical activity. *J. Phys. Chem. C*, **113**, 20974–20979.
66. Zang, L., Lange, C., Abraham, I., Storck, S., Maier, W.F., and Kisch, H. (1998) Amorphous microporous titania modified with platinum(IV) chloride: a new type of hybrid photocatalyst for visible light detoxification. *J. Phys. Chem. B*, **102**, 10765–10771.
67. Su, W., Chen, J., Wu, L., Wang, X., Wang, X., and Fu, X. (2008) Visible light photocatalysis on praseodymium(III)-nitrate-modified TiO_2 prepared by an ultrasound method. *Appl. Catal., B*, **77**, 264–271.
68. Mitoraj, D. and Kisch, H. (2008) The nature of nitrogen-modified titanium dioxide photocatalysts active in visible light. *Angew. Chem. Int. Ed.*, **47**, 9975–9978.
69. Zhou, X., Peng, F., Wang, H., Yu, H., and Fan, Y. (2011) Carbon nitride polymer sensitized TiO_2 nanotube arrays with enhanced visible light photoelectrochemical and photocatalytic performance. *Chem. Commun.*, **47**, 10323–10325.
70. Song, L., Qiu, R., Mo, Y., Zhang, D., Wei, H., and Xiong, Y. (2007) Photodegradation of Phenol in a polymer-modified TiO_2 semiconductor particulate system under the irradiation of visible light. *Catal. Commun.*, **8**, 429–433.
71. Wang, D., Wang, Y., Li, X., Luo, Q., An, J., and Yue, J. (2008) Sunlight photocatalytic activity of polypyrrole–TiO_2 nanocomposites prepared by 'in situ' method. *Catal. Commun.*, **9**, 1162–1166.
72. Xu, S.H., Li, S.Y., Wei, Y.X., Zhang, L., and Xu, F. (2010) Improving the photocatalytic performance of conducting polymer polythiophene sensitized

TiO_2 nanoparticles under sunlight irradiation. *React. Kinet. Mech. Catal.*, **101**, 237–249.

73. Liang, H. and Li, X. (2009) Visible-induced photocatalytic reactivity of polymer–sensitized titania nanotube films. *Appl. Catal., B*, **86**, 8–17.
74. Zhang, H., Zong, R., Zhao, J., and Zhu, Y. (2008) Dramatic visible photocatalytic degradation performances due to synergetic effect of TiO_2 with PANI. *Environ. Sci. Technol.*, **42**, 3803–3807.
75. Wang, F., Min, S., Han, Y., and Feng, L. (2010) Visible-light-induced photocatalytic degradation of methylene blue with polyaniline-sensitized TiO_2 composite photocatalysts. *Superlattices Microstruct.*, **48**, 170–180.
76. Wang, Y.Z., Zhong, M.Q., Chen, F., and Yang, J.T. (2009) Visible light photocatalytic activity of TiO_2/D-PVA for MO degradation. *Appl. Catal., B*, **90**, 249–254.
77. Liao, G., Chen, S., Quan, X., Chen, H., and Zhang, Y. (2010) Photonic crystal coupled TiO_2/polymer hybrid for efficient photocatalysis under visible light irradiation. *Environ. Sci. Technol.*, **44**, 3481–3485.
78. Song, L., Zhang, X., and Zeng, X. (2011) Application of poly(fluorene-co-bithiophene) as a novel sensitizer for TiO_2 in the photodegradation of phenol under irradiation of GaN LED clusters. *React. Kinet. Mech. Catal.*, **102**, 295–302.
79. Zhang, D., Qiu, R., Song, L., Eric, B., Mo, Y., and Huang, X. (2009) Role of oxygen active species in the photocatalytic degradation of phenol using polymer sensitized TiO_2 under visible light irradiation. *J. Hazard. Mater.*, **163**, 843–847.
80. Wang, Y.H., Hang, K., Anderson, N.A., and Lian, T.Q. (2003) Comparison of electron transfer dynamics in intramolecular and molecule-to-nanoparticle charge transfer complexes. *J. Phys. Chem. B*, **107**, 9434–9440.
81. Agrios, A.G., Gray, K.A., and Weitz, E. (2004) Narrow-band irradiation of a homologous series of chlorophenols on TiO_2: charge-transfer complex formation and reactivity. *Langmuir*, **20**, 5911–5917.
82. Paul, T., Miller, P., and Strathmann, T.J. (2007) Visible-light-mediated TiO_2 photocatalysis of fluoroquinolone antibacterial agents. *Environ. Sci. Technol.*, **41**, 4720–4727.
83. Li, M., Tang, P., Hong, Z., and Wang, M. (2008) High efficient surface-complex-assisted photodegradation of Phenolic compounds in single anatase titania under visible-light. *Colloid Surf., A*, **318**, 285–290.
84. Duncan, W.R. and Prezhdo, O.V. (2007) Theoretical studies of photoinduced electron Transfer in dye-sensitized TiO_2. *Annu. Rev. Phys. Chem.*, **58**, 143–184.
85. Wang, N., Zhu, L., Huang, Y., She, Y., Yu, Y., and Tang, H. (2009) Drastically enhanced visible-light photocatalytic degradation of colorless aromatic pollutants over TiO_2 via a charge-transfer-complex path: a correlation between chemical structure and degradation rate of the pollutants. *J. Catal.*, **266**, 199–206.
86. Park, Y., Singh, N.J., Kim, K.S., Tachikawa, T., Majima, T., and Choi, W. (2009) Fullerol–titania charge-transfer-mediated photocatalysis working under visible light. *Chem. Eur. J.*, **15**, 10843–10850.
87. Li, X.Z., Chen, C.C., and Zhao, J.C. (2001) Mechanism of photodecomposition of H_2O_2 on TiO_2 surfaces under visible light irradiation. *Langmuir*, **17**, 4118–4122.
88. Yu, H., Irie, H., Shimodaira, Y., Hosogi, Y., Kuroda, Y., Miyauchi, M., and Hashimoto, K. (2010) An efficient visible-light-sensitive Fe(III)-grafted TiO_2 photocatalyst. *J. Phys. Chem. C*, **114**, 16481–16487.
89. Serpone, N., Borgarello, E., and Grätzel, M. (1984) Visible light induced generation of hydrogen from H_2S in mixed semiconductor dispersions, improved efficiency through inter-particle electron transfer. *J. Chem. Soc., Chem. Commun.*, 342–344.

90. Robert, D. (2007) Photosensitization of TiO_2 by M_xO_y and M_xS_y nanoparticles for heterogeneous photocatalysis applications. *Catal. Today*, **122**, 20–26.
91. Liu, G., Wang, L., Yang, H.G., Cheng, H., and Lu, G.Q. (2010) Titania-based photocatalysts—crystal growth, doping and heterostructuring. *J. Mater. Chem.*, **20**, 831–843.
92. Li, G.S., Zhang, D.Q., and Yu, J.C. (2009) A new visible-light photocatalyst: CdS quantum dots embedded mesoporous TiO_2. *Environ. Sci. Technol.*, **43**, 7079–7085.
93. Lin, C.J., Yu, Y.H., and Liou, Y.H. (2009) Free-standing TiO_2 nanotube array films sensitized with CdS as highly active solar light-driven photocatalysts. *Appl. Catal., B*, **93**, 119–125.
94. Yang, L., Luo, S., Liu, R., Cai, Q., Xiao, Y., Liu, S., Su, F., and Wen, L. (2010) Fabrication of CdSe nanoparticles sensitized long TiO_2 nanotube arrays for photo-catalytic degradation of anthracene-9-carbonxylic acid under green monochromatic light. *J. Phys. Chem. C*, **114**, 4783–4789.
95. Xiao, G., Wang, X., Li, D., and Fu, X. (2008) $InVO_4$ -sensitized TiO_2 photocatalysts for efficient air purification with visible light. *J. Photochem. Photobiol., A*, **193**, 213–221.
96. Senevirathna, M.K.I., Pitigala, P.K.D.D.P., and Tennakone, K. (2005) Water photoreduction with Cu_2O quantum Dotson TiO_2 nano-particles. *J. Photochem. Photobiol., A*, **171**, 257–259.
97. Shu, X., An, Z., Wang, L., and He, J. (2009) Metal oxide-sensitized TiO_2 and $TiO_{2-x}N_x$ with efficient charge transport conduits. *Chem. Commun.*, 5901–5903.
98. Liu, H. and Gao, L. (2006) Preparation and properties of nanocrystalline α-Fe_2O_3-sensitized TiO_2 nanosheets as a visible light photocatalyst. *J. Am. Ceram. Soc.*, **89**, 370–373.
99. Ho, W., Yu, J.C., Lin, J., Yu, J., and Li, P. (2004) Preparation and photocatalytic behavior of MoS_2 and WS_2 nanocluster sensitized TiO_2. *Langmuir*, **20**, 5865–5869.
100. Lim, C.-S., Chen, M.-L., and Oh, W.-C. (2011) Synthesis of CdSe-TiO_2 photocatalyst and their enhanced photocatalytic activities under UV and visible light. *Bull. Korean Chem. Soc.*, **32**, 1657–1661.
101. Vogel, R., Hoyer, P., and Weller, H. (1994) Quantum-sized PbS, CdS, Ag_2S, Sb_2S_3, and Bi_2S_3 particles as sensitizers for various nanoporous wide-bandgap semiconductors. *J. Phys. Chem.*, **98**, 3183–3188.
102. Sun, W.-T., Yu, Y., Pan, H.-Y., Gao, X.-F., Chen, Q., and Peng, L.-M. (2008) CdS quantum dots sensitized TiO_2 nanotube-array photoelectrodes. *J. Am. Chem. Soc.*, **130**, 1124–1125.
103. Sambur, J.B. and Parkinson, B.A. (2010) CdSe/ZnS core/shell quantum dot sensitization of low index TiO_2 single crystal surfaces. *J. Am. Chem. Soc.*, **132**, 2130–2131.
104. Mali, S.S., Desai, S.K., Dalavi, D.S., Betty, C.A., Bhosale, P.N., and Patil, P.S. (2011) CdS-sensitized TiO_2 nanocorals: hydrothermal synthesis, characterization, application. *Photochem. Photobiol. Sci.*, **10**, 1652–1658.
105. Wu, L., Yu, J.C., and Fu, X.Z. (2006) Characterization and photocatalytic mechanism of nanosized CdS coupled TiO_2 nanocrystals under visible light irradiation. *J. Mol. Catal. A: Chem.*, **244**, 25–32.
106. Evans, J.E., Springer, K.W., and Zhang, J.Z. (1994) Femtosecond studies of interparticle. Electron transfer in a coupled CdS-TiO_2 colloidal system. *J. Chem. Phys.*, **101**, 6222–6225.
107. Wang, J.Y., Liu, Z.H., Zheng, Q., He, Z.K., and Cai, R.X. (2006) Preparation of photosensitized nanocrystalline TiO_2 hydrosol by nanosized CdS at low temperature. *Nanotechnology*, **17**, 4561–4566.
108. Fuji, H., Ohtaki, M., Eguchi, K., and Arai, H. (1998) Preparation and photocatalytic activities of a semiconductor composite of CdS embedded in a TiO_2 gel as a stable oxide semiconducting matrix. *J. Mol. Catal. A: Chem.*, **129**, 61–68.
109. Gao, B., Kim, Y.J., Chakraborty, A.K., and Lee, W.I. (2008) Efficient decomposition of organic compounds with $FeTiO_3/TiO_2$ heterojunction under

visible light irradiation. *Appl. Catal., B*, **83**, 202–207.

110. Xu, J., Wang, W., Sun, S., and Wang, L. (2012) Enhancing visible-light-induced photocatalytic activity By coupling with wide-band-gap semiconductor: a case study on Bi_2WO_6/TiO_2. *Appl. Catal., B*, **111–112**, 126–132.
111. Kowalska, E., Abe, R., and Ohtani, B. (2009) Visible light-induced photocatalytic reaction of gold-modified titanium(IV) oxide particles: action spectrum analysis. *Chem. Commun.*, **2**, 241–243.
112. Yu, J., Dai, G., and Huang, B. (2009) Fabrication and characterization of visible-light-driven plasmonic photocatalyst $Ag/AgCl/TiO_2$ nanotube-arrays. *J. Phys. Chem. C*, **113**, 16394–16401.
113. Jin, R., Cao, Y., Mirkin, C., Kelly, K., Schatz, G., and Zheng, J. (2001) Photoinduced conversion of silver nanospheres to nanoprisms. *Science*, **294**, 1901–1903.
114. Pichat, P. (1987) Surface properties, activity and selectivity of bifunctional powder photocatalysts. *New J. Chem.*, **11**, 135–140.
115. Sclafani, A., Mozzanega, M.-N., and Pichat, P. (1991) Effects of silver deposits on the photocatalytic activity of titanium dioxide samples for the dehydrogenation or oxidation of 2-propanol. *J. Photochem. Photobiol., A*, **59**, 181–189.
116. Naya, S., Inoue, A., and Tada, H. (2011) Visible-light activity enhancement of gold-nanoparticle-loaded titanium(IV) dioxide by preferential excitation of localized surface Plasmon resonance. *ChemPhysChem*, **12**, 2719–2723.
117. Kowalska, E., Mahaney, O.O.P., Abe, R., and Ohtani, B. (2010) Visible-light-induced photocatalysis through surface Plasmon excitation of gold on titania surfaces. *Phys. Chem. Chem. Phys.*, **12**, 2344–2355.
118. Chen, X., Zheng, Z., Ke, X., Jaatinen, E., Xie, T., Wang, D., Guo, C., Zhao, J., and Zhu, H. (2010) Supported silver nanoparticles as photocatalysts under ultraviolet and visible light irradiation. *Green Chem.*, **12**, 414–419.
119. Zhu, H., Chen, X., Zheng, Z., Ke, X., Jaatinen, E., Zhao, J., Guo, C., Xie, T., and Wang, D. (2009) Mechanism of supported gold nanoparticles as photocatalysts under ultraviolet and visible light irradiation. *Chem. Commun*, 7524–7526.
120. Furube, A., Du, L., Hara, K., Katoh, R., and Tachiya, M. (2007) Ultrafast Plasmon-induced electron transfer from gold nanodots into TiO_2 nanoparticles. *J. Am. Chem. Soc.*, **129**, 14852–14853.
121. Sakai, N., Fujiwara, Y., Takahashi, Y., and Tatsuma, T. (2009) Plasmon resonance-based generation of Cathodic photocurrent At electrodeposited gold nanoparticles coated with TiO_2 films. *ChemPhysChem*, **10**, 766–769.
122. Ide, Y., Matsuoka, M., and Ogawa, M. (2010) Efficient visible-light-induced photocatalytic activity on gold-nanoparticle-supported layered titanate. *J. Am. Chem. Soc.*, **132**, 16762–16764.
123. Szaciłowski, K., Macyk, W., Drzewiecka-Matuszek, A., Brindell, M., and Stochel, G. (2005) Bioinorganic photochemistry: frontiers and mechanisms. *Chem. Rev.*, **105**, 2647–2694.
124. Choi, W., Termin, A., and Hoffmann, M.R. (1994) The role of metal-ion dopants in quantum-sized TiO_2: correlation between photoreactivity and charge-carrier recombination dynamics. *J. Phys. Chem.*, **98**, 13669–13679.
125. Fuerte, A., Hernández-Alonso, M.D., Maira, A.J., Martínez-Arias, A., Fernández-García, M., Conesa, J.C., Soria, J., and Munuera, G. (2002) Nanosize Ti–W mixed oxides: effect of doping level in the photocatalytic degradation of toluene using sunlight-type excitation. *J. Catal.*, **212**, 1–9.
126. Anpo, M. and Takeuchi, M. (2003) Design and development of highly reactive titanium oxide photocatalysts operating under visible light irradiation. *J. Catal.*, **216**, 505–516.
127. Highfield, J.G. and Pichat, P. (1989) Photoacoustic study of the influence of platinum loading and bulk doping with chromium III ions on the reversible

photochromic effect in titanium dioxide. Correlation with photocatalytic properties. *New J. Chem.*, **13**, 61–66.

128. Li, G., Wang, X., Hu, X., Leung, C.W., Zhang, Z., and Yu, J.C. (2006) An ordered cubic *Im3m* mesoporous Cr–TiO_2 visible light photocatalyst. *Chem. Commun.*, 2717–2719.

129. Serpone, N., Lawless, D., Disdier, J., and Herrmann, J.-M. (1994) Spectroscopic, photoconductivity and photocatalytic studies of TiO_2 colloids: alone and with lattice-doped Cr(III), Fe(III), and V(V) cations. *Langmuir*, **10**, 643–652.

130. Gracia, F., Holgado, J.P., Caballero, A., and Gonzalez-Elipe, A.R. (2004) Structural, optical, and photoelectrochemical properties of M^{n+}-TiO_2 model thin film photocatalysts. *J. Phys. Chem. B*, **108**, 17466–17476.

131. Choi, J., Park, H., and Hoffmann, M.R. (2010) Effects of single metal-ion doping on the visible light photoreactivity of TiO_2. *J. Phys. Chem. C*, **114**, 783–792.

132. Sato, S. (1986) Photocatalytic activity of NO_x-doped TiO_2 in the visible light region. *Chem. Phys. Lett.*, **123**, 126–128.

133. Zhang, H., Chen, G., and Bahnemann, D.W. (2009) Photoelectrocatalytic materials for environmental applications. *J. Mater. Chem.*, **19**, 5089–5121.

134. Rehman, S., Ullah, R., Butt, A.M., and Gohar, N.D. (2009) Strategies of making TiO_2 and ZnO visible light active. *J. Hazard. Mater.*, **170**, 560–569.

135. Kumar, S.G. and Devi, L.G. (2011) Review on modified TiO_2 photocatalysis under UV/visible light: selected results and related mechanisms on interfacial charge carrier transfer dynamics. *J. Phys. Chem. A*, **115**, 13211–13241.

136. Dozzi, M.V., Ohtani, B., and Selli, E. (2011) Absorption and action spectra analysis of ammonium fluoride-doped titania photocatalysts. *Phys. Chem. Chem. Phys.*, **13**, 18217–18227.

137. Wang, Q., Chen, C.C., Ma, W.H., Zhu, H.Y., and Zhao, J.C. (2009) Pivotal role of fluorine in tuning band structure and visible-light photocatalytic activity of nitrogen-doped TiO_2. *Chem. Eur. J.*, **15**, 4765–4769.

138. Chen, C.C., Ma, W.H., and Zhao, J.C. (2010) Photocatalytic degradation Of organic pollutants by co-doped TiO_2 under visible light irradiation. *Curr. Org. Chem.*, **14**, 630–644.

139. Zuo, F., Wang, L., Wu, T., Zhang, Z., Borchardt, D., and Feng, P. (2010) Self-doped Ti^{3+} enhanced photocatalyst for hydrogen production under visible light. *J. Am. Chem. Soc.*, **132**, 11856–11857.

140. Ariga, H., Taniike, T., Morikawa, H., Tada, M., Min, B.K., Watanabe, K., Matsumoto, Y., Ikeda, S., Saiki, K., and Iwasawa, Y. (2009) Surface-mediated visible-light photo-oxidation on pure TiO_2(001). *J. Am. Chem. Soc.*, **131**, 14670–14672.

141. Li, R., Kobayashi, H., Guo, J., and Fan, J. (2011) Visible-light-driven surface reconstruction of mesoporous TiO_2: toward visible-light absorption and enhanced photocatalytic activities. *Chem. Commun.*, **47**, 8584–8586.

142. Cronemeyer, D.C. (1952) Electrical and optical properties of rutile single crystals. *Phys. Rev.*, **87**, 876–886.

143. Cronemeyer, D.C. (1959) Infrared absorption of reduced rutile TiO_2 single crystals. *Phys. Rev.*, **113**, 1222–1226.

144. Ghosh, A.K., Wakim, F.G., and Addiss, R.R. (1969) Photoelectronic processes in rutile. *Phys. Rev.*, **184**, 979–988.

145. Kuznetsov, V.N. and Serpone, N. (2009) On the origin of the spectral bands in the visible absorption spectra of visible-light-active TiO_2 specimens analysis and assignments. *J. Phys. Chem. C*, **113**, 15110–15123.

146. Justicia, I., Garcia, G., Battiston, G.A., Gerbasi, R., Ager, F., Guerra, M., Caixach, J., Pardo, J.A., Rivera, J., and Figueras, A. (2005) Photocatalysis in the visible range of sub-stoichiometric anatase films prepared by MOCVD. *Electrochim. Acta*, **50**, 4605–4608.

147. Justicia, I., Garcia, G., Vazquez, L., Santiso, J., Ordejon, P., Battiston, G., Gerbasi, R., and Figueras, A. (2005) Self-doped titanium oxide thin films for efficient visible light photocatalysis: an example: nonylphenol photodegradation. *Sens. Actuators, B*, **109**, 52–56.

148. Justicia, I., Ordejon, P., Canto, G., Mozos, J.L., Fraxedas, J., Battiston, G., Gerbasi, R., and Figueras, A. (2002) Designed self-doped titanium oxide thin films for efficient visible-light photocatalysis. *Adv. Mater.*, **14**, 1399–1402.

149. Tao, J., Luttrell, T., and Batzill, M. (2011) A two-dimensional phase of TiO_2 with a reduced bandgap. *Nat. Chem.*, **3**, 296–300.

150. Serpone, N., Emeline, A.V., Kuznetsov, V.N., and Ryabchuk, V.K. (2010) in *Environmentally Benign Photocatalysts: Applications of Titanium Oxide-Based Materials* (eds M. Anpo and P.V. Kamat), Springer, New York, Dordrecht and Heidelberg, London, pp. 35–112.

151. Emeline, A.V., Kuznetsov, V.N., Ryabchuk, V.K., and Serpone, N. (2008) Visible-light-active titania photocatalysts. The case of N-doped TiO_2s–properties and some fundamental issues. *Int. J. Photoenergy*. doi: 10.1155/ 2008/258394

152. Martyanov, I.N., Uma, S., Rodrigues, S., and Klabunde, K.J. (2004) Structural defects cause TiO_2-based photocatalysts to be active in visible light. *Chem. Commun.*, 2476–2477.

153. Komaguchi, K., Maruoka, T., Nakano, H., Imae, I., Oyama, Y., and Harima, Y. (2010) Electron-transfer reaction of oxygen species on TiO_2 nanoparticles induced by sub-band-gap illumination. *J. Phys. Chem. C*, **114**, 1240–1245.

154. Chen, X.B., Liu, L., Yu, P.Y., and Mao, S.S. (2011) Increasing solar absorption for photocatalysis with black hydrogenated titanium dioxide nanocrystals. *Science*, **331**, 746–750.

155. Yan, X., Ohno, T., Nishijima, K., Abe, R., and Ohtani, B. (2006) Is methylene blue an appropriate substrate for a photocatalytic activity test? A study with visible-light responsive titania. *Chem. Phys. Lett.*, **429**, 606–610.

9
Photoelectrocatalysis for Water Purification

Rossano Amadelli and Luca Samiolo

9.1
Introduction

The common feature of photoelectrocatalysis (PEC) and electrocatalysis is the application of a potential to affect or induce reactions at the interface. In the latter case, the electrode material is a conductor and one can exploit a relatively large potential window where different types of oxidation and/or reduction processes can be studied. In PEC, however, the occurrence of processes in a wide potential range can be explored only if electrodes are irradiated. This is because the electrode material is generally a semiconductor, which means that one will observe rectifying phenomena as the applied potential is varied over a wide range: reduction and oxidation in the dark will take place at n-type and p-type semiconductors, respectively.

Heterogeneous photocatalysis (PC) on semiconductors, on the other hand, is defined as a change in the rate of chemical reactions or their generation under the action of light in the absence of applied potential. In this respect, PEC at open circuit is the same as PC. The key common factor is a light-induced electron–hole charge separation while differences reside in the fact that separation of charges is assisted by an applied potential. In this respect, and probably correctly, this research field is referred to in several publications as electroassisted photocatalysis.

The seminal article by Fujishima and Honda published in 1972 [1] on the photoelectrochemical splitting of water at a TiO_2 photoelectrode is considered to be the origin of the impressive development of research on PC, although a bias was used. Seven years after the publication of this article, a brief review by Bard [2] clearly indicates PEC and PC as closely connected scientific fields: PC on semiconductor particle dispersions is already described in some detail including its potentiality for pollutants oxidation. Only less than a decade later, Hodes and Grätzel [3] published a comparative study on photoelectrochemistry at electrode and small particles.

Bard classified photoelectrochemical cells into three categories: (i) light is converted into energy, with no net change in the chemical composition of solution species (photovoltaic cells); (ii) light drives the overall cell reaction in a

Photocatalysis and Water Purification: From Fundamentals to Recent Applications, First Edition. P. Pichat.

nonspontaneous direction ($\Delta G > 0$) and the radiant energy is stored as chemical energy (photoelectrosynthetic cells); and (iii) a reaction is driven in a spontaneous ($\Delta G < 0$) direction and radiant energy is used to overcome the energy of activation rather than being stored as chemical energy (PC cells). In this connection, the article by Fujishima and Honda, by showing the PEC formation of hydrogen, can be considered as well to have led the way to a lasting interest in photoelectrochemical production of fuels [4]. An account of the historical development of photoelectrochemical sciences, published few years ago by Honda [5], does mention PEC applications among several others.

A major drawback of PC processes is poor efficiency mainly due to extensive recombination of photogenerated charges. After investigating various organic compounds, Serpone *et al.* [6] have estimated that the efficiency of the photocatalytic process is only 1–2%. Application of a potential can prevent charge recombination resulting in an extension of the lifetime of active holes. The benefits of electroassisting PC processes have been demonstrated, in fact, by quite a few published works, yet one can safely state that no detailed systematic study has been reported on the photoelectrochemical behavior of the oxidation of organic substances at film electrodes. Rather, support of photoelectrochemical measurements to heterogeneous PC has been put in evidence [7], and indeed, the combination of PC and electrochemical techniques has often proven to be useful both from practical and fundamental point of views [8, 9].

Although information on the use of photoelectrodes for water depollution is scattered across many publications, it emerges, especially from the recent literature, that PEC has its place among advanced oxidation methods [10, 11], and growing interest is witnessed by the very recent development of photoelectrochemical methods for air cleaning [12].

9.2
Photoeffects at Semiconductor Interfaces

This chapter briefly discusses some essential points of semiconductor electrochemistry. The fundamentals can be found in several books and reviews [13–19] and a condensed, very good introduction is given in Ref. [19].

Photoelectrochemistry as electrochemistry discusses the effects of electron transfer between phases. These effects extend to situations found in photoelectrosynthesis (e.g., production of hydrogen) and PEC at photoelectrodes or dispersed semiconductors and are strongly dependent on the charged structure of the interface. A charge interface is formed whenever two media of different conductivity are brought into contact. Thus, when a solid (metal, semiconductor, or insulator) is immersed in an electrolytic solution, its surface acquires a neat electrical charge that is balanced by a region of opposite charge in the solution adjacent to the solid surface. This separation of charge is known as the electrical double layer [20, 21]. It has a complex structure that is, however, simplified in concentrated electrolyte solutions in which its thickness is reduced to ~1 nm (the Helmholtz layer).

The charged interface behaves as a condenser and, accordingly, is characterized by a capacity, denoted as C_H for the Helmholtz layer. The potential drop across this layer can reach 1 V [20] and, in metal electrodes, it plays a key role in assisting electron transfer to and from the species in solution.

Adsorption of ions is among the factors that are at the origin of the appearance of charge on the surface of a solid immersed in an electrolyte solution. Anions are usually less strongly hydrated than cations and can move closer to the surface where they can be "specifically" adsorbed. This happens whenever the concentration of a species at the interface is greater than that that can be accounted for by electrostatic interactions.

For oxide electrodes, the pH-dependent dissociation/protonation of OH surface groups is a major factor determining the surface charge:

$$\mathrm{MOH} + \mathrm{H}^+ \rightarrow \mathrm{MOH_2}^+$$

$$\mathrm{MOH} + \mathrm{HO}^- \rightarrow \mathrm{MO}^- + \mathrm{H_2O}$$

Because of the fast protonation of these surface groups, a nearly Nernstian dependence on electrolyte pH is observed. Protons and hydroxyl ions are referred to as potential-determining ions, and the pH at which the oxide surface has no net charge is defined as the point of zero charge (pzc).

The lower concentration of free charge carriers in semiconductors compared to metals causes important phenomena on the solid side. A metal with a typical density of 10^{22} electrons per cubic centimeter and a charge of 10^{13} U cm^{-2} on the solution side will be compensated by a countercharge (q_{Me}) within only a fraction of an angstrom below the surface. In contrast, for a semiconductor having, for example, a free carrier density of 10^{17} cm^{-3} the countercharge (q_{sc}) would be found in a region of 1000 nm thickness.

Then a diffuse layer appears inside the semiconductors (space charge layer) whose thickness, W, ranges from 10 to 1000 nm depending on the surface charge and the free carrier concentration [3, 16]. It is described by a capacitance C_{sc} and a potential drop V_{sc} between the charged surface and the neutral semiconductor bulk. In terms of an equivalent electric circuit, C_{sc} is in series with C_H and the overall measured capacity ($1/C_{meas} = 1/C_{sc} + 1/C_H$) will be essentially equal to the smaller one, that is, C_{sc}.

The important consequence is that potential will drop inside the space charge layer without altering the rate interfacial reactions, that is, changes in the applied potential change the width of the space charge layer and the potential associated with it (V_{sc}) (Figure 9.1). For an ideal semiconductor, V_{sc} vanishes as the charge on the semiconductor becomes zero. The potential of the semiconductor corresponding to this condition is the flat band potential (V_{fb}), which plays the same role as the potential of zero charge for metals.

Despite the different role of applied potential in semiconductors, the Helmholtz layer is important because it influences the PEC biasing requirements and is directly affected by the chemisorption of charged species.

Upon irradiation with light of suitable wavelength, electrons are promoted to the conduction band leaving holes in the valence band. An appropriate choice of the

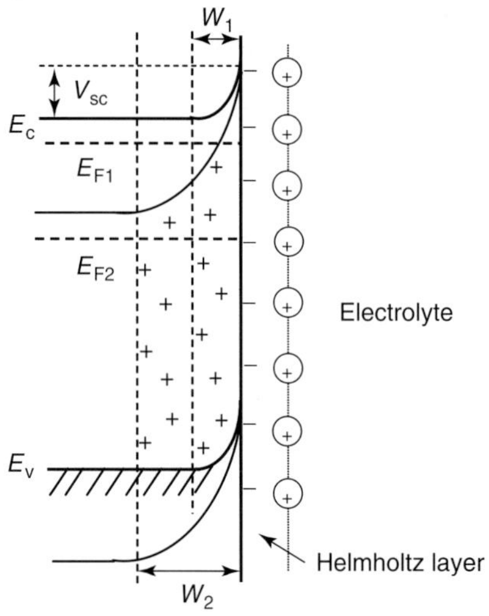

Figure 9.1 A not-to-scale image of electrical double-layer formation at an n-type semiconductor/electrolyte interface. A positive applied potential is shown to shift the position of the Fermi level from E_{F1} to E_{F2} and to increase the width of the space charge layer and the potential drop V_{sc} inside the semiconductor.

applied potential can deplete the surface of the majority carriers (electrons for n-type and holes for p-type semiconductors). These can be transferred through an external circuit and a current (photocurrent) can be measured. The photocurrent is an important parameter that directly reflects the efficiency of the photoelectrochemical system, and a more effective transport of photogenerated electrons within the electrode material will occur during a greater photocurrent response.

The photoelectrochemical behavior can often be complicated by the presence of surface states (ss). These are electronic energy levels located in the forbidden gap associated with dangling bonds that are caused by the termination of the semiconductor. Several species dissolved in solution, exhibiting a tendency to strongly adsorb onto the electrode surface, may be viewed as extrinsic ss [22].

The fundamental aspects of photoelectrochemical processes are investigated by different methods, several of which are based on analysis of measured photocurrents (I_{ph}) versus potential, time, and light irradiance [23–28]. The behavior of photocurrents as a function of irradiation wavelength (action spectra) [29] is particularly interesting because it can reveal a contribution of ss. It has been noted that often photocurrents do not show a sharp onset at the band gap threshold, but rather a tail into the sub-band-gap region that was ascribed to excitation from the occupied ss [29].

Important information is also obtained from cyclic voltammetry [30–33], open circuit photovoltage [34–38], and surface photovoltage [39]. Under strong irradiation, the open circuit photovoltage provides a means to determine the flat band potential (V_{fb}) [17] and the presence of ss [38].

Impedance measurements [40–43] provide important insights into the photoelectrochemical systems. One common use of impedance is the determination of the flat band potential from the dependence of capacitance on potential (Mott–Scottky relationship) [13–15], under conditions that lead to band bending. Analysis can often be complicated by the presence of ss [13, 17, 44, 45].

9.3 Water Depollution at Photoelectrodes

A number of studies demonstrated the potentialities of PEC oxidation processes in the degradation of organic compounds. However, despite a unanimous consensus on the high efficiency of the PEC approach, investigations are less numerous and significantly less systematic in comparison with the enormous amount of studies on depollution by PC [46–49].

The main classes of pollutants whose degradation was investigated by PEC are phenols [50–78], carboxylic acids [37, 43, 54, 79–92], surfactants [93–96] and, especially, dyes [97–130]. Several interesting papers have also demonstrated that PEC is an efficient approach to water disinfection [54, 131–139]. Various factors influence photooxidative reactions in both PC and PEC systems, including the light wavelength and irradiance, the semiconductor surface area, the photonic absorption spectrum of the photocatalyst, the efficiency of interfacial charge transfer, the nature of the electrolytes, and the reactor design. Hence, a precise control of the reaction conditions is critical in PEC as in other advanced oxidation methods.

In most publications, factors that are generally taken into consideration when analyzing the PEC processes are (i) electrode material and morphology, (ii) effect of applied potential, (iii) effect of pH, (iv) effect of oxygen, and (v) effect of electrolyte composition. In the brief comment that follows, it is assumed that the photocatalysts are n-type semiconductors unless otherwise stated.

9.3.1 Morphology and Microstructure

An inherent feature of PEC is the use of films on various conductive substrates. The use of immobilized photocatalysts instead of suspensions is an attractive possibility for PC applications too since it avoids the need for separating the catalyst at the end of the process (Chapters 5 and 6). An obvious disadvantage of films is the low active surface area, which decreases the adsorbed amounts of the pollutants to be converted.

The need to increase the films' surface area has stimulated intense research, and large effort has been put into developing nanocrystalline, porous films. In this context, research has concentrated on various forms of nanostructured films, their physical characteristics, and photoactivity.

A distinction between different types of nanostructured films is likely arbitrary. However, several high-area photoelectrodes used in water depollution can be roughly classified as nanoporous and mesoporous on the basis of particle interconnection to form spatially ordered networks. In the former, there is no regular packing of the particles and the films are "statistically porous" [140]; in the latter, the particles form an ordered structure with pores whose size can be regulated by preparation methods.

The photoelectrochemical behavior of nanoporous films has been extensively studied [25, 26, 29, 140–147] mainly for prospective application in dye-sensitized solar cells (DCS), but work has also been published in which the specific target is the photoelectrochemical oxidation of organic compounds [23–26]. Research overwhelmingly concerns TiO_2 as the photocatalyst, and it is generally concluded that structure is a fundamental factor determining the PEC activity. In particular, the performance of TiO_2 nanoporous electrodes is strongly influenced, among others, by electron mobility within the layer [142] that relies strongly on properties such as photocatalyst particle interconnectivity [145], and accordingly, the layer's thickness as well as the irradiation mode require optimization [146–148].

Nanoporous films are characterized by a considerable amount of ss and structural disorder at the contact between particles leading to reduced electron mobility. During their drift to the external circuit through the metal contact, free electrons can be trapped (and de-trapped) at ss, and the consequence is an increased likelihood of recombination phenomena that are revealed by low photocurrents. Electron transport via a surface trap-to-trap (T) hopping mechanism has been proposed by Ke *et al.* [149] on basis of electron paramagnetic (spin) resonance (EPR) data. Models of charge diffusion processes involving traps have been described [150, 151].

A complete analysis of a photoelectrochemical system involves the knowledge of both the kinetic parameters describing the bulk material and the interfacial properties [140, 152, 153]. In this context, an analysis of the factors affecting the photooxidation mechanism of simple organic compounds has been reported [23, 24].

The strongly ordered nanoscale structure of mesoporous films [142, 154], with well-interconnected particles, compares favorably with the relatively random structure of nanoporous films and offers the potential for improved electron transport, leading to higher photoefficiency. In addition, the structure of mesoporous electrodes, in particular, their thinner pore walls, allows a deeper penetration of light into the film.

A step further in the development of ordered nanostructures is the preparation of nanotubular (NT) electrodes [155–159]. Many reports indicated that TiO_2 nanotubes being formed on the Ti substrate by anodization exhibit striking properties when compared to nanoparticulate films [113, 160–164]. First, the TiO_2 nanotubes, with a regular pore structure, allow more efficient absorption of incident photons because of an increased light penetration depth and better light scattering. Second, the photogenerated electrons can be rapidly and efficiently transferred leading to a much reduced electron–hole recombination rate [4, 159].

9.3.2
Effect of Applied Potential

Conceptually, the role of potential is different for compact and nanostructured film electrodes. In the latter case, band bending is negligible or absent [25, 26, 142] since the typical particle diameter (5–50 nm) is much smaller than the thickness of the space charge layer for compact electrodes (Section 9.2). In general, for particles with radius less than thickness, W, ($\sim$1 μm), a full space charge may not develop [3].

In the cited publications on PEC depollution, it is generally concluded that an applied potential increases the pollutant's degradation rate [e.g., [63, 115]] and that best results can be obtained when the appropriate photoelectrolysis potential is chosen.

Photocurrents usually become constant (saturation) above a certain applied potential, when the interfacial kinetics are fast. The frequent observation of photocurrents leveling off at higher potentials can be attributed to the limitation of the hole capture process, and this also depends on film morphology and the nature of oxidizable species [152, 153]. In this context, it is also appropriate to recognize the role of electromigration of charged electroactive species, such as some dyes or organic acids, in the solution permeating the porous films. Application of a positive potential will attract or repel negatively and positively charged substrates, respectively [35, 142].

The oxidation of model organic species at nanoporous titanium dioxide thin films provides a good illustration of the effects of potential [165]. It was observed that as the applied positive potential is increased, the photooxidation rate of salicylic acid decreased to a low value, even for high concentrations, due to competing oxidation of water. Thus, the potential dependence of the net photocurrent due to oxidation of the organic species provides useful information on the mechanism, possibly involving parallel reactions.

9.3.3
Effect of pH

The solution pH influences the PEC process in many ways [110], including inducing changes in the flat band potential and the adsorption ability of the target compound.

Results of PEC depollution put in evidence the need to optimize pH depending on the system investigated, as for the applied potential. The change of surface charge with pH (Section 9.2) can be exploited to favor adsorption of the active species. In principle, the degradation of anionic substrates on, for example, TiO_2 is favored in the low pH range where the surface is positively charged and the adsorption of the substrates is stronger. This is indeed the case of the PEC degradation of several dyes, such as indigo carmine [130]. Likewise, degradation of some phenols decreases with increasing pH because they can dissociate and can be repelled by a negatively charged surface [56, 75].

In practice, however, there seems no good general rule of thumb. For example, the degradation rate of Reactive Orange 16 (anionic dye) is higher in alkaline solutions and in fact decreases if the solution pH decreases due to simultaneous $H_2O/OH^{\bullet -}$ oxidation [102]. Moreover, Yang *et al.* [63] reported that the efficiency of PEC degradation of 4-chlorophenol at TiO_2 electrodes is higher at pH 10, contrary to what might be expected. The authors proposed that despite repulsive forces between the phenolate anion and the surface, more OH radicals are formed at this alkaline pH. The conclusion is that the search for the most efficient degradation conditions must be found empirically from experiments at different solution pH values.

9.3.4
Effect of Oxygen

Oxygen is the typical electron acceptor in PC, and in additionally, oxygen, as well as its reduction products, can participate in the overall oxidative process of substrates [166]. In contrast, no role of oxygen is a priori expected in PEC since electrons are removed through an external circuit by an applied positive potential. In reality, results reported in the literature are controversial and this is seemingly due to an effect of the electrode microstructure.

Semiconductor particles in a nanostructured film are surrounded by the electrolyte, and photoelectrons may be captured by an electron acceptor in the solution on the way to the back contact [29]; indeed, quenching of anodic photocurrents, due to the presence of oxygen, is observed even at relatively large positive potentials [28]. It is noteworthy that oxygen can be formed as a result of water oxidation within nanoporous films even in nominally anaerobic solutions and might affect photocurrents.

From the point of view of mechanism, marginal effects of oxygen on dye degradation have been observed [114]. Likewise, the PEC degradation rate of 4-chlorophenol, at sufficiently positive potentials, was not influenced by oxygen [63]. Intermediates were the same whether or not oxygen was present, but notably, their relative concentrations changed in the two cases. This is taken by the authors as an indication that, in any case, oxygen still has some role in the overall degradation mechanism.

On a different level, an interesting approach is the use of one-compartment electrochemical cells in which oxidation of pollutants occurring at the photoanode are assisted by H_2O_2 electrogenerated at the cathode by O_2 reduction. Synergistic effects in this photoelectrochemical Fe^{3+} Fenton system bring about a conspicuous enhancement in the degradation efficiency of pollutants [53, 109, 167]. These type of experiments should be distinguished from those also called photoelectrocatalytic Fenton processes where the role of light is to prevent the formation of stable complexes between degradation products, for example, oxalate, and the Fenton reagent Fe^{3+} [168].

9.3.5 Electrolyte Composition

The salient point here is that several anions constituting the electrolyte can be converted to highly oxidizing radicals that can participate in the photodegradation process [105]. This is typically observed in chloride-containing media in which enhanced dye degradation is attributed to $Cl^{\bullet}$ radicals [100] or other species such as $Cl_2^{\bullet}$, $HClO^{\bullet}$, Cl_2, HClO, and ClO^- [122]. Similar results are observed in chlorine-mediated degradation of dyes at conductive electrodes [169].

On the other hand, the degradation of bisphenol A seems to be inhibited by chloride, which is explained by competitive adsorption and photooxidation of the anion [75], and the reaction was thus carried out in sodium sulfate at high potential. A sodium sulfate electrolyte was also found to be ideal in the PEC degradation of pentachlorophenol (PCP) [56]. It must be pointed out, however, that involvement of SO_4 radicals cannot be discounted. According to Solarska *et al.* [170], these radicals appear as plausible intermediates in the oxidation of dilute solutions of organic species containing sulfate ions and have also been detected at PbO_2 anodes in sulfuric acid (1 mol l^{-1}), even in the presence of oxygen evolution [171].

9.4 Photoelectrode Materials

As mentioned earlier, the target pollutants do not represent a large variety of organic compounds and that the degradation of selected simple pollutants such as oxalate, formic acid, and methanol is often examined to assess the photoactivity of photoelectrode materials. However, there are also cases in which the degradation of more complex pollutants is examined [172].

In quite a few publications, electrochemistry is frequently used to validate or complement data of PC investigations [7]. Given that materials performance is one of the main concern, the following sections discuss some of the more widespread photoelectrode constituents used in water depollution. A more detailed discussion is found in Chapters 5 and 6.

9.4.1 Titanium Dioxide

Much of the research on semiconductor PC or PEC focuses on the use of TiO_2, and the advantages of using TiO_2 is not discussed in this chapter, as this is the object of a vast literature [173, 174]. The mounting interest in NT photoelectrodes has been emphasized earlier, and within the work on PEC depollution, the overwhelming majority of recent publications deals with nanotubes. A very good discussion on the use of TiO_2 NT photoelectrodes for the degradation of phenol can be found in the paper by Liu *et al.* [64]. On the tide of the interest on nanotubes, PEC has gained a new impetus that likely has yet to run its full course (Figure 9.2).

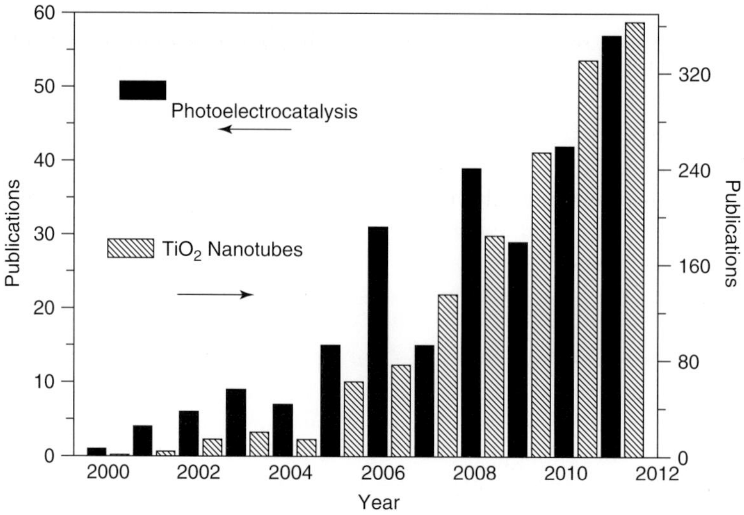

Figure 9.2 Number of publication in the years from 2000 to 2011 on photoelectrocatalysis and nanotubes. Confronted numbers are based on the Scifinder general search using photoelectrocatalysis and TiO_2 nanotubes as key words.

The fact remains that TiO_2 is a UV absorber and its modification by cation or anion doping has attracted enormous attention in an effort to improve PC activity as well as the visible light response [175].

9.4.1.1 **Cation Doping**

Achievements and prospective development in the field of metal-doped TiO_2 have been discussed in recent general [176–178] and specific reviews [179]. Literature reports on metal-doped TiO_2 reveal considerable discrepancy between the objectives and the actual results related to PC activity in general and in response to visible light in particular. Both positive and negative effects have been observed depending, among others, on the distribution of dopants in the TiO_2 lattice. In the case of cobalt-doped TiO_2, surface doping is more effective than bulk, uniform doping at degrading pollutants even under visible light irradiation [180].

Good PEC performance has been reported, for example, for TiO_2 doped with Co or Ni [181], Ce [182], and W [94]. Iron is a common dopant that has proved to be very efficient in enhancing the PC properties of TiO_2 in PEC applications too [131, 183].

9.4.1.2 **Nonmetal Doping**

Among nonmetal dopants, carbon and, especially, nitrogen enjoy high popularity because the resulting doped photocatalysts absorb visible light [49, 175]. Especially for nitrogen-doped TiO_2, there is some debate over the origin of the visible light response, although recent views appear to converge on the important role played by vacancies [184–187]. In any case, enhanced absorbance in the visible region does not seem to result in a high PEC activity for visible light irradiation [44,

160]. Preparation methods have a profound influence on the final properties, and from the results of a vast literature, one is seemingly led to conclude that ion implantation and nitridation in a stream of NH_3 at high temperature give the best results. Among a host of publications on nitrogen-doped TiO_2, very few are related, sometimes marginally, to water depollution by PEC [188, 189].

Improvement of PEC activity for pollutant abatement has been reported with other non-metal-dopant-codoped TiO_2, such as silicon [58], boron [60, 117], fluorine [190], and boron/fluorine [191]. Interestingly, visible light response is reported for the boron-doped oxide.

9.4.2 Other Semiconductor Photoelectrodes

9.4.2.1 Zinc Oxide and Iron Oxide

Although the optical properties of ZnO are similar to those of TiO_2, there are only a few examples of ZnO photoelectrodes [99]. The use of pure ZnO is limited by its instability as a function of pH and under irradiation [13].

Iron oxide (Fe_2O_3) is a very attractive material because of its visible light response; on the other hand, poor electron mobility and short hole diffusion length are its major limitations. The advent of nanotubes seems to be increasing its potential applications [160, 192].

9.4.2.2 Tungsten Trioxide

Tungsten trioxide is the second most used semiconductor next to TiO_2 for photoelectrochemical water purification [59, 98, 171, 193–195]. Its use in PC is limited by the position of the conduction band, which is too positive for oxygen reduction to occur. The flat band potential is 0.55–0.059 pH (V vs NHE) [194], while the pH-independent potential of the O_2/O_2^- couple is −0.16 V versus NHE [196].

Conversely, the PEC process does not need electron acceptors, and WO_3 photoelectrodes have been shown to be stable ($pH < 5$) and efficient for the degradation of various organic pollutants, adding that WO_3 response to visible light increases its application potentiality.

9.4.2.3 Bismuth Vanadate

Bismuth vanadate ($BiVO_4$), as WO_3, is poorly suitable for use in PC for degradation of a variety of organics [197, 198] because it is characterized by a too positive flat band potential for reduction of O_2 except in the alkaline pH range [198]. It shows, however, promising photoelectrochemical properties for a variety of organic degradation reactions, also for irradiation at wavelengths in the visible range [199–201].

9.4.3 Coupled Semiconductors

It was mentioned earlier that doping is one of the strategies that have been followed to improve the PC activity and visible light absorption. Other approaches that have been proposed include coupling of semiconductors with metals or other semiconductors. Surface modification with metal clusters has been long investigated [202] and has been found to improve PEC oxidations by inhibiting electron–hole recombination [86, 203].

Coupled, or hybrid semiconductors, are most broadly defined as a combination of an n-type semiconductor with a second one having n-type (n–n heterojunction) or p-type conductivity (p–n heterojunction). The basic idea is to choose semiconductors that have different flat band potentials and band gaps. Some typical situations are given in Figure 9.3 where semiconductor I, for example, TiO_2, is combined with semiconductor II, for example, WO_3, having a smaller band gap and that can be excited by visible light. Upon UV excitation (Figure 9.3a), electron will be transferred from I to II and the hole in the reverse direction. On the other hand, on visible light irradiation (Figure 9.3b), only II will be excited; holes can move to component I and electrons are accumulated in II. Efficient charge separation is possible, but enhanced PC activity relies on how fast holes are transferred to I in competition with recombination phenomena in II. Clearly, in the PEC case, an applied positive potential can efficiently remove electrons thereby increasing the oxidation power of the system.

For the heterojunction arrangement of Figure 9.3c, semiconductor II can inject photoelectrons into I both for UV and visible light absorption. Holes will accumulate in II, and let alone problems of material stability, their oxidizing power is too low

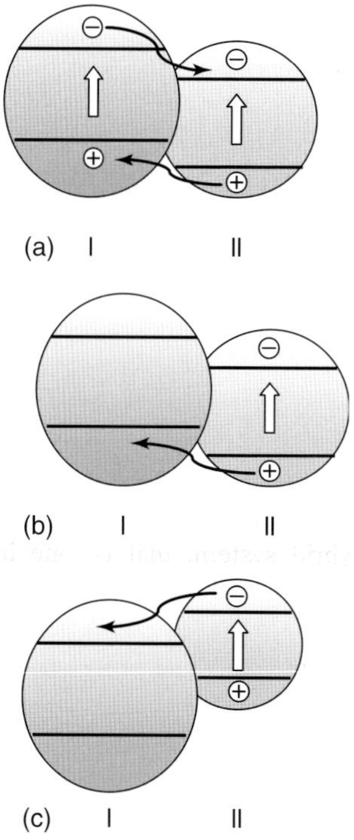

Figure 9.3 Some examples of n–n type heterojunctions under irradiation with UV (a) or visible light (b,c).

for the PC or PEC degradation of most recalcitrant pollutants. The depicted hybrid system is probably best suited for application in solar cells or as a photocathode.

Coupled semiconductors are discussed in detail elsewhere in this book and in the following sections, a few examples that are strictly relevant to PEC application are discussed. The semiconductor/carbon heterojunctions are not discussed.

9.4.3.1 n–n Heterojunctions

One should realize that the commonest case of n–n type junction is the mixed-phase anatase/rutile combination [204, 205]. It is not coincidental that one of the more active commercial photocatalysts, Degussa P25, is constituted of a 75 : 25 ratio of anatase to rutile.

The system WO_3/TiO_2 has also been much investigated and described in a number of publications reporting beneficial effects on the oxidation of organic compounds. A very good account is given in the recent review by Georgieva *et al.* [206]. In addition, some works on the characterization of these composite photoelectrodes are noteworthy [207, 208].

Photoelectrodes based on Bi_2O_3/TiO_2 hybrids are expected to show a similar behavior. Bi_2O_3 is close to WO_3 in terms of band gap and flat band potential [13] and thus Bi_2O_3/TiO_2 might be conforming to the case illustrated in Figure 9.3a,b. Despite the potential interest, there are few examples of PEC application, yet results are encouraging. Zhao *et al.* [78] have observed a good activity for the degradation of 2,4-dichlorophenol for both PC and PEC conditions, under UV and visible light irradiation. Interestingly, PEC gave the best results especially for visible light irradiation.

One of the earliest examples of positive effects of coupling TiO_2 with other oxides was given by Vinodgopal and Kamat [209] who found that the use of TiO_2/SnO_2 photoelectrodes greatly improved the photoelectrochemical degradation of 4-chlorophenol and acid orange with respect to SnO_2 and TiO_2 used separately. Later work on this composite material has confirmed synergistic effects on PEC degradation of pollutants [210, 211]. These works report enhanced efficiency in the degradation of orange II and 2,4-dichlorophenoxyacetic acid, respectively. Additionally, in both cases, redshifts in the absorption spectrum are observed. In one case, however, it is ascribed to the presence of carbonaceous residues; in the other, the preparation method and probably doping of SnO_2 with Sb leads to a redshift in light absorption [212], suggesting caution in the interpretation of the phenomenon.

The fact that in these works one of the component exhibits metallic conductivity draws attention to another interesting case of hybrid system, that is, one in which the PEC material simultaneously presents PC as well as electrocatalytic activity. Synergistic effects of irradiation are observed, for example, in the case of BiO_x–TiO_2/Ti [213] and PbO_2/TiO_2 [214] electrodes for the oxidation of phenol and oxalic acid, respectively.

Encouraging results for the degradation of phenols have been obtained with other composite film photoelectrodes including $ZnFe_2O_4/TiO_2$ [71] and more complex hybrids such as $Fe_2O_3/ZnO/TiO_2$ [52] and $ZnO/ZnWO_4/WO_3$ [124].

Ferrites (MFe_2O_4), in particular, are widely studied materials, as they combine visible light absorption and stability.

9.4.3.2 p–n Heterojunctions

These junctions also referred to as photochemical diodes and were proposed several years ago by Nozik *et al.* [215, 216] as an attractive approach to achieve efficient electron–hole separation that minimizes recombination phenomena. Like in conventional diodes, a field is built-in at the p–n interface. Under near-UV irradiation, the photogenerated electron–hole pairs are separated by the inner electric field: holes will be driven to the p-type component and electrons to the n-type one, and thence to the species in solution or to metal contacts (Figure 9.4).

A number of semiconductor p–n junctions have been used as PEC and, especially, PC catalysts. Some of the more common combinations include n-TiO_2/p-Cu_2O [217–219], n-TiO_2/p-NiO [220, 221], and n-TiO_2/p-ZnO [222]. Recent contributions focus particularly on p–n junctions that absorb visible light. Examples are p-$CaFe_2O_4$/n-Ag_3VO_4 [223], n-TiO_2/p-$CuInS_2$ [224], n-TiO_2/p-BiOI [225], and n-ZnO/p-$CaFe_2O_4$ [226]. It is confirmed that ferrites are a very interesting class of compounds that can be prepared as n-type or p-type; they combine, in addition, low band gaps and stability. A good example is provided by the work of Kim *et al.* [227] on the PC system p-$CaFe_2O_4$/n-$MgFe_2O_4$.

The results obtained are generally encouraging, although p–n heterojunctions are comparatively still less widely used as materials for decontamination by PEC [219, 224, 225].

The case of NiO/TiO_2 is worth a comment. NiO is considered to be a p-type semiconductor [228] and the junction is expected to operate as described in Figure 9.4. It was actually shown that irradiation of nickel oxide alone did not induce photoeffects [229], and concerning the mechanism, things can be seen from a different angle that does not necessarily imply formation of a p–n junction [230]. In this case, holes generated on TiO_2 are transferred to the NiO overlayer where they can oxidize Ni^{2+} to Ni^{3+} and the composite works as a redox energy storage material (Figure 9.5).

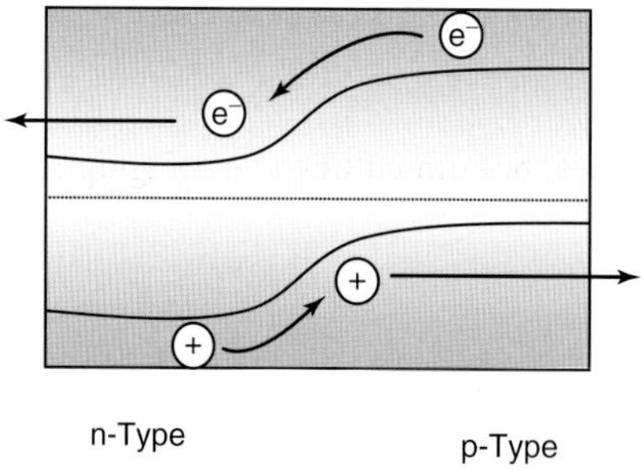

Figure 9.4 Schematic illustration of charge displacement in a photocatalytic p–n junction under irradiation.

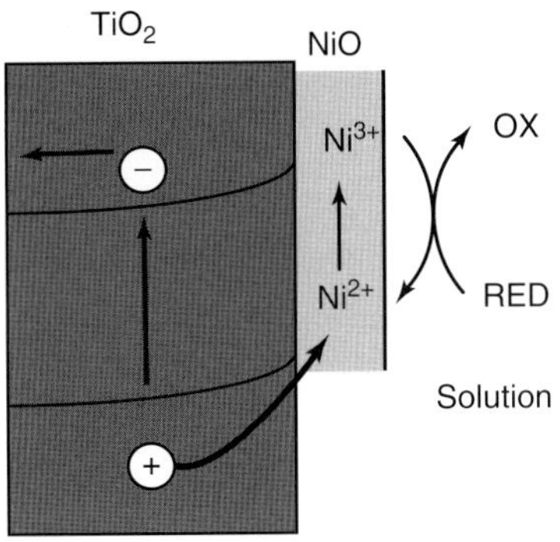

Figure 9.5 Model of the oxidative energy storage for TiO_2/NiO hybrid system under irradiation.

The oxidative energy stored can be exploited for oxidizing a number of organic and inorganic species such as alcohols, aldehydes, phenol, acetone, and iodide [230]. The search for analogous systems with tunable properties can probably lead to interesting developments, as shown recently in [38].

Finally, n-type semiconductors/boron-doped diamond (BDD) composites are considered to form a p–n junction [62, 231].

9.5 Electrodes Preparation and Reactors

As discussed in Section 9.3, preparation methods affect cardinal parameters, such as morphology, that determine the activity of photoelectrodes. An overview of some of the most common electrochemical methods for the preparation is given in Ref. [206]. In addition, one can add electrophoresis [232, 233] and occlusion methods [214, 234]. Special arrangements include interdigitated electrodes [143, 235, 236].

To date, sol–gel procedures are still particularly attractive for film fabrication and scaling-up since a variety of liquid precursors are available that can be easily applied on a substrate by dip-coating, spinning, or spraying [175, 237–243].

Poznyak *et al.* [29] concluded that electrodes nanostructured by sol–gel procedures are more active than compact or single-crystal ones. It is often difficult to compare the PEC behavior of electrodes prepared by different methods. In this sense, use of NT films offers more possibilities, but difficulties in scaling-up are often pointed out [72].

Research on photoelectrochemical reactors is still mostly limited to the description of lab-scale cells with some notable exceptions [72, 105, 128, 162, 244]. The use of three-dimensional packed-bed reactors might enhance the potential for scaling-up [105]. The rotating disk photoelectrocatalytic reactor (RDPR) may provide another interesting possibility [128, 162]. Its advantages include high PEC efficiencies combined with a simple design and good scaling-up perspectives. Titanium rotating disks can be anodically oxidized to grow TiO_2 nanotubes or

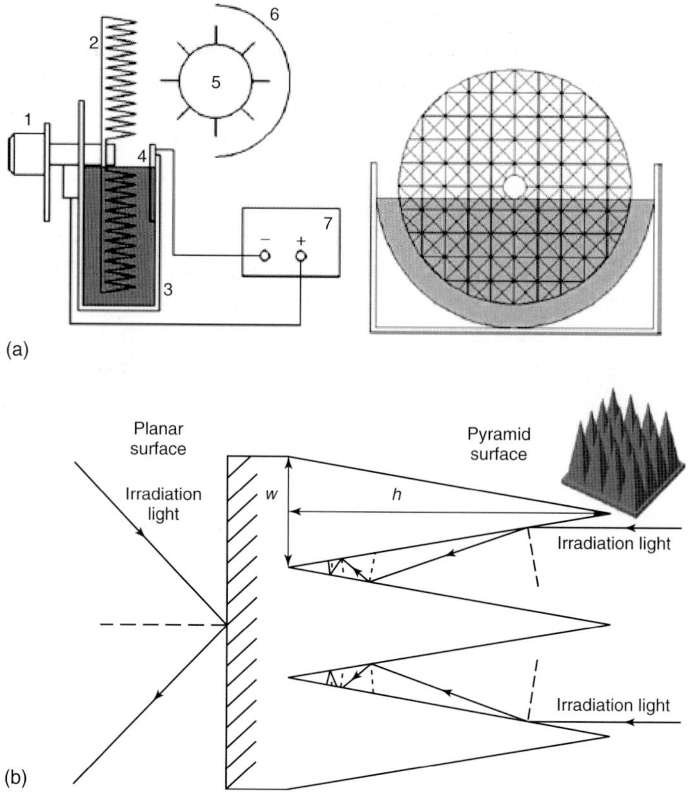

Figure 9.6 (a) Schematic view of a TiO_2/Ti rotating disk photoelectrochemical reactor: 1, motor; 2, TiO_2 anode; 3, electrolytic cell; 4, cathode; 5, lamp; 6, aluminum foil; and 7, power supply. (b) Details of the special pyramid electrode surface. (Source: Reprinted with permission from Ref. [245].)

coated with TiO_2 nanoparticles. It is noteworthy that in a comparative study on the photoelectrochemical degradation of methyl orange, nanotube electrodes showed sensibly higher efficiencies than the nanoparticulate ones [162].

One RDPR configuration [245] uses the so-called Ti pyramid-surface electrodes (Figure 9.6) that are fabricated by an electrical discharge linear cutting machine. These Ti disks, having a surface area of up to 260 cm^2, were then coated with TiO_2 by sol–gel and dip-coating processes to obtain photoelectrodes.

Interesting developments for application purposes may come from application of fuel-cell technology, as described in a number of recent publications [69, 246–248].

9.6 Conclusions

This chapter gives an account of water decontamination by PEC. It is a short overview of a research field where it is often difficult to discriminate between real

PEC and situations where electrochemistry is used as a support to substantiate results of PC studies. We have given a selection of results that hopefully can provide a reasonable picture of the field of application discussed.

Analysis of results shows that the electrode material, the morphology, the applied potential, the pH, the electrolyte composition, and the presence of oxygen strongly influence PEC performance.

For full screening purposes, one should compare photocurrents in terms of efficiency of one electrode toward different substrates or different electrode preparations for one substrate. The electrode structure turns out to play an important role, and for example, the development of NT films has given a big boost to PEC research, although there are voices of dissent on true benefits.

Results showed that the use of hybrid semiconductor materials is a promising approach to increase charge separation and lifetime of the charge carriers, thereby also enhancing interfacial charge transfer to adsorbed substrates.

In the overwhelming majority of publications, PEC and PC results on pollutant degradation are confronted and there is broad consensus that the former approach is more efficient [29, 162]. Then, quoting Egerton's words [72]: *"Does photoelectrocatalysis work?" may be "Yes", but the answer to the corresponding technological question is "No or Not yet."* Problems connected to reactor design are seen as a limiting factor but, in this respect, the use of photofuel cells, using conventional fuel-cell technology, might offer new perspectives.

References

1. Fujishima, A. and Honda, K. (1972) Electrochemical photolysis of water at a semiconductor electrode. *Nature*, **238**, 37–38.
2. Bard, A.J. (1979) Photoelectrochemistry and heterogeneous photo-catalysis at semiconductors. *J. Photochem.*, **10** (1), 59–75.
3. Hodes, G. and Grätzel, M. (1984) Photoelectrochemistry at semiconductor electrodes and small particles: a comparative study. *New J. Chem.*, **8** (8–9), 510–520.
4. Caramori, S., Cristino, V., Meda, L., Argazzi, R., and Bignozzi, C.A. (2011) in *Photocatalysis*, (ed. C.A. Bignozzi), Springer, Berlin, pp. 39–94.
5. Honda, K. (2004) Dawn of the evolution of photoelectrochemistry. *J. Photochem. Photobiol., A*, **166** (1–3), 63–68.
6. Serpone, N., Sauvé, G., Koch, R., Tahiri, H., Pichat, P., Piccinini, P., Pelizzetti E., and Hidaka H. (1996) Standardization protocol of process efficiencies and activation parameters in heterogeneous photocatalysis: relative photonic efficiencies ζ_r. *J. Photochem. Photobiol., A*, **94** (2–3), 191–203.
7. Taghizadeh, A., Lawrence, M.F., Miller, L., Anderson, M.A., and Serpone, N. (2000) (Photo)electrochemical behavior of selected organic compounds on TiO_2 electrodes. Overall relevance to heterogeneous photocatalysis. *J. Photochem. Photobiol., A*, **130** (2), 145–156.
8. Rajeshwar, K. (1995) Photoelectrochemistry and the environment. *J. Appl. Electrochem.*, **25** (12), 1067–1082.
9. Neumann-Spallart, M. (2007), Aspects of photocatalysis on semiconductors: photoelectrocatalysis. *Chimia*, **61** (12), 806–808.
10. Egerton, T.A. and Christensen, P.A. (2004) in *Advanced Oxidation Processes for Water and Wastewater Treatment*, (ed. S. Parsons), IWA Publishing, London, pp. 167–184.
11. Li, D. and Qu, J. (2009) The progress of catalytic technologies in water

purification: a review. *J. Environ. Sci.*, **21** (6), 713–719.
12. Georgieva, J., Armyanov, S., Poulios, I., and Sotiropoulos, S. (2009) An all-solid photoelectrochemical cell for the photooxidation of organic vapours under Ultraviolet and visible light illumination. *Electrochem. Commun.*, **11** (8), 1643–1646.
13. Morrison, S.R. (1980) *Electrochemistry at Semiconductor and Oxidized Metal Electrodes*, Plenum Publishing Corporation, New York, NY.
14. Sato, N. (1998) *Electrochemistry at Metal and Semiconductor Electrode*, Elsevier Science, Amsterdam.
15. Memming, R. (2001) *Semiconductor Electrochemistry*, Wiley-VCH Verlag GmbH, Weinheim.
16. Licht, S. (ed.) (2002) Semiconductor electrodes and photoelectrochemistry, in *Encyclopedia of Electrochemistry*, Vol. 6 (eds Bard A. J. and Stratmann M.) Wiley-VCH Verlag GmbH, Weinhiem.
17. Tan, X.M., Laibinis, P.E., Nguyen, S.T., Kesselman, J.M., Stanton, C.E., and Lewis, N.S. (1994) Principles and applications of semiconductor electrochemistry, in *Progress in Inorganic Chemistry*, Vol. 41 (ed. K.D. Karlin), John Wiley & Sons, Inc.
18. Pleskov, Y.V. and Gurevich, Y.Y. (1986) *Semiconductor Electrochemistry*, Consultants Bureau, New York.
19. Zhang, X.G. (2001) *Electrochemistry of Silicon and its Oxide*, Kluwer Academic Publishers, New York.
20. Albery, J. (1975) *Electrode Kinetics*, Clarendon Press, Oxford.
21. Bockris, J.O.M. and Reddy, A.K.N. (1977) *Modern Electrochemistry*, Vol. 2. Plenum Publishing Co., New York.
22. Grimley, T.B. (1960) Surface states associated with adsorbed atoms. *J. Phys. Chem. Solids*, **14**, 227–232.
23. Villarreal, T.L., Gómez, R., Neumann-Spallart, M., Alonso-Vante, N., and Salvador, P. (2004) Semiconductor photooxidation of pollutants dissolved in water: a kinetic model for distinguishing between direct and indirect interfacial hole transfer. I. Photoelectrochemical experiments with polycrystalline anatase electrodes under current doubling and absence of recombination. *J. Phys. Chem. B*, **108** (39), 15172–15181, and references therein.
24. Mora-Seró, I., Villarreal, T.L., Bisquert, J., Pitarch, A. Gómez, R., and Salvador, P. (2005) Photoelectrochemical behavior of nanostructured TiO_2 thin-film electrodes in contact with aqueous electrolytes containing dissolved pollutants: a model for distinguishing between direct and indirect interfacial hole transfer from photocurrent measurements. *J. Phys. Chem. B*, **109** (8), 3371–3380.
25. Solarska, R., Rutkowska I., Morand, R., and Augustynski, J. (2006) Photoanodic reactions occurring at nanostructured titanium dioxide films. *Electrochim. Acta*, **51** (11), 2230–2236.
26. Bilecka, I., Barczuk, P.J., and Augustynski, J. (2010) Photoanodic oxidation of small organic molecules at nanostructured TiO_2 anatase and rutile film electrodes. *Electrochim. Acta*, **55** (3), 979–984.
27. Hagfeldt, A., Lindström, H., Södergren, A., and Lindquist, S.E. (1995) Photoelectrochemical studies of colloidal TiO_2 films: the effect of oxygen studied by photocurrent transients. *J. Electroanal. Chem.*, **381** (1), 39–46.
28. Byrne, J.A. and Eggins, B.R. (1998) Photoelectrochemistry of oxalate on particulate TiO_2 electrodes. *J. Electroanal. Chem.*, **457** (1–2), 61–72.
29. Poznyak, S.K., Kokorin, A.I., and Kulak, A.I. (1998) Effect of electron and hole acceptors on the photoelectrochemical behaviour of nanocrystalline microporous TiO_2 electrodes. *J. Electroanal. Chem.*, **442** (1–2), 99–105.
30. Augustynski, J. (1988) Aspects of photoelectrochemical and surface behavior of titanium(IV) oxide. *Struct. Bond.*, **69**, 1–61.
31. Fabregat-Santiago, F., Mora-Seró, I., Garcia-Belmonte, G., and Bisquert J. (2003) Cyclic voltammetry studies of nanoporous semiconductors. Capacitive and reactive properties of nanocrystalline TiO_2 electrodes in aqueous electrolyte. *J. Phys. Chem. B*, **107** (3), 758–768.

32. Mintsouli, I., Philippidis, N., Poulios, I., and Sotiropolos, S. (2006) Photoelectrochemical characterization of thermal and particulate titanium dioxide electrodes. *J. Appl. Electrochem.*, **36** (4), 463–474.
33. Yu, H., Ma, J., Zhang, Y., Zhang, X., and Shi, W. (2011) Cyclic voltammetry studies of TiO_2 nanotube arrays Electrode: conductivity and reactivity in the presence of H^+ and aqueous redox systems. *Electrochim. Acta*, **56** (18), 6498–6502.
34. Mao, D., Kim, K.-J., and Frank, A.J. (1994) Open-circuit photovoltage and charge recombination at semiconductor/liquid interfaces. *J. Electrochem. Soc.*, **141** (5), 1231–1236.
35. McMurray, T.A., Byrne, J.A., Dunlop, P.S.M., and McAdams, E.T. (2005) Photocatalytic and electrochemically assisted photocatalytic oxidation of formic acid on TiO_2 films under UVA and UVB irradiation. *J. Appl. Electrochem.*, **35** (7–8), 723–731.
36. Krýsa, J., Zlámal, M., and Waldner, G. (2007) Effect of oxidisable substrates on the photoelectrocatalytic properties of thermally grown and particulate TiO_2 layers. *J. Appl. Electrochem.*, **37** (11), 1313–1319.
37. Byrne, J.A., Davidson, A., Dunlop, P.S.M., and Eggins, B.R. (2002) Water treatment using nano-crystalline TiO_2 electrodes. *J. Photochem. Photobiol., A*, **148** (1–3), 365–374.
38. Amadelli, R., Samiolo, L., Maldotti, A., Molinari, A., and Gazzoli, D. (2011) Selective photooxidation and photoreduction processes at TiO_2 surface-modified by grafted vanadyl. *Int. J. Photoenergy*, **2011**, Article ID 259453, 10 pp.
39. Kronik, L. and Shapira, Y. (2001) Surface photovoltage spectroscopy of semiconductor structures: at the crossroads of physics, chemistry and electrical engineering. *Surf. Interface Anal.*, **31** (10), 954–965.
40. Semenikhin, O.A., Kazarinov, V.E., Jiang, L., Hashimoto, K., and Fujishima, A. (1999) Suppression of surface recombination on TiO_2 anatase photocatalysts in aqueous solutions containing alcohol. *Langmuir*, **15** (11), 3731–3737.
41. Van de Lagemaat, J., Park, N.-G., and Frank, A.J. (2000) Influence of electrical potential distribution, charge transport, and recombination on the phtotopotential and photocurrent conversion efficiency of dye-sensitized nanocrystalline TiO_2 solar cells: a study by electrical impedance and optical modulation techniques. *J. Phys. Chem. B*, **104** (9), 2044–2052.
42. Liu, H., Li, H.Z., Leng, Y.J., and Li, W.Z. (2003) An alternative approach to ascertain the rate-determining steps of TiO_2 photoelectrocatalytic reaction by electrochemical impedance. *J. Phys. Chem. B*, **107** (34), 8988–8996.
43. Liu, H., Cheng, S., Wu, M., Wu, H., Zhang, J., Li, W., and Cao, C. (2000) Photoelectrocatalytic degradation of sulfosalycilic acid and its electrochemical impedance spectroscopy investigation. *J. Phys. Chem. B*, **104** (30), 7016–7020.
44. Samiolo, L., Valigi, M., Gazzoli, D., and Amadelli, A. (2010) Photo-electro catalytic oxidation of aromatic alcohols on visible light-absorbing nitrogen-doped TiO_2. *Electrochim. Acta*, **55** (26), 7788–7795.
45. Fabregat-Santiago, F., Garcia-Belmonte, G., Bisquert, J., Bogdanoff, P., and Zaban, A. (2003) Mott-Schottky analysis of nanoporous semiconductor electrodes in dielectric state deposited on SnO_2/(F) conducting substrate. *J. Electrochem. Soc.*, **150** (6), E293–E298.
46. Carp, O., Huisman, C.L., and Reller, A. (2004) Photoinduced reactivity of titanium dioxide. *Prog. Solid State Chem.*, **32** (1–2), 33–177.
47. Hashimoto, K., Irie, H., and Fujishima, A. (2005) TiO_2 Photocatalysis: a historical overview and future prospects. *Jpn. J. Appl. Phys.*, **44** (12), 8269–8285.
48. Agrios, A.G. and Pichat, P. (2005) State of the art and perspectives on materials and applications of photocatalysis over TiO_2. *J. Appl. Electrochem.*, **35** (7–8), 655–663.
49. Zhang, H., Chen, G., and Bahnemann, D.W. (2009) Photoelectrocatalytic materials for environmental applications. *J. Mater. Chem.*, **19** (29), 5089–5121.

50. Yan, X., Shi, H., and Wang, D. (2003) Photoelectrocatalytic degradation of phenol using a TiO_2/Ni thin-film electrode. *Korean J. Chem. Eng.*, **20** (4), 679–684.
51. Tada, H., Kokubu, A., Iwasaki, M., and Ito, S. (2004) Deactivation of the TiO_2 photocatalyst by coupling with WO_3 and the electrochemically assisted high photocatalytic activity of WO_3. *Langmuir*, **20** (11), 4665–4670.
52. Yang, S., Quan, X., Li, X., Liu, Y., Chen, S., and Chen, G. (2004) Preparation, characterization and photoelectrocatalytic properties of nanocrystalline Fe_2O_3/TiO_2, ZnO/TiO_2, and $Fe_2O_3/ZnO/TiO_2$ composite film electrodes towards pentachlorophenol degradation. *Phys. Chem. Chem. Phys.*, **6** (3), 659–664.
53. Xie, Y.-B. and Li, X.-Z. (2006) Degradation of bisphenol a in aqueous solution by H_2O_2-assisted photoelectrocatalytic oxidation. *J. Hazard. Mater.*, **138** (3), 526–533.
54. Egerton, T.A., Christensen, P.A., Kosa, S.A.M., Onoka, B., Harper, J.C., and Tinlin, J.R. (2006) Photoelectrocatalysis by titanium dioxide for water treatment. *Int. J. Environ. Pollut.*, **27** (1–3), 2–19.
55. He, C., Li, X.Z., Graham, N., and Wang, Y. (2006) Preparation of TiO_2/ITO and TiO_2/Ti photoelectrodes by magnetron sputtering for photocatalytic application. *Appl. Catal. Gen.*, **305** (1), 54–63.
56. Quan, X., Ruan, X., Zhao, H., Chen, S., and Zhao, Y. (2007) Photoelectrocatalytic degradation of pentachlorophenol in aqueous solution using a TiO_2 nanotube film electrode. *Environ. Pollut.*, **147** (2), 409–414.
57. Kim, D.W., Lee, S., Jung, H.S., Kim, J.Y., Shin, H., and Hong, K.S. (2007) Effects of heterojunction on photoelectrocatalytic properties of ZnO-TiO_2 films. *Int. J. Hydrogen Energy*, **32** (15), 3137–3140.
58. Su, Y., Chen, S., Quan, X., Zhao, H., and Zhang, Y. (2008) A silicon-doped TiO_2 nanotube arrays electrode with enhanced photoelectrocatalytic activity. *Appl. Surf. Sci.*, **255** (5, Pt. 1), 2167–2172.
59. Nissen, S., Alexander, B.D., Dawood, I., Tillotson, M., Wells, R.P.K., Macphee, D.E., and Killham, K. (2009) Remediation of a chlorinated aromatic hydrocarbon in water by photoelectrocatalysis. *Environ. Pollut.*, **157** (1), 72–76.
60. Li, J., Lu, N., Quan, X., Chen, S., and Zhao, H. (2008) Facile method for fabricating boron-doped TiO_2 nanotube array with enhanced photoelectrocatalytic properties. *Ind. Eng. Chem. Res.*, **47** (11), 3804–3808.
61. Liu, W., Quan, X., Cui, Q., Ma, M., Chen, S., and Wang, Z.-J. (2008) Ecotoxicological characterization of photoelectrocatalytic process for degradation of pentachlorophenol on titania nanotubes electrode. *Ecotoxicol. Environ. Saf.*, **71** (1), 267–273.
62. Qu, J. and Zhao, X. (2008) Design of BDD-TiO_2 hybrid electrode with P-N function for photoelectrocatalytic degradation of organic contaminants. *Environ. Sci. Technol.*, **42** (13), 4934–4939.
63. Yang, J., Dai, J., Chen, C., and Zhao, J. (2009) Effects of hydroxyl radicals and oxygen species on the 4-chlorophenol degradation by photoelectrocatalytic reactions with TiO_2-film electrodes. *J. Photochem. Photobiol., A*, **208** (1), 66–77.
64. Liu, Z., Zhang, X., Nishimoto, S., Jin, M., Tryk, D.A., Murakami, T., and Fujishima, A. (2008) Highly ordered TiO_2 nanotube arrays with controllable length for photoelectrocatalytic degradation of phenol. *J. Phys. Chem. C*, **112** (1), 253–259.
65. Wang, X., Zhao, H., Quan, X., Zhao, Y., and Chen, S. (2009) Visible light photoelectrocatalysis with salicylic acid-modified TiO_2 nanotube array electrode for p-nitrophenol degradation. *J. Hazard. Mater.*, **166** (1), 547–552.
66. Hou, Y., Li, X., Zhao, Q., Quan, X., and Chen, G. (2010) Electrochemically assisted photocatalytic degradation of 4-chlorophenol by $ZnFe_2O_4$-modified TiO_2 nanotube array electrode under

visible light irradiation. *Environ. Sci. Technol.*, **44** (13), 5098–5103.

67. Valova, E., Georgieva, J., Armyanov, S., Sotiropoulos, S., Hubin, A., Baert, K., and Raes, M. (2010) Morphology, structure and photoelectrocatalytic activity of TiO_2/WO_3 coatings obtained by pulsed electrodeposition onto stainless steel. *J. Electrochem. Soc.*, **157** (5), D309–D315.
68. Su, Y., Wu, J., Quan, X., and Chen, S. (2010) Electrochemically assisted photocatalytic degradation of phenol using silicon-doped TiO_2 nanofilm electrode. *Desalination*, **252** (1–3), 143–148.
69. Yuan, S.-J., Sheng, G.-P., Li, W.-W., Lin, Z.-Q., Zeng, R.J., Tong, Z.-H., and Yu, H.-Q. (2010) Degradation of organic pollutants in a photoelectrocatalytic system enhanced by a microbial fuel cell. *Environ. Sci. Technol.*, **44** (14), 5575–5580.
70. Oliveira, H.G., Nery, D.C., and Longo, C. (2010) Effect of applied potential on photocatalytic phenol degradation using nanocrystalline TiO_2 electrodes. *Appl. Catal. Environ.*, **93** (3–4), 205–211.
71. Hou, Y., Li, X.-Y., Zhao, Q.-D., Quan, X., and Chen, G.-H. (2010) Electrochemical method for synthesis of a $ZnFe_2O_4/TiO_2$ composite nanotube array modified electrode with enhanced photoelectrochemical activity. *Adv. Funct. Mater.*, **20** (13), 2165–2174.
72. Egerton, T.A. (2011) Does photoelectrocatalysis by TiO_2 work?. *J. Chem. Technol. Biotechnol.*, **86** (8), 1024–1031.
73. Frontistis, Z., Daskalaki, V.M., Katsaounis, A., Poulios, I., and Mantzavinos, D. (2011) Electrochemical enhancement of solar photocatalysis: degradation of endocrine disruptor bisphenol-a on Ti/TiO_2 films. *Water Res.*, **45** (9), 2996–3004.
74. Gu, Y., Zhang, Y., Zhang, F., Wei, J., Wang, C., Du, Y., and Ye, W. (2010) Investigation of photoelectrocatalytic activity of Cu_2O nanoparticles for p-nitrophenol using rotating ring-disk electrode and application for electrocatalytic determination. *Electrochim. Acta*, **56** (2), 953–958.
75. Brugnera, M.F., Rajeshwar, K., Cardoso, J.C., and Boldrin Zanoni, M.V. (2010) Bisphenol a removal from wastewater using self-organized TiO_2 nanotubular array electrodes. *Chemosphere*, **78** (5), 569–575.
76. Chai, S., Zhao, G., Li, P., Lei, Y., Zhang, Y., and Li, D. (2011) Novel sieve-like SnO_2/TiO_2 nanotubes with integrated photoelectrocatalysis: fabrication and application for efficient toxicity elimination of nitrophenol wastewater. *J. Phys. Chem. C*, **115** (37), 18261–18269.
77. Zheng, Q., Li, J., Chen, H., Chen, Q., Zhou, B., Shang, S., and Cai, W. (2011) Characterization and mechanism of the photoelectrocatalytic oxidation of organic pollutants in a thin-layer reactor. *Chin. J. Catal.*, **32** (6–8), 1357–1363.
78. Zhao, X., Liu, H., and Qu, J. (2011) Photoelectrocatalytic degradation of organic contaminants at Bi_2O_3/TiO_2 nanotube array electrode. *Appl. Surf. Sci.*, **257** (10), 4621–4624.
79. Kim, D.H. and Anderson, M.A. (1994) Photoelectrocatalytic degradation of formic acid using a porous titanium dioxide thin-film electrode. *Environ. Sci. Technol.*, **28** (3), 479–483.
80. Candal, R.J., Zeltner, W.A., and Anderson, M.A. (1998) TiO_2-Mediated photoelectrocatalytic purification of water. *J. Adv. Oxid. Technol.*, **3** (3), 270–276.
81. Krýsa, J. and Jirkovsky, J. (2002) Electrochemically assisted photocatalytic degradation of oxalic acid on particulate TiO_2 film in a batch mode plate photoreactor. *J. Appl. Electrochem.*, **32** (6), 591–596.
82. An, T., Xiong, Y., Li, G., Zha, C., and Zhu, X. (2002) Synergetic effect in degradation of formic acid using a new photoelectrochemical reactor. *J. Photochem. Photobiol., A*, **152** (1–3), 155–165.
83. He, C., Li, X., Xiong, Y., Zhu, X., and Liu, S. (2005) The enhanced PC and PEC oxidation of formic acid in aqueous solution using a Cu-TiO_2/ITO film. *Chemosphere*, **58** (4), 381–389.
84. An, T., Xiong, Y., Li, G., Zhu, X., Sheng, G., and Fu, J. (2006) Improving ultraviolet light transmission in a packed-bed photoelectrocatalytic reactor for removal of oxalic acid from

wastewater. *J. Photochem. Photobiol., A*, **181** (2–3), 158–165.

85. Egerton, T.A., Janus, M., and Morawski, A.W. (2006) New TiO_2/C sol–gel electrodes for photoelectrocatalytic degradation of sodium oxalate. *Chemosphere*, **63** (7), 1203–1208.
86. He, C., Xiong, Y., Shu, D., Zhu, X., and Li, X. (2006) Preparation and photoelectrocatalytic activity of Pt(TiO_2)-TiO_2 hybrid films. *Thin Solid Films*, **503** (1–2), 1–7.
87. Gan, W.Y., Lee, M.W., Amal, R., Zhao, H., and Chiang, K. (2008) Photoelectrocatalytic activity of mesoporous TiO_2 films prepared using the sol–gel method with tri-block copolymer as structure directing agent. *J. Appl. Electrochem.*, **38** (5), 703–712.
88. Shinde, P.S., Sadale, S.B., Patil, P.S., Bhosale, P.N., Brueger, A., Neumann-Spallart, M., and Bhosale, C.H. (2008) Properties of spray deposited titanium dioxide thin films and their application in photoelectrocatalysis. *Sol. Energy Mater. Sol. Cells*, **92** (3), 283–290.
89. Selcuk, H., Sene, J.J., and Anderson, M.A. (2003) Photoelectrocatalytic humic acid degradation kinetics and effect of pH, applied potential and inorganic ions. *J. Chem. Technol. Biotechnol.*, **78** (9), 979–984.
90. Selcuk, H., Sene, J.J., Sarikaya, H.Z., Bekbolet, M., and Anderson, M.A. (2004) An innovative photocatalytic technology in the treatment of river water containing humic substances. *Water Sci. Technol.*, **49** (4), 153–158.
91. Selcuk, H. and Bekbolet, M. (2008) Photocatalytic and photoelectrocatalytic humic acid removal and selectivity of TiO_2 coated photoanode. *Chemosphere*, **73** (5), 854–858.
92. Li, A., Zhao, X., Liu, H., and Qu, J. (2011) Characteristic transformation of humic acid during photoelectrocatalysis process and its subsequent disinfection byproduct formation potential. *Water Res.*, **45** (18), 6131–6140.
93. Hidaka, H., Ajisaka, K., Horikoshi, S., Oyama, T., Takeuchi, K., Zhao, J., and Serpone, N. (2001) Comparative assessment of the efficiency of TiO_2/OTE thin film electrodes fabricated by three deposition methods-photoelectrochemical degradation of the DBS anionic surfactant. *J. Photochem. Photobiol., A*, **138** (2), 185–192.
94. Gong, J., Yang, C., Pu, W., and Zhang, J. (2011) Liquid phase deposition of tungsten doped TiO_2 films for visible light photoelectrocatalytic degradation of dodecyl-benzenesulfonate. *Chem. Eng. J.*, **167** (1), 190–197.
95. Horikoshi, S., Satou, Y., Hidaka, H., and Serpone, N. (2001) Enhanced photocurrent generation and photooxidation of benzene sulfonate in a continuous flow reactor using hybrid TiO_2 thin films immobilized on OTE electrodes. *J. Photochem. Photobiol., A*, **146** (1–2), 109–119.
96. Hidaka, H., Asai, Y., Zhao, J., Nohara, K., Pelizzetti, E., and Serpone, N. (1995) Photoelectrochemical decomposition of surfactants on a TiO_2/TCO particulate film electrode assembly. *J. Phys. Chem.*, **99** (20), 8244–8248.
97. Pelegrini, R., Peralta-Zamora, P., de Andrade, A.R., Reyes, J., and Durán, N. (1999) Electrochemically assisted photocatalytic degradation of reactive dyes. *Appl. Catal. Environ.*, **22** (2), 83–90.
98. Luo, J. and Hepel, M. (2001) Photoelectrochemical degradation of naphthol blue black diazo dye on WO_3 film electrode. *Electrochim. Acta*, **46** (19), 2913–2922.
99. Quintana, M., Ricra, E., Rodriguez, J., and Estrada, W. (2002) Spray pyrolysis deposited zinc oxide films for photoelectrocatalytic degradation of methyl orange: influence of the pH. *Catal. Today*, **76** (2–4), 141–148.
100. Zhang, W., An, T., Xiao, X., Fu, J., Sheng, G., Cui, M., and Li, G. (2003) Photoelectrocatalytic degradation of reactive brilliant orange K-R in a new continuous flow photoelectrocatalytic reactor. *Appl. Catal. Gen.*, **255** (2), 221–229.
101. Zanoni, M.V.B., Sene, J.J., and Anderson, M.A. (2003) Photoelectrocatalytic degradation of remazol brilliant orange 3R on titanium dioxide thin-film electrodes. *J. Photochem. Photobiol., A*, **157** (1), 55–63.

102. Carneiro, P.A., Osugi, M.E., Sene, J.J., Anderson, M.A., and Boldrin Zanoni, M. (2004) Evaluation of color removal and degradation of a reactive textile azo dye on nanoporous TiO_2 thin-film electrodes. *Electrochim. Acta*, **49** (22–23), 3807–3820.
103. Chen, J., Liu, M., Zhang, J., Ying, X., and Jin, L. (2004) Photocatalytic degradation of organic wastes by electrochemically assisted TiO_2 photocatalytic system. *J. Environ. Manage.*, **70** (1), 43–47.
104. Su, Y.-F. and Chou, T.-C. (2005) Comparison of the photocatalytic and photoelectrocatalytic decolorization of methyl orange on sputtered TiO_2 thin films. *Z. Naturforsch., B: Chem. Sci.*, **60** (11), 1158–1167.
105. Zhang, W., An, T., Cui, M., Sheng, G., and Fu, J. (2005) Effects of anions on the photocatalytic and photoelectrocatalytic degradation of reactive dye in a packed-bed reactor. *J. Chem. Technol. Biotechnol.*, **80** (2), 223–229.
106. Yang, J., Chen, C., Ji, H., Ma, W., and Zhao, J. (2005) Mechanism of TiO_2-assisted photocatalytic degradation of dyes under visible irradiation: photoelectrocatalytic study by TiO_2-film electrodes. *J. Phys. Chem. B*, **109** (46), 21900–21907.
107. Egerton, T.A., Purnama, H., Purwajanti, S., and Zafar, M. (2006) Decolourization of dye solutions using photoelectrocatalysis and photocatalysis. *J. Adv. Oxid. Technol.*, **9** (1), 79–85.
108. Zainal, Z. and Lee, C.Y. (2006) Properties and photoelectrocatalytic behaviour of sol–gel derived TiO_2 thin films. *J. Sol-Gel Sci. Technol.*, **37** (1), 19–25.
109. Xie, Y.B. and Li, X.Z. (2006) Interactive oxidation of photoelectrocatalysis and electro-Fenton for azo dye degradation using TiO_2-Ti mesh and reticulated vitreous carbon electrodes. *Mater. Chem. Phys.*, **95** (1), 39–50.
110. Li, G., Qu, J., Zhang, X., Liu, H., and Liu, H. (2006) Electrochemically assisted photocatalytic degradation of orange II: influence of initial pH values. *J. Mol. Catal. A: Chem.*, **259** (1–2), 238–244.
111. Zainal, Z., Lee, C.Y., Hussein, M.Z., Kassim, A., and Yusof, N.A. (2007) Electrochemical-assisted photodegradation of mixed dye and textile effluents using TiO_2 thin films. *J. Hazard. Mater.*, **146** (1–2), 73–80.
112. Macedo, L.C., Zaia, D.A.M., Moore, G.J., and de Santana, H. (2007) Degradation of leather dye on TiO_2: a study of applied experimental parameters on photoelectrocatalysis. *J. Photochem. Photobiol., A*, **185** (1), 86–93.
113. Zlamal, M., Macak, J.M., Schmuki, P., and Krýsa, J. (2007) Electrochemically assisted photocatalysis on self-organized TiO_2 nanotubes. *Electrochem. Commun.*, **9** (12), 2822–2826.
114. Sohn, Y.S., Smith, Y.R., Misra, M., and Subramanian, V. (2008) Electrochemically assisted photocatalytic degradation of methyl orange using anodized titanium dioxide nanotubes. *Appl. Catal. Environ.*, **84** (3–4), 372–378.
115. Osugi, M.E., Zanoni, M.V.B., Chenthamarakshan, C.R., de Tacconi, N.R., Woldemariam, G.A., Mandal, S.S., and Rajeshwar, K. (2008) Toxicity assessment and degradation of disperse azo dyes by photoelectrocatalytic oxidation on Ti/TiO_2 nanotubular array electrodes. *J. Adv. Oxid. Technol.*, **11** (3), 425–434.
116. Zhang, Z., Yuan, Y., Liang, L., Cheng, Y., Shi, G., and Jin, L. (2008) Preparation and photoelectrocatalytic activity of ZnO nanorods embedded in highly ordered TiO_2 nanotube arrays electrode for azo dye degradation. *J. Hazard. Mater.*, **158** (2–3), 517–522.
117. Su, Y., Han, S., Zhang, X., Chen, X., and Lei, L. (2008) Preparation and visible-light-driven photoelectrocatalytic properties of boron-doped TiO_2 nanotubes. *Mater. Chem. Phys.*, **110** (2–3), 239–246.
118. Zhou, Z., Zhu, L., Li, J., and Tang, H. (2009) Electrochemical preparation of TiO_2/SiO_2 composite film and its high activity toward the photoelectrocatalytic degradation of methyl orange. *J. Appl. Electrochem.*, **39** (10), 1745–1753.
119. Heikkila, M., Puukilainen, E., Ritala, M., and Leskela, M. (2009) Effect of thickness of ALD grown TiO_2 films

on photoelectrocatalysis. *J. Photochem. Photobiol., A*, **204** (2–3), 200–208.

120. Fu, J.F., Zhao, Y.Q., Xue, X.D., Li, W.C., and Babatunde, A.O. (2009) Multivariate-parameter optimization of acid blue-7 wastewater treatment by Ti/TiO_2 photoelectrocatalysis via the Box–behnken design. *Desalination*, **243** (1–3), 42–51.

121. Monteiro Paschoal, F.M., Anderson, M.A., and Zanoni, M.V.B. (2009) Simultaneous removal of chromium and leather dye from simulated tannery effuent by photoelectrochemistry. *J. Hazard. Mater.*, **166** (1), 531–537.

122. Osugi, M.E., Rajeshwar, K., Ferraz, E.R.A., de Oliveira, D.P., Araújo, Â.R., and Zanoni, M.V.B. (2009) Comparison of oxidation efficiency of disperse dyes by chemical and photoelectrocatalytic chlorination and removal of mutagenic activity. *Electrochim. Acta*, **54** (7), 2086–2093.

123. Li, J., Wang, J., Huang, L., and Lu, G. (2010) Photoelectrocatalytic degradation of methyl orange over mesoporous film electrodes. *Photochem. Photobiol. Sci.*, **9** (1), 39–46.

124. Wang, Y., Cai, L., Li, Y., Tang, Y., and Xie, C. (2010) Structural and photoelectrocatalytic characteristic of $ZnO/ZnWO_4/WO_3$ nanocomposites with double heterojunctions. *Physica E*, **43** (1), 503–509.

125. Wang, W.-Y., Yang, M.-L., and Ku, Y. (2010) Photoelectrocatalytic decomposition of dye in aqueous solution using nafion as an electrolyte. *Chem. Eng. J.*, **165** (1), 273–280.

126. Zhang, Y., Zhao, G., Lei, Y., Li, P., Li, M., Jin, Y., and Lv, B. (2010) CdS-encapsulated TiO_2 nanotube arrays lidded with ZnO nanorod layers and their photoelectrocatalytic applications. *ChemPhysChem*, **11** (16), 3491–3498.

127. Zhao, X., Liu, H., and Qu, J. (2010) Photoelectrocatalytic degradation of organic contaminant at hybrid BDD-$ZnWO_4$ electrode. *Catal. Commun.*, **12** (2), 76–79.

128. Zhang, A., Zhou, M., Liu, L., Wang, W., Jiao, Y., and Zhou, Q. (2010) A novel photoelectrocatalytic system for organic contaminant degradation on a TiO_2 nanotube (TNT)/Ti electrode. *Electrochim. Acta*, **55** (18), 5091–5099, and refs therein.

129. Zhanga, W., Bai, J., and Fu, J. (2011) Photoelectrocatalytic degradation of methyl orange on porous TiO_2 film electrode in NaCl solution. *Adv. Mater. Res.*, **213**, 15–19.

130. Guaraldo, T.T., Pulcinelli, S.H., and Zanoni, M.V.B. (2011) Influence of particle size on the photoactivity of Ti/TiO_2 thin film electrodes, and enhanced photoelectrocatalytic degradation of indigo carmine dye. *J. Photochem. Photobiol., A*, **217** (1), 259–266.

131. Egerton, T.A., Kosa, S.A.M., and Christensen, P.A. (2006) Photoelectrocatalytic disinfection of E. coli suspensions by iron doped TiO_2. *Phys. Chem. Chem. Phys.*, **8** (3), 398–406.

132. Li, G., An, T., Nie, X., Sheng, G., Zeng, X., Fu, J., Lin, Z., and Zeng, E.Y. (2007) Mutagenicity assessment of produced water during photoelectrocatalytic degradation. *Environ. Toxicol. Chem.*, **26** (3), 416–423.

133. Yu, H., Quan, X., Zhang, Y., Ma, N., Chen, S., and Zhao, H. (2008) Electrochemically assisted photocatalytic inactivation of Escherichia coli under visible light using a $ZnIn_2S_4$ film electrode. *Langmuir*, **24** (14), 7599–7604.

134. Marugán, J., Christensen, P., Egerton, T., and Purnama, H. (2009) Synthesis, characterization and activity of photocatalytic sol–gel TiO_2 powders and electrodes. *Appl. Catal. Environ.*, **89** (1–2), 273–283.

135. Baram, N., Starosvetsky, D., Starosvetsky, J., Epshtein, M., Armon, R., and Ein-Eli, Y. (2009) Enhanced inactivation of E. coli bacteria using immobilized porous TiO_2 photoelectrocatalysis. *Electrochim. Acta*, **54** (12), 3381–3386.

136. Fraga, L.E., Anderson, M.A., Beatriz, M.L.P.M.A., Paschoal, F.M.M., Romao, L.P., and Zanoni, M.V.B. (2009) Evaluation of the photoelectrocatalytic method for oxidizing chloride and simultaneous removal of microcystin toxins in surface waters. *Electrochim. Acta*, **54** (7), 2069–2076.

137. Philippidis, N., Nikolakaki, E., Sotiropoulos, S., and Poulios, I. (2010) Photoelectrocatalytic inactivation of E. coli XL-1 blue colonies in water. *J. Chem. Technol. Biotechnol.*, **85** (8), 1054–1060.
138. Shinde, S.S., Bhosale, C.H., and Rajpure, K.Y. (2011) Photocatalytic activity of sea water using TiO_2 catalyst under solar light. *J. Photochem. Photobiol., B*, **103** (2), 111–117.
139. Li, G., Liu, X., Zhang, H., An, T., Zhang, S., Carroll, A.R., and Zhao, H. (2011) In situ photoelectrocatalytic generation of bactericide for instant inactivation and rapid decomposition of gram-negative bacteria. *J. Catal.*, **277** (1), 88–94.
140. Neumann, B., Brezesinsky, T., Smarsly, B., and Tributsch, H. (2010) Tayloring the photocatalytical activity of anatase TiO_2 thin film electrodes by three-dimensional mesoporosity. *Solid State Phenom.*, **162**, 91–113.
141. Stride, J.A. and Tuong, N.T. (2010) Controlled synthesis of titanium dioxide nanostructures. *Solid State Phenom.*, **162**, 261–294.
142. Solarska, R., Augustynski, J., and Sayama, K. (2006) Viewing nanocrystalline TiO_2 photoelectrodes as three-dimensional electrodes: effect of the electrolyte upon the photocurrent efficiency. *Electrochim. Acta*, **52** (2), 694–703.
143. Solarska, R., Rutkowska, I., and Augustynski, J. (2008) Unusual photoelectrochemical behaviour of nanocrystalline TiO_2 films. *Inorg. Chim. Acta*, **361** (3), 792–797.
144. Sadale, S.B., Chaqour, S.M., Gorochov, O., and Neumann-Spallart, M. (2008) Photoelectrochemical and physical properties of tungsten trioxide films obtained by aerosol pyrolysis. *Mater. Res. Bull.*, **43** (6), 1472–1479.
145. Ofir, A., Dor, S., Grinis, L., Zaban, A., Dittrich, T., and Bisquert, J. (2008) Porosity dependence of electron percolation in nanoporous TiO_2 layers. *J. Chem. Phys.*, **128** (6), 064703, (9 pages).
146. Bartkova, H., Kluson, P., Bartek, L., Drobek, M., Cajthaml, T., and Krysa, J. (2007) Photoelectrochemical and photocatalytic properties of titanium (IV) oxide nanoparticulate layers. *Thin Solid Films*, **515** (24), 8455–8460.
147. Waldner, G. and Krysa, J. (2005) Photocurrents and degradation rates on particulate TiO_2 layers. Effect of layer thickness, concentration of oxidisable substance and illumination direction. *Electrochim. Acta*, **50** (22), 4498–4504.
148. Hitchman, M.L. and Tian, F. (2002) Studies of TiO_2 thin films prepared by chemical vapour deposition for photocatalytic and photoelectrocatalytic degradation of 4-chlorophenol. *J. Electroanal. Chem.*, **538–539** (1), 165–172.
149. Ke, S.C., Wang, T.C., Wong, M.S., and Gopal, N.O. (2006) Low temperature kinetics and energetics of the electron and hole traps in irradiated TiO_2 nanoparticles as revealed by EPR spectroscopy. *J. Phys. Chem. B*, **110** (24), 11628–11634.
150. Bisquert, J., Cahen, D., Hodes, G., Rühle, S., and Zaban, A. (2004) Physical chemical principles of photovoltaic conversion with nanoparticulate, mesoporous dye-sensitized solar sells. *J. Phys. Chem. B*, **108** (24), 8106–8118.
151. Bisquert, J. (2004) Chemical diffusion coefficient of electrons in nanostructured semiconductor electrodes and dye-sensitized solar cells. *J. Phys. Chem. B*, **108** (7), 2323–2332.
152. Jiang, D., Zhao, H., Jia, Z., Cao, J., and John, R. (2001) Photoelectrochemical behaviour of methanol oxidation at nanoporous TiO_2 film electrodes. *J. Photochem. Photobiol., A*, **144** (2–3), 197–204.
153. Jiang, D., Zhao, H., Zhang, S., and John, R. (2003) Characterization of photoelectrocatalytic processes at nanoporous TiO_2 film electrodes: photocatalytic oxidation of glucose. *J. Phys. Chem. B*, **107** (46), 12774–12780.
154. Gan, W.Y., Zhao, H., and Amal, R. (2009) Photoelectrocatalytic activity of mesoporous TiO_2 thin film electrodes. *Appl. Catal. Gen.*, **354** (1–2), 8–16.
155. Grimes, C.A. and Mor, G.K. (2009) *TiO_2 Nanotube Arrays Synthesis, Properties, and Applications*. Springer, Dordrecht.

156. Grimes, C.A. (2007) Synthesis and application of highly ordered arrays of TiO_2 nanotubes. *J. Mater. Chem.*, **17** (15), 1451–1457.
157. Roy, P., Berger, S., and Schmuki, P. (2011) TiO_2 Nanotubes: synthesis and applications. *Angew. Chem. Int. Ed.*, **50** (13), 2904 – 2939.
158. Liang, H., Li, X., and Nowotny, J. (2010) Photocatalytical properties of TiO_2 nanotubes. *Solid State Phenom.*, **162**, 295–328.
159. Zhang, H., Zhao, H., Zhang, S., and Quan, X. (2008) Photoelectrochemical manifestation of photoelectron transport properties of vertically aligned nanotubular TiO_2 photoanodes. *ChemPhysChem*, **9** (1), 117–123.
160. Shankar, K., Basham, J.I., Allam, N.K., Varghese, O.K., Mor, G.K., Feng, X., Paulose, M., Seabold, J.A., Choi, K.-S., and Grimes, C.A. (2009) Recent advances in the use of TiO_2 nanotube and nanowire arrays for oxidative photoelectrochemistry. *J. Phys. Chem. C*, **113** (16), 6327–6359.
161. Xie, Y. (2006) Photoelectrochemical application of nanotubular titania photoanode. *Electrochim. Acta*, **51** (17), 3399–3406.
162. Zhang, A., Zhou, M., Han, L., and Zhou, Q. (2010) Combined potential of three types of TiO_2 nanotube (TNT)/Ti and nanoparticle (TNP)/Ti photoelectrodes. *Appl. Catal. Gen.*, **385** (1–2), 114–122, and references therein.
163. Xie, Y. and Fu, D. (2010) Photoelectrocatalysis reactivity of independent titania nanotubes. *J. Appl. Electrochem.*, **40** (7), 1281–1291.
164. Dale, G.R., Hamilton, J.W.J., Dunlop, P.S.M., and Byrne, J.A. (2009) Electrochemically assisted photocatalysis on anodic titania nanotubes. *Curr. Top. Electrochem.*, **14**, 89–97.
165. Li, M. C. and Shen, J.N. (2006) Photoelectrochemical oxidation behavior of organic substances on TiO_2 thin-film electrodes. *J. Solid State Electrochem.*, **10** (12), 980–986.
166. Pichat, P. (2003) in: M.A. Tarr (Ed.), *Chemical Degradation Methods for Wastes and Pollutants: Environmental and Industrial Applications*. Marcel Dekker, Inc., New York, Basel, 77–119.
167. Khataee, A.R. and Zarei, M. (2011) Photocatalysis of a dye solution using immobilized ZnO nanoparticles combined with photoelectrochemical process. *Desalination*, **273** (2–3), 453–460.
168. Boye, B., Dieng, M.M., and Brillas, E. (2003) Anodic oxidation, electro-Fenton and photoelectro-fenton treatments of 2,4,5-trichlorophenoxyacetic acid. *J. Electroanal. Chem.*, **557** (1), 135–146.
169. Rajkumar, D. and Kim, J.G. (2006) Oxidation of various reactive dyes with in situ electro-generated active chlorine for textile dyeing industry wastewater treatment. *J. Hazard. Mater.*, **136** (2), 203–212.
170. Solarska, R., Santato, C., Jorand-Sartoretti, C., Ulmann, M., and Augustynski, J. (2005) Photoelectrolytic oxidation of organic species at mesoporous tungsten trioxide film electrodes under visible light illumination. *J. Appl. Electrochem.*, **35** (7–8), 715–721.
171. Amadelli, R., Maldotti, A., Molinari, A., Danolov, F.I., and Velichenko, A.B. (2002) Influence of the electrode history and effects of the electrolyte composition and temperature on O_2 evolution at β-PbO_2 anodes in acid media. *J. Electroanal. Chem.*, **534** (1), 1–12.
172. Xin, Y., Liu, H., Han, L., and Zhou,Y. (2011) Comparative study of photocatalytic and photoelectrocatalytic properties of alachlor using different morphology TiO_2/Ti photoelectrodes. *J. Hazard. Mater.*, **192** (3), 1812– 1818.
173. Henderson, M.A. (2011) A surface science perspective on TiO_2 photocatalysis. *Surf. Sci. Rep.*, **66** (6–7), 185–297.
174. Fujishima, A., Zhang, X., Tryk, D.A. (2008) TiO_2 Photocatalysis and related surface phenomena. *Surf. Sci. Rep.*, **63** (12), 515–582.
175. Ismail, A.A. and Bahnemann, D.W. (2011) Mesoporous titania photocatalysts: preparation, characterization and reaction mechanisms. *J. Mater. Chem.*, **21** (32), 11686–11707.

176. Tryba, B. (2008) Increase of the photocatalytic activity of TiO_2 by carbon and iron modifications. *Int. J. Photoenergy*, Article ID 721824, 15 pp.
177. Zaleska A. (2008) Doped-TiO_2: a review. *Recent Patents Eng.*, **2** (3), 157–164.
178. Choi, J., Park, H., and Hoffmann, M.R. (2010) Effects of single metal-ion doping on the visible-light photoreactivity of TiO_2. *J. Phys. Chem. C*, **114** (2), 783–792.
179. Rauf, M.A., Meetani, M.A., and Hisaindee, S. (2011) An overview on the photocatalytic degradation of azo dyes in the presence of TiO_2 doped with selective transition metals. *Desalination*, **276** (1–3), 13–27.
180. Amadelli, R., Samiolo, L., Maldotti, A., Molinari, A., Valigi, M., and Gazzoli, D. (2008) Preparation, characterisation, and photocatalytic behaviour of Co-TiO_2 with visible light response. *Int. J. Photoenergy*, Article ID 853753, 9 pp.
181. Esquivel, K., García, J., Ma, G., Rodríguez, F.J., Vega González, M., Escobar-Alarcón, L., Ortiz-Frade, L., and Godínez, L.A. (2011) Titanium dioxide doped with transition metals ($M_xTi1\text{-}xO_2$,M: Ni, Co): synthesis and characterization for its potential application as photoanode. *J. Nanopart. Res.*, **13** (8), 3313–3325.
182. Wang, P., Cao, M., Ao, Y., Wang, C., Hou, J., and Qian, J. (2011) Investigation on Ce-doped TiO_2-coated BDD composite electrode with high photoelectrocatalytic activity under visible light irradiation. *Electrochem. Commun.*, **13** (12), 1423–1426.
183. Xu, Z. and Yu, J. (2011) Visible-light-induced photoelectrochemical behaviors of Fe-modified TiO_2 nanotube arrays. *Nanoscale*, **3** (8), 3138–3144.
184. Emeline, A.V., Kuznetsov, V.N., Rybchuk, V.K., and Serpone, N. (2008) Visible-light-active titania photocatalysts: the case of N-doped TiO_2s—properties and some fundamental issues. *Int. J. Photoenergy*, **2008**, 19 pp., Article ID 258394, and references therein.
185. Peng, F., Cai, L., Yu, H., Wang, H., and Yang, J. (2008) Synthesis and characterization of substitutional and interstitial nitrogen-doped titanium dioxides with visible light photocatalytic activity. *J. Solid State Chem.*, **181** (1), 130–136.
186. Zhou, X., Peng, F., Wang, H., Yu, H., and Yang, J. (2011) Effect of nitrogen-doping temperature on the structure and photocatalytic activity of the B,N-doped TiO_2. *J. Solid State Chem.*, **184** (1), 134–140.
187. Feng, C., Wang, Y., Jin, Z., Zhang, J., Wu, S.Z., and Zhang, J. (2008), Photoactive centers responsible for visible-light photoactivity of N-doped TiO_2. *New J. Chem.*, **32** (6), 1038–1047.
188. Yan, G., Zhang, M., Hou, J., and Yang, J. (2011) Photoelectrochemical and photocatalytic properties of N + S co-doped TiO_2 nanotube array films under visible light irradiation. *Mater. Chem. Phys.*, **129** (1–2), 553–557.
189. Han, L., Xin, Y., Liu, H., Ma, X., and Tang, G. (2010) Photoelectrocatalytic properties of nitrogen doped TiO_2/Ti photoelectrode prepared by plasma based ion implantation under visible light. *J. Hazard. Mater.*, **175** (1–3), 524–531.
190. Wu, G., Wang, J., Thomas, D.F., and Chen, A. (2008) Synthesis of F-doped flower-like TiO_2 nanostructures with high photoelectrochemical activity. *Langmuir*, **24** (7), 3503–3509.
191. Su, Y., Zhang, X., Han, S., Chen, X., and Lei, L. (2007) F–B-codoping of anodized TiO_2 nanotubes using chemical vapor deposition. *Electrochem. Commun.*, **9** (9), 2291–2298.
192. Zhang, Z., Hossain, M.F., and Takahashi, T. (2010) Self-assembled hematite (α-Fe_2O_3) nanotube arrays for photoelectrocatalytic degradation of azo dye under simulated solar light irradiation. *Appl. Catal. Environ.*, **95** (3–4), 423–429.
193. Waldner, G., Brüger, A., Gaikwad, N.S., and Neumann-Spallart, M. (2007) WO_3 Thin films for photoelectrochemical purification of water. *Chemosphere*, **67** (4), 779–784.
194. Hepel, M. and Hazelton, S. (2005) Photoelectrocatalytic degradation of diazo dyes on nanostructured

WO_3 electrodes. *Electrochim. Acta*, **50** (25–26), 5278–5291.

195. Xu, Z., Bai, X., Wei, M., Tong, Y., Gao, Y., Chen, J., Zhou, J., and Wang, Z.L. (2011) Three-dimensional WO_3 nanostructures on carbon paper: photoelectrochemical property and visible light driven photocatalysis. *Chem. Commun.*, **47** (20), 5804–5806.

196. Sawyer, D.T. (1991) *Oxygen Chemistry*, Oxford University Press Inc., New York.

197. Saison, T., Chemin, N., Chaneac, C., Durupty, O., Ruaux, V., Mariey, L., Mauge, F., Beaunier, P., and Joliver, J.P. (2011) Bi_2O_3, $BiVO_4$, and Bi_2WO_6: impact of surface properties on photocatalytic activity under visible light. *J. Phys. Chem. C*, **115** (13), 5657–5666.

198. Castillo, N., Ding, L., Heel, A., Graule, T., and Pulgarin, C. (2010) On the photocatalytic degradation of phenol and dichloroacetate by $BiVO_4$. *J. Photochem. Photobiol., A*, **216** (2–3), 221–227.

199. Zhang, X., Chen, S., Quan, X., and Zhao, H. (2009) Preparation and characterization of $BiVO_4$ film electrode and investigation of its photoelectrocatalytic (PEC) ability under visible light. *Sep. Purif. Technol.*, **64** (3), 309–313.

200. Zhanga, X., Quana, X., Chena, S., and Zhang, Y. (2010) Effect of Si doping on photoelectrocatalytic decomposition of phenol of $BiVO_4$ film under visible light. *J. Hazard. Mater.*, **177** (1–3), 914–917.

201. Zhou, B., Qu, J., Zhao, X., and Liu, H. (2011) Fabrication and photoelectrocatalytic properties of nanocrystalline monoclinic $BiVO_4$ thin-film electrode. *J. Environ. Sci.*, **23** (1), 151–159.

202. Allongue, P. (1992) in *Modern Aspects of Electrochemistry*, Vol. 23 (eds. B.E. Conway, J.O. Bockris, and R.E. White), Plenum Press, New York, pp. 239–314.

203. He, C., Li, X.Z., Graham, N., and Xiong, Y. (2005) Photoelectrocatalytic degradation of bisphenol a in aqueous solution using a Au-TiO_2/ITO film. *J. Appl. Electrochem.*, **35** (7–8), 741–750.

204. Sun, B. and Smirniotis, P.G. (2003) Interaction of anatase and rutile TiO_2 particles in aqueous photooxidation. *Catal. Today*, **88** (1–2), 249–259.

205. Li, G., Chen, L., Graham, M.E., and Gray, K.A. (2007) A comparison of mixed phase titania photocatalysts prepared by physical and chemical methods: the importance of the solid-solid interface. *J. Mol. Catal. A: Chem.*, **275** (1–2), 30–35.

206. Georgieva, J., Valova, E., Armyanov, S., Philippidis, N., Poulios, I., and Sotiropoulos, S. (2012) Bi-component semiconductor oxide photoanodes for the photoelectrocatalytic oxidation of organic solutes and vapours: a short review with emphasis to TiO_2–WO_3 photoanodes. *J. Hazard. Mater.*, **211-212**, 30–46.

207. Somasundaram, S., Tacconi, N., Chenthamarakshan, C.R., Rajeshwar, K., and de Tacconi, N.R. (2005) Photoelectrochemical behavior of composite metal oxide semiconductor films with a WO_3 matrix and occluded Degussa P 25 TiO_2 particles. *J. Electroanal. Chem.*, **577** (1), 167–177.

208. Higashimoto, S., Ushiroda, Y., and Azuma, M. (2008) Electrochemically assisted photocatalysis of hybrid WO_3/TiO_2 films: effect of the WO_3 structures on charge separation behavior. *Top. Catal.*, **47** (3–4), 148–154.

209. Vinodgopal, K. and Kamat, P.V. (1995) Electrochemically assisted photocatalysis using nanocrystalline semiconductor thin films. *Sol. Energy Mater. Sol. Cells*, **38** (1–4), 401–410.

210. Chen, L.-C., Tsai, F.-R., Fang, S.-H., and Ho, Y.-C. (2009) Properties of sol–gel SnO_2/TiO_2 electrodes and their photoelectrocatalytic activities under UV and visible light illumination. *Electrochim. Acta*, **54** (4), 1304–1311.

211. Li, P., Zhao, G., Li, M., Cao, T., Cui, X., and Li, D. (2012) Design and high efficient photoelectric-synergistic catalytic oxidation activity of 2D macroporous SnO_2/1D TiO_2 nanotubes. *Appl. Catal. Environ.*, **111–112**, 578–585.

212. Fan, C.M., Hua, B., Wang, Y., Liang, Z.H., Hao, X.G., Liu, S.B., and Sun, Y.P. (2009) Preparation of Ti/SnO_2–Sb_2O_4 photoanode by electrodeposition and dip coating for PEC oxidations. *Desalination*, **249** (2), 736–741.

213. Park, H., Bak, A., Ahn, Y.Y., Choi, J., Hoffmannn, M.R. Photoelectrochemical performance of multi-layered BiO_x–TiO_2/Ti electrodes for degradation of phenol and production of molecular hydrogen in water. *J. Hazard. Mater.*, doi: 10.1016/j.jhazmat.2011.05.009 (in press).
214. Amadelli, R., Samiolo, L., Velichenko, A.B., Knysh, V.A., Luk'yanenko, T.V., and Danilov, F.I. (2009) Composite PbO_2–TiO_2 materials deposited from colloidal electrolyte: electrosynthesis, and physicochemical properties. *Electrochim. Acta*, **54** (22), 5239–5245.
215. Nozik, A.J. (1977) Photochemical diodes. *Appl. Phys. Lett.*, **30** (11), 567–569.
216. Fornarini, L., Nozik, A.J., and Parkinson, B.A. (1984) The energetics of p/n photoelectrolysis cells. *J. Phys. Chem.*, **88** (15), 3238–3243.
217. Zhang, Y.-G., Ma, L.-L., Li, J.-L., and Yu, Y. (2007) In situ Fenton reagent generated from TiO_2/Cu_2O composite film: a new way to utilize TiO_2 under visible light irradiation. *Environ. Sci. Technol.*, **41** (17), 6264–6269, and references therein.
218. Hou, Y., Li, X., Quan, X., and Chen, G. (2009) Photoelectrocatalytic activity of a Cu_2O-loaded self-organized highly oriented TiO_2 nanotube array electrode for 4-chlorophenol degradation. *Environ. Sci. Technol.*, **43** (3), 858–863.
219. Yang, L., Luo, S., Li, Y., Xiao, Y., Kang, Q., and Cai, Q. (2010) High efficient photocatalytic degradation of p-nitrophenol on a unique Cu_2O/TiO_2 p-n heterojunction network catalyst. *Environ. Sci. Technol.*, **44** (19), 7641–7646.
220. Lin, J., Shen, J., Wang, R., Cui, J., Zhou, W., Hu, P., Liu, D., Liu, H., Wang, J., Boughton, R.I., and Yue, Y. (2011) Nano-p–n junctions on surface-coarsened TiO_2 nanobelts with enhanced photocatalytic activity. *J. Mater. Chem.*, **21** (13), 5106–5113.
221. Chen, C.-J., Liao, C.-H., Hsu, K.-C., Wu, Y.-T., and Wu, J.C.S. (2011) P–N junction mechanism on improved NiO/TiO_2 photocatalyst. *Catal. Commun.*, **12** (14), 1307–1310.
222. Chen, S., Zhao, W., Liu, W., and Zhang, S. (2008) Preparation, characterization and activity evaluation of p–n junction photocatalyst p-ZnO/n-TiO_2. *Appl. Surf. Sci.*, **255** (5), 2478–2484.
223. Chen, S., Zhao, W., Liu, W., Zhang, H., Yu, X., and Chen, Y. (2009) Preparation, characterization and activity evaluation of p–n junction photocatalyst p-$CaFe_2O_4$/n-Ag_3VO_4 under visible light irradiation. *J. Hazard. Mater.*, **172** (2–3), 1415–1423 and references therein.
224. Liu, R., Liu, Y., Liu, C., Luo, S., Teng, Y., Yang, L., Yang, R., and Cai, Q. (2011) Enhanced photoelectrocatalytic degradation of 2,4-dichlorophenoxyacetic acid by $CuInS_2$ nanoparticles deposition onto TiO_2 nanotube arrays. *J. Alloys Compd.*, **509** (5), 2434–2440.
225. Dai, G., Yu, J., and Liu, G. (2011) Synthesis and enhanced visible-light photoelectrocatalytic activity of $p-n$ junction BiOI/TiO_2 nanotube arrays. *J. Phys. Chem. C*, **115** (15), 7339–7346.
226. Chen, S., Zhao, W., Liu, W., Zhang, H., and Yu, X. (2009) Preparation, characterization and activity evaluation of p–n junction photocatalyst p-$CaFe_2O_4$/n-ZnO. *Chem. Eng. J.*, **155** (1–2), 466–473, and references therein.
227. Kim, H.G., Borse, P.H., Jang, J.S., Jeong, E.D., Jung, O.-S., Suhd, Y.J., and Lee, J.S. (2009) Fabrication of $CaFe_2O_4$/$MgFe_2O_4$ bulk heterojunction for enhanced visible light photocatalysis. *Chem. Commun.*, **2009** (39), 5889–5891.
228. Bandara, J., Divarathne, C.M., and Nanayakkara, S.D. (2004) Fabrication of n–p junction electrodes made of n-type SnO_2 and p-type NiO for control of charge recombination in dye-sensitized solar cells. *Sol. Energy Mater. Sol. Cells*, **81** (4), 429–437.
229. Kostecki, R., Richardson, T., and McLarnon, F. (1998) Photochemical and photoelectrochemical behavior of a novel TiO_2/$Ni(OH)_2$ electrode. *J. Electrochem. Soc.*, **145** (7), 2380–2385.

230. Takahashi, Y. and Tatsuma, T. (2005) Oxidative energy storage ability of a TiO_2-$Ni(OH)_2$ bilayer photocatalyst. *Langmuir*, **21** (26), 12357–12361, and references therein.

231. Zhao, X., Qu, J., Liu, H., Qiang, Z., Liu, R., and Hu, C. (2009) Photoelectrochemical degradation of anti-inflammatory pharmaceuticals at Bi_2MoO_6–boron-doped diamond hybrid electrode under visible light irradiation. *Appl. Catal. Environ.*, **91** (1–2), 539–545.

232. Wu, K. and Zhitomirsky, I. (2011) Electrophoretic deposition of ceramic nanoparticles. *Int. J. Appl. Ceram. Technol.*, **8** (4), 920–927.

233. Cao, G.Z. (2004) Growth of oxide nanorod arrays through sol electrophoretic deposition. *J. Phys. Chem. B*, **108** (52), 19921–19931.

234. Chenthamarakshan, C.R., de Tacconi, N.R., Shiratsuchi, R., and Rajeshwar, K. (2003) Tungsten trioxide-titanium dioxide composite films prepared by occlusion electrosynthesis in a nickel matrix. *J. Electroanal. Chem.*, **553**, 77–85.

235. Shemer, G. and Paz, Y. (2011) Interdigitated electrophotocatalytic cell for water purification. *Int. J. Photoenergy*, **2011**, 7 pp. Article ID 596710.

236. Neumann-Spallart, M. (2011) Photoelectrochemistry on a planar, interdigitated electrochemical cell. *Electrochim. Acta*, **56** (24), 8752–8757.

237. Chen, X. and Mao, S.S. (2007) Titanium dioxide nanomaterials: synthesis, properties, modifications, and applications. *Chem. Rev.*, **107** (7), 2891–2959.

238. Mohammadi, M.R., Fray, D.J., and Mohammadi, A. (2008) Sol–gel nanostructured titanium Dioxide: controlling the crystal structure, crystallite size, phase transformation, packing and ordering. *Microporous Mesoporous Mater.*, **112** (1–3), 392–402.

239. Huang, C.-H., Yang, Y.-T., and Doong, R.-A. (2011) Microwave-assisted hydrothermal synthesis of mesoporous anatase TiO_2 via sol–gel process for dye-sensitized solar cells. *Microporous Mesoporous Mater.*, **142** (2–3), 473–480.

240. Vinu, A., Mori, T., and Ariga, K. (2006) New families of mesoporous materials. *Sci. Technol. Adv. Mater.*, **7** (8), 753–771.

241. Brinker, C.J., Hurd, A.J., Schunk, P.R., Frye, G.C., and Ashley, C.S. (1992) Review of sol–gel thin film formation. *J. Non-Cryst. Solids*, **147–148**, 424–436.

242. Bradley, D.C., Mehotra, R.C., Rothwell, I.P., and Singh, A. (2001) *Alkoxo and Aryloxo Derivatives of Metals*, Academic Press, San Diego, CA.

243. Huang, Y., Zheng, H., Ball, I., and Luo, Z. (2001) Advances in Sol–Gel Technology. Ceramic Industry Magazine (Dec. 2001).

244. Fernandez-Ibañez, P., Malato, S., and Enea, O. (1999) Photoelectrochemical reactors for the solar decontamination of water. *Catal. Today*, **54** (2–3), 329–339.

245. Li, K., Yang, C., Wang, Y.-L., Jia, J.-P., Xu, Y.-L., and He, Y. (2012) A high-efficient rotating disk photoelectrocatalytic (PEC) reactor with macro light harvesting pyramid-surface electrode. *AIChE J.*, **58** (8), 2448–2455, and references therein.

246. Lianos, P. (2011) Production of electricity and hydrogen by photocatalytic degradation of organic wastes in a photoelectrochemical cell: the concept of the photofuelcell: a review of re-emerging research field. *J. Hazard. Mater.*, **185** (2–3), 575–590.

247. Liu, Y., Li, J., Zhou, B., Lv, S., Li, X., Chen, H., Chen, Q., and Cai, W. (2012) Photoelectrocatalytic degradation of refractory organic compounds enhanced by a photocatalytic fuel cell. *Appl. Catal. Environ.*, **111** (1), 485–491.

248. Macphee, D., Wells, R., Kruth, A., Todd, M., Elmorsi, T., Smith, C., Pokrajac, D., Strachan, N., Mwinyhija, M., Scott-Emuakpor, E., Nissen, S., and Killham K. (2010) A visible light driven photoelectrocatalytic fuel cell for clean-up of contaminated water supplies. *Desalination*, **251** (1–3), 132–137.

Part III
Effects of Photocatalysis on Natural Organic Matter and Bacteria

10
Photocatalysis of Natural Organic Matter in Water: Characterization and Treatment Integration

Sanly Liu, May Lim, and Rose Amal

10.1
Introduction

Natural organic matter (NOM) is a collective term assigned to a broad group of organic materials that is produced from the decomposition of living materials and synthetic activities of microorganisms [1, 2]. NOM consists of an extraordinarily complex mixture of compounds; the composition and properties of which may vary with source (origin), age, fate, and season [3]. Organic compounds which constitute NOM include aquatic humic substances, carboxylic acids, phenols, amino acids, proteins, hydrocarbons, carbohydrates, and trace organic compounds. The physical and chemical properties of NOM are of great interest to scientists and engineers because of its prevalence in the environment, its many ecosystem functions, and its impact on engineering processes such as the operation of water treatment plant.

NOM can impart color to water and therefore reduce the aesthetic qualities of drinking water. It also serves as a substrate for microbial growth in the distribution system [4] and is able to form stable salts and complexes with metal ions [5, 6]. NOM can exert significant oxidant demand, thereby increasing the disinfectant dose required during drinking water treatment. NOM also interferes with water treatment processes; it causes membrane fouling and competes with other pollutants for activated carbon adsorption sites [7, 8]. Thus, effective removal of NOM is of considerable interest to those concerned with drinking water quality.

Furthermore, NOM needs to be eliminated before the chlorine disinfection process in a drinking water treatment plant. The reaction of NOM in the presence of chlorine will produce a variety of chlorinated disinfection by-products (DBPs), such as trihalomethanes (THMs), haloacetic acids (HAAs), haloacetonitriles, and haloketones [9], some of which are potentially hazardous to human health [10]. The trend in regulation is to set lower limits for permissible DBPs levels and to expand the number of DBPs controlled. This in turn drives the search for alternative disinfectants or advanced treatment methods for precursor (organic matter) removal. However, the use of alternative disinfectants can still potentially generate a new set of DBPs [11] and as such the effective removal of organic matter, which serves as DBP precursors, continues to be the preferred control strategy.

Photocatalysis and Water Purification: From Fundamentals to Recent Applications, First Edition. P. Pichat.

The treatment of water for potable use has traditionally focused on the removal of NOM by conventional treatment methods, such as coagulation and later by advanced treatment process, such as membrane filtration, ion exchange/adsorption, and ozonation/biodegradation. Recently, extensive research has focused on the TiO_2 mediated photocatalytic oxidation of NOM and humic acids (HAs) as model compounds of NOM. The initial work in this field was carried out by Bekbolet and Ozkosemen [12], who studied the effectiveness of photocatalytic treatment on the degradation of HA. The oxidation of NOM via photocatalysis is often investigated as it has the potential to mineralize the organic matter instead of simply concentrating or transferring the organic material from one phase to another, as is the case for conventional treatment methods, such as coagulation, ion exchange, or membrane filtration. Photocatalysis has been envisaged to obtain potable water in remote places by use of portable water treatment units.

The key to the successful removal of NOM is the in-depth characterization of its physical and chemical properties, allowing a better understanding of the removal mechanism. A combination of synergistic treatments can then be applied to optimize NOM removal. Nevertheless, despite several decades of extensive functional group and elemental analysis research, the exact chemical structure of NOM remains elusive and not well defined [13]. This is due to the large number of component molecules combined with the numerous types of linkages that bind these molecules together.

This article reviews some of the most widely used characterization methods to monitor the photocatalytic oxidation of NOM, including changes to the NOM structure after treatment. Data on the by-products of NOM photocatalytic degradation as well as its resultant DBPs are examined. Furthermore, a critical review is made on the issues related to the large scale application of TiO_2 photocatalysis for drinking water purification and the need to combine photocatalysis with other water treatment technologies. Specific examples on the integration of photocatalysis with other treatment methods are also briefly discussed.

10.2
Monitoring Techniques

At present, there is no analytical method for the unique identification of NOM due to its complex and heterogeneous nature. Instead, numerous analytical methods and combinations of approaches are needed to obtain meaningful information about the chemical and structural properties responsible for the color, solubility, and other physical and chemical properties of NOM. The following section discusses several characterization methods that have been used widely in the literature and/or practice to monitor the transformation of NOM and humic substances as a result of the photocatalytic oxidation process.

Until recently, the removal of NOM and HAs following photocatalytic oxidation was usually quantified in terms of organic carbon measurements and UV absorbance (typically at 254 nm). These two analyses have been widely applied as the procedures are cheap, quick, and simple, requiring neither intensive labor nor sophisticated equipment.

10.2.1 Total Organic Carbon

Total organic carbon (TOC) is defined as the carbon bound in a variety of organic compounds in water and wastewater. A related definition of organic carbon is dissolved organic carbon (DOC), which is the fraction of TOC that passes through a 0.45 μm pore diameter filter. Organic carbon concentration is used as a composite parameter in evaluating the efficiency of water treatment processes and the quantity of the organic matter. Although organic carbon analysis is adequate for a rapid assessment of the efficiency of the photocatalytic oxidation, they do not elucidate the chemistry of the process. In addition, the total organic content analysis does not provide any information about the removal of individual contaminants. However, removing the TOC or DOC from a drinking water supply reduces the concentration of potentially hazardous unidentified organic compounds. Organic carbon concentration measurement is also particularly useful in monitoring the extent of mineralization of NOM in oxidative treatment processes. Even after extended photocatalytic treatment, complete mineralization of NOM cannot be achieved [14, 15]. This might be due to the presence of refractory compounds in the original water sample and/or by-products of oxidation.

10.2.2 UV–vis Spectroscopy

NOM absorbs strongly in the UV region because of the presence of multiple bonds and unshared electron pairs in the organic molecules. Aromatic groups or molecules with conjugated double bonds, in particular, are deemed to be the major chromophores that absorb light in the UV region. The typical UV–vis absorption spectrum of NOM has the general trend of decreasing absorbance as the wavelength increases. The spectrum is featureless in the sense that there is no well-defined maxima or minima because of the overlaps in the absorption bands of various chromophoric groups [16]. As a result, many researchers have limited the monitoring of the absorbance at a single wavelength to provide an indication of water quality. Absorption wavelengths at 250, 254, 270, 280, 300, 365, 400, 436, and 465 nm, and absorption ratios such as Abs250/Abs365 and Abs465/Abs665 have been used to characterize NOM [16].

The UV absorbance at 254 nm (UV_{254}), in particular, is widely used as a surrogate parameter for DOC content of water samples, while absorbance measurements at 460 nm are generally used as an indicator of color. Also, UV absorbance is not only proportional to the concentration but also to the molar absorptivities of the chromophores that absorb light at the detection wavelength.

Liu *et al.* [17] monitored the changes in the absorbance spectrum of natural water samples during UV/TiO_2 treatment with time. With increasing reaction time, an overall decrease in UV absorbance was observed in most cases, particularly at wavelengths greater than 240 nm. At the end of 150 min treatment, the residual organic matter only absorbed at wavelengths less than 250 nm.

In photocatalysis treatment, the UV_{254} values usually decrease at a faster rate compared to the DOC concentration, which suggests that the chromophores in NOM are rapidly degraded by the photocatalytic oxidation process into intermediate products that did not absorb UV as strongly. These intermediate products then undergo a series of slower complex reactions before mineralization [18]. It can be further deduced that aromatic and olefinic moieties in the NOM structure are more strongly affected by photocatalytic degradation processes than aliphatic structures.

Another parameter called specific UV absorbance ($SUVA_{254}$) at 254 nm, which is the ratio of the UV absorbance at 254 nm to the DOC content, has been used as a measure for the apparent molecular weight (MW) and the aromaticity of NOM [16]. SUVA is analogous to molar absorptivity, as it provides an "average" molar absorptivity for all the molecules contributing to the DOC. As a general rule, SUVA values of $\geq 4\ m^{-1}(mg\ l^{-1})^{-1}$ indicate that the DOC is largely composed of humic substances, whereas SUVA values of less than $3\ m^{-1}(mg\ l^{-1})^{-1}$ imply that the organic matter is of lower MW, poor in aromatic content, and more hydrophilic [19]. Weishaar *et al.* [20] reported a strong correlation between $SUVA_{254}$ and the aromaticity of the NOM, as determined by ^{13}C nuclear magnetic resonance (^{13}C NMR), for a large number of aquatic organic matter isolates.

SUVA has also been found to be a good indicator to the potential of DBPs formation [21, 22]. Significant reduction in SUVA after dark adsorption on TiO_2 has been reported by Wei *et al.* [23], indicating the preferred adsorption of hydrophobic humic substances on TiO_2. When the light was turned on, the SUVA values of the HA solution continuously decreased with reaction time. Uyguner and Bekbolet [16] also reported significant decrease of SUVA after 60 min of photocatalytic degradation compared to that of raw HA, which implies a reduction in the degree of aromaticity in relation to the removal of organic carbon.

Although UV_{254} has been widely used as a potential surrogate measure for DOC, it is important to exercise caution and not use UV_{254} as a sole parameter to measure the efficiency of the photocatalytic oxidation process, or any oxidation process for that matter. A key limitation of UV spectroscopy for NOM characterization is that it only detects compounds of NOM that absorb UV light at that specified wavelength, and therefore might underestimate the total amount of organic matter. In photocatalysis treatment, oxidation changes the structure of NOM, and thereby reduces the UV_{254} absorbance. However, it does not mean that the dissolved organic matter has been removed. It can only be deduced that the intermediate oxidation products do not absorb significantly at 254 nm. In most cases, significant amount of organic carbon still remained after extended photocatalysis.

The determination of UV absorbance of NOM can also be subjected to interferences caused by turbidity of the sample (in the form of suspended solids) in the solution that can both scatter and absorb the light. In addition, inorganic species (typically present in natural fresh waters) that absorb light in the near-UV region up to 230 nm, such as nitrate, can interfere with the UV absorbance of NOM [24]. As there is no interaction between nitrate and the organic matter [20], the contribution of nitrate toward the UV absorbance could theoretically be subtracted, leaving only the UV absorbance of NOM.

10.2.3
Fluorescence Spectroscopy

Fluorescence is a phenomenon of light emission by certain molecules during their energy release from the excitation state to the ground state. The fluorescent intensity of NOM, similar to that of many organic compounds, can be related to its molecular structures [3] and environmental factors such as pH, temperature, the presence of electrolytes, and quenching by chelation with metal ions or other organic molecules [25]. The fluorescence bands of NOM are broad and rather featureless and the intensity generally decreases with increasing molecular size [26, 27]. In aromatic compounds, the presence of electron withdrawing groups such as COOH decreases the fluorescence intensity, while the presence of electron donating groups, such as –OH and $-NH_2$ increases the fluorescence intensity [27].

The traditional approach for obtaining fluorescence spectra of NOM is fluorescence emission spectroscopy, where emission is scanned over a range of wavelengths for a fixed excitation wavelength [28]. Further development resulted in the application of the synchronous fluorescence scanning (SFS) technique, which scans both the excitation and emission wavelengths, with emission wavelength usually chosen at an offset of generally 12–60 nm from the excitation wavelength [29, 30]. SFS proved useful in displaying the similarities and differences in the composition of NOM from different sources [31]. In recent days, technological advances have allowed the introduction of the excitation emission matrix (EEM) fluorescence spectroscopy in the study of NOM characteristics [32]. The principal of EEMS is that excitation, emission, and fluorescence intensity can be scanned over a range of wavelengths synchronously, generating a 3D plot of fluorescence excitation wavelength, emission wavelength, and intensity. The interpretation of a fluorescence EEM is, however, extremely challenging because of the thousands of wavelength-dependent intensities [33].

Eggins *et al.* [18] used fluorescence spectroscopy to evaluate the photocatalytic oxidation of HA. The fluorescent emission at 400 nm was found to initially increase in intensity before rapidly decreasing. In addition, they found that the rapid decrease in the chromophores absorbing at 254 and 400 nm could be correlated to the corresponding increase in the fluorescent components. Cho and Choi [34] reported the drastic quenching of fluorescence intensity of HA within 1 h irradiation. They postulated that as the fluorescence is largely due to $\pi^* \rightarrow \pi$ transitions in HA molecules, its rapid extinction should be ascribed to the destruction of aromatic structures of HA. Uyguner and Bekbolet [16] monitored the changes in the synchronous scan excitation spectra of HAs during photocatalytic oxidation. A gradual decrease of fluorescence intensity in the 450–600 nm wavelength region was reported, which was attributed to the degradation of the high MW components and formation of lower molecular size fractions during photocatalysis. Furthermore, the characteristic peak of HA was significantly quenched even prior to irradiation, indicating that fluorescent moieties in HA readily adsorb onto the TiO_2 surface.

Wei *et al.* [23] reported a decrease in peak intensity of fluorescence EEM spectra following dark adsorption with no change in the peak location, which indicates that there was no formation of new compounds and that TiO_2 adsorption did not change the chemical structure of HA. After photocatalysis, the peak that matched the excitation and emission wavelengths reported for HA-like fluorescence decreased significantly, while a new peak, which has been reported as a protein-like peak, appeared. This indicates that the by-product of HA oxidation has a protein-like structure.

10.2.4 Molecular Size Fractionation

Recent reports have focused on fractionating NOM based on molecular size to get an insight into the changes in the molecular size of the organics during the photocatalytic degradation process. Molecular size distribution characterization of NOM is valuable for the optimization of water treatment because some treatment processes preferentially remove organic molecules of a certain MW range.

High-performance size exclusion chromatography (HPSEC), an analytical technique to obtain the size distribution of NOM, has found wide application in water treatment optimization, as it is a rapid analytical method, is relatively inexpensive, and requires a small sample volume with minimal pretreatment. Moreover, it can measure the whole distribution of MW, rather than only the average MW. HPSEC is a separation technique based on molecular size by elution through beds of porous beads. Molecules that are too large to penetrate the pores of the beads are excluded and passed through the column with the solvent, while molecules that penetrate the beads are temporarily retained, and thus are separated from the larger molecules [35].

Several detection methods are available for the determination of MW distribution by HPSEC. The most commonly used detection method in HPSEC characterization is UV–vis spectrophotometry. However, as NOM moieties possess no obvious maximum absorption at the wavelength range in the UV–vis spectrum, HPSEC chromatograms with UV absorbance detection were usually reported at one particular wavelength in the range of 210–280 nm. UV–vis detectors are rather selective; they only detect analytes that absorb at the operating wavelength. Other detectors that have been used with HPSEC include DOC [36], refractive index [37], fluorescence [38], mass spectrometry [39], and multiangle laser light scattering [40]. Different detectors may yield different MW distributions, because each detector measures different properties of NOM molecules. Multiple detectors can be integrated to the HPSEC system, which permits a wider characterization of size, structure, and functionality, and provides insight into NOM type and source [41].

HPSEC has been used to investigate the adsorption extent of different DOC fractions onto the TiO_2 surface. Most of the adsorbed DOCs consist of larger MW fraction, with very little material adsorbed from the small MW fraction [42]. Murray *et al.* [43] postulated that lower MW compounds could not be readily adsorbed onto

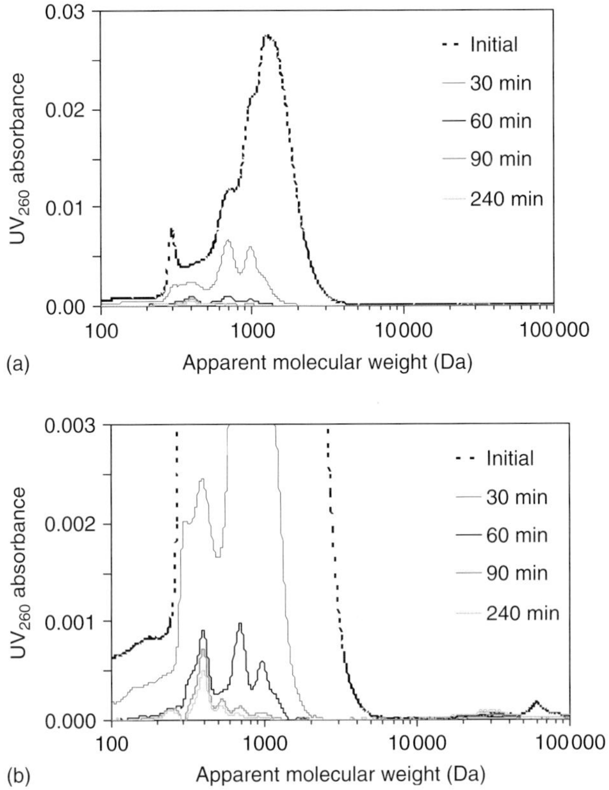

Figure 10.1 (a) HPSEC chromatograms of water samples treated using UVA/TiO_2 pH 7 after various periods of irradiation [15]. (b) Enlarged image of (a).

the TiO_2 surface as they lacked enough reactive functional groups. Liu *et al.* [15] showed that NOM with apparent MWs of 1–4 kDa were preferentially degraded, forming lower MW organic compounds (Figure 10.1). As the reaction proceeded, the NOM MW distribution continued shifting toward a lower MW. Doll and Frimmel [42] reported that new fractions of degraded products appeared at the typical retention time of aliphatic di- and monocarboxylic acids, which indicates the formation of low-MW acids after photocatalytic oxidation. Other authors have reported the formation of new peaks, which is direct evidence that large NOM compounds were decomposed into smaller molecules by the TiO_2 photocatalytic process [44, 45]. Liu *et al.* [17] showed that after photocatalysis, the organic molecules were transformed into compounds that absorb weakly at the typical detection wavelength of 250–260 nm, using HPSEC analysis with multiwavelength UV detection. In addition, the multiwavelength HPSEC results also revealed that photocatalytic oxidation yields by-products with a low aromaticity and low MW. In the same work, Liu *et al.* [17] identified low-MW acids and neutral compounds to be the residual organics after photocatalytic oxidation from HPSEC with the DOC detector.

10.2.5
Resin Fractionation

The fractionation approach is useful for evaluating the reactivity and chemical properties of NOM, particularly with respect to reactions in the water treatment processes or in the formation of DBPs. It is, however, important to note that the fractions obtained are operationally defined, rather than structurally defined.

More recently, advancements in chromatographic separation techniques has allowed NOM to be separated into discrete fractions based on hydrophobicity and charge. NOM fractionation with resin can be categorized into two major groups: analytical and preparative fractionation procedure. While analytical fractionation only aims to obtain the quantitative distribution of NOM in all the fractions, the preparative fractionation procedure attempts to collect enough fractions for further characterization and reactivity studies.

Leenheer and Huffman [46] were the first to report the analytical fractionation procedure for aquatic organic carbon into six fractions based on its acid/base/neutral and hydrophobic/hydrophilic characteristics. The analytical fractionation procedure was not satisfactory as a preparative procedure because of the irreversible adsorption of hydrophilic acids on the strong base anion exchange resin. Therefore, a preparative DOC fractionation procedure was then developed by Leenheer [47]. The fractionation procedure involved passing a NOM sample through a series of resin adsorbents (nonionic Amberlite XAD-8 resin, Bio-Rad AG-MP 50 cation exchange resin, and Duolite A-7 anion exchange resin) followed by back elution of the resins with chemicals to obtain the fractions. Other workers have developed various methods of fractionation using a different series of resins and/or elution procedures, which also led to different names or definitions for each isolated fraction [48–53].

Liu *et al.* [15] determined the hydrophobicity/hydrophilicity of the intermediates and by-products of TiO_2 photocatalytic oxidation of NOM with the resin fractionation technique. They found that oxidation changed the affinity of the bulk organic character from predominantly hydrophobic to more hydrophilic. The transformation in polarity of the organic compound after photocatalytic oxidation would affect the downstream treatment processes, as most conventional water treatment methods remove the hydrophobic fraction preferentially. The hydrophilic organics are more likely to be removed by biodegradation [54] and therefore might promote microbiological regrowth in the distribution system.

10.2.6
Infrared Spectroscopy

IR spectra of NOM contain a variety of bands that are diagnostic of functional groups. It is especially useful for the study of functional groups that form polar bonds, but the technique cannot provide useful information concerning the structural units to which such functional groups are attached. Fourier transform infrared (FTIR) has a number of advantages over the traditional infrared techniques, such as greater sensitivity through increased energy throughput and a higher signal to noise ratio [55]. However, as samples need to be mounted on a potassium bromide (KBr) pellet, with hygroscopic properties, it is difficult to eliminate the interference of water bands in the IR spectra. An alternative technology which overcomes the problem of the KBr pellet is diffuse reflectance infrared Fourier transform spectroscopy (DRIFTS) [55].

FTIR spectroscopy has been used to provide specific information on the types of molecular chemical bonding of NOM functional groups with TiO_2 surface. Wiszniowski *et al.* [56] studied the adsorption of HAs at three different pHs using FTIR DRIFTS technique and found that at an acidic pH when TiO_2 is positively charged, HAs are adsorbed on TiO_2 mainly by carboxylate groups.

10.3
By-products from the Photocatalytic Oxidation of NOM and its Resultant Disinfection By-Products (DBPs)

One of the major concerns in any advanced oxidation process is the formation of undesirable chemical products as reaction intermediates. It is not unusual to observe several degradation intermediates with toxicity greater than the parent compound. Although a number of specific by-products of NOM oxidation with ozone or UV have been reported, little is known about the by-products formed during photocatalytic oxidation. The characterization of such by-products is a prerequisite for the application of this technology in the field of drinking water treatment. Obviously, differences in the raw water DOC composition may result in different by-products following photocatalytic oxidation and also reactivity with chlorine disinfectant.

Liu *et al.* [15] reported that aldehydes and ketones were typical intermediate products formed during photocatalytic oxidation of NOM in surface waters and these compounds would be oxidized to form carboxylic acids as the reaction proceeded. Areerachakul *et al.* [14] confirmed the cleavage of large MW humic substances by solid-phase microextraction/gas chromatography with a flame ionization detector (SPME/GC FID). However, due to the complex nature of the by-products, GC FID did not allow identification of individual oxidation intermediates of the humic substances. Earlier, Wiszniowski *et al.* [56] followed the formation and disappearance of the by-products of photocatalytic oxidation using high-performance liquid chromatography with a UV detector. Although the by-products were not identified, they were able to observe the formation of two new peaks (P2 and P3) in the chromatogram with increased irradiation time (Figure 10.2). Moreover, the increase of the intensity of the original peak (P1) before a decrease with prolonged irradiation indicates that molar structures of HA were rearranged. With prolonged treatment, one of the new peaks (P2) decreased, while the other peak seemed to be associated with photocatalytically resistant products or nonadsorbed products.

When microbes are present in the water together with NOM, photocatalytic oxidation kills and lyses the cell. The inactivation of microbes, in addition to the

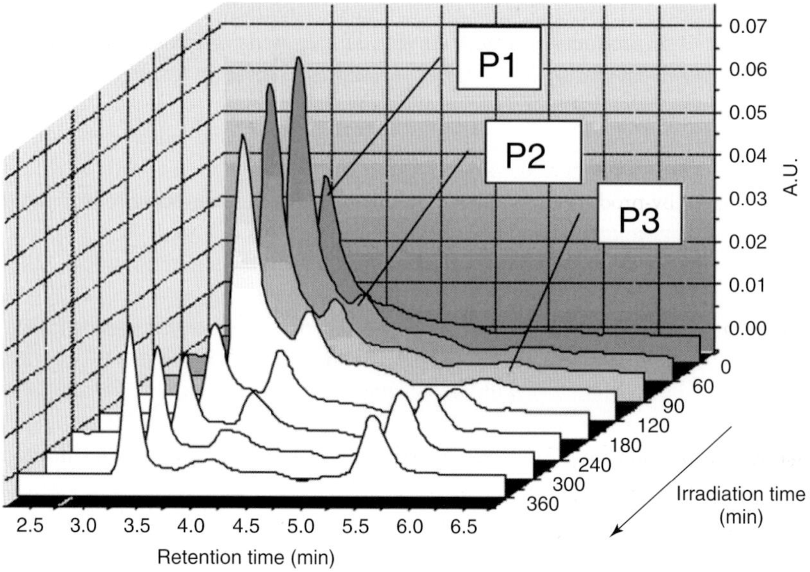

Figure 10.2 High-performance liquid chromatograms displaying the formation and disappearance of by-products from the photocatalytic degradation of humic acids with TiO_2 [56].

oxidation of NOM is a secondary benefit of photocatalysis, leading to a reduction in the chlorine dose needed for disinfection treatment. However, cell lysis may lead to the release of cell-bound particulates and dissolved organic matter, which can contribute to the DBP precursor pool.

In addition to organic carbon by-products, inorganic by-products can be formed during photocatalytic oxidation of NOM. Research into the fate and transformation of dissolved organic nitrogen and organic phosphorus during photocatalytic oxidation has lagged far behind that of the larger organic carbon pool. Dissolved organic nitrogen and organic phosphorus analysis is more analytically challenging to quantify than DOC. Some investigations have shown that nitrogen-containing molecules are mineralized mostly into inorganic nitrogens, such as nitrite (easily oxidized to nitrate by photocatalysis), nitrate, and ammonium ions [57, 58], while organic phosphorus compounds are degraded into phosphate ions or orthophosphate [59].

To evaluate toxicity of the by-products after photocatalytic oxidation, bioassays such as those involving *Vibrio fischeri*, *Daphnia magna*, and *Pseudokirchneriella subcapitata* may be carried out. However, these by-products typically occur at low concentrations and microbial assays may not be sensitive enough to characterize their toxicity [60]. Using *D. magna* test, Bekbolet *et al.* [61] reported that the toxicity of water after photocatalysis were in accordance with the DBPs distribution and concentration, rather than NOM removal efficiency.

The hydroxyl radicals produced as a result of TiO_2 photocatalytic oxidation has a high oxidation potential and leads to largely unselective reactions. If elevated concentration of bromide ions are present in the water, then the bromide ions can be photocatalytically oxidized to bromine species (HOBr, Br_2) [62]; the formation of brominated by-products in irradiated TiO_2 suspensions is a possibility that has to be considered. The formation of bromoform was reported in irradiated titanium dioxide suspensions in the presence of both DOC and bromide ions [63, 64]. Brominated DBPs pose a greater health risk than their chlorinated counterparts [65]. There may be considerable variability in the extent and time scale of formation of brominated by-products depending on the quality of the raw water to be treated (DOC and bromide ion concentrations) and on the reaction parameters, such as the light intensity. Therefore, the formation of brominated by-products should be considered and tested on a case-by-case basis.

Photocatalytic oxidation may also form by-products that have different reactivity with chlorine disinfectants. The concentration of THMs was generally reported to decrease in water treated with photocatalytic oxidation [66, 15, 67], and the oxidation by-products were sometimes found to be as reactive as the untreated organic compounds [66]. However, Gerrity *et al.* [68] have shown that limited photocatalysis has the potential to increase the trihalomethane formation potential (THMFP) after chlorination if the organic matter is not completely mineralized. Liu *et al.* [69] has also reported increases in specific THMFP at shorter treatment times during photocatalytic oxidation, which indicates that the intermediate products formed during photocatalytic oxidation may be more reactive with chlorine than the parent compound. It may be deduced then that the oxidation process fragments large aromatic compounds, thereby exposing more reactive sites to attack by chlorine at short photocatalytic treatment times. However, with further treatment, those sites were oxidized and became less reactive with chlorine.

In the studies of Liu *et al.* [45], THMFP of water after a short photocatalysis treatment was still relatively high and only decreased after prolonged treatment, although DOC, UV absorbance, and the intensity of the HPSEC spectrum indicated similar low concentrations of organics for both samples. This suggested that the commonly used analytical techniques for NOM characterization may not accurately predict THMFP trends for water treated with photocatalysis, or any advanced oxidation processes. This would be a significant concern for the commercial application of photocatalytic oxidation and therefore implies the need for continuous monitoring of the DBP formation potential, and downstream treatment such as granular activated carbon or biological treatment to remove these reactive DBP precursors before any disinfection stage.

The formation potential of another class of DBPs called haloacetic acid during photocatalytic treatment of NOM has not been extensively studied. Bekbolet *et al.* [61] compared the formation potential of four HAA species: chloroacetic acid, dichloroacetic acid, trichloroacetic acid, and dibromoacetic acid in raw water taken from different regions in Istanbul (Turkey) and their corresponding photocatalytically treated water. They found that the formation of trichloroacetic acid was enhanced after photocatalysis. Liu *et al.* [15] reported an increase in the specific trihaloacetic acid formation potential ($THAA_5FP$) at the initial stage of the photocatalytic oxidation, which indicates the formation of photocatalytic intermediates that had a greater tendency to form HAAs. However, with prolonged irradiation, specific $THAA_5FP$ reverted back to its initial value and remained relatively constant. The increase in the specific $THAA_5FP$ was attributed to an increase in hydrophilic substituents during the photocatalytic oxidation process, which has been reported to be the major precursor of HAAs [70].

In addition to THMs and HAAs, Bekbolet *et al.* [61] also investigated the formation potential of other DBPs, including chloral hydrate, haloacetonitriles, haloketones, and chloropicrin. The compounds trichloroacetonitrile, 1,3-dichloropropanone, and chloropicrin were not detected in the samples of raw water. Minor amounts of other DBPs were detected with the exception of chloral hydrate, which displayed initially high levels but diminished after photocatalytic treatment. Comparing the DBP formation potential of raw and treated waters, lower DBP formation potentials were reported for water after TiO_2 photocatalysis. Although bromoacetonitrile was not detected in the raw water samples, it was detected in the water following TiO_2 photocatalytic treatment.

Richardson *et al.* [71] observed a single organic DBP, tentatively identified as 3-methyl-2,4-hexanedione in ultrafiltered natural water treated with TiO_2 photocatalysis. When chlorine was used as a secondary treatment after TiO_2 photocatalysis, several chlorinated and brominated DBPs were formed, among them were some halomethanes and several halonitriles. Most of the halogenated DBPs were the same as those observed when the ultrafiltered natural water was chlorinated. However, one by-product, tentatively identified as dihydro-4,5-dichloro-2(3*H*)furanone, was formed only by a combination of photocatalysis and chlorine disinfection [71].

10.4
Hybrid Photocatalysis Technologies for the Treatment of NOM

Although photocatalytic oxidation of NOM is a promising technology for NOM degradation and removal of the DBPs precursor, commercial application of this technology is still limited. This is due to several practical issues, which is discussed below. Photocatalytic processes are usually studied in a slurry of fine particles of the TiO_2 semiconductor to maximize the available surface area for adsorption and oxidation. Nevertheless, the use of a photocatalyst in a slurry form requires an additional treatment step to remove the fine particles from the treated water, which increases the cost of the treatment. In recent years, attempts have been made to immobilize the catalyst on a suitable support, such as a material with a high surface area that settles readily [72–74] or a magnetic material [75–77]. Mounting TiO_2 nanoparticles onto magnetic platforms shows potential in solving this separation problem as they can be retained and recycled via magnetic separation [78, 79, 76, 80]. However, from an engineering point of view, the attachment of TiO_2 nanoparticles on those supports must be strong enough to hold for long operation periods.

The photocatalytic oxidation process has a low throughput, while requiring high energy to mineralize the organic matter. This is compounded by the fact that the mineralization of the organic matter usually takes a long time, rendering it impractical for treatment plant applications. A well-designed photoreactor configuration is also important to ensure optimal light distribution throughout the reactor for catalyst activation. Recent advances in photonic technology, such as the invention of UV light-emitting diode (LED), have provided promising solutions for transmitting UV into any photoreactor configurations.

The high energy requirement for UV activation has driven significant research to exploit solar energy as a source of light for activation and to render TiO_2 photocatalytically active beyond its absorption threshold of 390 nm. The UV region constitutes only approximately 4% of the energy available within the solar spectrum [81]. Thus, considerable research is being carried out to extend the photocatalyst response into the visible part of the solar spectrum. Several approaches have been reported in the literature, which include the use of narrow-band-gap semiconductors as sensitizers [82–84], metal ion doping into the band gap structure [85, 86], nonmetal doping [87–89], and preparation of a mixed oxide with iron component [90]. However, the reaction rates achieved with the newly developed visible light photocatalysts are lower than their UV-activated counterparts [91]. Therefore, optimization is urgently required so these benefits can be maximized.

The presence of turbidity in the water can interfere with TiO_2 photocatalysis as it blocks the penetration of light needed to activate the photocatalyst [92]. Therefore, the application of separation steps such as coagulation, sedimentation, and filtration before the application of photocatalytic reactions is recommended. In addition, the presence of inorganic ions in the water, such as chloride, sulfate, phosphate, carbonate, or bicarbonate, impedes the photocatalytic degradation rate due to hydroxyl radical scavenging or competitive adsorption on the surface of the catalyst [93–96]. For water with high alkalinity and inorganic ions content, mineralization will require prolonged reaction time, which may not be economically attractive for the commercial application of the technology. Photocatalysis may also increase the amount of biodegradable organic carbon (BOC) from NOM oxidation, promoting downstream microbiological regrowth [97]. Therefore, downstream treatment may be necessary to improve the biological stability of the water.

TiO_2 nanoparticles can aggregate severely into supramicron particles when added to water. Aggregation of nanoparticles in water may decrease the photocatalytic oxidation rate due to the decrease in surface area. The aggregation of TiO_2 photocatalyst can then influence adsorption of organic molecules, light scattering and photon absorption, charge carrier dynamics, and so on [98, 91].

The adsorption of NOM on the surface of the photocatalyst is an important factor for achieving high degradation rates [99]. NOM adsorption on TiO_2 is pH dependent due to the electrostatic interaction between negative charges of NOM and positive charges on surfaces of the oxide. Zeta potential values of TiO_2 decreased with increasing pH, and TiO_2 has positive charges on their surfaces in relatively low pH ranges but negative charges in relatively high pH ranges, showing zero point of charge at a pH of 5.9 [100]. pH of the water has to be adjusted to the acidic range in order to promote the adsorption of NOM onto the surface of the TiO_2 photocatalyst and then readjusted after treatment, which therefore increased the economic cost.

Because of these issues, photocatalytic oxidation should be used as a polishing step or combined with other treatment technologies. A combination of photocatalysis with other treatment technologies increases the overall treatment efficiency by decreasing the reaction time and cost of the treatment (in terms of light energy). Beneficial effects of combined treatment are commonly reported from laboratory studies [43, 101, 14, 102], which suggest potential large scale application via process integration rather than single technology processing.

Murray *et al.* [43] compared the adsorption pilot trials using TiO_2 pellets with the ferric sulfate coagulation treatment. They concluded that TiO_2 adsorption alone is not able to achieve the same removal as coagulation and therefore ruled out the potential of TiO_2 adsorption as a single stage replacement of the existing coagulation treatment. Rather, Murray *et al.* [43] suggested that TiO_2 adsorption could be performed before coagulation process to reduce the required coagulant dose. Following TiO_2 adsorption, coagulation with ferric sulfate was reported to reduce UV_{254} and DOC by 60 and 79% respectively, leading to a combined reduction of 86% DOC and 94% UV_{254}.

As was aforementioned, photocatalytic oxidation changes the structural properties of NOM constituents, and thus affecting both the adsorption and coagulation characteristics of NOM. When photocatalysis was used as a preoxidation method before coagulation, a decrease in $color_{436}$ and UV_{254} removal efficiency of approximately less than 15% was observed [103]. Changes in the molecular sizes, functional groups, hydrophobicities, and charge densities after photocatalysis are thought to be the factors responsible for the reductions in the removal efficiencies during coagulation. However, when photocatalysis was used as a polishing step after coagulation treatment, DOC removal was enhanced by about 12–32% compared to coagulation alone. Shon *et al.* [104] also found that the overall organic removal efficiency was improved when photocatalysis was performed after coagulation. Coagulation as a pretreatment before photocatalysis has another added advantage, as it can also remove turbidity from the water, therefore reducing the interference to the photocatalytic oxidation reaction.

Combining photocatalysis with activated carbon adsorption significantly improves removal efficiency in comparison to that treated with photocatalysis alone. Shon *et al.* [104] studied the effects of coupling adsorption with photocatalysis. When powdered activated carbon adsorption and TiO_2 photocatalysis were performed simultaneously, organic removal was improved significantly. Areerachakul *et al.* [14] investigated the degradation of humic substances in water using a coupled photocatalysis/powdered activated carbon adsorption system. DOC removal of the combined treatment improved by a factor of 2–3 times compared to TiO_2 alone. Xue *et al.* [99] immobilized nanosized TiO_2 on granular activated carbon and reported that the degradation of HA on the composite material was facilitated by the synergistic relationship between the surface adsorption characteristics and photocatalysis.

Photocatalysis can also be used as a pretreatment to enhance the biodegradability of NOM [56, 105] as photocatalysis can break long organic molecules into smaller organics, which are typically amenable to further biological treatment. The readily biodegradable compounds are promptly utilized as substrates by microorganisms in the biologically activated carbon filtration.

The use of membrane systems for drinking water treatment is rapidly growing. NOM is a major membrane foulant, which results in a reduction of the membrane permeability, and hence a decline in membrane permeate flux or increase in applied pressure, which leads to higher operating costs. A number of investigations have sought to integrate the photocatalytic process with membrane filtration. Hybrid photocatalysis/membrane processes have been widely tested for the removal of NOM from water because membranes can retain the catalyst in the system. Incorporation of photoactive nanomaterials makes the membranes "reactive" instead of being a simple physical barrier. In addition, immobilization of TiO_2 photocatalyst on the membrane surface has also been reported to reduce membrane fouling by organic or biological matter due to the inactivation of bacteria and degradation of organic matter by TiO_2 [106–108]. Different configurations have been reported: – membrane filtration for suspended catalyst recycling [109], filtration for suspended catalyst recycling and NOM removal [110], immobilized photocatalyst on membranes [111], and membranes made of photocatalyst materials [112]. Schematic diagrams for the configurations of photocatalytic membrane reactor are presented in Figure 10.3.

Although the immobilization of TiO_2 nanoparticles on the membrane has generated a lot of interests, concerns regarding possible release of nanoparticles into the water cannot be dismissed completely. Retention of nanoparticles within the membrane is critical not only because of the cost associated with the loss of the nanomaterials, but also, and more importantly, because of the potential impacts of nanomaterials on human health and ecosystems [113].

10.5 Conclusions

A heterogeneous photocatalytic process is a potentially promising technology for NOM oxidation, provided certain limitations can be overcome. Its application has not been tested beyond the laboratory scale for a number of reasons. First, a comprehensive understanding of the by-product of NOM oxidation is vital if TiO_2 photocatalysis that is to be used on a large scale drinking water treatment plant. It is possible that the intermediate or by-product of NOM oxidation is more harmful than NOM itself. Advances in the area of rapid monitoring of by-products are crucial because the changes in the concentration or the nature of the dissolved organics in the source water, or variation in the photocatalytic treatment time can affect the nature of the by-products. Moreover, the potential of the by-product to react with disinfectants needs to be continually assessed and monitored to meet the increasingly strict DBPs regulations.

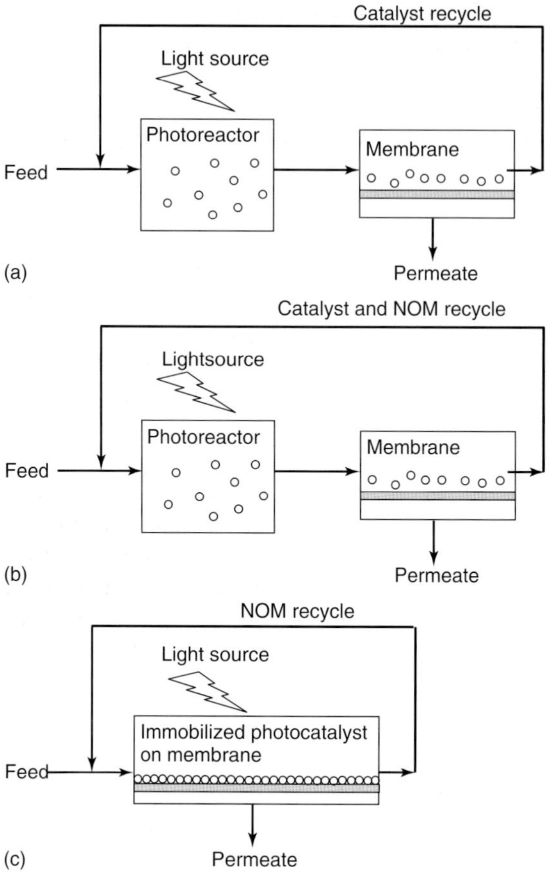

Figure 10.3 Photocatalytic membrane reactor configurations: (a) membrane filtration for suspended catalyst recycling [109], (b) filtration for suspended catalyst recycling and NOM removal [110], and (c) immobilized photocatalyst on membranes [111].

The present treatment cost of photocatalytic systems are higher than those of conventional water treatment technologies, but the efforts being made in the design of more efficient systems with improved catalyst activity will establish this technology as a cleaner and cost-effective alternative. In general, photocatalysis cannot be applied as a stand-alone process to replace any existing treatment processes. However, the potential of using photocatalytic systems to treat NOM in commercial/industry-scale applications will increase if it is integrated with other treatment process. Coupling TiO_2 photocatalytic treatment with other conventional treatment processes can provide enhanced treatment efficiency with potential overall savings in energy and chemical use. Synergistic effects can be achieved through coupling heterogeneous photocatalysis with membrane separation, adsorption, coagulation, and biodegradation to enhance overall NOM removal efficiency.

References

1. Aiken, G.R., McKnight, D.M., Wershaw, R.L., and MacCarthy, P. (1985) *Humic Substances in Soil, Sediment, and Water: Geochemistry, Isolation, and Characterization*, John Wiley & Sons, Inc., New York.
2. Suffet, I.H. and MacCarthy, P. (1989) *Aquatic Humic Substances: Influence on Fate and Treatment of Pollutants*, American Chemical Society, Washington, DC.
3. Thurman, E.M. (1985) *Organic Geochemistry of Natural Waters*, Martinus Nijhoff, Dordrecht.
4. Volk, C.J., Bell, K., Ibrahim, E., Verges, D., Amy, G., and LeChevallier, M. (2000) Impact of enhanced and optimized coagulation on removal of organic matter and its biodegradable fraction in drinking water. *Water Res.*, **34**, 3247–3257.
5. Tipping, E. and Hurley, M.A. (1992) A unifying model of cation binding by humic substances. *Geochim. Cosmochim. Acta*, **56**, 3627–3641.
6. Jones, M.N. and Bryan, N.D. (1998) Colloidal properties of humic substances. *Adv. Colloid Interface Sci.*, **78**, 1–48.
7. Pelekani, C. and Snoeyink, V.L. (1999) Competitive adsorption in natural water: role of activated carbon pore size. *Water Res.*, **33**, 1209–1219.
8. Zularisam, A.W., Ismail, A.F., and Salim, R. (2006) Behaviours of natural organic matter in membrane filtration for surface water treatment–a review. *Desalination*, **194**, 211–231.
9. Nikolau, A., Kostopoulou, M., and Lekkas, T. (1999) Organic by-products of drinking water chlorination. *Global Nest: Int. J.*, **1**, 143–156.
10. USEPA (2005) Economic Analysis for the Stage 2 Disinfectants and Disinfection Byproducts Rule, EPA 815-R-05-010.
11. Richardson, S.D. (2005) New disinfection by-product issues: emerging DBPs and alternative routes of exposure. *Global Nest: Int. J.*, **7**, 43–60.
12. Bekbolet, M. and Ozkosemen, G. (1996) A preliminary investigation on the photocatalytic degradation of a model humic acid. *Water Sci. Technol.*, **33**, 189–194.
13. Leenheer, J.A. (2007) Progression from model structures to molecular structures of natural organic matter components. *Ann. Environ. Sci.*, **1**, 57–68.
14. Areerachakul, N., Vigneswaran, S., Kandasamy, J., and Duangduen, C. (2008) The degradation of humic substance using continuous photocatalysis systems. *Sep. Sci. Technol.*, **43**, 93–112.
15. Liu, S., Lim, M., Fabris, R., Chow, C., Drikas, M., and Amal, R. (2008) TiO_2 Photocatalysis of natural organic matter in surface water: impact on trihalomethane and haloacetic acid formation potential. *Environ. Sci. Technol.*, **42**, 6218–6223.
16. Uyguner, C.S. and Bekbolet, M. (2005) Evaluation of humic acid photocatalytic degradation by UV-vis and fluorescence spectroscopy. *Catal. Today*, **101**, 267.
17. Liu, S., Lim, M., Fabris, R., Chow, C.W.K., Drikas, M., Korshin, G., and Amal, R. (2010) Multi-wavelength spectroscopic and chromatography study on the photocatalytic oxidation of natural organic matter. *Water Res.*, **44**, 2525–2532.
18. Eggins, B.R., Palmer, F.L., and Byrne, J.A. (1997) Photocatalytic treatment of humic substances in drinking water. *Water Res.*, **31**, 1223.
19. Edzwald, J.K. and Van Benschoten, J.E. (1990) Aluminium coagulation of natural organic matter, in *Chemical Water and Waste Water Treatment* (eds H.H. Hahn and R. Klute), Springer-Verlag, Berlin.
20. Weishaar, J.L., Aiken, G.R., Bergamaschi, B.A., Fram, M.S., Fujii, R., and Mopper, K. (2003) Evaluation of specific ultraviolet absorbance as an indicator of the chemical composition and reactivity of dissolved organic carbon. *Environ. Sci. Technol.*, **37**, 4702–4708.
21. Croue, J.P., Korshin, G.V., Leenheer, J.A., and Benjamin, M.M. (2000) Characterisation of Natural Organic Matter

in Drinking Water, AWWA Research Foundation, American Water Works Association, Denver, Colorado.

22. Kitis, M., Karanfil, T., Kilduff, J.E., and Wigton, A. (2001) The reactivity of natural organic matter to disinfection by-products formation and its relation to specific ultraviolet absorbance. *Water Sci. Technol.*, **43**, 9–16.
23. Wei, Y., Chu, H., Dong, B., and Li, X. (2011) Evaluation of humic acid removal by a flat submerged membrane photoreactor. *Chin. Sci. Bull.*, **56**, 3437–3444.
24. Korshin, G.V., Li, C.-W., and Benjamin, M.M. (1997) Monitoring the properties of natural organic matter through UV spectroscopy: a consistent theory. *Water Res.*, **31**, 1787–1795.
25. Stevenson, F.J. (1994) *Humus Chemistry: Genesis, Composition, Reactions*, John Wiley & Sons, Inc., New York.
26. Stewart, A.J. and Wetzel, R.G. (1980) Fluorescence: absorbance ratios-a molecular-weight tracer of dissolved organic matter. *Limnol. Oceanogr.*, **25**, 559–564.
27. Senesi, N. (1990) Molecular and quantitative aspects of the chemistry of fulvic acid and its interactions with metal ions and organic chemicals: part II. The fluorescence spectroscopy approach. *Anal. Chim. Acta*, **232**, 77–106.
28. Lakowicz, R. (1999) *Principles of Fluorescence Spectroscopy*, Kluwer Academic and Plenum Publishers, New York.
29. Miano, T.M. and Senesi, N. (1992) Synchronous excitation fluorescence spectroscopy applied to soil humic substances chemistry. *Sci. Total. Environ.*, **117–118**, 41–51.
30. Wu, F. and Liu, C. (2006) Humic substance, in *Chromatographic Analysis of the Environment*, 3rd edn (ed L.M.L. Nollet), CRC Press, Boca Raton, FL.
31. Galapate, R.P., Baes, A.U., Ito, K., Mukai, T., Shoto, E., and Okada, M. (1998) Detection of domestic wastes in Kurose river using synchronous fluorescence spectroscopy. *Water Res.*, **32**, 2232–2239.
32. Holbrook, R.D., DeRose, P.C., Leigh, S.D., Rukhin, A.L., and Heckert, N.A. (2006) Excitation–emission matrix fluorescence spectroscopy for natural organic matter characterization: a quantitative evaluation of calibration and spectral correction procedures. *Appl. Spectrosc.*, **60**, 791–799.
33. Chen, W., Westerhoff, P., Leenheer, J.A., and Booksh, K. (2003) Fluorescence excitation-emission matrix regional integration to quantify spectra for dissolved organic matter. *Environ. Sci. Technol.*, **37**, 5701–5710.
34. Cho, Y. and Choi, W. (2002) Visible light-induced reactions of humic acids on TiO_2. *J. Photochem. Photobiol., A*, **148**, 129–135.
35. De Nobili, M. and Chen, Y. (1999) Size exclusion chromatography of humic substances: limits, perspectives and prospectives. *Soil Sci.*, **164**, 825–833.
36. Huber, S.A. and Frimmel, F.H. (1996) Size-exclusion chromatography with organic carbon detection (LC-OCD): a fast and reliable method for the characterization of hydrophilic organic matter in natural waters. *Vom Wasser*, **86**, 277–290.
37. Conte, P. and Piccolo, A. (1999) High pressure size exclusion chromatography (HPSEC) of humic substances: molecular sizes, analytical parameters, and column performance. *Chemosphere*, **38**, 517–528.
38. Nagao, S., Matsunaga, T., Suzuki, Y., Ueno, T., and Amano, H. (2003) Characteristics of humic substances in the kuji river waters as determined by high-performance size exclusion chromatography with fluorescence detection. *Water Res.*, **37**, 4159–4170.
39. Persson, L., Alsberg, T., Kiss, G., and Odham, G. (2000) On-line size-exclusion chromatography/electrospray ionisation mass spectrometry of aquatic humic and fulvic acids. *Rapid Commun. Mass Spectrom.*, **14**, 286–292.
40. Wagoner, D.B., Christman, R.F., Cauchon, G., and Paulson, R. (1997) Molar mass and size of Suwannee river natural organic matter using multi-angle laser light scattering. *Environ. Sci. Technol.*, **31**, 937–941.
41. Amy, G. and Her, N. (2004) Size exclusion chromatography (SEC) with

multiple detectors: a powerful tool in treatment process selection and performance monitoring. *Water Sci. Technol. Water Supply*, **4**, 19–24.

42. Doll, T.E. and Frimmel, F.H. (2005) Photocatalytic degradation of carbamazepine, clofibric acid and iomeprol with P25 and hombikat UV100 in the presence of natural organic matter (NOM) and other organic water constituents. *Water Res.*, **39**, 403–411.
43. Murray, C.A., Goslan, E.H., and Parsons, S.A. (2007) TiO_2/UV: single stage drinking water treatment for NOM removal? *J. Environ. Eng. Sci.*, **6**, 311–317.
44. Huang, X., Leal, M., and Li, Q. (2008) Degradation of natural organic matter by TiO_2 photocatalytic oxidation and its effect on fouling of low-pressure membranes. *Water Res.*, **42**, 1142–1150.
45. Liu, S., Lim, M., Fabris, R., Chow, C., Chiang, K., Drikas, M., and Amal, R. (2008) Removal of humic acid using TiO_2 photocatalytic process–fractionation and molecular weight characterisation studies. *Chemosphere*, **72**, 263–271.
46. Leenheer, J.A. and Huffman, E.W.D. Jr., (1976) Classification of organic solutes in water by using macroreticular resins. *U.S. Geol. J. Res.*, **4**, 737–751.
47. Leenheer, J.A. (1981) Comprehensive approach to preparative isolation and fractionation of dissolved organic carbon from natural waters and wastewaters. *Environ. Sci. Technol.*, **15**, 578–587.
48. Aiken, G.R., McKnight, D.M., Thorn, K.A., and Thurman, E.M. (1992) Isolation of hydrophilic organic acids from water using nonionic macroporous resins. *Org. Geochem.*, **18**, 567–573.
49. Malcolm, R. and MacCarthy, P. (1992) Quantitative evaluation of XAD-8 and XAD-4 resins used in tandem for removing organic solutes from water. *Environ. Int.*, **18**, 597–607.
50. Croue, J.P., Martin, B., Deguin, A., and Legube, B. (1994) Isolation and Characterisation of Dissolved Hydrophobic and Hydrophilic Organic Substances of a Reservoir Water. Natural Organic Matter in Drinking Water, American Water Works Association, Denver.
51. Bolto, B., Abbt-Braun, G., Dixon, D., Eldridge, R., Frimmel, F., Hesse, S., King, S., and Toifl, M. (1999) Experimental evaluation of cationic polyelectrolytes for removing natural organic matter from water. *Water Sci. Technol.*, **40**, 71–79.
52. Leenheer, J.A. (2004) Comprehensive assessment of precursors, diagenesis, and reactivity to water treatment of dissolved and colloidal organic matter. *Water Sci. Technol. Water Supply*, **4**, 1–9.
53. Chow, C.W.K., Fabris, R., and Drikas, M. (2004) A rapid fractionation technique to characterise natural organic matter for the optimisation of water treatment processes. *J. Water Supply Res Technol.*, **53**, 85–92.
54. Croue, J.P. (1999) Isolation, fractionation, characterization and reactive properties of natural organic matter. Proceedings of the 18th Federal AWWA Convention, Adelaide, Australia.
55. McDonald, S., Bishop, A.G., Prenzler, P.D., and Robards, K. (2004) Analytical chemistry of freshwater humic substances. *Anal. Chim. Acta*, **527**, 105–124.
56. Wiszniowski, J., Robert, D., Surmacz-Gorska, J., Miksch, K., and Weber, J.-V. (2002) Photocatalytic decomposition of humic acids on TiO_2: part I: discussion of adsorption and mechanism. *J. Photochem. Photobiol., A*, **152**, 267–273.
57. Takeda, K. and Fujiwara, K. (1996) Characteristics on the determination of dissolved organic nitrogen compounds in natural waters using titanium dioxide and platinized titanium dioxide mediated photocatalytic degradation. *Water Res.*, **30**, 323–330.
58. Onar, A.N. and Akdemir, N. (2007) Effect of complexation on photocatalytic degradation of dissolved organic nitrogen compounds. *Anal. Lett.*, **40**, 2949–2958.
59. Malato, S., Blanco, J., Richter, C., Milow, B., and Maldonado, M. (1999) Solar photocatalytic mineralization

of commercial pesticides: methamidophos. *Chemosphere*, **38**, 1145–1156.
60. Rizzo, L. (2011) Bioassays as a tool for evaluating advanced oxidation processes in water and wastewater treatment. *Water Res.*, **45**, 4311–4340.
61. Bekbolet, M., Uyguner, C.S., Selcuk, H., Rizzo, L., Nikolaou, A.D., Meric, S., and Belgiorno, V. (2005) Application of oxidative removal of NOM to drinking water and formation of disinfection by-products. *Desalination*, **176**, 155–166.
62. Herrmann, J.M. and Pichat, P. (1980) Heterogeneous photocatalysis. Oxidation of halide ions by oxygen in ultraviolet irradiated aqueous suspension of titanium dioxide. *J. Chem. Soc., Faraday Trans. 1: Phys. Chem. Condens. Phases*, **76**, 1138–1146.
63. Tercero Espinoza, L.A., Rembor, M., Matesanz, C.A., Heidt, A., and Frimmel, F.H. (2009) Formation of bromoform in irradiated titanium dioxide suspensions with varying photocatalyst, dissolved organic carbon and bromide concentrations. *Water Res.*, **43**, 4143–4148.
64. Tercero Espinoza, L.A. and Frimmel, F.H. (2008) Formation of brominated products in irradiated titanium dioxide suspensions containing bromide and dissolved organic carbon. *Water Res.*, **42**, 1778–1784.
65. Richardson, S.D. (2003) Disinfection by-products and other emerging contaminants in drinking water. *Trends Anal. Chem.*, **22**, 666–684.
66. Philippe, K.K., Hans, C., MacAdam, J., Jefferson, B., Hart, J., and Parsons, S.A. (2010) Photocatalytic oxidation, GAC and biotreatment combinations: an alternative to the coagulation of hydrophilic rich waters? *Environ. Technol.*, **31**, 1423–1434.
67. Hand, D.W., Perram, D.L., and Crittenden, J.C. (1995) Destruction of DBP precursors with catalytic oxidation. *J. Am. Water Works Assoc.*, **87**, 84–96.
68. Gerrity, D., Mayer, B., Ryu, H., Crittenden, J., and Abbaszadegan, M. (2009) A comparison of pilot-scale photocatalysis and enhanced coagulation for disinfection byproduct mitigation. *Water Res.*, **43**, 1597–1610.
69. Liu, S., Lim, M., Fabris, R., Chow, C., Drikas, M., and Amal, R. (2010) Comparison of photocatalytic degradation of natural organic matter in two Australian surface waters using multiple analytical techniques. *Org. Geochem.*, **41**, 124–129.
70. Hwang, C.J., Sclimenti, M.J., Bruchet, A., Croué, J.P., and Amy, G.L. (2001) DBP yields of polar NOM fractions from low humic waters. Proceedings of Water Quality Technology Conference 2001, AWWA, Denver, CO.
71. Richardson, S.D., Thruston, A.D. Jr.,, Collette, T.W., Patterson, K.S., Lykins, B.W. Jr.,, and Ireland, J.C. (1996) Identification of TiO_2/UV disinfection byproducts in drinking water. *Environ. Sci. Technol.*, **30**, 3327–3334.
72. Aronson, B.J., Blanford, C.F., and Stein, A. (1997) Solution-phase grafting of titanium dioxide onto the pore surface of mesoporous silicates: synthesis and structural characterization. *Chem. Mater.*, **9**, 2842–2851.
73. Chong, M.N., Jin, B., Chow, C.W.K., and Saint, C. (2010) Recent developments in photocatalytic water treatment technology: a review. *Water Res.*, **44**, 2997–3027.
74. Wang, H., Yang, B., and Zhang, W.J. (2010) Photocatalytic degradation of methyl orange on Y zeolite supported TiO_2. *Adv. Mat. Res.*, **129**, 733–737.
75. Lee, S., Drwiega, J., Wu, C.Y., Mazyck, D., and Sigmund, W.M. (2004) Anatase TiO2 nanoparticle coating on barium ferrite using titanium bis-ammonium lactato dihydroxide and its use as a magnetic photocatalyst. *Chem. Mater.*, **16**, 1160–1164.
76. Watson, S., Beydoun, D., and Amal, R. (2002) Synthesis of a novel magnetic photocatalyst by direct deposition of nanosized TiO_2 crystals onto a magnetic core. *J. Photochem. Photobiol., A*, **148**, 303–313.
77. Fu, W., Yang, H., Chang, L., Li, M., and Zou, G. (2006) Anatase TiO_2 nanolayer coating on strontium ferrite

nanoparticles for magnetic photocatalyst. *Colloids Surf., A Physicochem. Eng. Asp.*, **289**, 47–52.
78. Beydoun, D., Amal, R., Low, G.K.C., and McEvoy, S. (2000) Novel photocatalyst: titania-coated magnetite. Activity and photodissolution. *J. Phys. Chem. B*, **104**, 4387–4396.
79. Beydoun, D., Amal, R., Scott, J., Low, G., and McEvoy, S. (2001) Studies on the mineralization and separation efficiencies of a magnetic photocatalyst. *Chem. Eng. Technol.*, **24**, 745–748.
80. Watson, S., Scott, J., Beydoun, D., and Amal, R. (2005) Studies on the preparation of magnetic photocatalysts. *J. Nanopart. Res.*, **7**, 691–705.
81. Ohtani, B. (2008) Preparing articles on photocatalysis—beyond the illusions, misconceptions, and speculation. *Chem. Lett.*, **37**, 216–229.
82. Bessekhouad, Y., Robert, D., and Weber, J. (2004) Bi_2S_3/TiO_2 And CdS/TiO_2 heterojunctions as an available configuration for photocatalytic degradation of organic pollutant. *J. Photochem. Photobiol., A*, **163**, 569–580.
83. Ho, W. and Yu, J.C. (2006) Sonochemical synthesis and visible light photocatalytic behavior of CdSe and $CdSe/TiO_2$ nanoparticles. *J. Mol. Catal. A: Chem.*, **247**, 268–274.
84. Liu, J., Yang, R., and Li, S. (2006) Preparation and characterization of the TiO_2-V_2O_5 photocatalyst with visible-light activity. *Rare Met.*, **25**, 636–642.
85. Castro, C.A., Centeno, A., and Giraldo, S.A. (2011) Iron promotion of the TiO_2 photosensitization process towards the photocatalytic oxidation of azo dyes under solar-simulated light irradiation. *Mater. Chem. Phys.***129**(3), 1176–1183.
86. Kim, S., Hwang, S.J., and Choi, W. (2005) Visible light active platinum-ion-doped TiO2 photocatalyst. *J. Phys. Chem. B*, **109**, 24260–24267.
87. Hong, X., Wang, Z., Cai, W., Lu, F., Zhang, J., Yang, Y., Ma, N., and Liu, Y. (2005) Visible-light-activated nanoparticle photocatalyst of iodine-doped titanium dioxide. *Chem. Mater.*, **17**, 1548–1552.
88. In, S., Orlov, A., Berg, R., García, F., Pedrosa-Jimenez, S., Tikhov, M.S., Wright, D.S., and Lambert, R.M. (2007) Effective visible light-activated B-doped and B, N-codoped TiO_2 photocatalysts. *J. Am. Chem. Soc.*, **129**, 13790–13791.
89. Hsu, S.W., Yang, T.S., Chen, T.K., and Wong, M.S. (2007) Ion-assisted electron-beam evaporation of carbon-doped titanium oxide films as visible-light photocatalyst. *Thin Solid Films*, **515**, 3521–3526.
90. Tada, H., Jin, Q., Nishijima, H., Yamamoto, H., Fujishima, M., Okuoka, S., Hattori, T., Sumida, Y., and Kobayashi, H. (2011) Titanium (IV) dioxide surface –modified with iron oxide as a visible light photocatalyst. *Angew. Chem.*, **123**, 3563–3567.
91. Friedmann, D., Mendive, C., and Bahnemann, D. (2010) TiO_2 For water treatment: parameters affecting the kinetics and mechanisms of photocatalysis. *Appl. Catal., B*, **99**, 398–406.
92. Sharma, P., Durga Kumari, V., and Subrahmanyam, M. (2008) Photocatalytic degradation of isoproturon herbicide over TiO_2/Al-MCM-41 composite systems using solar light. *Chemosphere*, **72**, 644–651.
93. Bahnemann, D., Cunningham, J., Fox, M.A., Pelizzetti, E., Pichat, P., and Serpone, N. (1994) in *Aquatic and Surface Photochemistry* (eds G.R. Helz, R.G. Zepp, and D.G. Crosby) Chapter 21, Lewis Publisher, pp. 261–316.
94. Bekbölet, M., Boyacioglu, Z., and Özkaraova, B. (1998) The influence of solution matrix on the photocatalytic removal of color from natural waters. *Water Sci. Technol.*, **38**, 155–162.
95. Wiszniowski, J., Robert, D., Surmacz-Gorska, J., Miksch, K., Malato, S., and Weber, J.-V. (2004) Solar photocatalytic degradation of humic acids as a model of organic compounds of landfill leachate in pilot-plant experiments: influence of inorganic salts. *Appl. Catal., B*, **53**, 127.
96. Kim, M.J., Choo, K.H., and Park, H.S. (2010) Photocatalytic degradation of seawater organic matter using a submerged membrane reactor. *J. Photochem. Photobiol., A*, **216**, 215–220.

97. Jinhui, Z. and Wei, C. (2009) UV-TiO2 photocatalytic disinfection and its influence on drinking water biological stability. IEEE 3rd International Conference on Bioinformatics and Biomedical Engineering, Beijing, 2009, pp. 1–5.

98. Egerton, T.A. and Tooley, I.R. (2004) Effect of changes in TiO_2 dispersion on its measured photocatalytic activity. *J. Phys. Chem. B*, **108**, 5066–5072.

99. Xue, G., Liu, H., Chen, Q., Hills, C., Tyrer, M., and Innocent, F. (2011) Synergy between surface adsorption and photocatalysis during degradation of humic acid on TiO_2/activated carbon composites. *J. Hazard. Mater.*, **186**, 765–772.

100. Jaffrezic-Renault, N., Pichat, P., Foissy, A., and Mercier, R. (1986) Study of the effect of deposited Pt particles on the surface charge of TiO2 aqueous suspensions by potentiometry, electrophoresis and labelled ion adsorption. *J. Phys. Chem.*, **90**, 2733–2738.

101. Uyguner, C.S., Bekbolet, M., and Selcuk, H. (2007) A comparative approach to the application of a physico-chemical and advanced oxidation combined system to natural water samples. *Sep. Sci. Technol.*, **42**, 1405–1419.

102. Zou, L. and Zhu, B. (2008) The synergistic effect of ozonation and photocatalysis on color removal from reused water. *J. Photochem. Photobiol., A*, **196**, 24–32.

103. Uyguner, C.S., Suphandag, S.A., Kerc, A., and Bekbolet, M. (2007) Evaluation of adsorption and coagulation characteristics of humic acids preceded by alternative advanced oxidation techniques. *Desalination*, **210**, 183–193.

104. Shon, H.K., Vigneswaran, S., Ngo, H.H., and Kim, J.H. (2005) Chemical coupling of photocatalysis with flocculation and adsorption in the removal of organic matter. *Water Res.*, **39**, 2549–2558.

105. Bekbolet, M., Cecen, F., and Ozkosemen, G. (1996) Photocatalytic oxidation and subsequent adsorption characteristics of humic acids. *Water Sci. Technol.*, **34**, 65–72.

106. Li, J.H., Xu, Y.Y., Zhu, L.P., Wang, J.H., and Du, C.H. (2009) Fabrication and characterization of a novel TiO_2 nanoparticle self-assembly membrane with improved fouling resistance. *J. Membr. Sci.*, **326**, 659–666.

107. Ciston, S., Lueptow, R.M., and Gray, K.A. (2008) Bacterial attachment on reactive ceramic ultrafiltration membranes. *J. Membr. Sci.*, **320**, 101–107.

108. Damodar, R.A., You, S.J., and Chou, H.H. (2009) Study the self cleaning, antibacterial and photocatalytic properties of TiO_2 entrapped PVDF membranes. *J. Hazard. Mater.*, **172**, 1321–1328.

109. Fu, J., Ji, M., and An, D. (2005) Fulvic acid degradation using nanoparticle TiO_2 in a submerged membrane photocatalysis reactor. *J. Environ. Sci.*, **17**, 942–945.

110. Molinari, R., Borgese, M., Drioli, E., Palmisano, L., and Schiavello, M. (2002) Hybrid processes coupling photocatalysis and membranes for degradation of organic pollutants in water. *Catal. Today*, **75**, 77–85.

111. Syafei, A.D., Lin, C.F., and Wu, C.H. (2008) Removal of natural organic matter by ultrafiltration with TiO_2-coated membrane under UV irradiation. *J. Colloid Interface Sci.*, **323**, 112–119.

112. Zhang, X., Du, A.J., Lee, P., Sun, D.D., and Leckie, J.O. (2008) TiO_2 Nanowire membrane for concurrent filtration and photocatalytic oxidation of humic acid in water. *J. Membr. Sci.*, **313**, 44–51.

113. Pichat, P. (2010) A brief survey of the potential health risks of TiO_2 particles and TiO_2-containing photocatalytic or non-photocatalytic materials. *J. Adv. Oxid. Technol.*, **13**, 238–246.

11
Waterborne *Escherichia coli* Inactivation by TiO_2 Photoassisted Processes: a Brief Overview

Julián Andrés Rengifo-Herrera, Angela Giovana Rincón, and Cesar Pulgarin

11.1
Introduction

The most common used techniques in use now for water disinfection are chlorination, heating, and ozonation. However, in the case of chlorination, there is a negative effect: the appearance of trihalomethanes (THMs) as by-products of the reaction between chlorine with organic matter [1–3]. Other methods, for example, ozonation, are either expensive or involve a high consumption of electric energy; moreover after ozonation, chlorination is required in order to avoid bacteria regrowing. Solar disinfection (SODIS) is a simple technology developed to inactivate bacteria on water; in this case, SODIS makes a good use of the synergistic effect of the UV-component of the sunlight irradiation (5–6%) and the ambient heat created by the incoming sunlight to inactivate pathogen microorganisms [4, 5]. However, this process sometimes leads to microbial regrowth and is not effective in the case of bacterial strains such as *Salmonella typhimurium* [6]. In this respect, the use of photochemical processes with or without photocatalysts in solution such as advanced oxidation processes (AOPs), photo-Fenton, heterogeneous, and other supported photocatalysis seem a promising alternative to chlorination and could also be used to enhance the efficiency of the SODIS technology, which is now in use [7, 8].

In 1985, Matsunaga *et al.* [9] reported the first use of heterogeneous photocatalysis by TiO_2 as a sterilization process to inactivate *Escherichia coli*, *Lactobacillus acidophilus*, and *Saccharomyces cerevisiae* cells, and this approach generated a few thousand studies during the past three decades.

TiO_2 exists in three crystalline forms: anatase, rutile, and brookite. Indeed, the photocatalytic activities of various anatase and rutile samples overlap because other parameters influence these activities. The allegation about the superiority of anatase per se is based on the observation that, at least until now, the most active TiO_2 samples are anatase specimens [10], but it is not demonstrated that anatase is intrinsically more active. When TiO_2 nanoparticles are illuminated with UV light below 385 nm, an electron (e^-) is promoted from the valence band (VB) to the conduction band (CB), leaving in the VB a hole (h^+). These charge

Photocatalysis and Water Purification: From Fundamentals to Recent Applications, First Edition. P. Pichat.

carriers (e^-/h^+) can migrate to the TiO_2 surface being trapped on surface defects, dangling bonds or impurities. Once the charge carriers are trapped, they undergo electronic transfer; the surface-trapped h^+ and e^- can react with either electron donors or acceptors. Water molecules or hydroxyl groups adsorbed on the TiO_2 surface behave as electron donors leading to surface adsorbed hydroxyl radicals ($OH^•$) with high oxidative power (1.8 V vs NHE). Trapped electrons can react with oxygen molecules previously adsorbed on TiO_2 surfaces producing superoxide radical ($O_2^{•-}$) [11–15].

$$O_2^{•-} + H^+ \rightarrow HO_2^{•} \quad (pK_a = 4.8)\ [15] \tag{11.1}$$

$$HO_2^{•} + O_2^{•-} + H^+ \rightarrow H_2O_2 + O_2 \quad (k = 1 \times 10^8 M^{-1}s^{-1})\ [15] \tag{11.2}$$

$$HO_2^{•} + HO_2^{•} \rightarrow H_2O_2 + O_2 \quad (k = 8.6 \times 10^5 M^{-1}s^{-1})\ [15] \tag{11.3}$$

$$O_2^{•-} + O_2^{•-} + 2H^+ \rightarrow H_2O_2 + O_2 \quad (k << 100 M^{-1}s^{-1})\ [15] \tag{11.4}$$

Recently, it has been reported that anatase particles produce singlet oxygen ($^1O_2$) under UV light, through the oxidation of $O_2^{•-}$ radicals by VB holes [16, 17] (Chapter 1).

$$O_2^{•-} + h_{VB}^+ \rightarrow {}^1O_2 \tag{11.5}$$

In summary, irradiation of TiO_2 nanoparticles by UV light leads to the production of reactive oxygen species (ROS) such as $OH^•$, $O_2^{•-}$, $^1O_2$, and H_2O_2, which are all highly oxidative species capable of oxidizing/abating microorganisms. These species are able to oxidize the cell constituents: proteins, lipids, and nucleic acids.

11.2 Physicochemical Aspects Affecting the Photocatalytic *E. coli* Inactivation

The bulk and surface characteristics of TiO_2, its concentration, the intensity of the incoming light [18, 19], the presence of anions or organic substances [20, 21], pH [20], and the available oxygen concentration [22] play a major role in the inactivation of bacteria. The specific TiO_2 bulk and surface properties [23] like crystalline structure, isoelectric point (IEP), particle size, and aggregates size play a determinant role on the photocatalytic inactivation of microorganisms in solution.

11.2.1 Effect of Bulk Physicochemical Parameters

11.2.1.1 Effect of TiO_2 Concentration and Light Intensity

The TiO_2 concentrations used for photocatalytic *E. coli* inactivation range from 0.1 to 3 g l^{-1}. This concentration depends on *E. coli* strains, the kind of light source, and the reactor geometry used in different studies. However, in summary, at a high TiO_2 concentration (>2 g l^{-1}), the photocatalytic bacteria inactivation decreases due to the restricted (short) light penetration into the suspension decreasing the photocatalyst activity. The concomitant action of UV light alone on the bacteria (direct photolysis) is also diminished. In contrast, light intensity also plays an important role, as it induces charge separation on TiO_2 particles leading to cell damage. It has been reported that increasing the light intensity increases the photocatalytic bacteria inactivation, but when light irradiation is applied intermittently, the photocatalytic cell inactivation decreases or does not proceed to completion in reasonable times [18, 24–26].

11.2.1.2 Simultaneous Presence of Anions and Organic Matter

The simultaneous presence of anions and organic matter can affect negatively the photocatalytic disinfection. The effect of anions such as HCO_3^-/CO_3^{2-}, HPO_4^{2-}, $H_2PO_4^-$, SO_4^{2-}, Cl^- on the photocatalytic waterborne bacteria inactivation has been reported. Anions can scavenge photoinduced ROS or h^+_{VB} [27, 28] and can be preferentially adsorbed on TiO_2 surfaces blocking the active sites where ROS generation or charge carriers tapping takes place [29]; these results have been discussed in [30]. Adsorption of some anions might affect positively or negatively the interactions between cell and TiO_2 [31]. The ion toxicity depends on its concentration and constitutes a supplementary stress for the bacteria affecting both their metabolism and the osmotic strength of solutions. The turbidity and optical absorption increases in the presence of anions and influences the light penetration and consequently diminishes the light absorption on both bacteria and TiO_2.

The composition and the concentration of organic matter in water bodies such as humic compounds in surface water, groundwater, and wastewater influence the absorption of light affecting the light absorption by the organic chromophores as well as by the long-lived organic intermediates generated during the photocatalytically induced degradation of pathogens. Organic compounds present in these waters can either positively affect the photocatalytic disinfection by the enhancement of ROS production or negatively by competing with bacteria for the photogenerated oxidative species [20, 32]. This competition will depend on the chemical nature of the organic compound and its interaction with TiO_2 surfaces.

The effect of the simultaneous presence of organic molecules and bacteria was studied in the photocatalytic processes, using *E. coli* and resorcinol and hydroquinone as the biological and chemical targets, respectively. It was found that the presence of resorcinol affected, to a larger extent, the inactivation of *E. coli* compared to hydroquinone. Resorcinol seems to interact preferentially with TiO_2 by forming inner-sphere complexes, that is, chemisorptions at acidic and neutral pH values; thus the chemisorbed resorcinol can react rapidly with shallow trapped holes competing efficiently either for photoinduced holes or ROS. Hydroquinone molecules do not interact strongly with TiO_2 surfaces leading only to a minor photocatalytic inactivation. These results reveal that if there is a strong interaction between chemical substances and TiO_2 surfaces, the photocatalytic inactivation of bacteria/pathogens will be negatively affected. Photocatalytic oxidation of chemical substances produced during the degradation process could also produce toxic by-products or photosensitizers, which may affect positively or negatively the inactivation process [21].

11.2.1.3 pH Influence

It was recently reported that the photocatalytic *E. coli* inactivation using Degussa-Evonik TiO_2 P25 in Milli-Q water did not vary between pH 4.0 and 9.0 [20], showing that pH range affects the bacteria survival besides the properties of the TiO_2 used. For instance, modification of the pH during illumination by successive acid addition (without TiO_2) resulted in a greater diminution of bacteria compared with the no acidified system. This suggests that photocatalytic disinfection results from the accumulative and possible synergistic effect of three factors: pH modification, light, and photogenerated oxidative species in solution/suspension.

The solution pH modifies the surface charge of TiO_2 and the net charge of the outer bacteria membrane. The IEP of anatase TiO_2 particles is close to 6.0–7.0 [23, 33]. When the initial pH of solution is below Z-potential, anatase particles become slightly positive charged, whereas when the pH is beyond this value, the surface particles will be negatively charged. *E. coli* cells have also a Z-potential around 4.3 [34]; thus under physiological pH conditions (pH $\sim$ 7), an interaction between slightly positive charged surface particles of TiO_2 and negatively charged cell membranes might take place affecting the inactivation process [31]. On a strictly electrostatic basis, the *E. coli* abatement should be much less favored at pH values higher than 7.0 and below 4.3, as *E. coli* cells and the surface of TiO_2 are negatively or positively charged creating repulsion between cells and the TiO_2 particles.

In addition, a change in the initial pH of solution containing chemical substances and bacteria would change not only the equilibrium (i.e., amphoteric) of certain anions and the TiO_2 itself but also the ionic strength and modify the interactions between TiO_2 and bacteria and consequently the response of bacteria to the light irradiation.

11.2.1.4 Oxygen Concentration

As mentioned above, dissolved oxygen (DO) molecules are acceptor of electrons from the CB of TiO_2 and are transformed into superoxide anion radicals. This species is disproportionate generating H_2O_2 (Eqs. (11.1–11.5)) or singlet oxygen ($^1O_2$). High concentrations of DO led to a decrease of charge carrier recombination with concomitant increase of the ROS concentration. It was observed that the presence of oxygen enhanced the inactivation process. The effect of oxygen on bacterial photocatalytic inactivation is crucial in water containing a large amount of organic matter. For practical reasons, oxygenation of waters seems to be essential to prevent the inhibition of bacterial inactivation [22]. Wei *et al.* [35] reported that bactericidal activity increased with the concentration of DO in TiO_2 suspensions under solar irradiation. Added aeration exerted a dramatic effect on the bacterial inactivation of irradiated TiO_2 suspensions when changing the gas added from 100% O_2 to 100% N_2. Note that SODIS depends on the DO concentration even in the absence of TiO_2 [36, 37].

11.2.2
Physicochemical Characteristics of TiO_2

Gumy *et al.* [23] reported the *E. coli* inactivation in suspension scanning 13 different TiO_2 with different crystalline structure, IEP, specific surface area (S), primary size particle, and aggregation size. Except for two mixed crystalline phase anatase–rutile materials from Degussa, P25, and P25 TN90, all photocatalysts were pure anatase. The particle size of the TiO_2 samples varied from about 5 to 700 nm and with BET (Brunauer-Emmett-Teller) areas ranging from 9 to 335 $m^2\ g^{-1}$. IEP and aggregation size of the materials in suspensions were measured by electroacoustic Z-potential measurements and evaluated at different pH values. The aggregate size appeared to be dependent on the pH value of the suspension. The IEP of the TiO_2 samples ranged from acidic (IEP, 3) to neutral values.

Interestingly, S, particle size, and aggregate size seemed not to be correlated with *E. coli* inactivation. However, surface charges appeared to be crucial for the TiO_2 and bacteria interaction. Indeed, negatively charged TiO_2, that is, TiO_2 with acidic IEP, were not efficient in the *E. coli* inactivation kinetics probably due to the electrostatic repulsions with the *E. coli* outer membrane. This phenomenon was observed while investigating the photocatalysis at different pH values. In contrast, mixed anatase–rutile showed higher inactivating activity than anatase alone. In this case, the transfer of photogenerated electrons is supposed to be enhanced [38]. These results suggest that the preparation method, superficial charge, the size of the aggregates, and TiO_2 crystalline phases, together, play an important role during the photocatalytic bacteria inactivation. Further optimization of these parameters may lead to an increase in bacterial inactivation (Table 11.1).

11.3
Using of N-Doped TiO_2 in Photocatalytic Inactivation of Waterborne Microorganisms

An important drawback of TiO_2 for applications in photocatalysis is that its band gap is rather large (3.0–3.2 eV). Thus, the overall photocatalytic efficiency of TiO_2 is limited to only a small fraction of the solar spectrum corresponding to the UV region ($\lambda < 385$ nm), which accounts for ~5% of the incident solar energy. During the past years, the synthesis of visible-light-response TiO_2 absorbing 45% of the solar radiation has received so much attention. Several approaches have been adressed used to achieve this goal [39] being the N-doping the method that has recently received more attention and has a potential to be used as an effective visible light photocatalyst [40] (Chapters 4 and 5).

It has reported the existence of two kinds of N-doping on TiO_2, namely, substitutional, where O atoms from the surface are replaced by N atoms, and interstitial, where N atoms could be bonded to surface O atoms, the former shows XPS (X-ray Photo-electron spectroscopy) N 1s signals at 396–397 eV and the latter at 399–400 eV, respectively. Depending of the preparation method, it is possible to obtain interstitial or substitutional N-doped TiO_2 [41].

Table 11.1 Resume of effect of physicochemical aspects on TiO_2 photocatalytic *E. coli* inactivation.

Physicochemical aspects	Effect on photocatalytic *E. coli* inactivation
Light intensity	Enhancement of bacteria inactivation is correlated with light intensity.
Light intermittence	Continuous light irradiation is more efficient than intermittent irradiation.
Optimal TiO_2 concentration	The concentration depends of light intensity, type reactor, and bacteria concentration.
Presence of anions	Leads to negative and positive effects such as blocking of active sites on TiO_2, scavenging of photoinduced ROS and h^+_{VB}, and affect negatively or positively TiO_2–bacteria interactions.
Presence of organic matter	Positive effect generating extra ROS by photosensitized reaction. Negative effect by UV-light-absorbing compounds and competition with bacteria cells by ROS.
pH	pH alters the bacteria survival, isoelectric point of TiO_2 and cells, and TiO_2–bacteria interactions.
O_2 concentration	High O_2 concentrations leads to the decreasing of e^-/h^+ recombination and concomitant increasing of ROS concentration.
Physicochemical characteristics of TiO_2	Presence of mixed TiO_2 rutile–anatase phase showed high photocatalytic bacteria inactivation. TiO_2 IEP plays a key role on bacteria–TiO_2 interactions.

DFT (Density Functional Theory) calculations and EPR (Electronic Paramagnetic Resonance) experiments have demonstrated that N-doping could induce the formation of mid-gap electronic states partially filled beyond the VB and responsible for the visible light absorption at wavelengths up to 400 nm. Experimental and theoretical works show that N-doping might reduce the energy formation of oxygen vacancies (V_o) in TiO_2 (4.2 eV in pure anatase TiO_2 and 0.6 eV in N-doped anatase TiO_2) [42]. V_o defects in TiO_2 play an important role as adsorption sites for molecular oxygen and as trapping sites of photogenerated electrons [43].

Few studies on N-doped TiO_2 prepared by sol–gel [44–49] doping commercial TiO_2 powders [50–52] with different N-precursors such as thiourea, urea, and amines have addressed the inactivation of *E. coli* cells using natural or simulated sunlight and artificial visible light.

Wu *et al.* [47] prepared Ag, interstitial-N-codoped TiO_2 particles by sol–gel method using tetramethylammonium as N-precursor, this material led to photocatalytic *E. coli* inactivation under visible light irradiation. The authors show evidence for $OH^•$ radicals that may be responsible of *E. coli* inactivation. This work is controversial because it has been found that in nonmetallic doping, the holes photoinduced on mid-gap states should not have the mobility and oxidative power enough to produce $OH^•$ radicals under visible light irradiation [53, 54].

Recently, studies reported by our group [51, 52] showed evidence about a simple mechanical mixing of thiourea with commercial anatase TiO_2 (Tayca TKP 102) leads interstitial N-doped anatase TiO_2 with visible light absorption due to N-doping and oxygen vacancies (Electronic spin resonance (ESR) signal with g = 2.003). Time-resolved diffuse reflectance (TRDR) technique shows that absorption of visible light photons induces electron promotion to the TiO_2 CB and leads to photocatalytic *E. coli* inactivation when the powders were irradiated with blue light ($400 > \lambda < 500$).

ESR-spin trapping measurements revealed evidence that visible-light-doped anatase TiO_2 particles did not produce $OH^{\bullet}$ radicals. It was found that $^1O_2$ would be the ROS generated by visible light irradiation. The mechanism of $^1O_2$ photosensitization by TiO_2 nanoparticles is closely related to the formation of superoxide radicals ($O_2^{\bullet-}$) due to dioxygen reduction by solvated e^-_{CB} [17]. Thus formed $O_2^{\bullet-}$ species could undergo a further oxidation to form singlet oxygen. This oxidation is thermodynamically favored as E° ($O_2^{\bullet-}/^1O_2$) = 0.34 V versus NHE [15].

This is probably due to the photoexcitation of electrons from localized mid-gap N or S levels to the CB generating localized hole on N or S levels. Reduction of dioxygen by e^-_{CB} is thermodynamically favored producing superoxide radical (E° ($O_2/O_2^{\bullet-}$) = −0.28 V vs NHE at pH = 7). The redox potential of the TiO_2 CB is not modified by nonmetallic doping [53]. Thus, it was suggested that N, S-codoped commercial TiO_2 powders could excite electrons from mid-gap levels to the CB by visible light absorption. These electrons could react with the molecular oxygen previously adsorbed on oxygen vacancies producing superoxide radicals. This latter radical may be oxidized by the holes trapped on the mid-gap N, S-states leading to the singlet oxygen-mediated *E. coli* inactivation (Figure 11.1), as it is well known that $^1O_2$ can participate in lipid peroxidation (LPO) reactions toxic to microorganisms (Figure 11.2).

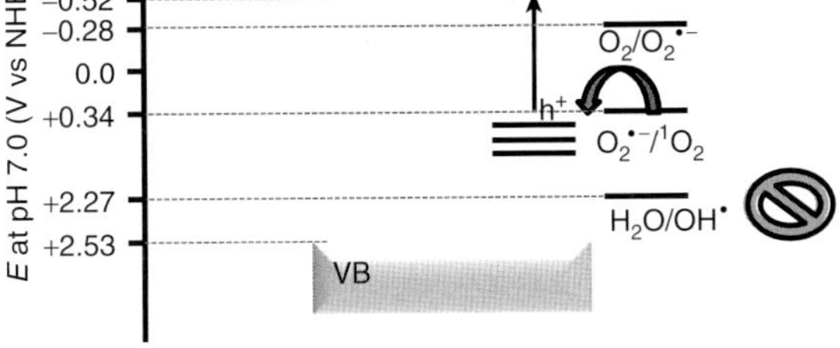

Figure 11.1 Scheme shows that singlet-oxygen generation under visible light irradiation seems to be thermodynamically favored in N, S-codoped TiO_2.

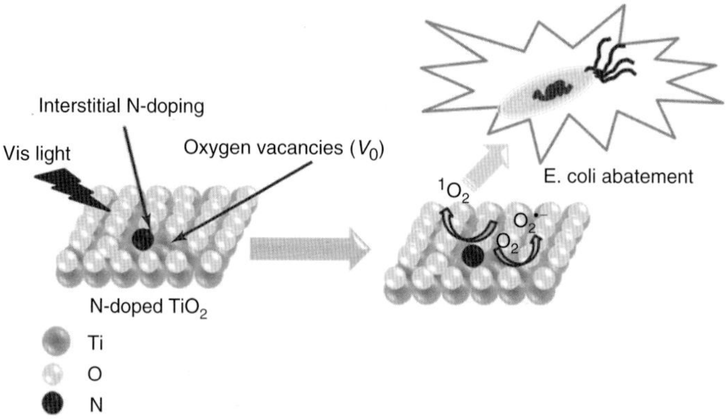

Figure 11.2 Inactivation of *E. coli* cells by N, S-codoped TiO_2 under visible light irradiation.

11.4 Biological Aspects

Different studies reported in the literature have found that the photocatalytic inactivation processes on TiO_2 particles depends mainly on two biological parameters discussed in the following sections.

11.4.1 Initial Bacterial Concentration

The initial bacteria population is an important parameter during the evaluation of the photocatalytic water disinfection efficiency. Examples of photocatalytic inactivation starting at four different initial *E. coli* concentrations confirmed that longer time is required for bacterial inactivation for a high initial bacterial concentration because under these circumstances dead cells and their excreted intracellular components might compete for the photogenerated ROS or screen the light penetration protecting the remaining bacteria in suspension. Moreover, bacterial inactivation follows a first order kinetic behavior [18, 54–56]. However, although at high initial bacterial concentration longer times are required to reach total cells inactivation, the initial deactivation rate expressed in (CFU min^{-1} ml^{-1}) is 10^4 times higher for an initial concentration of 10^7 CFU ml^{-1} than for an initial concentration of 10^2 CFU ml^{-1}.

11.4.2 Physiological State of Bacteria

The influences of the growth state of *E. coli* on UV disinfection have been documented [57, 58]. There are a number of studies reporting that the stationary phase of the bacteria is more resistant to inactivation than exponential phase cells. The stationary-phase response to environmental changes involves the synthesis of a

Table 11.2 Resume of effect of biological aspects on TiO_2 photocatalytic *E. coli* inactivation.

Biological aspects	Effect on photocatalytic *E. coli* inactivation
Initial bacterial concentration	Time required to reach a total cell inactivation increases with the initial bacterial concentration
Physiological state of bacteria	Bacteria in stationary phase were inactivated in lesser extent than those in exponential growth phase

set of proteins conferring *E. coli* resistance to heat shock, oxidation (e.g., by UV light), hyperosmolarity, acidity, and nutrient scarcity [59, 60]. Bacterial inactivation not only depends on the type of bacteria but also on their growth phase [56] (Table 11.2).

11.5 Proposed Mechanisms Suggested for Bacteria Abatement by Heterogeneous TiO_2 Photocatalysis

Only few studies on the mechanisms leading to cell death or cell inactivation have been reported so far [9, 25, 26, 31, 60–69]. The mechanism is not fully understood and some explanations are controversial. The first killing mechanism proposed on the literature, implies an oxidation of the intracellular coenzyme A (CoA), which inhibits the cell respiration and subsequently causes cell death as a result of a direct contact between TiO_2 and the target cell [9]. The second killing mode suggests that bacterial death is caused by a significant disorder in the cell permeability and by the decomposition of the cell walls [26]. It is suggested that the cell wall damages take place prior to the cytoplasmic membrane damage [70, 71]. Photocatalytic treatment progressively increases the cell permeability and subsequently allows the efflux of intracellular constituents, which eventually lead to cell death.

Indeed the mechanism of photocatalytic cell abatement is still under discussion. Often the synergetic effect of light and ROS is neglected and the mechanisms proposed takes into account only the effect of $OH^{\bullet}$ radicals induced by photocatalysis. However, there is experimental evidence showing that the UV-A light (UV-C and UV-B wavelengths will not be taken into account because these wavelengths are not abundant on the terrestrial surface) could affect the *E. coli* antioxidant defenses and degrade intracellular proteins containing iron allowing its release into the cell [68–70]. Here, we suggest the existence of synergic effects responsible of the *E. coli* abatement owing to three processes: photocatalytic ROS production, TiO_2 nanoparticles under dark conditions, and the applied UV-A light.

11.5.1 Effect of UV-A Light Alone and TiO_2 in the Dark

UV-A light does not affect directly the DNA; however, several studies have reported that antioxidant enzymes such as superoxide dismutase (SOD) and catalase might be deactivated by UV-A irradiation [69]. Thus the intracellular presence of $O_2^{\bullet-}$ and H_2O_2 cannot be eliminated. Furthermore iron-containing proteins can also undergo partial degradation releasing Fe(II) ions into the cell [70, 71].

On the other hand, it has been proposed [31, 68] that simple contact or adsorption between TiO_2 nanoparticles and bacteria might produce damage to the cell envelope integrity and more precisely to the outer membrane permeability because TiO_2 is an abrasive material, and this fact could lead to the penetration of deleterious substances. This interaction can be favored at neutral pH, attractive interactions between negatively charged outer cell membrane and slightly positive charged TiO_2 surfaces (Section 11.2.1.3). Thus, it has been suggested that dark interactions between TiO_2 nanoparticles might participate in the photocatalytic bacteria inactivation.

11.5.2 Cell Inactivation by Irradiated TiO_2 Nanoparticles

In Section 11.1, it was mentioned that irradiation of TiO_2 nanoparticles leads to the generation of ROS such as $OH^{\bullet}$ and $O_2^{\bullet-}$ radicals and $^1O_2$, H_2O_2. Often the bactericidal effect of irradiated TiO_2 has been correlated to the attack of $OH^{\bullet}$ radicals on the bacterial outer membrane initiating LPO. Thus, only the hydroxyl radicals yielded on irradiated TiO_2 surfaces and close to the bacteria wall might be responsible for the outer membrane damage. However, due to the high reactivity and short lifetime of $OH^{\bullet}$ radicals induced by TiO_2 irradiation, they cannot penetrate into the cell and attack vital parts of the bacteria, such as the DNA [72]. Although photocatalytically induced $O_2^{\bullet-}$ could also play a role in LPO reactions, at pH values close to neutrality, the superoxide radical-anion undergoes rapid disproportionate reactions yielding H_2O_2 (Eq. (11.4)); thus its direct effect on the photoinduced TiO_2 bactericidal inactivation could be neglected. However, superoxide radicals could play an indirect role by producing H_2O_2 that can alter the permeability of cell outer membrane and penetrate into the cell [71]. The other ROS produced by heterogeneous photocatalysis on TiO_2 is the singlet oxygen $^1O_2$, which is an ROS with a lifetime longer than the $OH^{\bullet}$ radical (~microseconds vs nanoseconds) [72]; $^1O_2$ can react with fatty acids and proteins composing the cell membrane and leads to the initiation of LPO reactions on outer-membranes damaging also the respiratory enzymes along the suppression of the ATP formation [73, 74].

11.6 Conclusion

The photocatalytic *E. coli* inactivation on TiO_2 involves damages on the cell outer membrane (loss of selective permeability) caused by its interaction with TiO_2 surfaces and the attack of photoinduced surface adsorbed $OH^{\bullet}$ radicals and $^1O_2$. The former cannot penetrate into the cell because of its short diffusion length and short lifetime. But, intracellular damages may occur by effect of UV-A irradiation

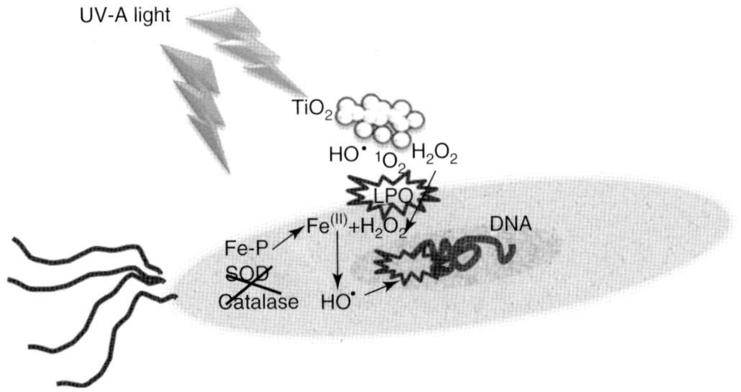

Figure 11.3 Scheme of photocatalytic *E. coli* inactivation by TiO_2 nanoparticles irradiated by UV-A light.

on the antioxidant enzymes (SOD and catalase) and iron-containing proteins. Thus, photocatalytically induced H_2O_2 with a longer lifetime may penetrate into the cell and participate in intracellular Haber–Weiss reactions with Fe(II) ions previously released by the UV-A-induced destruction of iron-containing proteins, yielding intracellular $OH^{\bullet}$ radicals attacking the DNA or RNA causing the cell death [6, 70, 75, 76] (Figure 11.3).

References

1. Cooper, W.J., Cadavid, E., Nickelsen, M.G., Lin, K.J., Kurucz, C.N., Waite, T.D. (1993) Removing THMs from drinking-water using high energy electron-beam irradiation. *J. Am. Water Work Assn.* **85**, 106–112.
2. Richardson, S.D., Termes, T.A. (2005) Water analysis: emerging contaminants and current issues. *Anal. Chem.* **79**, 3807–3838.
3. Bond, T., Goslan, E.H., Parsons, S.A., Jefferson, B. (2011) Treatment of disinfection by-products precursors. *Environ. Technol.* **32**, 1–25.
4. Summer, R., Cabaj, A.,Haider, T. (1996) Microbial effect of reflected UV radiation devices for water disinfection. *Water Res.* **34**, 173–177.
5. Mc Guigan, K., Joyce, T., Conroy, R., Gillespie, J., Elmore-Meegan, M. (1998) Solar disinfection of drinking water contained in transparent plastic bottles: characterizing the bacteria inactivation process. *J. Appl. Microbiol.* **84**, 1138–1148.
6. Berney, M., Weilenmann, H-U., Simonetti, A., Egli, T. (2006) Efficacy of solar disinfection of *Escherichia coli, Shigella flexneri, Salmonella typhimurium* and *Vibrio cholerae*. *J. Appl. Microbiol.* **101**, 828–836.
7. Sciacca, F., Rengifo-Herrera, J.A., Whété, J., Pulgarin, C. (2010) Dramatic enhancement of solar disinfection (SODIS) of wild *Salmonella* sp. in PET bottles by H_2O_2 addition on natural water of Burkina Faso containing dissolved iron. *Chemosphere* **78**, 1186–1191.
8. Gamage, J., Zhang, Z.(2010) Applications of photocatalytic disinfection. *Int. J. Photoenergy*. Art No. 764870.
9. Matsunaga, T., Tomoda, R., Nakajima, T., Wake, H. (1985) Photoelectrochemical sterilization of microbial cells by semiconductor powders. *FEMS Microbiol. Lett.* **29**, 211–214.
10. Friedmann, D., Mendive, C., Bahnemann, D. (2010) TiO_2 for water treatment: parameters affecting the

kinetics and mechanisms of photocatalysis. *Appl. Catal. B: Environ.* **99**, 398–406.

11. Linsebigler, A.L., Lu, G., Yates. J.T. (1995) Photocatalysis on TiO_2 surfaces: principles, mechanisms and selected results. *Chem. Rev.* **95**, 735–758.
12. Fujishima, A., Zhang, X. (2006) Titanium dioxide photocatalysis: present situation and future approaches. *C.R.Chimie* **9**,750–760.
13. Fujishima, A., Zhang, X., Tryk, D.A. (2008) TiO_2 photocatalysis and related surface phenomena. *Surf. Sci. Rep.* **63**, 515–582.
14. Yates, J.T. (2009) Photochemistry on TiO_2: mechanism behind the surface chemistry. *Surf. Sci.* **603**, 1605–1612.
15. Sawyer, D.T., Valentine, J.S. (1981) How super is superoxide? *Acc. Chem. Res.* **14**, 393–400.
16. Nosaka, Y., Daimon, T., Nosaka, A.Y., Murakami, Y. (2004) Singlet oxygen formation in photocatalytic TiO_2 aqueous suspension. *Phys. Chem. Chem. Phys.* **6**, 2917–2918.
17. Daimon, T., Nosaka, Y. (2007) Formation and behavior of singlet molecular oxygen in TiO_2 photocatalysis studied by detection of near-infrared phosphorescence. *J. Phys. Chem. C* **111**, 4420–4424.
18. Bennabu, A.K., Derriche, Z., Felix, C., Moules, V., Lejeune, P., Guillard, C. (2007) Photocatalytic inactivation of *Escherichia coli*. Effect of concentration of TiO_2 and microorganism and intensity of UV irradiation. *Appl. Catal. B-Environ.* **76**, 257–263.
19. Rincón, A.G., Pulgarin, C. (2003) Photocatalytic inactivation of *E. coli*: effect of (continuous-intermittent) light intensity and of (suspended-fixed) TiO_2 concentration. *Appl. Catal. B: Environ.* **44**, 263–284.
20. Rincón, A-G, Pulgarin, C. (2004) Effect of pH, inorganic ions, organic matter and H_2O_2 on E. coli K-12 inactivation by TiO_2: implications in solar water solar disinfection. *Appl. Catal. B: Environ.* **51**, 283–302.
21. Moncayo-Lasso, A., Mora-Arismendi, L.E., Rengifo-Herrera, J.A., Sanabria, J., Benitez, N., Pulgarin, C. (2011) The detrimental influence of bacteria (*E. coli, Shigella, Salmonella*) on the degradation of organic compounds (and vice-versa) in TiO_2 photocatalysis and near neutral photo-Fenton processes under simulated solar light. *Photoch. Photobiol. Sci.* in press.
22. Rincón, A.G., Pulgarin, C. (2005) Use of coaxial photocatalytic reactor (CAPHORE) in the TiO_2 photo-assisted treatment of mixed *E. coli* and *Bacillus sp.* and bacterial community present in wastewater. *Catal. Today* **101**, 331–344.
23. Gumy, D., Morais, C., Bowen, P., Pulgarin, C., Giraldo, S., Hadju, R., Kiwi, J. (2006) Catalytic activity of commercial TiO_2 powders for the abatement of the bacteria (*E. coli*) under solar simulated light: influence of the isoelectric point. *Appl. Catal. B: Environ.* **63**, 76–84.
24. Blake, D.M., Maness, P.C., Huang, Z., Wolfrum, E.J., Huang, J., Jacoby, W.A. (1999) Application of the photocatalytic chemistry of titanium dioxide to disinfection and the killing of cancer cells. *Sep. Purif. Meth.* **28**, 1–50.
25. Maness, P.C., Smolinski, S., Blake, D.M., Huang, Z., Wolfrum, E.J., Jacoby, W.A. (1999) Bactericidal activity of photocatalytic TiO_2 reaction: toward an understanding of its killing mechanism. *Appl. Environ. Microb.* **65**, 4094–4098.
26. Saito,T., Iwase, T., Horie, J., Morioka, T. (1992) Mode of photocatalytic bactericidal action of powdered semiconductor TiO_2 on mutants *Streptococci*. *J. Photoch. Photobiol. B* **14**, 369–379.
27. Chen, H.Y., Zahraa, O., Bouchy, M. (1997) Inhibition of the adsorption and photocatalytic degradation of an organic contaminant in an aqueous suspension of TiO_2 by inorganic ions. *J. Photoch. Photobiol. A* **108**, 37–44.
28. Abdullah, M., Low, G.K.C., Mattheus, R.W. (1990) Effects of common inorganic anions on rates of photocatalytic degradation of organic carbon over illuminated titanium dioxide. *J. Phys. Chem.* **94**, 6820–6825.
29. Poulina, A., Mikhailova, S.S. (1995) Influence of impurities on adsorption interaction between surfactants and titanium dioxide. *Colloid. J.* **57**, 116–117.

30. Bahnemann, D., Cunningham, J., Fox, M.A., Pelizzetti, E., Pichat, P., Serpone, N. in *Aquatic and Surface Photochemistry*, Chapter 21 (eds G.R. Helz, R.G. Zepp, D.G. Crosby), Lewis Publisher, pp. 261–316, 1994.
31. Gogniat, G., Thyssen, M., Denis, M., Pulgarin, C., Dukan, S. (2006) The bactericidal effect of TiO_2 photocatalysis involves adsorption onto catalyst and the loss of membrane integrity. *FEMS Microbiol. Lett.* **258**, 18–24.
32. Rincon, A.G., Pulgarin, C., Adler, N., Peringer, P. (2001) Interaction between E. coli inactivation and DBP-precursors-dihydroxybenzene isomers-in the photocatalytic process of drinking-water disinfection with TiO_2. *J. Photoch. Photobiol. A* **139**, 233–241.
33. Jaffrezic-Renault, N., Pichat, P., Foissy, A., Mercier, R. (1986) Effect of deposited Pt particles on the surface charge of TiO_2 aqueous suspensions by potentiometry, electrophoresis and labeled ion adsorption. *J. Phys. Chem.* **90**, 2733–2738.
34. Fang, L., Cai, P., Chen, W., Liang, W., Hong, Z., Huang, Q. (2009) Impact of cell wall structure on the behavior of bacterial cells in the binding of cooper and cadmium. *Colloid Surf. A* **347**, 50–55.
35. Reed, R.H. (2004) The inactivation of microbes by sunlight: solar disinfection as a water treatment process. *Adv. Appl. Microbiol.* **54**, 333–365.
36. Reed, R.H., Mani, S.K., Meyer,V. (2000) Solar photo-oxidative disinfection of drinking water: preliminary field observations. *Lett. Appl. Microbiol.* **30**, 432–436.
37. Hurum, D.C., Agrios, A.G., Crist, S.E., Gray, K.A., Rajh, T., Thurnauer, M.C. (2006) Probing reaction mechanisms in mixed phase TiO_2 by EPR. *J. Electron Spectrosc.* **150**, 155–163.
38. Chatterjee, D., Dasgupta, S. (2005) Visible light induced photocatalytic degradation of organic pollutants. *J. Photochem. Photobiol. C* **6**, 186–205.
39. Zhang, J., Wu, Y., Xing, M., Leghari, S.A.K., Sajjad, S. (2010) Development of modified N-doped TiO_2 photocatalyst with metals, non metals and metal oxides. *Energy Environ. Sci.* **3**, 715–726.
40. Livraghi, S., Chierotti, E., Giamello, E., Magnacca, G., Paganini, G., Cappelletti, G., Bianchi, C.L. (2008) Nitrogen doped titanium dioxide active in photocatalytic reactions with visible light: a multi-technique characterization of differently prepared materials. *J. Phys. Chem. C* **112**,17244–17252.
41. Di Valentin, C., Pacchioni, G., Selloni, A., Livraghi, S., Giamello, E. (2005) Characterization of paramagnetic species in N-doped TiO_2 powders by EPR spectroscopy and DFT calculations. *J. Phys. Chem. B.* **109**, 11414–11419.
42. Thompson, T.L., Yates, J.T. (2006) Surface science studies on the photoactivation of TiO_2-new photochemical processes. *Chem. Rev.* **106**, 4428–4453.
43. Li, Y., Li, J., Qiu, X., Burda, C. (2007) Bactericidal activity of nitrogen-doped metal oxide nanocatalysts and the influence of bacterial extracellular polymeric substances (EPS). *J. Photochem. Photobiol. A.* **190**, 94–100.
44. Li, Y., Li, J., Qiu, X., Burda, C. (2006) Novel TiO_2 nanocatalysts for wastewater purification: tapping energy from the sun. *Water Sci. Technol.* **54**, 47–54.
45. Li, Q., Xie, R., Li, Y.W., Mintz, E., Shang, J.K. (2007) Enhanced visible-light induced photocatalytic disinfection of *E. coli* by carbon-sensitized nitrogen-doped titanium oxide. *Environ. Sci. Technol.* **41**, 5050–5056.
46. Cheng, C.C., Sun, D.S., Chu, W.-C., Tseng, Y.-H., Ho, H.-C., Wang, J.-B., Chung, P.-H., Chen, J.-H., Tsai, P., Lin, N.T., Yu, M.S., Chang, H.-H. (2009) The effects of the bactericidal interaction with visible –light responsive titania photocatalyst on the bactericidal performance. *J. Biomed. Sci.***16**, Art. No. 7.
47. Wu, P., Xie, R., Imlay, K., Shang, J.K. (2010) Visible-light-induced bactericidal activity of titanium dioxide co-doped with nitrogen and silver. *Environ. Sci. Technol.* **44**, 6992–6997.
48. Wong, M.-S., Chu, W.-C., Sun, D.-S., Huang, H.-S., Chen, J.-S., Tasi,

P.-J., Lin, N.-T., Yu, M.-S., Hsu, S.-T., Wang,S.-H., Chang, H.-H. (2006) Visible-light-induced bactericidal activity of a nitrogen-doped titanium photocatalyst against human pathogens. *Appl. Environ. Microbiol.* **72**, 6111–6116.

49. Rengifo-Herrera, J.A., Mielczarski, E., Mielczarski, J., Castillo, N.C., Kiwi, J., Pulgarin, C. (2008) *Escherichia coli* inactivation by N, S-co-doped commercial TiO_2 powders under UV and visible light. *Appl. Catal. B* **84**,448–456.

50. Rengifo-Herrera, J.A., Pierzchala, K., Sienkiewicz, A., Forró, L., Kiwi, J., Pulgarin, C. (2009) Abatement of organics and *Escherichia coli* by N, S co-doped TiO_2 under UV and visible light. Implications of the formation of singlet oxygen ($^1O_2$) under visible light. *Appl. Catal. B: Environ.* **88**, 398–406.

51. Rengifo-Herrera, J.A., Pierzchala, K., Sienkiewicz, A., Forró, L., Kiwi, J., Moser, J.E., Pulgarin, C. (2010) Synthesis, characterization and photocatalytic activities of nanoparticulate N,S-co-doped TiO_2 having different surface-to-volume ratios. *J. Phys. Chem. C.* **114**, 2717–2723.

52. Mrowetz, M.; Balcerski, W.; Colussi, A. J.; Hoffmann; M. R. (2004) Oxidative power of nitrogen-doped TiO_2 photocatalysts under visible illumination. *J. Phys. Chem. B* **108**, 17269–17263.

53. Tachikawa, T., Tojo, S., Kawai, K., Endo, M., Fujitsuka, M., Ohno, T., Nishijima, K., Miyamoto, Z., Majima T. (2004) Photocatalytic oxidation reactivity of holes in the sulfur- and –carbon-doped TiO_2 powders studied by time-resolved diffuse reflectance spectroscopy. *J. Phys. Chem. B* **108**, 19299–19306.

54. Wei, C., Lin, W.Y., Zainal, Z., Williams, N.E., Zhu, K., Kruzic, A.P., Smith, R.L., Rajeshwar, K. (1994) Bactericidal activity of TiO_2 photocatalyst in aqueous-media-toward a solar-assisted water disinfection system. *Environ. Sci. Technol.* **28**, 934–938.

55. Huang, N., Xiao, Z., Huang, D., Yuan, C. (1998) Photochemical disinfection of Escherichia coli with a TiO_2 colloid solution and a self-assembled TiO_2 thin film. *Supramol. Sci.* **5**, 559–564.

56. Rincón, A.G., Pulgarin, C. (2004) Bactericidal action of illuminated TiO_2 on pure *Escherichia coli* and natural bacterial consortia: post-irradiation events in the dark and assessment of the effective disinfection time. *Appl. Catal. B: Environ.* **49**, 99–112.

57. Kadavy, D.R., Shaffer, J.J., Lott, S.E.,Wolf, T.A., Bolton, C.E., Gallimore, W.H., Martin, E.L., Nickerson, K.W., Kokjohn, T.A. (2000) Influence of infected cell growth state on bacteriophage reactivation levels. *Appl. Environ. Microbiol.* **66**, 5206–5012.

58. Lewis, C.K. Burt-Maxcy, R. (1984) Effect of physiological age on radiation resistance of some bacteria that are highly radiation resistant. *Appl. Environ. Microbiol.* **47**, 915–918.

59. Child, M., Strike, P., Pickup, R., Edwards, C. (2002) *Salmonella typhimurium* displays cyclical patterns of sensitivity to UV-C killing during prolonged incubation in the stationary phase of growth. *FEMS Microbiol. Lett.* **213**, 81–85.

60. Murno, P.M., Flatatau, G.N., Clement, L.R., Gauthier, M.L. (1995) Influence of RpoS (KatF) sigma factor on maintenance of viability and culturability of *Escherichia coli* and *Salmonella typhimurium* in seawater. *Appl. Environ. Microbiol.* **61**, 1853–1858.

61. Ireland, J.C., Klostermann, P., Rice, E.W., Clark, R.M. (1993) Inactivation of *Escherichia coli* by titanium dioxide photocatalytic oxidation. *Appl. Environ. Microbiol.* **59**, 1668–1670.

62. Kikuchi, Y., Sunada, K., Iyoda, T., Hashimoto, K., Fujishima, A. (1997) Photocatalytic bactericidal effect of TiO_2 thin-films- dynamic view of the active oxygen species responsible of the effect. *J. Photoch. Photobiol. A* **106**, 51–56.

63. Huang, Z., Maness, P.C., Blake, D.M., Wolfrum, E.J., Smolinski, S.L., Jacoby, W.A. (2000) Bactericidal mode of titanium dioxide photocatalysis. *J. Photochem. Photobiol. A* **130**, 163–170.

64. Salih, F.M. (2002) Enhancement of solar inactivation of *Escherichia coli* by titanium dioxide photocatalytic oxidation. *J. Appl. Microbiol.* **92**, 920–926.

65. Sunada, K., Watanabe, T., Hashimoto, K. (2003) Studies of photokilling of bacteria on TiO_2 thin film. *J. Photochem. Photobiol. A* **156**, 227–233.

66. Dalrymple, O.K., Stefanakos, E., Trotz, M.A., Goswami, Y. (2010) A review of the mechanisms and modeling of photocatalytic disinfection. *Appl. Catal. B: Environ.* **98**, 27–38.

67. Markowska-Szczupak, A., Ulfig, K., Morawski, A.W. (2011) The application of titanium dioxide for deactivation of bioparticulates: an overview. *Catal. Today* **169**, 249–257.

68. Piegeot-Rémy, S., Simonet, F., Errazuriz-Cerda, E., Lazzaroni, J.C., Atlan, D., Guillard, C. (2011) Photocatalysis and disinfection of wáter: identification of potential bacterial targets. *Appl. Catal. B: Environ.* **104**, 390–398.

69. Kapuscinski, R.B., Mitchell, R. (1981) Solar irradiation induces sublethal injury in *Escherichia coli* in seawater. *Appl. Environ. Microbiol.* **41**, 670–674.

70. Imlay, J.A. (2003) Pathways of oxidative damage. *Annu. Rev. Microbiol.* **57**, 395–418.

71. Kruszewski, M. (2003) Labile iron pool: the main determinant of cellular response to oxidative stress. *Mutat. Res.-Fund. Mol. Mech. Mutagen* **531**, 81–92.

72. Cheng, Z., Li, Y. (2007) What is responsible for the initiating chemistry of iron-mediated lipid peroxidation: an update. *Chem. Rev.* **107**, 748–766.

73. Tatsuzawa, H., Maruyama, T., Misawa, N., Fujimori, K., Hori, K., Sano, Y., Kambayashi, Y., Nakano, M. (1998) Inactivation of bacterial respiratory chain enzymes by singlet oxygen. *FEBS Lett.* **439**, 329–333.

74. Valduga, G., Bertolini, G., Reddi, E., Jori, G. (1993) Effect of extracellularly generated singlet oxygen on Gram-positive and Gram- negative bacteria. *J. Photochem. Photobiol. B.* **21**, 81–86.

75. Imlay, J.A. (2008) Cellular defenses against superoxide and hydrogen peroxide. *Annu. Rev. Biochem.* **77**, 755–776.

76. Halliwell, B., Gutteridge, J.M. (1992) Biologically relevant metal ion-dependent hydroxyl radical generation. *FEBS Lett.*, **307**, 108–112.

Part IV
Modeling. Reactors. Pilot plants

12
Photocatalytic Treatment of Water: Irradiance Influences

David Ollis

12.1
Introduction

12.1.1
Chapter Topics

Photocatalysis is an example of indirect photochemistry, in which a nonreactant is photoexcited and subsequent thermal chemistry is involved in reactant conversion. We discuss examples of irradiance dependent photocatalyzed kinetics. In each case, a simple kinetic model is used to illustrate a key irradiance influence. The examples considered are listed below.

Section	Section title
12.2	Reaction order in irradiance: Influence of electron–hole recombination and the high irradiance penalty
12.3	Langmuir–Hinshelwood kinetic form: Equilibrated adsorption
12.4	Pseudo-steady-state analysis: Nonequilibrated adsorption
12.5	Mass transfer influence with immobilized photocatalysts
12.6	Controlled periodic illumination: Attempt to beat recombination
12.7	Solar-driven photocatalysis: Nearly constant nUV irradiance
12.8	Mechanism of hydroxyl radical attack: Same irradiance dependence
12.9	Simultaneous homogeneous and heterogeneous photochemistry
12.10	Dye-photosensitized oxidation
12.11	Interplay between fluid residence times and irradiance profiles
12.12	Quantum yield, photonic efficiency, and electrical energy per order

12.1.2
Photon Utilization Efficiency

At the modest irradiances of conventional light sources, single photon excitation is the rule, so the excitation process is uniformly first order in all cases considered here,

Photocatalysis and Water Purification: From Fundamentals to Recent Applications, First Edition. P. Pichat.

whether of photocatalysis, photosensitized reaction, or photolysis. The interplay of electron–hole recombination (de-excitation), reaction, diffusion, and mass transfer complicates the reaction rate description, and produces rate equations in which the apparent reaction order in irradiance may lie between unity and zero. As measures of photon utilization efficiency all involve a reaction rate divided by a term containing irradiance (I) to the first power, it follows that these efficiencies vary as irradiance to the powers between 0 (no penalty for increasing irradiance) and −1 (severe penalty for increased irradiance). We revisit this topic in the concluding section.

12.2 Reaction Order in Irradiance: Influence of Electron–Hole Recombination and the High Irradiance Penalty

A classical 1979 paper by Egerton and King [1] showed that the photocatalyzed oxidation of neat liquid isopropanol exhibited reactions rates which were first order in near-UV (nUV) irradiance, I, up to about 1 mW cm^{-2}, and transitioned into a half order dependence at higher irradiance values (Figure 12.1). This study over a remarkable 5 orders of magnitude in irradiance has remained a bedrock result. Numerous subsequent papers have observed first order or half order kinetics. Furthermore, as the transition region in Figure 12.1 typifies a number of photocatalysis reports, intermediate values of the irradiance exponent have also

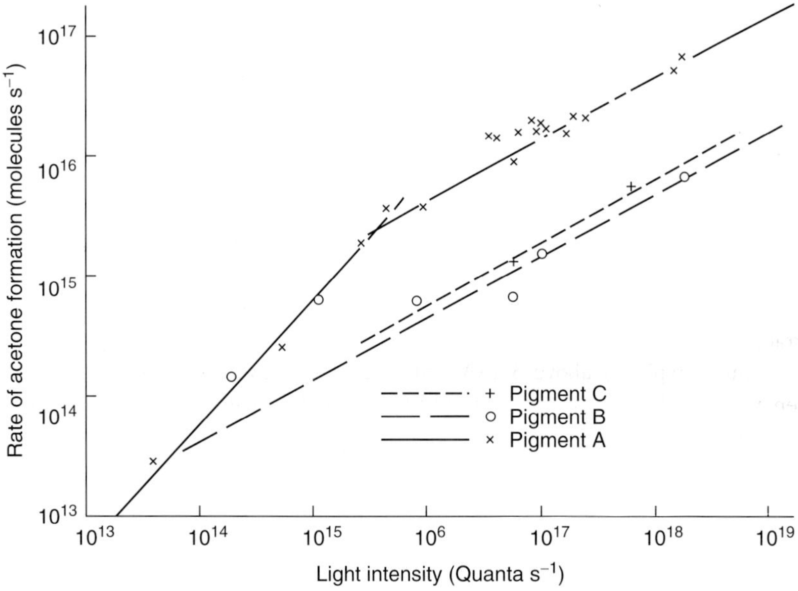

Figure 12.1 Rate of acetone formation from neat isopropanol under near-UV illumination of TiO_2 slurry. (Source: Egerton and King [1], reprinted by permission.)

been reported, most frequently where only a narrow (less than 10-fold) variation in irradiance was explored.

12.3 Langmuir–Hinshelwood (LH) Kinetic Form: Equilibrated Adsorption

This classical law of thermal heterogeneous catalysis invokes equilibrated reactant adsorption, followed by a slow, irreversible catalyzed reaction, as represented in Eqs. (12.1) and (12.2):

$$A + S \underset{\text{(fast)}}{\overset{k_1}{\rightarrow}} AS \underset{\text{(fast)}}{\overset{k_{-1}}{\rightarrow}} A + S \tag{12.1}$$

$$AS \underset{\text{(slow)}}{\overset{k_r}{\rightarrow}} \text{Products} \tag{12.2}$$

We may ask if this form might be valid in photocatalysis as well. A typical photon flux is 1 mW cm^{-2}, which provides the order of 1 incident photon per surface atom per second. As aqueous-phase photocatalysis has apparent quantum efficiencies routinely of the order of one-tenth to several percent, the apparent turnover frequency, F, is one event per 30–1000 s, which is slow compared to the A + S collision kinetic rate of the order of 10^8 s^{-1} in liquid phases. In consequence, Eq. (12.1) is assumed to be in equilibrium, and Eq. (12.2) contains the irradiance dependence, typically of the form given by Eq. (12.3):

$$k_r = k_{ro} I^n \quad \text{where} \quad \frac{1}{2} < n < 1.0 \tag{12.3}$$

We may say that $n = 1$ corresponds to reaction dominance and $n = 1/2$ corresponds to electron–hole recombination dominance.

In most photocatalysis equations used to date, this simple form suffices. Exceptions are discussed in Section 12.4, pseudo-steady-state assumption.

A figure of merit in photochemistry is the quantum yield (reaction events per photon absorbed), or, where absorption data are not available, the photoefficiency (PE) (reaction events per incident photon). In either case, these figures of merit exhibit an irradiance dependence proportional to the power $(n - 1)$. Thus, low irradiances give constant photoefficiencies independent of irradiance, while high irradiances, typically above 5 mW cm^{-2} for TiO_2, give an inverse square root dependence. This simple kinetic fact has prevented the use of strong light sources, or light concentrators, from achieving economically practical importance.

The classical equilibrium coverage, $Ø_e$, on a uniform surface by an adsorbing reactant A is given by Eq. (12.4):

$$Ø_e = \frac{K_a C}{1 + K_a C} \tag{12.4}$$

where K_a = adsorption equilibrium constant = k_1/k_{-1} and $Ø_e$ = equilibrium surface coverage at reactant concentration C.

The reaction rate for the slow step (Eq. (12.2)) is then given by Eq. (12.5):

$$r = \text{Rate} = k_r \, Ø_e = k_{ro} \, I^n \left(\frac{K_a C}{1 + K_a C} \right) \tag{12.5}$$

Literature values of n are routinely reported, which lie between ½ and 1.0.

At constant irradiance, a reciprocal rate versus reciprocal concentration plot is given by Eq. (12.6):

$$\frac{1}{r} = \left(\frac{1}{k_{ro} I^n} \right) \left(\frac{1 + K_a C}{K_a C} \right) = \left(\frac{1}{k_r} \right) \left(1 + \left(\frac{1}{K_a C} \right) \right) \tag{12.6}$$

which allows evaluation of the rate constant, k_r (intercept = $1/k_r$) and equilibrium adsorption constant K_a (slope = $1/(k_r\ K_a)$).

Early aqueous-phase studies demonstrated linear plots for $1/r$ versus $1/C_o$ for a wide range of reactants (Figure 12.2a,b).

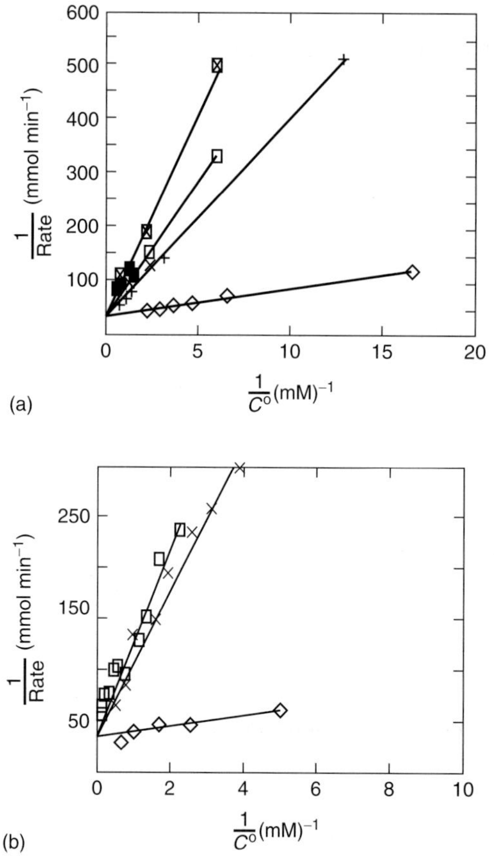

Figure 12.2 Reciprocal initial rates versus reciprocal initial concentration for (a) chloroform (squares), dichloromethane (×), perchloroethylene (diamonds), and dichloroacetic acid (+). (b) 1,2-dichloroethane (squares), 1,2-dichloroethane, and vinyl chloride (triangles) [2]. (Source: Reprinted by permission of Academic Press.)

Although the influence of concentration on rate was frequently measured, irradiance influences were studied less frequently, so the intricacies of irradiance dependence still continue to be revealed. These early studies summarized by Turchi and Ollis [2] provided an additional feature. Reciprocal plots for many reactants showed the expected linearity, but produced, for a given catalyst, reactor, and lamp, vertical intercepts which were essentially identical, independent of reactant structure (Figure 12.2a,b). This result implies that the maximum rate is given by Eq. (12.7):

$$\text{Rate (max)} = k_{\text{r}} = k_{\text{ro}} I^n \tag{12.7}$$

Thus, the maximum photocatalytic rate is typically independent of reactant structure and depends only on incident irradiance, I, and the particular photocatalyst at hand. This asymptotic reaction rate behavior is not found for conventional, thermal heterogeneous catalysts. The implication is clear that flooding the photocatalyst surface with all participating reactants ($1/C \rightarrow 0$) results in an irradiance – limited reaction rate. Under these conditions, the PE for reaction may be near unity, as observed early by Egerton and King with neat liquid isopropanol oxidation [1], and later by Xu and Langford with neat liquid acetophenone [3].

12.4 Pseudo-Steady-State Analysis: Nonequilibrated Adsorption

The validity of the Langmuir–Hinshelwood (LH) rate form in Section 12.3 above rests on the assumption of rapid adsorption and desorption steps, compared to an assumed slow reaction step (Eq. (12.2)) on the active catalyst surface. As photocatalysis generates highly reactive intermediates such as holes (h^+) and oxygen radicals (OH, OH_2, O_2^-, etc.) with short lifetimes, we may expect this LH assumption to be invalid, especially at ambient temperatures where an activated desorption step may be quite slow [4].

Application of the pseudo-steady-state hypothesis avoids this LH-equilibrated adsorption assumption [4]. Assuming for simplicity that OH radicals are the dominant active species produced by illuminated TiO_2, a transient surface balance on coverage by reactant A yields Eq. (12.8).

$$\begin{array}{llll} \dfrac{d\varnothing_{\text{A}}}{dt} = k_1 C_{\text{A}} \left(1 - \varnothing_{\text{A}}\right) & - k_{-1} \varnothing_{\text{A}} & - k_{\text{r}}(\text{OH})\, \varnothing_{\text{A}} & \\ \text{Rate} = \text{Adsorption} & - \text{Desorption} & - \text{Reaction} & (12.8) \end{array}$$

We assume that the steady-state hydroxyl concentration (OH) depends on irradiance according to I^n, where $\frac{1}{2} < n < 1.0$ as observed by Egerton and King [1].

Thus, at steady state where $d\varnothing_{\text{A}}/dt = 0$, we have

$$0 = k_1 C_{\text{A}} \left(1 - \varnothing_{\text{A}}\right) - k_{-1} \varnothing_{\text{A}} - k_{\text{ro}} b\, I^n \varnothing_{\text{A}} \tag{12.9}$$

and the steady-state coverage is Eq. (12.10)

$$\text{Ø}_A = \frac{k_1 C_A}{k_1 C_A + k_{-1} + k_{ro} b I^n} \tag{12.10}$$

The reaction rate may again be written as

$$r = \text{Rate} = k_r \text{Ø}_A = k_{ro}\,(\text{OH})\,\text{Ø}_A$$

Assuming that the steady-state hydroxyl (or other active oxidant) concentration depends on irradiance as I^n, we have finally

$$\begin{aligned} r &= \frac{k_{ro} b I^n k_1 C_A}{k_1 C_A + k_{-1} + k_{ro} b I^n} \\ &= \frac{k_{ro} b I^n C_A}{C_A + \left(\frac{k_{-1} + k_{ro} b I^n}{k_1}\right)} \\ &= \frac{k_r C_A}{C_A + K_d\,(\text{apparent})} \end{aligned} \tag{12.11}$$

where K_d(apparent) is the apparent desorption equilibrium constant (= $1/K_a$(apparent)).

The pseudo-steady-state rate Eq. (12.11) has the same dependence on concentration as does the equilibrium formulation, Eq. (12.5). However, now the apparent rate constant k_r and the dissociation constant K_d (= $1/K_a$) both depend on irradiance, as indicated in Eq. (12.12a,b).

$$k\,(\text{app}) = k_r b I^n = k_{LH} \tag{12.12a}$$

$$K_d\,(\text{apparent}) = \frac{k_{-ro} + k_r b I^n}{k_1} \tag{12.12b}$$

For example, photocatalyzed phenol degradation studied by Emeline *et al.* [5] was shown to exhibit k_{LH} and K_d (= K_a^{-1}) values that were each dependent on irradiance to the first power (Figure 12.3a,b) [5].

We note that a plot of K_d versus k_{LH} for a series of irradiance values will give a slope = $1/k_1$ and intercept = k_{-1}/k_1, thus allowing evaluation of the fundamental adsorption and desorption rate constants [4]. For data of Emeline *et al.*, Figure 12.3 shows that $K_d \to 0$ as irradiance $\to 0$, thus indicating that $k_{-1} = 0$ or very small.

This analysis was also found to be consistent with results for phenol [4], 4-chlorophenol [6, 7], and methyl viologen [8]. Our paper [3] also found that k_{-1} is nearly zero, as might be expected for activated desorption at ambient temperatures.

Thus, the major kinetic steps in the limit of very small k_{-1} are

$$\text{A} \xrightarrow[k_1]{} \text{A}_{ads} \xrightarrow[k_r]{} \text{Products} \tag{12.13}$$

so that the reaction appears to consist of two irreversible steps in series. This result confirms our original assumption that the pseudo-steady-state analysis is needed, because adsorption equilibrium does not operate under typical illumination conditions.

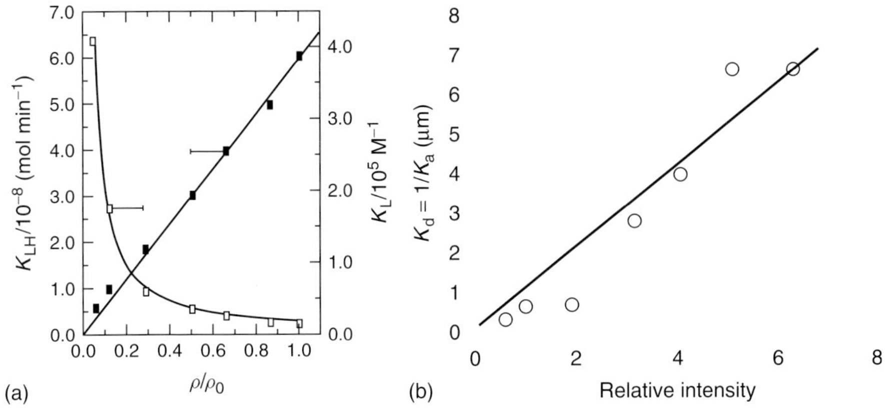

Figure 12.3 (a) Reaction rate constant, k_{LH} and (b) apparent desorption equilibrium constant (inverse of apparent binding constant, K_a in (a)) versus irradiance [5]. (Source: (a) reprinted by permission of American Chemical Society.)

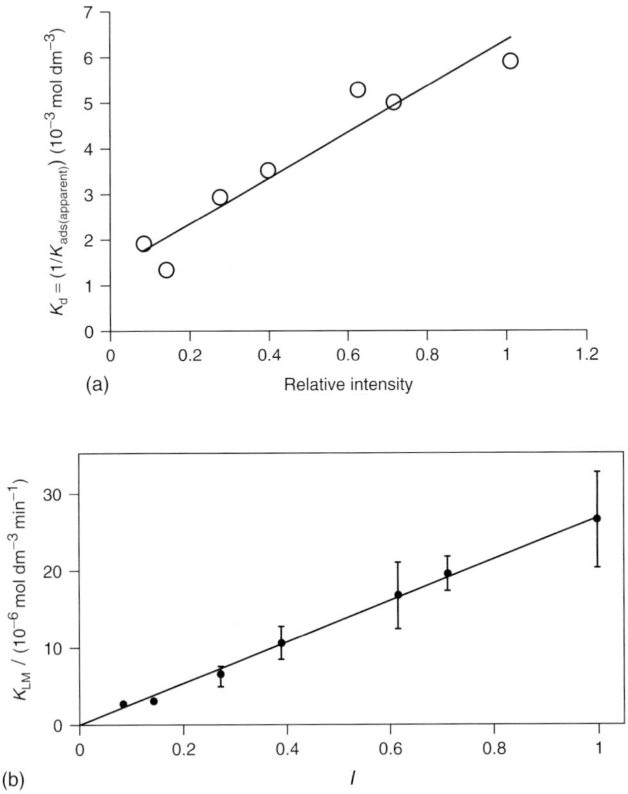

Figure 12.4 (a,b) Apparent desorption equilibrium constant K_d and rate constant k_{LH} versus irradiance [6, 7]. (Source: Reprinted by permission of American Chemical Society.)

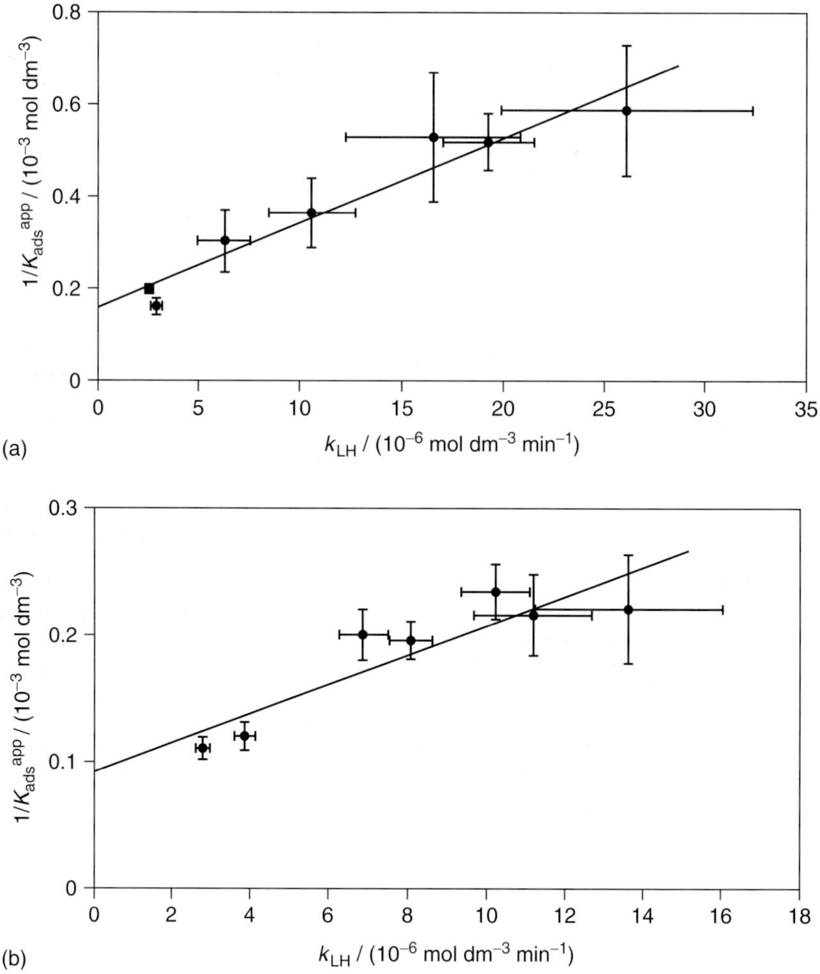

Figure 12.5 Plot of K_d versus k_{LH} for 4-chlorophoenol-photocatalyzed oxidation: (a) TiO_2 film and (b) TiO_2 slurry. (Source: Mills *et al.* [6, 7], reprinted by permission of American Chemical Society.)

Subsequent results by Mills *et al.* [6, 7] with 4-chlorophenol validate this dependence of k_{LH} and K_d on irradiance (Figure 12.4a,b) and linearity in plots of K_d versus k_{LH} (Figure 12.5) for both TiO_2 dispersions (Figure 12.5a,b) and films (Figure 12.5a). For this system, evaluation of the fundamental rate constants produces the following results:

k_1	=	0.054 min^{-1} (film) or 0.087 min^{-1} (slurry)
k_{-1}	=	8.6 μM min^{-1} or 8.0 μM min^{-1}
k_{LH}	=	0–27 μM min^{-1} (film) or 0–14 μM min^{-1} (slurry).

Here again, the equilibrium adsorption assumption fails, because the k_{LH} range is of the same order as k_{-1}, hence the pseudo-steady-state analysis is required.

More detailed analytical models have been proposed by Emeline *et al.* [5] and others. However, the commonality in all cases is that the irradiance dependence of the apparent dissociation and adsorption equilibrium constants implies the need for a pseudo-steady-state analysis, because the equilibrium adsorption model is invalid.

12.5 Mass Transfer and Diffusion Influences at Steady Conditions

In well-dispersed slurry photocatalysts, the particle–particle distance is very small, and dissolved reactants and molecular oxygen, hydrogen peroxide, and so on, can collide frequently with TiO_2 surfaces, so that diffusional mass transfer is never rate limiting or kinetically important.

However, for TiO_2 agglomerates, which may reach sizes of one-tenth to several micrometers, internal diffusion within particle agglomerates may be limiting (Section 12.6). For immobilized photocatalysts, where the distance from bulk solution to the nearest photocatalyst surface may be millimeters to centimeters, external mass transfer resistance may be important: an active photocatalyst may exhaust near surface reactant concentration, so that the observed rate is largely determined by mass transfer (physical) rather than photocatalytic (chemical) rate.

For a thick, porous photocatalyst film immobilized on a solid support, both external mass transfer and internal reactant diffusion may be influential, especially if illumination arrives from the back (support side) of the film, rather than the front (bulk solution) side.

As found in all areas of heterogeneous catalysis, at steady state, the rate of external mass transfer must equal the total rate of reaction in the catalyst. For model convenience, we consider an example where the internal diffusion is unimportant, or the photocatalyst is nonporous, and the reaction rate is first order in reactant C. Then the rate of mass transfer from bulk solution (b) to catalyst surface (S) equals the surface rate:

$$k_m\left(C_b - C_S\right) = k_{LH} C_S \tag{12.14}$$

Solving for surface concentration C_S in terms of the bulk concentration, C_b, gives

$$C_S = \frac{k_m C_b}{k_m + k_{LH}} \tag{12.15}$$

and the corresponding rate equation (assuming first order) is given by

$$r = k_{LH} C_S = \frac{C_b}{1/k_{LH} + 1/k_m} \tag{12.16}$$

Given the assumption in Section 12.2 above that $k_{LH} = k_{ro}\, bI^n$, we have

$$\text{Rate} = r = \frac{C_b}{1/k_m + 1/k_{ro} bI^n} \tag{12.17}$$

In consequence, at low irradiances, $k_m \gg k_{LH}$, and the rate simplifies to

$$r = k_{LH}\text{Ø}_A = k_{ro}bI^n \qquad (12.18)$$

At sufficiently high irradiances, and thus surface rate, C_S will approach zero and the rate observed will be determined by the mass transfer coefficient, k_m:

$$r = k_m\left(C_b - C_S\right) = k_m C_b \ \left(\text{as } C_S \geq 0\right) \qquad (12.19)$$

The relative contribution of mass transfer conductance, k_m, versus catalytic rate conductance, k_{LH}, is given by the Damkohler ratio:

$$\text{Da} = \frac{k_{LH}}{k_m} = \frac{k_{ro}bI^n}{k_m} \qquad (12.20)$$

As lamp power and/or use of solar concentrators increases irradiance, I, the fluid flow rate must be increased to keep $\text{Da} < 0.1$ (free of mass transfer effect), or else the reaction will become more limited by mass transfer as irradiance, I, increases.

It is also clear that such mass transfer resistance will depress the quantum yield, or a surrogate ratio, PE, that of reaction rate divided by photon flux.

$$\text{PE} = \frac{\text{Rate}}{\text{irradiance}} = \frac{r}{I} = \frac{\left(C_b/I\right)}{1/k_m + 1/k_{ro}bI^n} \qquad (12.21)$$

In the limit of high irradiance, the reaction becomes mass transfer limited and the PE decreases as I^{-1}.

$$\text{PE} = \frac{k_m C_b}{I} \qquad (12.22)$$

At low irradiances, reaction rate is the dominant resistance, and reactant is plentiful at the catalyst surface. Here we expect $n = 1$, so the rate simplifies to k_{ro} C_b I, and the PE is independent of I.

$$\text{PE} = k_r C_b \qquad (12.23)$$

As noted earlier, the reaction order dependence on irradiance may exhibit three regimes [9]:

Low I	Rate varies as I^1 QE (quantum efficiency) and PE = constant
Intermediate I	Rate varies as $I^{1/2}$ (recombination dominant) QE and PE vary as $I^{-1/2}$
High I	Rate independent of I QE and PE vary as I^{-1}

For slow fluid motion where k_m is small and the reactant arrives by pure diffusion in water, the mass transfer limit may be reached before achievement of recombination kinetic dominance.

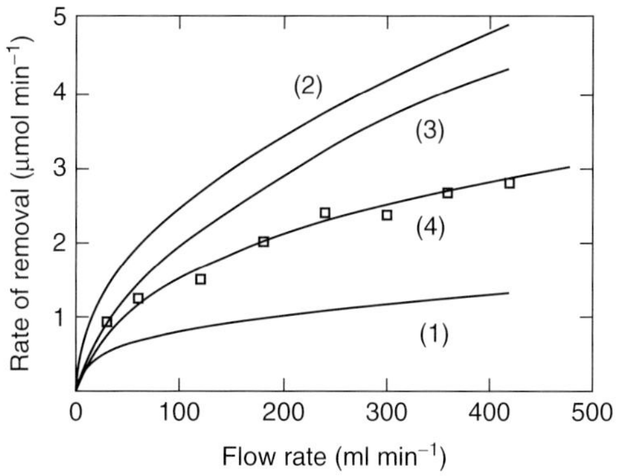

Figure 12.6 Rate of photocatalyzed oxidation of salicylic acid in a helical quartz reactor; catalyst immobilized on internal wall [11]. (Source: Reprinted by permission of American Chemical Society.)

An example of mass transfer influence is evident in early work of Matthews [10]. Figure 12.6 displays the observed rate of photocatalyzed aqueous-phase oxidation of salicylic acid in a helical quartz reactor which had a photocatalyst film on its interior wall. The calculated curves of reaction rate versus flow rate represent purely mass transfer limited behavior for flow in a straight tube (1), in a coiled tube (2), and in a coiled tube where the bulk reactant concentration decreases from inlet to outlet (3) [11]. Fitting the curve (3) equation with a photochemical reaction term allowed achievement of curve (4) [5]. This good model fit shows that at Matthew's irradiance, his data were entirely mass transfer dominated at flow rates below 75 ml min^{-1}. Even at the maximum experimental flow rate of 400 ml min^{-1}, mass transfer was 75% of the total resistance. Thus immobilization of photocatalyst adds a further photonic penalty to a photocatalyzed reaction, with corresponding decrease in quantum or photonic efficiency. In contrast, use of a slurry reactor provides uniform distribution of catalyst throughout the reactor, so that mass transfer influence is negligible because the reactant–catalyst distance is small everywhere.

12.6
Controlled Periodic Illumination: Attempt to Beat Recombination

As steady-state studies have routinely shown, increasing irradiances leads to a decrease in photonic efficiency or quantum yield, however the figure of merit is defined. As electron discharge to dioxygen (molecular oxygen) is frequently the assumed slow step in photocatalysis, as demonstrated earlier by Gerischer and Heller [12a,b], consideration of a dark period to allow electron discharge has arisen. These authors demonstrated that illuminated titania particles acquire a long-lived negative charge, which could be collected when TiO_2 particles in an illuminated region are diffused into a dark region and discharged on a collector electrode.

As undesired recombination is a second order process, the reduction of excess electrons by periodic light and dark intervals, named *controlled periodic illumination* or *CPI*, was thought to contribute to an enhanced PE, first reported by Sczechowki *et al.* [13]. This transient situation was modeled by Upadhya and Ollis [14] for the case of oxygen deficiency and formate abundance, finding that "model simulation provides trends of PE with variations in the length of light and dark periods which resemble the PE trends" observed by Szechowski [12].

However, subsequent reports by Buechler *et al.* [15, 16] for both slurry and immobilized titania photocatalysts concluded that "the periodic illumination effects reported for this reaction, and for other photocatalytic oxidations resulted instead from intraparticle diffusion in flocculated particles, mass transport of oxygen to the catalyst surface, slow or weak adsorption of formate ion, or a combination of these processes." In all cases, utilization of a dark period allows for increased dissolved oxygen and reactant (formate ion) within the titania agglomerates. Upon re-illumination, an initial rate higher than that observed at steady-state results briefly, until the steady state is again restored, typically after about 3 s.

This conclusion was verified by Cornu *et al.* [17] who explored periodic illumination of highly dispersed (sonicated), dilute slurries (which should contain no substantial acid-induced agglomerates), finding that the reaction rate per absorbed photon was the same for constant or periodic illumination. We conclude that CPI does not enhance rate versus steady state obtained in the absence of internal diffusion or external mass transfer effects, but if enhancement occurs, it indicates the presence of such physical resistances.

The level of photon flux, or irradiance, does cause a change in the operating kinetic regime. Thus Buechler *et al.* [15, 16] reported using CPI with a rotating disk of immobilized titania, finding that "as the light irradiance was increased from 0.05 to 5 mW cm^{-2}, the PE for continuous illumination experiments decreased from 80 to 5%. At a light time of 0.6 s and a dark time of 2.0 s and a light irradiance of 5.5 mW cm^{-2}, the PE increased from 5% during continuous illumination experiments to 20% with CPI. However, at low irradiances ($I < 0.5$ mW cm^{-2}), CPI did not increase the PE. Analysis of the results indicates that the reactor is oxygen diffusion-limited at light irradiances above 0.5 mW cm^{-2} when air is used as the oxidant. At irradiances below 0.3 mW cm^{-2} the reaction is photon limited . . . " and the intrinsic chemical kinetics are observable. "At light irradiances between 0.3 and 0.5 mW cm^{-2}, the reaction is controlled by both surface kinetics and diffusional limitations" [16].

We conclude by reference to Figure 12.7 [15, 16] (Case I) that small single particles (Figure 12.7a) dispersed in dilute slurries do not show diffusion effects [17], whereas large agglomerates (Figure 12.7b) or in immobilized titania films (Figure 12.7c) will evidence such mass transfer disguises [15–18].

12.7
Solar-Driven Photocatalysis: Nearly Constant nUV Irradiance

Ahmed [19] examined solar photocatalysis in a tubular slurry photoreactor. Their initial rate data versus time of day (Figure 12.8) showed a nearly constant rate over the summer time interval between 2.5 h after sunrise and before sunset, in

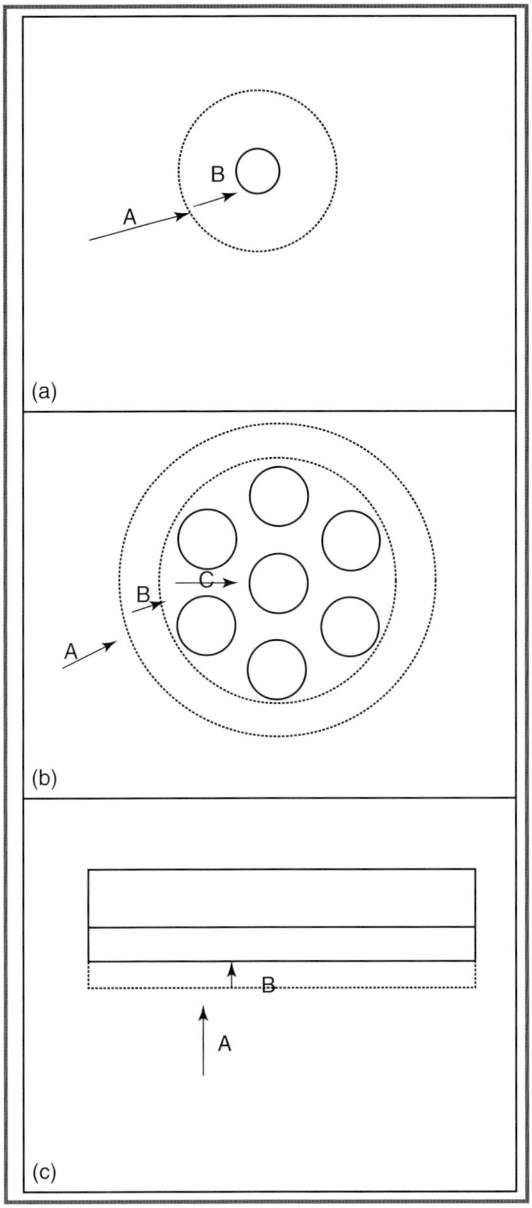

Figure 12.7 Mass transport through well-mixed bulk phase (A), external film (B), and internal porous agglomerate (C): (a) uniformly dispersed particles, (b) flocculated particles, and (c) rough film of TiO_2 supported on SS [15]. (Source: Reprinted by permission of American Chemical Society.)

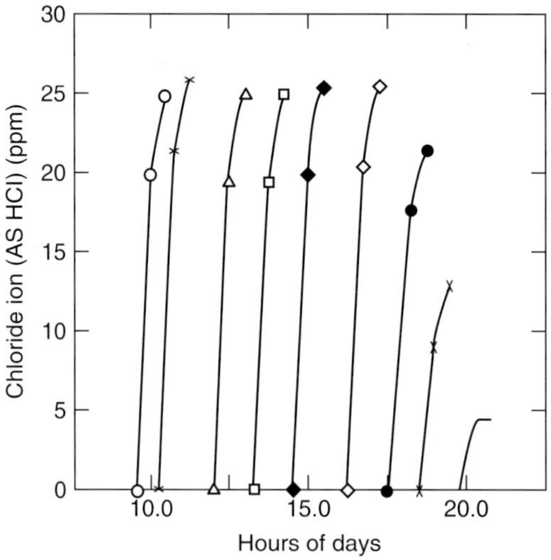

Figure 12.8 Initial trichloroethylene (TCE)-photocatalyzed degradation versus time of summer day [19]. (Source: Reprinted by permission of American Chemical Society.)

agreement with pyroheliometer data which reveals only about a 10% variation of total scattered (blue to nUV) sunlight over this interval [19].

This important result suggests that solar concentrators are not necessarily an efficient configuration, inasmuch as the scattered (blue) sunlight cannot be concentrated, whereas direct sunlight, which is nearly free of nUV, can. Examination of the fraction of sunlight which is scattered versus wavelength indicates that half or more of the nUV (300–400 nm) photons arrive via scattered paths.

12.8
Mechanism of Hydroxyl Radical Attack: Same Irradiance Dependence

The exact spatial location of hydroxyl radical (or other photoproduced active oxidant species) may vary. Accordingly, Turchi and Ollis [2] earlier proposed four possibilities for the dark reaction of hydroxyl radicals (or other active oxidant) with reactant, which are listed in the following.

1) Reaction between adsorbed OH and adsorbed reactant R (LH mechanism)
2) Reaction between adsorbed OH and free R (Eley–Rideal mechanism)
3) Reaction between free OH and adsorbed R (Eley–Rideal mechanism), and
4) Reaction between free OH and free R (solution kinetics).

Each of these is a dark reaction, uninfluenced by irradiance. Hence, variation of rate with irradiance reveals only the rate of oxidant generation, but not the spatial location of its ultimate disappearance.

12.9 Simultaneous Homogeneous and Heterogeneous Photochemistry

Photocatalytic treatment of water is predicated on the arrival of photons at the photocatalyst. In many waters, the solution absorbance may be high, and preclude economic use of photocatalysis. In a related problem, both homogeneous photolysis of reactant and photocatalytic oxidation may occur simultaneously.

If we consider for simplicity a steady-state flow reactor with a uniform catalyst slurry moving in direction z and illuminated at one wall with irradiance I_o, then the equations in Section 12.11.2 below describe the general case involving reactant photolysis as well as photocatalysis and oxidation due to photolyzable oxidant (hydrogen peroxide, ozone, etc.). This general case was explored by Turchi [20]. For ease of presentation, he defined R as the ratio of all reaction rates divided by the rate predicted if no catalyst were present. The general results are shown in Figure 12.9 for values of photocatalyst absorbance A_c ranging from zero to infinity.

The conclusion [20] is "that for high optical density solutions, a homogeneous photoreaction ought to be used first (e.g., photolysis of reactant or attack by H_2O_2 photolysis) in order to reduce solution absorbance to a value of $A_h C_R = 0.1$. Subsequent photocatalyst addition is more efficient than continued operation in its absence. Alternatively, many low-level contaminant streams will have an initial value of homogeneous absorbance $A_h C_R$ that is less than 0.1; here, immediate photocatalyst addition provides the best results" [19].

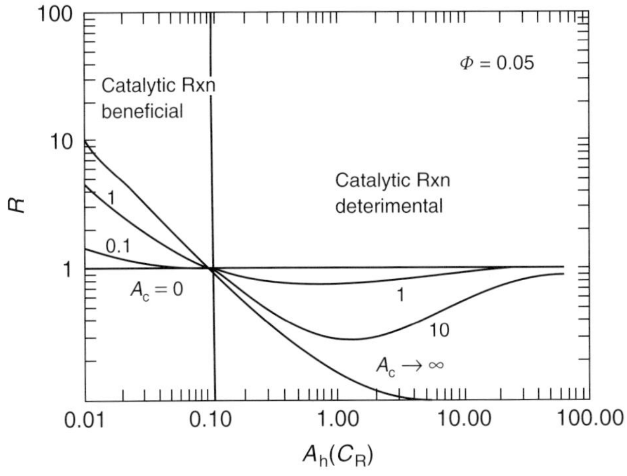

Figure 12.9 Ratio R (reaction with catalyst to total reaction without catalyst) versus solution homogeneous absorbance, A_h. For $R < 1.0$, catalyst addition is beneficial; for $R > 1$, detrimental. Curves drawn for various heterogeneous catalyst absorbance values, A_c [20].

12.10
Dye-Photosensitized Auto-Oxidation

Organic dyes have strong absorption bands in the visible light range. In the presence of TiO_2, dyes in solution may adsorb on the oxide surface and undergo photosensitized degradation (Figure 12.10). Here the photoexcited dye S* may inject an electron into the TiO_2 conduction band [21] and the resulting dye cation can be oxidatively bleached, resulting in decolorization. While O_2^- radicals were generated, hydroxyl radicals were not, as evidenced by the failure of photosensitized bleaching to produce dye mineralization.

A photosensitization example for Acid Orange 7 (AO7) degradation [22] using visible light appears in Figure 12.11. Here dye sensitized degradation occurs, with parallel destruction of azo linkages, naphthalene rings and benzene rings

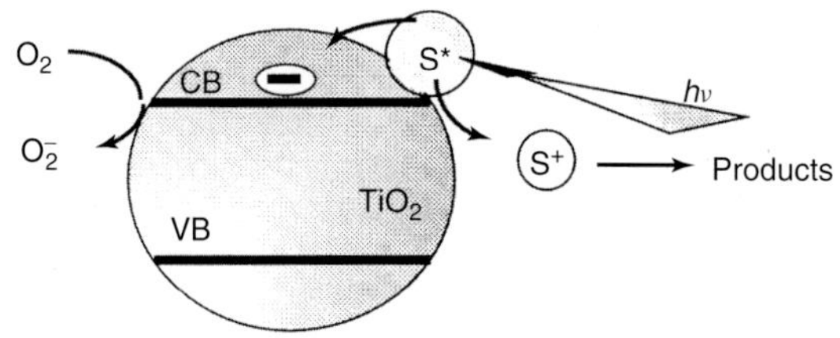

Figure 12.10 Photoexcited dye S* injects an electron into the conduction band of TiO_2, and resulting dye cation S^+ is subsequently oxidized, resulting in permanent bleaching of color [21]. (Source: Reprinted by permission of American Chemical Society.)

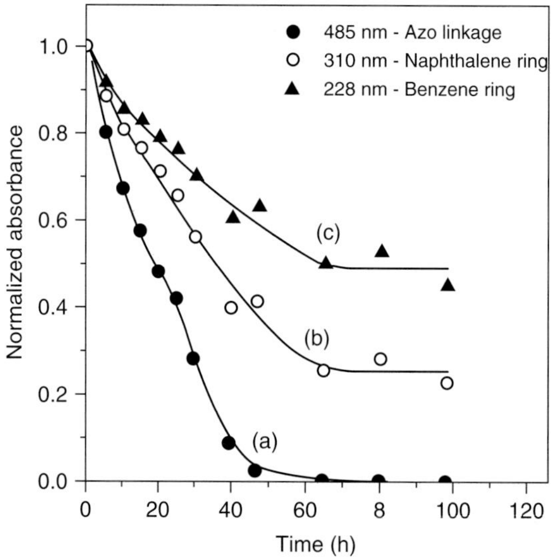

Figure 12.11 Absorbance bands of azo naphthalene rings and benzene rings in dye AO7 versus time. Photosensitized degradation halts upon exhaustion of azo linkages [22]. (Source: Reprinted by permission of Elsevier.)

in the dye molecule. Upon exhaustion of diazo linkages, the degradation reaction halts: no further destruction of benzene and naphthalene rings occurs. Traces of hydrogen peroxide are produced, but generation of this oxidant also halts upon azo exhaustion. Thus, photosensitization of dye can produce complete decolorization, but not mineralization [20].

Photosensitized degradation of dyes may help decolorize dye water streams but its presence complicates use of dyes to demonstrate visible light activated photocatalysis. As these dyes also have strong nUV absorbance, the same remark applies to dye studies with nUV (300–400 nm) light sources.

12.11 Interplay between Fluid Residence Times and Irradiance Profiles

12.11.1 Batch Reactors

In a well-stirred batch reactor every fluid element has the same residence time, and theoretically the same averaged exposure to illumination. Many aqueous-phase studies have utilized catalyst slurry concentrations of the order of 1 g l^{-1}, which were found to give nearly complete light absorption Thus a reactor with 1 cm ID tubing could have an optical density of 3, hence the TiO_2 absorption coefficient, ß, is estimated as

$$ßC_{cat}L = 3,$$

$$\text{so } ß = \frac{3}{C_{cat}L} = \frac{3}{\left(1\ \text{g}\,\text{l}^{-1}\right)\times 1\ \text{l}\ 10^{-3}\text{cm}^3 \times 1\ \text{cm}^3 4.23\ \text{g}^{-1}} \times 1\ \text{cm} = 13\ 000\ \text{cm}^{-1}.$$

Cassano and Alfano used integrating sphere measurements to find averaged absorption coefficient values (300–365 nm) of 55 000 cm^{-1} (Degussa P25), 35 000 cm^{-1} (Aldrich), and 23 000 cm^{-1} (Hombikat) [23].

Time averaged uniform illumination has been routinely achieved in stirred flasks as well as recirculating batch reactors [24–26]. In these small systems, reaction time is slow (minutes to hours) versus recirculation time (seconds), and the recirculating system is treatable as a single well-stirred tank with uniform concentrations. The complete mixing assumption also assures that every particle see the same time averaged photon flux.

12.11.2 Flow Reactors

Fluid flow velocity profiles in continuous flow reactors are illustrated in Figure 12.12 showing (Figure 12.12a) laminar, parabolic flow in a slit (Figure 12.12b), laminar, parabolic flow in a falling film, and (Figure 12.12c) turbulent flow. For laminar flow reactors, the nonuniform residence time for each differential fluid layer creates concentrations which vary with both axial distance (direction of solid arrows) and distance from the illumination source (flow boundary).

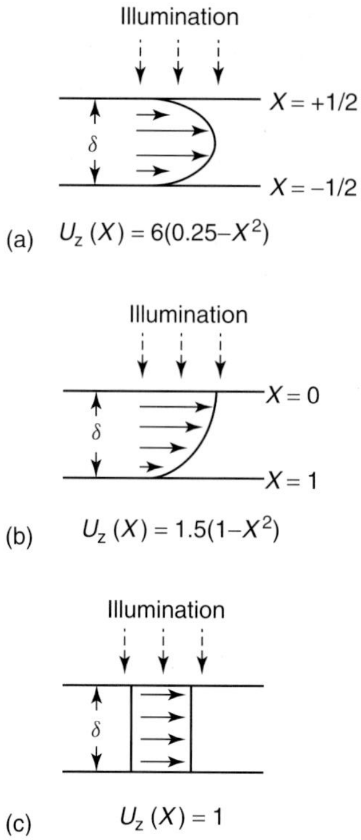

Figure 12.12 Flow reactor velocity profiles for (a) laminar flow in slit, (b) laminar falling film flow, and (c) turbulent flow [20]. (Source: Reprinted by permission of American Institute of Chemical Engineers.)

As a simple reactor model, all three configurations may be modeled by a single Eq. (12.24) below [20]:

$$U(x)\ \frac{\partial C}{\partial z} = ß\frac{\partial^2 C}{\partial X_2} - (\mathrm{Da})\ CI^n(x) \tag{12.24}$$

where C = reactant concentration, I = irradiance, z = axial distance, and x = distance from wall where $I(x = 0) = I_o$ (source), $U(x)$ = velocity profile, and Da = the Dahmkohler number = k_{ro}/k_m.

The irradiance profile is the Beer–Lambert form:

$$\frac{\partial I}{\partial x} = -ßC_c I(x) \tag{12.25}$$

where ß = molar absorption coefficient of photocatalyst and C_c = catalyst concentration.

This equation represents multiple configurations:

1) no mixing, laminar slit flow (ß = 0)(Figure 12.12)
2) no mixing, laminar falling film flow (ß = 0) (Figure 12.12)
3) no mixing, plug (turbulent) flow (ß = 0) (Figure 12.12).
4) complete lateral mixing, ß = infinity, all profiles equivalent [20].

Lateral mixing (ß > 0) always provides better reactant conversion than unmixed flows (ß = 0).

Inspection of Figure 12.12 yields an intuitive result: the most efficient configuration is the falling film, because illumination irradiance is greatest where flow is fastest (Figure 12.12b), whereas Figure 12.12a is less efficient as the faster flowing middle of the slit receives less than maximum possible illumination. Configuration Figure 12.12c also appears efficient except that turbulent giving the flat velocity profile shown requires high flow rates Re = Reynolds number = density $\times$ velocity $\times$ X/viscosity $>$ 2000, with resulting low residence time per pass.

When a photolyzable oxidant (e.g., H_2O_2) is also present, the balance equation must include two sinks for photons: the photocatalyst and the photolyzable oxidant, and a separate mass balance equation for oxidant is also required [20].

The literature on photoreactor design is substantial and is not the focus of this chapter. The reader can find detailed recent analyzes and summaries by groups of Cassano *et al.* [23] in Argentina and Li Puma *et al.* [27] in England. Suffice it to say that such global design equations depend for their accuracy on a proper form of the reaction rate equation, including irradiance influences, which has been the focus of this chapter.

12.12 Quantum Yield, Photonic Efficiency, and Electrical Energy per Order

Photocatalysis quantum yields are typically of the order of 0.5–5%. A careful study by Sun and Bolton [28] using an integrating sphere reported QE = 0.040 $\pm$ 0.003 for photocatalyzed generation of hydroxyl radicals at pH = 7. Their sonicated Aldrich titania showed that QE varies inversely with the square root of irradiance in the photon flux range of 1.3–1.8 $\times$ 10^{-8} Einstein s^{-1}. The PE of a photocatalytic process is the ratio of observed rate divided by the incident photon rate. As not all photons are absorbed from an incident source, this PE is always less than the QE for hydroxyl radical generation. Thus, the approximately 4% quantum yield measured for an Aldrich titania is the upper limit for this material.

For economic comparisons, a figure of merit defined by Bolton *et al.* [29] as an IUPAC standard is the electrical energy per order (EEO) that gives the power required to reduce the concentration of contaminant by 1 order of magnitude. Under dilute conditions where the degradation reaction is first order, this value can be calculated from the following formulas:

$$E_{EO}\left(\text{kW h m}^{-3} - \text{order}\right) = \frac{Pt\ 1000}{V \log\left({}^{C_i}/{}_{C_f}\right)} \quad \text{(batch system)} \tag{12.26}$$

$$E_{EO}\left(\text{kW h m}^{-3} - \text{order}\right) = \frac{P}{F \log {}^{C_i}/{}_{C_f}} \quad \text{(flow system)} \tag{12.27}$$

where P = electrical power for light source, V = reactor liquid volume, t = time (h), F = flow rate ($m^3\ h^{-1}$), and C_i and C_f are initial and final contaminant concentrations (batch) or inlet and exit concentrations (flow), respectively. Use of these formulas allows estimation of the electricity operating costs for any desired degree of (first order) contaminant removal.

12.13 Conclusions

Semiconductor photocatalysis requires reaction initiation by absorption of light, hence the local irradiance of band gap photons has a dominant influence on reaction rate, and on attempts to construct credible kinetic models for this photochemistry. Pseudo-steady-state treatment is shown superior to the simpler equilibrated adsorption assumption; however, at fixed light irradiance, the LH form still fits most kinetic results. Future research reports should always contain irradiance measurements, as well as evaluation of model kinetic parameters. The observed kinetics may be influenced also by mass transfer limitations of sparingly soluble oxygen or target contaminants at low concentrations, by transient illumination schedules, by simultaneous photosensitization and/or photolysis of target molecules, and by fluid residence time distributions within a photoreactor. Each of these phenomena may strongly influence the observed (apparent) kinetics photocatalyzed reactions, hence understanding their relative importance will continue to require our attention.

References

1. Egerton, T. and King, C. (1979) Influence of light irradiance on photoactivity in TiO_2 pigmented systems. *J. Oil Colour Chem. Assoc.*, **62**(10), 386–391.
2. Turchi, C. and Ollis, D. (1990) Photocatalyti degradation of organic water contaminants: mechanisms involving hydroxyl radical attack. *J. Catal.*, **122**, 178–192.
3. Xu, Y. and Langford, C. (2000) Variation of langmuir adsorption constant determined for TiO_2 photocatalyzed degradation of acetophenone under different light irradiance. *J. Photochem. Photobio. A: Chem.*, **133**, 67–71.
4. Ollis, D.F. (2005) Kinetics of liquid phase photocatalyzed reactions: an illuminating approach. *J. Phys.Chem. B*, **100**, 2439–2444.
5. Emeline, A., Ryabchuk, V., and Serpone, N. (2000) Factors affecting the efficiency of a photocatalyzed process in aqueous-metal oxide dispersions- prospects of distinguishing between two kinetic models. *J. Photochem. Photobiol. A, Chem.*, **133**, 89.
6. Mills, A., Wang, J., and Ollis, D. (2006) Kinetics of liquid phase semiconductor photoassisted reactions: supporting observations for a pseudo-steady state model. *J. Phys. Chem., B*, **110**, 14386–14390.
7. Mills, A., Wang, J., and Ollis, D. (2006) Dependence of the kinetics of liquid phase photocatalyzed reactions on oxygen concentration and light irradiance. *J. Catal.*, **243**, 1–6.
8. Martyanov, I. and Savinov, E. (2000) Photocatalytic steady-state methylviologen oxidation in air-saturated TiO_2 aqueous suspension: Initial photonic efficiency and initial oxidation rate as a function of methylviologen concentration and light irradiance. *J. Photochem. Photobiol. A, Chem.*, **134**, 219–226.
9. Ollis, D., Pelizzeti, E., and Serpone, N. (1991) Photocatalyzed destruction of water contaminants. *Environ. Sci. Technol.*, **25**, 1522–1529.
10. Matthews, R.W. (1987) Photooxidation of organic impurities in water using thin-films of titanium dioxide. *J. Phys. Chem.*, **91**, 3328–3333.
11. Turchi, C. and Ollis, D. (1988) Photocatalytic reactor design: an example of mass-transfer limitations with immobilized catalyst. *J. Phys. Chem.*, **92**, 6852–6853.
12. (a) Gerischer, H. and Heller, A., Photocatalytic oxidation of organic molecules at TiO_2 particles by sunlight in aerated

water, *J. Electrochem. Soc.*, 1992, **139**, 113–118. (b) Gerischer, H. and Heller, A., The role of oxygen in photooxidation of organic molecules on semiconductor particles, *J. Phys. Chem.*, 1991, **98**, 5261–5267.

13. (a) Sczechowski, J. Koval, C. and Noble, R., Improved photoefficiencies for TiO_2 photocatalytic reactors through use of controlled periodic illumination, *Trace. Met. Anal.* 1993, 645–650. (b) Foster, N., Koval, C., Sczechowski, J. et al. Investigation of controlled periodic illumination effects oh photooxidation processes with titanium dioxide films using rotating ring photoelectrochemistry, *J. Electroanal. Chem.*, 1996, **406**, 213–219.
14. Upadhya, S. and Ollis, D. (1997) Simple photocatalysis model for photoefficiency enhancement via controlled periodic illumination. *J. Phys. Chem. B*, **101**(14), 2625–263.
15. Buechler, K., Zawistowski, T., Noble, R., and Koval, C. (2001) Investigation of the mechanism of the controlled periodic illumination effect in TiO_2 photocatalysis. *Ind. Eng. Chem.*, **40**, 1097–1102.
16. Buechler, K., Nam, Sawistowski, T., Noble, R., and Koval, C. (1999) Design and evaluation of a novel controlled periodic illumination reactor to study photocatalysis. *Ind. Eng. Chem. Res.*, **38**, 1258–1263.
17. Cornu, C., Colussi, A., and Hoffmann, M. (2003) Time scales and pH dependences of the redox processes determining the photocatalytic efficiency of TiO_2 nanoparticles from periodic illumination experiments in the stochastic regime. *J. Phys. Chem. B*, **107**, 3156–3160.
18. Ollis, D.F. (2005) Kinetic disguises in heterogeneous photocatalysis. *Top. Catal.*, **35**, 217–223.
19. Ahmed, S. and Ollis, D.F. (1984) Solar photoassisted catalytic decomposition of the chlorinated hydrocarbons trichloroethylene and trichloromethane. *Sol. Energy*, **12**(5), 597–601.
20. Ollis, D. and Turchi, C. (1990) Heterogeneous photocatalysis for water purification: contaminant mineralization kinetics and elementary reactor analysis. *Environ. Prog.*, **9**(4), 229–234.
21. Nasr, C., Vinodgopal, K., Fisher, L., Hotchandani, S., Chattopadhyay, A., and Kamat, P. (1996) Environmental photochemistry on semiconductor surfaces. Visible light induced degradation of a textile diazo dye, naphthol blue black, on TiO_2 nanoparticles. *J. Phys. Chem.*, **100**, 8436–8442.
22. Stylidi, M., Kodearides, D., and Verykios, X. (2004) Visible light-induced photocatalytic degradation of acid orange 7 in aqueous TiO2 suspensions. *Appl. Catal., B*, **47**, 189–201.
23. Cassano, A. and Alfano, O. (2000) Reaction engineering of suspended solid heterogeneous photocatalytic reactors. *Catal. Today*, **58**, 167–197.
24. Pruden, A. and Ollis, D. (1983) Photoassisted heterogeneous catalysis. The degradation of trichloroethylene in water. *J. Catal.*, **82**(2), 404–417.
25. Pruden, A. and Ollis, D. (1983) Degradatiin of chloroform by photoassisted heterogeneous catalysis in dilute aqueous suspensions of titanium dioxide. *Environ. Sci. Tech.*, **17**(10), 628–631.
26. Ollis, D.F., Hsiao, C.-Y., Budiman, L., and Lee, C.L. (1984) Heterogeneous photocatalysis; conversions of percloroethylene, dichloroethane, chloroacetic acids, and chlorobenzens. *J. Catal.*, **88**, 89–96.
27. Colina-Marquez, J., Fiderman, M., and Li Puma, G. (2010) Radiation absorption and optimization of solar photocatalytic reactors for environmental applications. *Environ. Sci. Technol.*, **44**(13), 5112–5120.
28. Sun, L. and Bolton, J. (1996) Determination of the quantum yield for the photochemical generation of hydroxyl radicals in TiO2 suspensions. *J. Phys. Chem.*, **100**, 4127–4134.
29. Bolton, J., Bircher, K., Tumas, W., and Tolman, C. (2001) Figures-of-merit for the technical development and application of advanced oxidation technologies for both electric- and solar-driven systems. *Pure. Appl. Chem.*, **73**, 627–637.

13
A Methodology for Modeling Slurry Photocatalytic Reactors for Degradation of an Organic Pollutant in Water

Orlando M. Alfano, Alberto E. Cassano, Rodolfo J. Brandi, and María L. Satuf

13.1
Introduction and Scope

Photocatalysis is a reaction that it is said to be known since 1929 [1], when it was discovered the flaking of paints and the degradation of fabrics containing titanium dioxide under solar UV radiation. However, it was not until 1970 when Fujishima and Honda published, first in a Japanese journal and later, with much more diffusion, in Nature [2] and in spite of some limitations and bias, the possibility of water splitting employing TiO_2 and UV light. This publication coincided with the petroleum crisis and, up to the year 1980, produced a vertiginous growth of the scientific literature on the topic until its absence of economic feasibility was demonstrated. The gas-phase photocatalytic oxidation of alkanes [3, 4] and the formation of superoxide [4] over TiO_2 were independently reported at that time. A new thrust of similar extraordinary development occurred in 1977 when Frank and Bard [5], as consequence of previous studies in hydrogen production, found the possibility of using the oxidative properties of the catalyst to destroy organic compounds. This fact guided its application to environmental pollution abatement. These studies, initially aimed at the treatment of contaminated water, were later also directed to the treatment of air pollution. Despite the intensity and diffusion of the work [6], there were no major practical advances and by the end of the 1990s the subject began to decline again. But almost at the same time, two new lines of work emerged, particularly applied to large external surfaces. On the one hand, the oxidative properties of titanium dioxide was used to achieve self-cleaning surfaces by means of solar light [7–9] and, on the other hand, it was sought to use the capacity of the aforementioned oxide combined with ultraviolet radiation, to transform surfaces into highly hydrophilic textures, so rains contribute to their cleaning [10]. These uses had a reasonable commercial success. In what seems to be the beginning of a more recent era of photocatalysis, new work oriented toward the use of visible light would extend all its possibilities of application from those conceived from 1970 to the present [11, 12]. The goal is to modify the catalyst properties by adding chemical elements so it can be activated by the visible spectrum of the sun.

Photocatalysis and Water Purification: From Fundamentals to Recent Applications, First Edition. P. Pichat.

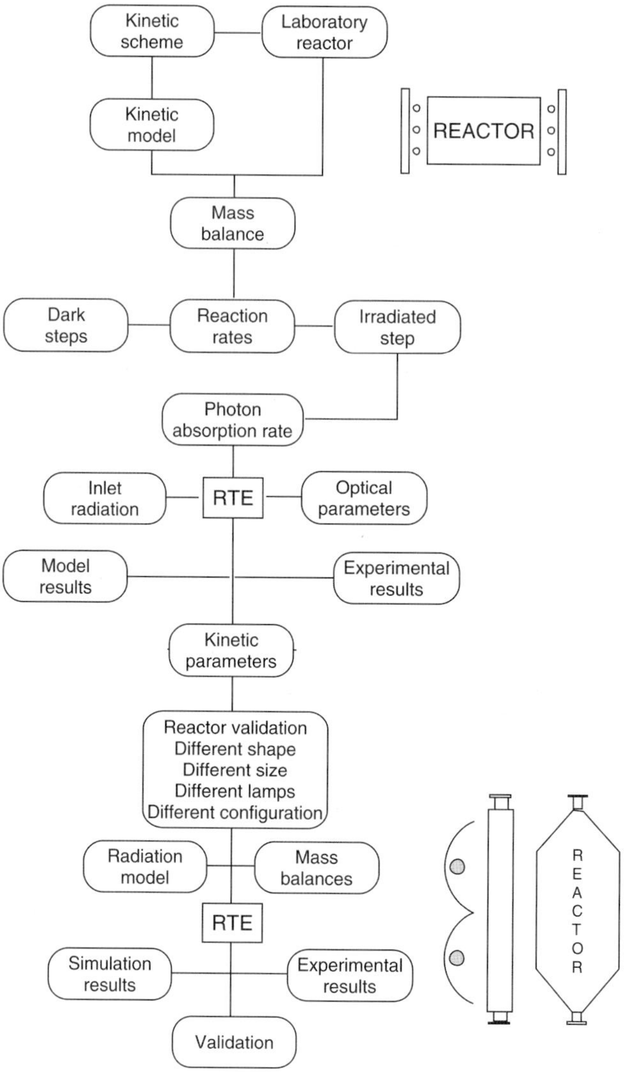

Figure 13.1 Flow sheet of the reactor modeling and scaling-up methodology.

Since 1993, our group has been dedicated to the study of photocatalytic processes, developing procedures to obtain quantitative reaction kinetic models [13, 14] and reactor scaling-up [15]. The idea is to move from especially designed laboratory experiments to the commercial size apparatus in one step. To achieve this objective, it is necessary to analyze both equipment with rigorous mathematical modeling, involving the solution of the complete radiative transfer equation (RTE) [15]. In Figure 13.1, a conceptual flow sheet describes the methodology. To work in the laboratory reactor, one starts with a kinetic scheme and derives a kinetic model. To analyze the results, it is necessary to have a mass balance. For the irradiated

Table 13.1 Dimensions and main characteristics of the laboratory- and bench-scale reactors.

Description	Laboratory-scale reactor	Bench-scale reactor
Shape	Cylindrical	Rectangular, flat plate
Main dimensions	Inner diameter = 8.6 cm Length = 5.0 cm	Width = 18.0 cm Length = 34.0 cm Thickness = 1.2 cm
Reactor volume (cm^3)	290.4	734.4
Total system volume (cm^3)	1000	5000
Number of lamps	8	2
Nominal input power (W)	4 (per lamp)	40 (per lamp)
Emission range (nm)	300–400	310–410
Radiation flux (100%) (einstein cm^{-2} s^{-1})	7.55×10^{-9}	6.40×10^{-9}
Lamp diameter (cm)	1.6	3.8
Lamp arc length (cm)	13.6	60

Source: Adapted from Ref. [16], Copyright 2007, with permission from Elsevier.

steps, the photon absorption rate must be known. Here, it is indispensable to have information about the radiation that arrives at the reactor wall and the optical properties of the catalyst (absorption coefficient, scattering coefficient, and the scattering phase function). With this data, the results of the kinetic experiments are compared with those obtained from the simulation of the kinetic model resorting to a nonlinear, multiparameter estimation to render the kinetic constants. For scaling-up purposes (changing the reactor shape, size, lamps, and configuration), the same approach must be followed, directly applying the kinetic equations to the new reactor and solving the radiation model. There is only one limitation to this practice: the lamps spectral wavelength distribution of both systems must be very similar.

This chapter presents in detail the modeling and scaling-up of a slurry reactor, from laboratory-scale to bench-scale size, for the photocatalytic degradation of a model pollutant: 4-chlorophenol (4-CP, IUPAC name: 4-chlorophenol) [16]. Dimensions and main characteristics of both reactors are summarized in Table 13.1.

13.2 Evaluation of the Optical Properties of Aqueous TiO_2 Suspensions

The evaluation of the radiation field inside a slurry reactor constitutes a central step in the study of photocatalytic reactions. The complexity of this task lies in the simultaneous existence of radiation absorption and scattering. A rigorous approach to obtain the spatial and directional distributions of radiation intensities is the application of the RTE to the system under study. To solve the RTE, three optical properties of the catalyst suspensions are required: the volumetric absorption coefficient κ_λ, the volumetric scattering coefficient σ_λ, and the phase function for scattering *p*. This section describes an experimental method to measure and calculate the above mentioned properties [17]. This method involves spectrophotometric measurements of absorbance, diffuse reflectance and diffuse transmittance of TiO_2 suspensions, the evaluation of the radiation field in the spectrophotometer sample cell, and the application of a nonlinear optimization program to adjust model predictions to experimental data. The catalyst evaluated was TiO_2 Aldrich (99.9+% anatase, cat. 23203–3, lot 10908DZ).

13.2.1
Spectrophotometric Measurements of TiO_2 Suspensions

To prepare the aqueous suspensions for measurements, the TiO_2 powder was dried in an oven at 150 °C for 12 h and then dispersed in ultrapure water. The samples were sonicated during 1 h and kept under magnetic stirring (between 10 and 30 min) until measurement. The catalyst concentrations ranged from 0.2×10^{-3} to 2.0×10^{-3} g cm^{-3}.

The extinction coefficient β_λ, defined as the sum of κ_λ and σ_λ, was obtained from absorbance spectrophotometric measurements of TiO_2 suspensions (ABS_λ), under specially designed conditions to minimize the collection of the scattered rays by the detector [18]: $\beta_\lambda = 2.303\ ABS_\lambda/L$, where L represents the cell path length. The specific extinction coefficients β^*_λ (i.e., per unit catalyst mass concentration C_m) can be obtained by applying a standard linear regression to the plot of β_λ versus C_m. The spectral diffuse reflectance (R_λ) and transmittance (T_λ) measurements of the suspensions were made by an Optronic OL Series 750 spectroradiometer equipped with an OL 740–70 integrating sphere reflectance attachment. The integrating sphere, coated with polytetrafluoroethylene (PTFE), has two openings in the wall for reflecting samples: the sample port and the comparison port. The detector is positioned on a port mounted on the top of the integrating sphere. The OL 740–70 attachment also contains a filter holder compartment for transmittance measurements. The integrating sphere configurations for measurements are schematically shown in Figure 13.2. For diffuse reflectance measurements, a pressed PTFE reflectance standard was used as the reference in the comparison port, and 100% reflectance reading was obtained placing another PTFE standard in the sample port (Figure 13.2a). To measure the reflectance of the sample, the quartz cell with the TiO_2 suspension was placed in the sample port with its back covered by a light trap, keeping the PTFE standard in the comparison port (Figure 13.2b). The light trap consists of a solid cube painted black that captures the light which passes through the sample, so it is not reflected back into the integrating sphere. For diffuse transmittance measurements, PTFE reflectance standards were placed in the comparison and sample ports, and 100% transmittance reading was obtained against air (Figure 13.2c). To measure the transmittance of the sample, the cell with the catalyst suspension was placed in the filter holder compartment (Figure 13.2d).

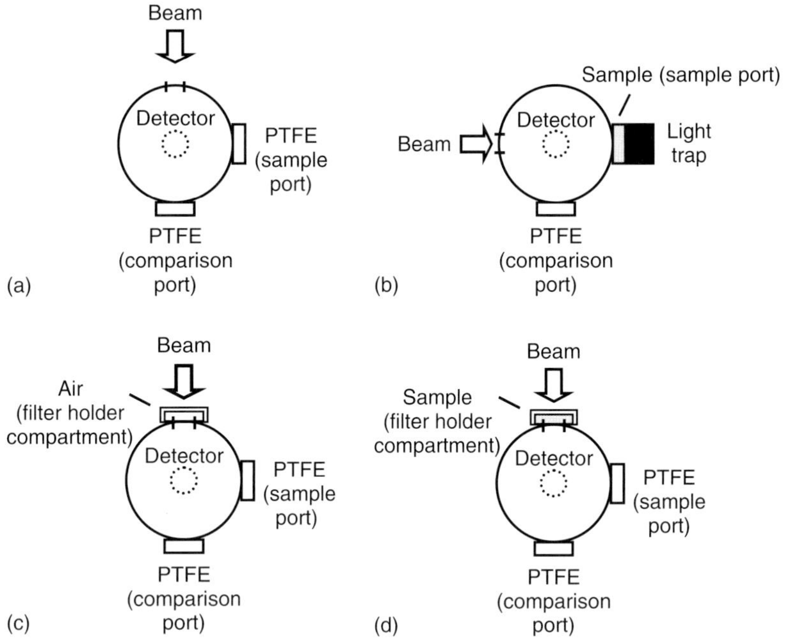

Figure 13.2 Integrating sphere configurations: (a,b) reflectance measurements and (c,d) transmittance measurements. (Source: Reprinted with permission from Ref. [17]. Copyright 2005 American Chemical Society.)

13.2.2
Radiation Field in the Spectrophotometer Sample Cell

The RTE describes the radiation intensity I_λ at any position along a ray path through a medium. This equation can be applied to evaluate the radiation field in the heterogeneous system constituted by TiO_2 particulate suspensions in water. For a medium with absorption and scattering (no emission is considered), the RTE can be written as

$$\frac{dI_\lambda(s,\Omega)}{ds} + \beta_\lambda I_\lambda(s,\Omega) = \frac{\sigma_\lambda}{4\pi}\int_{\Omega'=0}^{4\pi} I_\lambda\left(s,\Omega'\right)p\left(\Omega' \to \Omega\right)\, d\Omega' \tag{13.1}$$

The right-hand side of Eq. (13.1) represents the gain of radiant energy along the direction Ω due to the incoming scattering from all directions (Ω'). The phase function $p(\Omega' \to \Omega)$ represents the probability that the incident radiation from direction Ω' will be scattered and incorporated into direction Ω. This term gives the RTE its integro-differential nature and is responsible for most of the difficulties associated with its numerical integration. The rectangular spectrophotometer cell, employed for reflectance and transmittance measurements, can be represented as an infinite plane parallel medium with azimuthal symmetry (Figure 13.3). Then, a one-dimensional, one-directional radiation transport model was applied to solve the RTE in the cell; that is, [19]

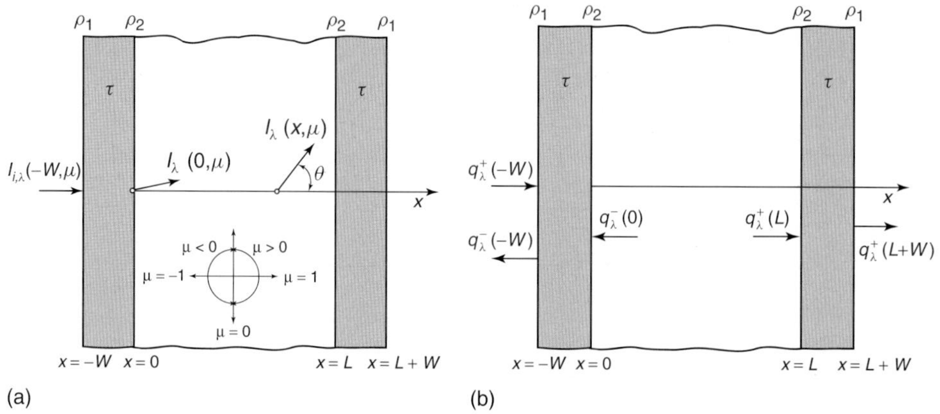

Figure 13.3 Schematic representation of the spectrophotometer cell. (a) Coordinate system for the one-dimensional, one-directional radiation model. (b) Inlet and outlet radiative fluxes. (Source: Reprinted with permission from Ref. [17]. Copyright 2005 American Chemical Society.)

$$\frac{\mu}{\beta_\lambda}\frac{\partial I_\lambda(x,\mu)}{\partial x} + I_\lambda(x,\mu) = \frac{\omega_\lambda}{2}\int_{\mu'=-1}^{1} I_\lambda\left(x,\mu'\right)p\left(\mu,\mu'\right)\,\mathrm{d}\mu' \tag{13.2}$$

where $\omega_\lambda = \sigma_\lambda/\beta_\lambda$ is the spectral albedo, defined as the ratio of the scattering coefficient to the extinction coefficient; x is the axial coordinate; μ is the direction cosine of the ray for which the RTE is written ($\mu = cos\theta$); and μ', the cosine of an arbitrary ray before scattering. Considering the quartz walls of the cell as specularly reflecting surfaces, the boundary conditions of Eq. (13.2) take the following form (Figure 13.3a):

$$I_\lambda(0,\mu) = \Gamma_{W2,\lambda} I_\lambda(0,-\mu) + \Upsilon_{W,\lambda} I_{i,\lambda}(-W,\mu) \quad (\mu > 0) \tag{13.3}$$

$$I_\lambda(L,\mu) = \Gamma_{W2,\lambda} I_\lambda(L,-\mu) \quad (\mu < 0) \tag{13.4}$$

where $\Upsilon_{W,\lambda}$ is the global wall transmission coefficient and $\Gamma_{W2,\lambda}$ is the global wall reflection coefficient corresponding to the radiation that arrives from the internal side of the cell [17].

The choice of the phase function p represents an important step in any calculation where multiple scattering is involved. Actually, in a well defined physical problem, the phase function is given, not chosen. Nevertheless, complicated functions lead to very time consuming computations. It is then a common practice to choose p so that it preserves the main characteristics of the actual function and still renders the multiple-scattering computations manageable. A very useful function is the Henyey and Greenstein phase function, which can be expressed as [20]

$$\mathrm{p}_{\text{H-G},\lambda}\left(\mu_0\right) = \frac{\left(1-\mathrm{g}_\lambda{}^2\right)}{\left(1+\mathrm{g}_\lambda{}^2-2\mathrm{g}_\lambda\mu_0\right)^{3/2}} \tag{13.5}$$

where g_λ is the dimensionless asymmetry factor and μ_0 is the cosine of the scattering angle θ_0 between the directions of the incident and scattered rays. The H-G function is then determined by a single free parameter (g_λ) that varies smoothly from isotropic ($g_\lambda = 0$) to a narrow forward peak ($g_\lambda = 1$) or to a narrow backward peak ($g_\lambda = -1$). The knowledge of g_λ alone suffices to obtain solutions of multiple-scattering problems with a high degree of accuracy, making the $p_{H\text{-}G,\lambda}$ ideal for calculations [21].

To solve the RTE, the Discrete Ordinate Method [22] was applied. This method transforms the RTE into a set of algebraic equations that can be solved numerically. We are facing an inverse analysis of radiative transfer, where the optical properties ω_λ and g_λ will be determined from a set of measured radiation quantities.

13.2.3 Parameter Estimation

In order to compare the model predictions with the experimental measurements of diffuse reflectance and transmittance, we should interpret the results in terms of net radiative fluxes. The spectral net radiative flux, $q_\lambda(x)$, for the one-dimensional, one-directional model with an azimuthal symmetry, can be written as:

$$q_\lambda(x) = 2\pi \int_{\mu=-1}^{1} I_\lambda(x,\mu)\,\mu\,d\mu \tag{13.6}$$

Accordingly, considering Figure 13.3b, diffuse reflectance and transmittance values can be interpreted as the ratio between net radiative fluxes:

$$R_\lambda = \frac{q_\lambda^-(-W)}{q_\lambda^+(-W)} = \frac{2\pi \int_{\mu=-1}^{0} I_\lambda(-W,\mu)\,\mu\,d\mu}{2\pi \int_{\mu=0}^{1} I_{i,\lambda}(-W,\mu)\,\mu\,d\mu} \tag{13.7}$$

$$T_\lambda = \frac{q_\lambda^+(L+W)}{q_\lambda^+(-W)} = \frac{2\pi \int_{\mu=0}^{1} I_\lambda(L+W,\mu)\,\mu\,d\mu}{2\pi \int_{\mu=0}^{1} I_{i,\lambda}(-W,\mu)\,\mu\,d\mu} \tag{13.8}$$

Reflectance and transmittance model predictions are calculated with the values of I_λ obtained from the solution of the RTE. However, these values represent the radiation intensities at the inner walls of the cell ($x = 0$ and $x = L$). Therefore, to compare theoretical values with experimental results, one must account for the effect of the cell walls and obtain the corresponding values at $x = -W$ and $x = L + W$. The approach used to calculate the intensities outside the sample cell is similar to the one applied for the boundary conditions of the RTE:

$$I_\lambda(-W,\mu) = \Gamma_{W1,\lambda} I_{i,\lambda}(-W,-\mu) + \Upsilon_{W,\lambda} I_\lambda(0,\mu) \quad (\mu < 0) \tag{13.9}$$

$$I_\lambda(L+W,\mu) = \Upsilon_{W,\lambda} I_\lambda(L,\mu) \quad (\mu > 0) \tag{13.10}$$

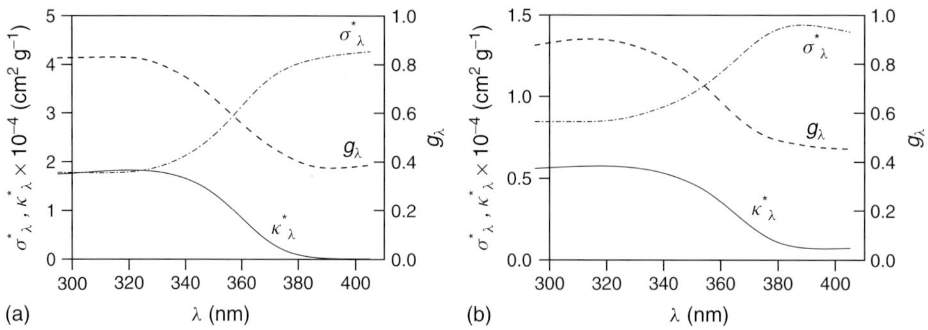

Figure 13.4 Spectral distribution of the catalyst optical properties: broken and dotted lines, σ_λ^*; solid line, κ_λ^*; broken line, g_λ; (a) natural pH (≅ 6.5) and (b) pH 2.5. (Source: Adapted with permission from Ref. [26]. Copyright 2007 American Chemical Society.)

where $\Gamma_{W1,\lambda}$ represents the global wall reflection coefficient corresponding to the radiation that arrives from the external side of the cell.

A nonlinear multiparameter regression procedure (a modified Levenberg–Marquardt method) was applied to adjust theoretical values to experimental information. The optimization program renders the values of ω_λ and g_λ that minimize the differences between model predictions and experimental data. Then, for each concentration of TiO_2 and each wavelength, the volumetric scattering and absorption coefficients were obtained as $\sigma_\lambda = \beta_\lambda \omega_\lambda$ and $\kappa_\lambda = \beta_\lambda - \sigma_\lambda$. The specific absorption and scattering coefficients ($\kappa_\lambda^* = \kappa_\lambda / C_m$, $\sigma_\lambda^* = \sigma_\lambda / C_m$), calculated per unit catalyst mass concentration, are obtained by applying a standard linear regression on the data κ_λ versus C_m and σ_λ versus C_m, respectively. The values of the dimensionless asymmetry factor, for each wavelength, are obtained by calculating the average value of g_λ for all the C_m tested in the experimental work.

The optical properties of aqueous catalytic suspensions are strongly dependent on the pH of the medium. The values of the extinction coefficient at pH 2.5 are approximately three times lower than the ones obtained at natural pH, with the consequent decrease in the absorption and scattering coefficients. These variations in the optical properties can be attributed to changes in the positions of the valence and conduction bands of the semiconductor [23], modification of the rheological properties of TiO_2 suspensions [24], or changes in the rate of the electron–hole recombination [25]. Figure 13.4 depicts the spectral distribution of g_λ, σ_λ^*, κ_λ^*, at the natural pH of the TiO_2 suspensions (pH 6.5), and at pH 2.5.

13.3
Radiation Model

The evaluation of the radiation field is essential in the modeling and design of photocatalytic reactors. The estimated radiation field expressed in terms of the local volumetric rate of photon absorption (LVRPA) can be introduced into kinetic expressions, thus obtaining equations independent of the experimental irradiation conditions. In this section, the setup of the laboratory device and the procedure employed in the experiments of Sections 13.4 and 13.5 are introduced. Then, the radiation model inside the laboratory-scale reactor is described in detail. A one-dimensional, one-directional radiation field model is proposed and solved to obtain the LVRPA.

13.3.1
Experimental Set Up and Procedure

The reactor is cylindrical, made of stainless steel, with an inner wall of Teflon. It is operated inside a batch recycling system, as shown in Figure 13.5. The reactor has two circular, flat windows made of borosilicate glass. In front of each window, a set of four UV lamps (TL 4 W/08 Black Light UVA lamps from Phillips, with an emission peak at about 350 nm), was placed. Ground glass plates, situated between the lamps and the reactor windows, were used to produce diffuse inlet radiation. The irradiation level was varied using optical filters between the lamps and the reactor. The optical filters were constructed on red sensitive, 0.1 mm thickness, polyester films, for recorders with He–Ne laser (HNm from AGFA Alliance Recording), by printing different tones of gray with the aid of a software. The storing tank was equipped with a sampling valve, a gas inlet for oxygen supply, and a water-circulating jacket to ensure isothermal conditions during the reaction time. The pump flow rate was adjusted in order to provide good mixing conditions, low conversion per pass in the reactor, and uniform concentration of the catalyst. The system was maintained under overpressure of pure oxygen to guarantee the renewal of the oxygen consumed by the chemical reaction. Each run lasted 8 h and the samples were taken every hour. The concentration of 4-CP in the samples

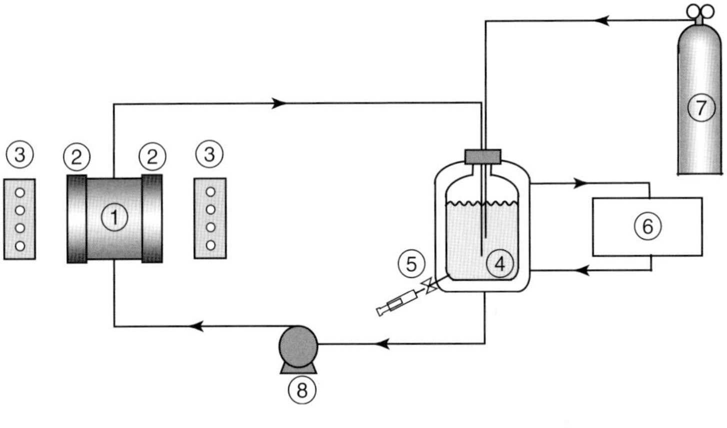

Figure 13.5 Schematic representation of the experimental set up.1, reactor; 2, windows; 3, lamps; 4, tank; 5, sampling valve; 6, thermostatic bath; 7, oxygen supply; and 8, pump. (Source: Reprinted with permission from Ref. [26]. Copyright 2007 American Chemical Society.)

was measured by HPLC using a Waters chromatograph equipped with an LC-18 Supelcosil reversed phase column (Supelco). The eluent was a ternary mixture of water (containing 1% v/v acetic acid), methanol, and acetonitrile (60 : 30 : 10)% v/v, pumped at a rate of 1 ml min^{-1}. UV detection of 4-CP was performed at 280 nm.

13.3.2
Radiation Field Inside the Photoreactor

To simplify the radiation model, it can be assumed that the greatest changes in the spatial distribution of the radiation field are mainly produced along the x coordinate axis (Figure 13.6). Therefore, the radiation propagation is modeled with only one spatial variable. On the other hand, due to the arrangement of the lamps and the presence of ground glass, it can be assumed that the radiation arriving at the reactor windows is diffuse with azimuthal symmetry. When the boundary conditions for the RTE are azimuthally symmetric, the radiation can be modeled with one angular variable [27]. Under these assumptions, Eq. (13.2) can be used to obtain the radiation intensity.

The optical effects of the reactor windows, that is, absorption, refraction, and reflection, are taken into account to obtain the boundary conditions of the RTE. At $x = 0$, radiation intensities in the forward direction ($\mu > 0$) are the result of two contributions: (i) the transmitted portion of the radiation arriving from outside of the reactor and (ii) the reflected portion of the radiation coming from the suspension. Although the radiation incident on the external face of the reactor window is diffuse, due to refraction, the angular directions of the intensities entering the suspension ($I_{0,\lambda}$) are between 0 and the critical angle θ_c, given by $\theta_c = \sin^{-1}\frac{n_a}{n_s}$. By applying a similar analysis at $x = L_R$, the boundary conditions of the RTE take the following form:

$$I_\lambda(0,\mu) = I_{0,\lambda} + \Gamma_{W,\lambda}(-\mu)\, I_\lambda(0,-\mu) \quad 1 > \mu > \mu_c \tag{13.11}$$

$$I_\lambda(0,\mu) = \Gamma_{W,\lambda}(-\mu)\, I_\lambda(0,-\mu) \quad \mu_c > \mu > 0 \tag{13.12}$$

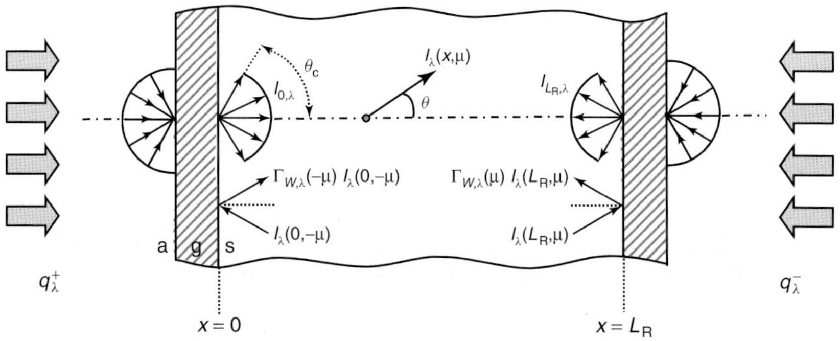

Figure 13.6 Coordinate system for the one-dimensional, one-directional radiation model. (Source: Reprinted with permission from Ref. [26]. Copyright 2007 American Chemical Society.)

$$I_\lambda\left(L_R,-\mu\right)=I_{L_R,\lambda}+\Gamma_{W,\lambda}\left(\mu\right)I_\lambda\left(L_R,\mu\right)\quad 1>\mu>\mu_c \tag{13.13}$$

$$I_\lambda\left(L_R,-\mu\right)=\Gamma_{W,\lambda}\left(\mu\right)I_\lambda\left(L_R,\mu\right)\quad \mu_c>\mu>0 \tag{13.14}$$

where $\mu_c = \cos\theta_c$ and $\Gamma_{W,\lambda}$ represents the global window reflection coefficient.

The Net Radiation Method [20] was adapted to compute the global window reflection coefficient as a function of μ:

$$\Gamma_{W,\lambda}\left(\mu\right)=\rho_{s-g}\left(\mu\right)+\frac{\left[1-\rho_{s-g}\left(\mu\right)\right]\left[1-\rho_{g-s}\left(\mu^*\right)\right]\rho_{g-a}\left(\mu^*\right)\tau_\lambda^2\left(\mu^*\right)}{1-\rho_{g-s}\left(\mu^*\right)\rho_{g-a}\left(\mu^*\right)\tau_\lambda^2\left(\mu^*\right)} \tag{13.15}$$

where ρ_{s-g} is the suspension–glass reflectivity, ρ_{g-s}, the glass–suspension reflectivity, ρ_{g-a} is the glass–air reflectivity, τ_λ represents the internal glass window transmittance, and μ^* is the cosine of the refracted angle in the glass.

The interface reflectivities were calculated by using Snell's law and Fresnel's equations. The values of $I_{0,\lambda}$ and $I_{L_R,\lambda}$ were obtained from experimental potassium ferrioxalate actinometric measurements of the radiation fluxes coming from the lamps and entering the reacting space at $x = 0$ and $x = L_R$, and from the spectral distribution of the lamp emission [26].

The solution of the RTE was obtained by applying the Discrete Ordinate Method. The LVRPA for polychromatic radiation is calculated from the values of $I_\lambda(x,\mu)$ as

$$e^a\left(x\right)=2\pi\int_\lambda \kappa_\lambda\int_{\mu=-1}^{1} I_\lambda\left(x,\mu\right)\mathrm{d}\mu\,\mathrm{d}\lambda \tag{13.16}$$

In Figure 13.7, the LVRPA profiles for different TiO_2 concentrations are shown. Photon absorption presents strong variations along the x-axis. For $C_m \geq 0.25 \times 10^{-3}$ g cm^{-3} (Figure 13.7a), more than 90% of the radiation is absorbed in the first 0.5 cm of the reactor. Smoother profiles are obtained as long as C_m decreases, as shown in Figure 13.7b.

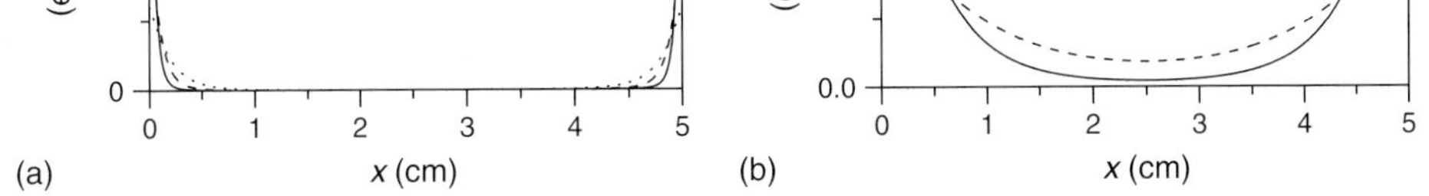

Figure 13.7 LVRPA profiles for different catalyst concentrations. (a) Solid line, 1.0×10^{-3} g cm^{-3}; broken line, 0.5×10^{-3} g cm^{-3}; dotted line, 0.25×10^{-3} g cm^{-3}. (b) Solid line, 0.1×10^{-3} g cm^{-3} and broken line, 0.05×10^{-3} g cm^{-3}. (Source: Reprinted with permission from Ref. [26]. Copyright 2007 American Chemical Society.)

13.4 Quantum Efficiencies of 4-Chlorophenol Photocatalytic Degradation

A way to evaluate the performance of photocatalytic processes and to compare experimental results obtained from different devices is to calculate efficiency parameters. Among them, the quantum yield (Φ_λ) and the quantum efficiency (η) are two of the most frequently used. For a given reaction, these parameters may present variations according to the type of catalyst employed, the operating conditions of the experiments (radiation wavelength, catalyst loading, initial reactant concentration, range of incident intensities, effects of reactions products, pH, etc.). When the energy source is monochromatic, Φ_λ is applied, whereas η is used for polychromatic radiation. The quantum efficiency η can be defined as the ratio of the number of reactant molecules degraded during a given time t to the total number of photons absorbed over the employed spectral range, during the same period of time [28, 29]. In terms of rates, η can be evaluated as

$$\eta_i = \frac{[\text{reaction rate of species } i]}{[\text{rate of photon absorption }]} \tag{13.17}$$

13.4.1 Calculation of the Quantum Efficiency

In the case of the degradation of 4-CP, $\eta_{4\text{-CP}}$ can be expressed as the ratio of the initial volumetric rate of 4-CP degradation $r_{4\text{-}CP}(\mathbf{x},t_0)$ to the LVRPA $e^a(\mathbf{x})$, both averaged over the reactor volume (V_R) [26]:

$$\eta_{4\text{-CP}} = \frac{\langle r_{4\text{-CP}}(\mathbf{x}, t_0)\rangle_{V_R}}{\langle e^a(\mathbf{x})\rangle_{V_R}} \tag{13.18}$$

The 4-CP initial reaction rate can be evaluated from the mass balance in the system. Under the following assumptions (i) perfectly mixed system, (ii) no mass transport limitations, (iii) only heterogeneous chemical reaction occurs, and (iv) differential conversion per pass in the reactor, the 4-CP mass balance takes the following form [14]

$$\varepsilon_L \frac{dC_{4\text{-CP}}(t)}{dt}\Big|_{Tk} = -\frac{V_R}{V_T}\langle r_{4\text{-CP}}(\mathbf{x}, t)\rangle_{V_R} \tag{13.19}$$

where ε_L is the liquid holdup ($\varepsilon_L \cong 1$), $C_{4\text{-CP}}$ is the molar concentration of 4-CP, t denotes reaction time, Tk refers to the tank, and $\langle r_{4\text{-CP}}(\mathbf{x}, t)\rangle_{V_R}$ is the 4-CP reaction rate averaged over the reactor volume. To calculate the initial reaction rate from experimental data, Eq. (13.19) can be used as follows:

$$\langle r_{4-CP}(x, t_0)\rangle_{VR} = -\varepsilon_L \frac{V_T}{V_R}\lim_{t\to t_0}\left(\frac{C_{4-CP}(t) - C_{4-CP}(t_0)}{t - t_0}\right)_{Tk} = -\varepsilon_L \frac{V_T}{V_R} m \tag{13.20}$$

Each experimental curve was fitted by an exponential function. The initial slope m in Eq. (13.20) represents the value of the first derivative of the exponential function at $t = t_0$.

For the one-dimensional model, the average value of $e^a(x)$ in the reactor volume, necessary to calculate the quantum efficiency, can be computed as

$$\langle e^a(x) \rangle_{V_R} = \frac{1}{V_R} \int_{V_R} e^a(x)\, dV \qquad (13.21)$$

13.4.2 Experimental Results

The quantum efficiency of the degradation of 4-CP was analyzed considering the effect of initial pH of the reacting solution. A set of experimental runs were performed with a catalyst loading of 0.5×10^{-3} g cm^{-3} and six different initial values of pH (between 2.5 and 10). In all experiments, the initial concentration of 4-CP was 1.4×10^{-7} mol cm^{-3}. Significant variations of the reaction rate were observed when the pH was modified, resulting higher rates under acidic conditions. This behavior could be explained by the pH dependence of the superficial charge of the catalyst particles. At acidic conditions, TiO_2 surface carries a net positive charge, whereas chlorophenols and their intermediates products remain neutral. Thus, these conditions could facilitate the adsorption of organic molecules and increase the rate of disappearance of the pollutant [30]. Different reaction mechanisms under acidic or alkaline conditions could also explain the reaction rate variations with pH [23]. As mentioned in Section 13.2, important changes in the optical properties of the TiO_2 suspensions also occur when the pH varies. Therefore, the pH alters both the value of the LVRPA and the reaction rate, producing significant variations in the values of $\eta_{4\text{-CP}}$.

In Figure 13.8, the values of $\eta_{4\text{-CP}}$ as a function of pH are represented. At low pH values, the quantum efficiency is high. By increasing the pH, the quantum efficiency decreases strongly to reach a minimum at pH $\cong$ 6.5 which is the natural

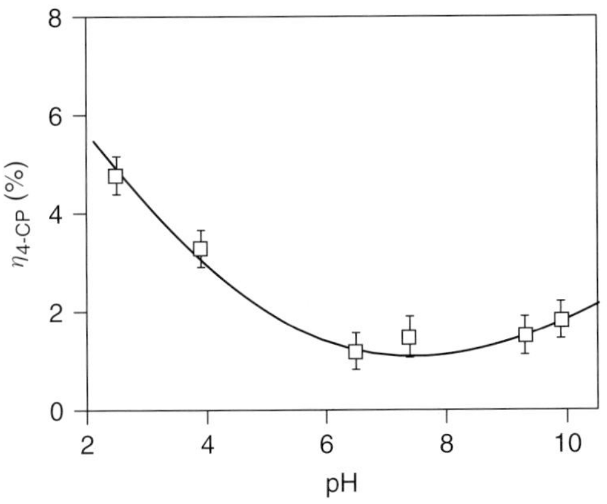

Figure 13.8 Quantum efficiency of the 4-CP degradation versus pH (bars represent 95% confidence interval). (Source: Reprinted with permission from Ref. [26]. Copyright 2007 American Chemical Society.)

value of the reacting solution. Under alkaline conditions, the quantum efficiency increases slightly.

13.5
Kinetic Modeling of the Pollutant Photocatalytic Degradation

In the present section, intrinsic kinetic expressions for the photocatalytic degradation of 4-CP are derived, as well as equations to represent the formation and disappearance of the main intermediate products: 4-chlorocatechol (4-CC, IUPAC name: 4-chlorobenzene-1,2-diol) and hydroquinone (HQ, IUPAC name: benzene-1,4-diol). The kinetic equations are based on mechanistic reaction steps and explicitly include the radiation effects [31].

13.5.1
Mass Balances

The mass balance presented in Eq. (13.19) can be written in a more general form as

$$\varepsilon_L \frac{dC_i(t)}{dt}\Big|_{Tk} = \frac{V_R}{V_T} a_v \langle \nu_i \, r(\mathbf{x}, t) \rangle_{A_R} \tag{13.22}$$

where C_i is the molar concentration of compound i = 4-CP, 4-CC, HQ (mol cm^{-3}); a_v denotes the catalytic surface area per unit suspension volume (cm^{-1}); ν_i is the stoichiometric coefficient; and $\langle r(\mathbf{x}, t) \rangle_{A_R}$ is the superficial reaction rate averaged over the catalytic reaction area (mol cm^{-2} s^{-1}).

The primary oxidation products of 4-CP are 4-CC, HQ, and benzoquinone (BQ, IUPAC name: cyclohexadienedione) [32–34] (Chapter 2). As shown in Section 13.4, the 4-CP removal efficiency is improved under acidic conditions. Consequently, experiments were performed at pH 2.5. At this pH, the oxidation via 4-CC represents the main pathway for the degradation of 4-CP. In second place appears the formation of HQ. Although BQ was detected during the experiments, its concentration was too low to permit quantification. Therefore, this intermediate product is not considered in the kinetic model. Then, the mass balances for 4-CP, 4-CC, and HQ, with the corresponding initial conditions, are

$$\begin{aligned} \varepsilon_L \frac{dC_{4\text{-CP}}(t)}{dt} &= -\frac{V_R}{V_T} a_v \left\{ \langle r_{4\text{-CP},1}(x, t) \rangle_{A_R} + \langle r_{4\text{-CP},2}(x, t) \rangle_{A_R} \right\} \\ C_{4\text{-CP}}(t = 0) &= C_{4\text{-CP},0} \end{aligned} \tag{13.23}$$

$$\begin{aligned} \varepsilon_L \frac{dC_{4\text{-CC}}(t)}{dt} &= \frac{V_R}{V_T} a_v \left\{ \langle r_{4\text{-CP},1}(x, t) \rangle_{A_R} - \langle r_{4\text{-CC}}(x, t) \rangle_{A_R} \right\} \\ C_{4\text{-CC}}(t = 0) &= 0 \end{aligned} \tag{13.24}$$

$$\begin{aligned} \varepsilon_L \frac{dC_{HQ}(t)}{dt} &= \frac{V_R}{V_T} a_v \left\{ \langle r_{4\text{-CP},2}(x, t) \rangle_{A_R} - \langle r_{HQ}(x, t) \rangle_{A_R} \right\} \\ C_{HQ}(t = 0) &= 0 \end{aligned} \tag{13.25}$$

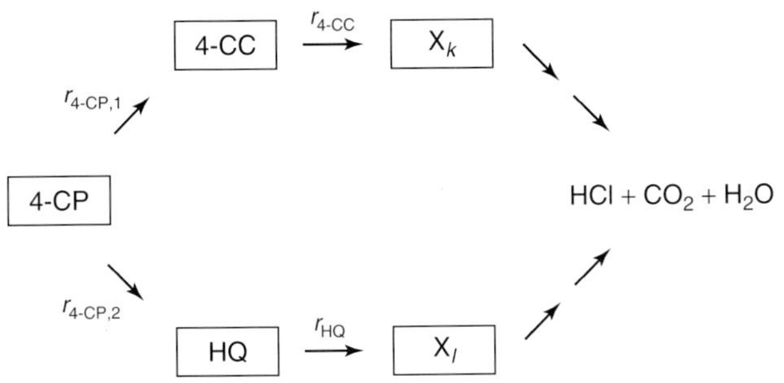

Figure 13.9 Reaction network of the photocatalytic degradation of 4-CP. (Source: Reprinted from Ref. [31], Copyright 2008, with permission from Elsevier.)

Two parallel reaction pathways are postulated for the degradation of 4-CP (Figure 13.9) [32]: $r_{4\text{-CP},1}$ represents the degradation rate to give 4-CC, whereas $r_{4\text{-CP},2}$ is the rate that leads to the formation of HQ. $r_{4\text{-CC}}$ and r_{HQ} denote the degradation rates of 4-CC and HQ, respectively. X_k and X_l represent secondary organic intermediate products.

13.5.2
Kinetic Model

The kinetic model proposed for the photocatalytic degradation of 4-CP is based on the reaction sequence summarized in Table 13.2 [35, 36]. X_i represents the primary organic intermediates of the reaction (i.e., 4-CC and HQ), that are further degraded by oxidation with the $OH^\bullet$ radical, and Y_i represents inorganic radicals and species that compete with the organic substrate for the existing $OH^\bullet$.

The derivation of the reaction rate expressions for the photodegradation of 4-CP, 4-CC, and HQ were based on the following assumptions: (i) the hydroxyl

Table 13.2 Kinetic steps for the 4-CP photocatalytic degradation.

Step	Reaction
Activation	$TiO_2 + h\nu \rightarrow TiO_2 + e^- + h^+$
Recombination	$e^- + h^+ \rightarrow \text{heat}$
Electron trapping	$e^- + O_{2,ads} \rightarrow O_2^{\bullet -}$
Hole trapping	$h^+ + H_2O_{ads} \rightarrow OH^\bullet + H^+$
	$h^+ + OH^-_{ads} \rightarrow OH^\bullet$
Hydroxyl attack	$4\text{-CP}_{ads} + OH^\bullet \rightarrow X_i$
	$X_{i,ads} + OH^\bullet \rightarrow X_j$
	$Y_{i,ads} + OH^\bullet \rightarrow Y_j$

Source: Reprinted from Ref. [31], Copyright 2008, with permission from Elsevier.

radical attack is the main responsible for the degradation of 4-CP and the organic intermediate products; (ii) photocatalytic reactions occur at the surface of the catalyst particles among adsorbed species; (iii) dynamic equilibrium is achieved between the bulk and the adsorbed concentrations of water, oxygen, organic compounds, and inorganic species [37]; (iv) molecular oxygen and organic compounds are considered to adsorb on different sites of the TiO_2 particle [38], and on the other hand, since adsorption sites for the organics are assumed to be the same, a competitive adsorption mechanism for 4-CP and its main intermediates is postulated; (v) the kinetic micro-steady-state approximation is applied for the concentration of electrons, holes, and hydroxyl radicals [35]; and (vi) the rate of electron–hole generation is proportional to the LVRPA [13]. Additional assumptions have been made to simplify the model: (i) there are no mass transport limitations, (ii) oxygen concentration is constant and in excess with respect to the stoichiometric demand, and (iii) the concentration of water molecules and hydroxide ions on the catalytic surface remains constant. On the basis of these considerations, the reaction rate expressions are

$$r_{4\text{-CP},1}(\mathbf{x},t) = \frac{\alpha_{2,1} C_{4\text{-CP}}(t)}{1+\alpha_3 C_{4\text{-CP}}(t) + \alpha'_1 C_{4\text{-CC}}(t) + \alpha'_2 C_{\text{HQ}}(t)} \times \left(-1+\sqrt{1+\frac{\alpha_1}{a_v} e^{a}(\mathbf{x})}\right) \tag{13.26}$$

$$r_{4\text{-CP},2}(\mathbf{x},t) = \frac{\alpha_{2,2} C_{4\text{-CP}}(t)}{1+\alpha_3 C_{4\text{-CP}}(t) + \alpha'_1 C_{4\text{-CC}}(t) + \alpha'_2 C_{\text{HQ}}(t)} \times \left(-1+\sqrt{1+\frac{\alpha_1}{a_v} e^{a}(\mathbf{x})}\right) \tag{13.27}$$

$$\mathrm{r}_{4-\text{CC}}(\mathbf{x},\mathrm{t}) = \frac{\alpha_4 C_{4-CC}(t)}{1+\alpha_3 C_{4-CP}(t) + \alpha'_1 C_{4-CC}(t) + \alpha'_2 C_{HQ}(t)} \times \left(-1+\sqrt{1+\frac{\alpha_1}{\mathrm{a_v}} \mathrm{e^a}(\mathbf{x})}\right) \tag{13.28}$$

$$r_{\text{HQ}}(\mathbf{x},t) = \frac{\alpha_5 C_{\text{HQ}}(t)}{1+\alpha_3 C_{4\text{-CP}}(t) + \alpha'_1 C_{4\text{-CC}}(t) + \alpha'_2 C_{\text{HQ}}(t)} \times \left(-1+\sqrt{1+\frac{\alpha_1}{a_v} e^{a}(\mathbf{x})}\right) \tag{13.29}$$

where α_i and α'_i are kinetic parameters.

13.5.3 Kinetic Parameters Estimation

In order to obtain the values of the kinetic parameters, experimental runs were performed at different levels of TiO_2 concentration: $(0.05, 0.1, 0.5, 1.0) \times 10^{-3}$ g cm^{-3};

and incident radiation: 30, 65, and 100%. The pH of the reacting suspensions was 2.5 and the initial 4-CP concentration in all experiments was 1.4×10^{-7} mol cm^{-3}.

The Levenberg–Marquardt optimization algorithm was applied to estimate the kinetic constants involved in Eqs. (13.26–13.29). The experimental concentrations of 4-CP, 4-CC, and HQ, obtained from samples at different reaction times (a sample was taken from the tank every hour), were compared with simulation results. The optimization procedure renders the values of the parameters that minimize the differences between the model predictions and the experimental data. Under the operating conditions of the experiments, it was found that the terms $\alpha_3 C_{4\text{-}CP}(t)$, $\alpha'_1 C_{4\text{-}CC}(t)$, and $\alpha'_2 C_{HQ}(t)$ were much lower than 1, and thus could be neglected. Consequently, the final kinetic expressions employed for the parameters estimation were the following:

$$r_{4\text{-CP},1}(\mathbf{x},t) = \alpha_{2,1} C_{4\text{-CP}}(t)\left(-1+\sqrt{1+\frac{\alpha_1}{a_\text{v}} e^\text{a}(\mathbf{x})}\right) \tag{13.30}$$

$$r_{4\text{-CP},2}(\mathbf{x},t) = \alpha_{2,2} C_{4\text{-CP}}(t)\left(-1+\sqrt{1+\frac{\alpha_1}{a_\text{v}} e^\text{a}(\mathbf{x})}\right) \tag{13.31}$$

$$r_{4\text{-CC}}(\mathbf{x},t) = \alpha_4 C_{4\text{-CC}}(t)\left(-1+\sqrt{1+\frac{\alpha_1}{a_\text{v}} e^\text{a}(\mathbf{x})}\right) \tag{13.32}$$

$$r_{\text{HQ}}(\mathbf{x},t) = \alpha_5 C_{\text{HQ}}(t)\left(-1+\sqrt{1+\frac{\alpha_1}{a_\text{v}} e^\text{a}(\mathbf{x})}\right) \tag{13.33}$$

The values of the five kinetic parameters, with the corresponding 95% confidence interval, are reported in Table 13.3.

Figure 13.10 shows experimental results and model predictions for 100% incident radiation and $C_\text{m} = 0.5 \times 10^{-3}$ g cm^{-3}. The degradation of 4-CP is complete after 8 h of treatment. The concentrations of 4-CC and HQ initially rise, reach a maximum around the second hour, and then decrease until they also disappear at the end of the run.

Considering all the experiments, good agreement was obtained between model predictions and experimental data, with a root mean square error (RMSE) of 14.4%.

Table 13.3 Estimated kinetic parameters.

Parameter	α_1	$\alpha_{2,1}$	$\alpha_{2,2}$	α_4	α_5
Value	1.09×10^{11}	9.43×10^{-6}	2.18×10^{-6}	1.29×10^{-5}	9.21×10^{-6}
Confidence interval (95%)	$\pm 0.07 \times 10^{11}$	$\pm 0.26 \times 10^{-6}$	$\pm 0.46 \times 10^{-6}$	$\pm 0.08 \times 10^{-5}$	$\pm 0.96 \times 10^{-6}$
Units	s cm^2 einstein^{-1}	cm s^{-1}	cm s^{-1}	cm s^{-1}	cm s^{-1}

Source: Reprinted from Ref. [31], Copyright 2008, with permission from Elsevier.

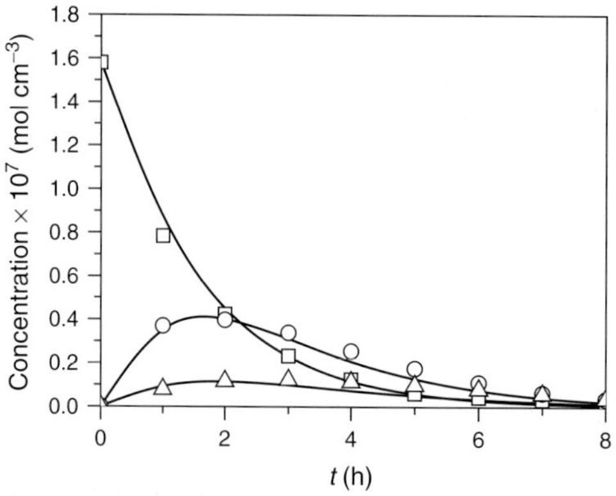

Figure 13.10 Experimental and predicted concentrations of 4-CP, 4-CC, and HQ versus time for 100% incident radiation and $C_m = 0.5 \times 10^{-3}$ g cm^{-3}. Experimental data: □, 4-CP; ○, 4-CC; and Δ, HQ. Model results: solid lines. (Source: Reprinted from Ref. [31], Copyright 2008, with permission from Elsevier.)

13.6
Bench-Scale Slurry Photocatalytic Reactor for Degradation of 4-Chlorophenol

The rigorous methodology previously described is used in this section to model a very different photocatalytic reactor in size, shape, and operating conditions. The kinetic model of the 4-CP degradation as well as its main intermediate compounds obtained in the laboratory-scale reactor will be applied to study the behavior of a bench-scale slurry reactor for the same photocatalytic reaction.

13.6.1
Experiments

A schematic representation of the experimental device is shown in Figure 13.11. The bench-scale reactor was operated inside a closed recycle system. It was built with two parallel borosilicate glasses and illuminated from one side by two 40 W Philips TLK40/09N tubular lamps. These UV actinic lamps, having a diameter of 38 mm, were placed at the focal axis of parabolic reflectors. The system was also equipped with a storage tank that has an oxygen supply and a liquid sampling valve, a heat exchanger, and a centrifugal pump. Additional details about the experimental device and procedure can be found in [16].

13.6.2
Reactor Model

13.6.2.1 Radiation Model

The monochromatic radiation intensity I_λ (x,z,Ω), as a function of position and direction, can be obtained from the solution of the RTE applied to the bench-scale photocatalytic reactor. To do this, a number of assumptions were made:

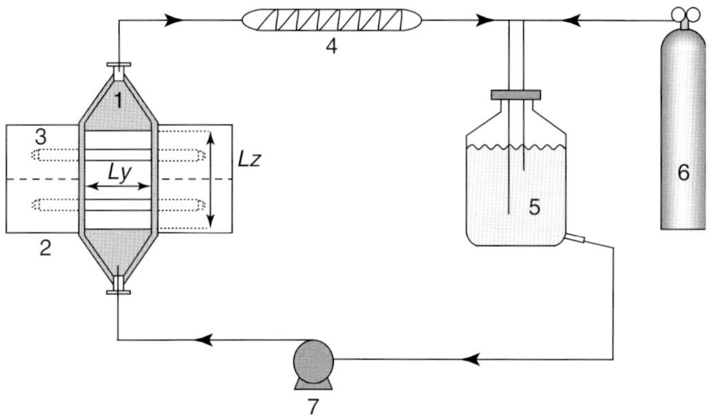

Figure 13.11 Schematic representation of the bench-scale experimental device: 1, front view of the reactor; 2, reflectors; 3, lamps; 4, heat exchanger; 5, tank; 6, oxygen supply; and 7, pump. (Source: Reprinted from Ref. [16], Copyright 2007, with permission from Elsevier.)

(i) radiation uniformity along the *y*-direction is assumed because the lamp length is considerably larger than the reactor width (Figure 13.11); thus, a two-dimensional model along the *x*- and *z*-coordinates is adopted to describe the radiation field, (ii) radiation propagation is modeled using the spherical coordinates (θ,ϕ), and (iii) the emission term of the RTE was neglected considering that the reactor is operated at ambient temperature (298 K).

Therefore, the RTE for each spatial position and direction inside the reacting space is given by the following integro-differential equation [39, 40]:

$$\mu \frac{\partial I_\lambda (x,z,\Omega)}{\partial x} + \eta \frac{\partial I_\lambda (x,z,\Omega)}{\partial z} + \beta_\lambda I_\lambda (x,z,\Omega)$$
$$= \frac{\sigma_\lambda}{4\pi} \int_{\Omega'=4\pi} I_\lambda \left(x,z,\Omega'\right) p\left(\Omega' \to \Omega\right) \mathrm{d}\Omega' \qquad (13.34)$$

In this equation, the direction cosines with respect to the *x*-axis ($\mu = \cos\phi \sin\theta$) and *z*-axis ($\eta = \sin\phi \sin\theta$) have been employed.

The boundary conditions at the irradiated, opposite, superior, and inferior walls of the reactor, necessary to solve the RTE, are

$$I_\lambda \left(x=0,\ z,\ \Omega=\Omega^i\right) = \chi \text{ (properties of the emitting system and the wall)} \qquad (13.35)$$

$$I_\lambda \left(x=\mathrm{Lx}, z,\ \Omega=\Omega^i\right) = \chi \text{ (properties of the arriving radiation and the wall)} \qquad (13.36)$$

$$I_\lambda \left(x, z=0,\ \Omega=\Omega^i\right) = 0 \qquad (13.37)$$

$$I_\lambda \left(x, z=\mathrm{Lz}, \Omega=\Omega^i\right) = 0 \qquad (13.38)$$

Here Ω^i represents the directions of radiation intensity entering the reactor.

The Discrete Ordinate Method [22, 39] was employed to solve the complete radiation system (Eqs. (13.34–13.38)). The spectral volumetric extinction (β_λ) and scattering (σ_λ) coefficients and the Henyey–Greenstein phase function ($p_{H\text{-}G,\lambda}$), evaluated in Section 13.2, were used to solve the RTE. Once the radiation intensities are calculated, the LVRPA involved in the reaction rate expressions for the pollutant degradation can be obtained by means of the following integration:

$$e^{a}(\mathbf{x}) = \int_{\lambda} \kappa_{\lambda} \int_{\Omega} I_{\lambda}(x, z, \Omega)\, d\Omega\, d\lambda \qquad (13.39)$$

13.6.2.2 Reaction Rates

The reaction rate expressions for each compound of the reacting system (4-CP, 4-CC, HQ) and the corresponding kinetic parameters were obtained from the laboratory-scale reactor studied in Section 13.5.

13.6.2.3 Mass Balances in the Tank and Reactor

The complete reactor model can be written considering separate mass balances for the reactor and the storage tank. For the tank, well stirred conditions and transient state were assumed. Thus, the resulting mass balance equation and initial condition in the tank, for a compound i (with i = 4-CP, 4-CC, HQ), are given by

$$\varepsilon_{L} \frac{dC_{i}(t)}{dt}\Big|_{Tk} = \frac{1}{\tau_{Tk}} \left[C^{i}_{i,Tk}(t) - C^{o}_{i,Tk}(t)\right] \qquad (13.40)$$

$$t = 0, C^{i}_{i,Tk} = C_{i,0} \qquad (13.41)$$

In Eq. (13.40) τ_{Tk} is the mean residence time in the storage tank (40 s).

The local mass balances in the reactor for 4-CP, 4-CC, HQ and the corresponding initial and boundary conditions, can be derived under the following assumptions: (i) direct photolysis is neglected, (ii) the chemical reaction takes place only at the solid–liquid interface, (iii) the fluid flow is laminar, (iv) diffusion in the z-direction is negligible when compared to convection in that direction, and (v) there is symmetry in the y-direction. Under these conditions, we have

$$\frac{\partial C_{i,R}}{\partial t} + v_{z} \frac{\partial C_{i,R}}{\partial z} - D_{i} \frac{\partial^{2} C_{i,R}}{\partial x^{2}} - r_{i} = 0 \qquad (13.42)$$

$$C_{i,R}(x, z, t = 0) = C_{i,0} \qquad (13.43)$$

$$C_{i,R}(x, z = 0, t) = C^{o}_{i,Tk}(t) \qquad (13.44)$$

$$\frac{\partial C_{i,R}}{\partial x}\Big|_{x=0} = 0 \quad \textit{and} \quad \frac{\partial C_{i,R}}{\partial x}\Big|_{x=Lx} = 0 \qquad (13.45)$$

Solving the system of partial differential equations with their initial and boundary conditions (Eqs. (13.42–13.45)), the average concentrations of compound i at the reactor outlet can be calculated by

$$\left\langle C_{i,R}(x, z = Lz, t)\right\rangle_{A_R} = \frac{\displaystyle\int_{x=0}^{x=Lx} C_{i,R}(x, z = Lz, t)\, v_{z}(x)\, dx}{\displaystyle\int_{x=0}^{x=Lx} v_{z}(x)\, dx} = C^{i}_{i,Tk}(t) \qquad (13.46)$$

Notice that the outlet concentration of the reactor given by the Eq. (13.46) is equal to the inlet concentration for the mass balance in the tank, $C^{i}_{i,Tk}(t)$.

13.6.3 Results

In this section, the performance of the bench-scale slurry reactor is studied. Firstly, the computed results derived from inserting the kinetic model and estimated parameters in the bench-scale reactor model are presented; then, these computed results are compared with experimental data.

Figure 13.12 shows a 3D plot of predicted concentrations of 4-CP (Figure 13.12a) and 4-CC (Figure 13.12b) at three reaction times: $t = 2$ s, 30 min, and 60 min, for runs performed with a TiO_2 mass concentration of 1.0×10^{-3} g cm^{-3} [41]. At the beginning of the reaction, the 4-CP concentration is uniform at the reactor inlet ($z = 0$). After that, the photocatalytic degradation takes place near the irradiated wall ($x = 0$) and the average pollutant concentration gradually decreases. These simulations are in agreement with the LVRPA spatial distribution: high radiation absorption near the irradiated wall ($x = 0$) and negligible radiation absorption close to the opposite wall ($x = 1.2$ cm). As might be expected, at longer reaction times ($t = 30$ and 60 min), the average 4-CP concentration decreases. On the other hand, for $t = 0$ min, the 4-CC concentration profiles is null at the reactor inlet ($z = 0$). For longer reaction times, the 4-CC concentration progressively increases close to the irradiated wall due to the conversion of 4-CP (Figure 13.12b). Then, at $t = 30$ min, a competition of formation and degradation of 4-CC is shown and finally, at $t = 60$ min, the degradation of 4-CC is detected. This photocatalytic process continues until the complete degradation of 4-CP and 4-CC is reached.

Finally, Figure 13.13 shows the experimental and predicted conversions of 4-CP at different reaction times and for all the runs performed with the bench-scale

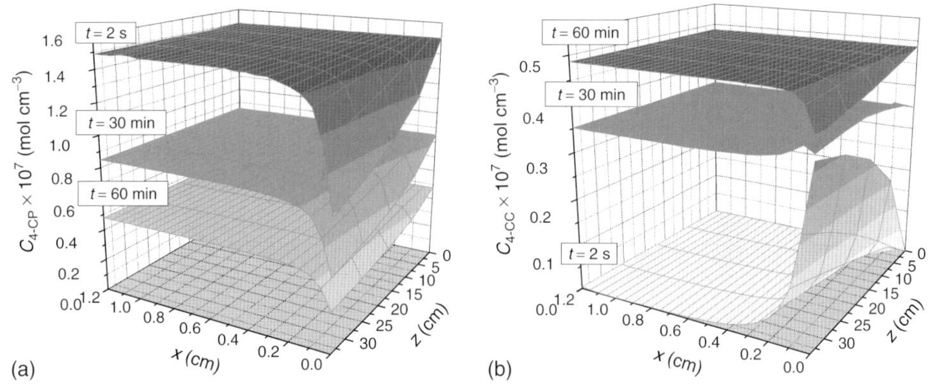

Figure 13.12 Spatial distribution of concentrations in the reaction space at different irradiation times. (a) 4-CP. (b) 4-CC. $C_m = 1.0 \times 10^{-3}$ g cm^{-3}. (Source: Reprinted from Ref. [41] with permission from the copyright holders, IWA Publishing.)

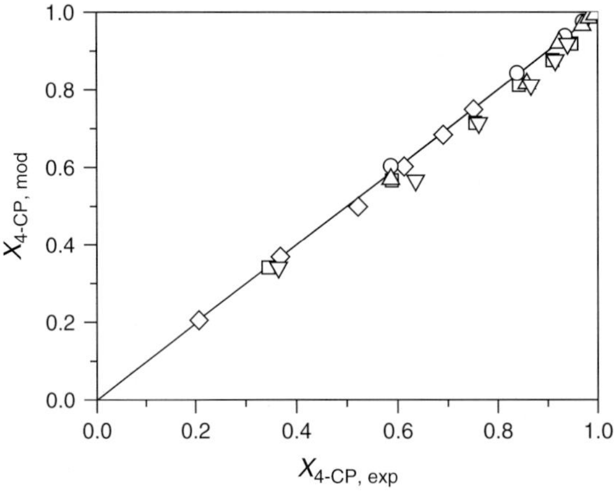

Figure 13.13 Experimental versus predicted conversions of 4-CP. Symbols correspond to different runs: ◇, $C_{4\text{-CP},0} = 1.4 \times 10^{-7}$ mol cm^{-3} and $C_m = 0.05 \times 10^{-3}$ g cm^{-3}; □, $C_{4\text{-CP},0} = 1.4 \times 10^{-7}$ mol cm^{-3} and $C_m = 0.1 \times 10^{-3}$ g cm^{-3}; ○, $C_{4\text{-CP},0} = 1.4 \times 10^{-7}$ mol cm^{-3} and $C_m = 0.5 \times 10^{-3}$ g cm^{-3}; Δ, $C_{4\text{-CP},0} = 1.4 \times 10^{-7}$ mol cm^{-3} and $C_m = 1.0 \times 10^{-3}$ g cm^{-3}; and ∇, $C_{4\text{-CP},0} = 0.7 \times 10^{-7}$ mol cm^{-3} and $C_m = 0.1 \times 10^{-3}$ g cm^{-3}. (Source: Reprinted from Ref. [16], Copyright 2007, with permission from Elsevier.)

reactor. The symbols indicate the operating conditions employed for each one of the experimental runs. Good agreement can be observed when the simulated results are compared with the experimental data. Taking into account all the experimental runs, the RMSE based only on the experimental and predicted concentrations for 4-CP and 4-CC, was 9.91%. HQ was not considered to calculate the RMSE due to the low concentrations observed during the experimental work (<1.5 mg l^{-1}).

13.7 Conclusions

A methodology has been proposed for modeling and scaling-up of a photocatalytic slurry reactor, from laboratory to bench scale, to study the degradation of a model organic pollutant in water: the 4-CP. To achieve this objective, it was necessary to analyze both equipment using a rigorous mathematical modeling, involving the solution of the complete RTE. Firstly, the kinetic expressions for the degradation of the organic pollutants and the corresponding kinetic parameters were determined using a laboratory photoreactor. Then, these kinetic expressions were employed to study a bench-scale, slurry photocatalytic reactor. From the reported results, it is possible to conclude that the proposed methodology has shown to be appropriate to simulate the performance of a photocatalytic slurry reactor of different shape, size, radiation emitting system, and configuration.

Acknowledgments

The authors are grateful to Universidad Nacional del Litoral (UNL), Consejo Nacional de Investigaciones Científicas y Técnicas (CONICET), and Agencia Nacional de Promoción Científica y Tecnológica (ANPCyT) for the financial support. They also thank Antonio C. Negro for his valuable help during the experimental work and Claudia M. Romani for her technical assistance.

References

1. Hashimoto, K., Irie, H. and Fujishima, A. (2005) TiO_2 Photocatalysis: a historical overview and future prospects. *Jpn. J. Appl. Phys.*, **44**, 8269–8285.
2. Fujishima, A. and Honda, K. (1972) Electrochemical photolysis of water at a semiconductor electrode. *Nature*, **238**, 37–38.
3. Formenti, M., Juillet, F., and Teichner, S.J. (1970) *Compt. Rend. Acad. Sci. C*, **270**, 138–141.
4. Formenti, M., Juillet, F., Meriaudeau, P., and Teichner, S.J. (1971) Heterogeneous photocatalysis for partial oxidation of paraffins. *Chem. Technol.*, **1**, 680–686.
5. Frank, N.S. and Bard, A.J. (1977) Heterogeneous photocatalytic oxidation of cyanide and sulfite in aqueous solutions. *J. Phys. Chem.*, **81**, 1484–1488.
6. Hoffmann, M.R., Martin, S.T., Choi, W., and Bahnemann, D.W. (1995) Environmental applications of semiconductor photocatalysis. *Chem. Rev.*, **95**, 69–96.
7. Watanabe, T., Kitamura, A., Kojima, E., Nakayama, C., Hashimoto, K., and Fujishima, A. (1993) Photocatalytic activity of TiO_2 thin film under room light, in *Photocatalytic Purification and Treatment of Water and Air* (eds D.F. Ollis and H. Al-Ekabi), Elsevier, Amsterdam.
8. Paz, Y., Luo, Z., Rabenberg, L., and Heller, A. (1995) Photooxidative self-cleaning transparent titanium dioxide films on glass. *J. Mater. Res.*, **7**, 49–63.
9. Negishi, N., Iyoda, T., Hashimoto, K., and Fujishima, A. (1995) Preparation of transparent TiO_2 thin film photocatalyst and its photocatalytic activity. *Chem. Lett.*, 841–842.
10. Wang, R., Hashimoto, K., Fujishima, A., Chikuni, M., Kojima, E., Kitamura, A., Shimohigoshi, M., and Watanabe, T. (1997) Light-induced amphiphilic surfaces. *Nature*, **388**, 431–432.
11. Sato, S. (1986) Photocatalytic activity of NO_x-doped TiO_2 in the visible light region. *Chem. Phys. Lett.*, **123**, 126–128.
12. Anpo, M., Ichihashi, Y., and Yamashita, H. (1997) Design of highly active titanium oxide photocatalysts and second-generation titanium oxide photocatalysts operating under visible rays for the decomposition of NOx. *Petrotech Tokyo*, **20**, 66–72.
13. Alfano, O.M., Cabrera, M.I., and Cassano, A.E. (1997) Photocatalytic reactions involving hydroxyl radical attack. I. reaction kinetics formulation with explicit photon absorption effects. *J. Catal.*, **172**, 370–379.
14. Cabrera, M.I., Negro, A.C., Alfano, O.M., and Cassano, A.E. (1997) Photocatalytic reactions involving hydroxyl radical attack: II. kinetics of the decomposition of trichloroethylene using titanium dioxide. *J. Catal.*, **172**, 380–390.
15. Alfano, O.M. and Cassano, A.E. (2009) in *Advances in Chemical Engineering* (eds H. De Lasa and B. Serrano-Rosales), Elsevier, pp. 229–287.
16. Satuf, M.L., Brandi, R.J., Cassano, A.E., and Alfano, O.M. (2007) Scaling-up of slurry reactors for the photocatalytic degradation of 4-chlorophenol. *Catal. Today*, **129**, 110–117.
17. Satuf, M.L., Brandi, R.J., Cassano, A.E., and Alfano, O.M. (2005) Experimental method to evaluate the optical properties of aqueous titanium dioxide suspensions. *Ind. Eng. Chem. Res.*, **44**, 6643–6649.
18. Cabrera, M.I., Alfano, O.M., and Cassano, A.E. (1996) Absorption and

scattering coefficients of titanium dioxide particulate suspensions in water. *J. Phys. Chem.*, **100**, 20043–20050.

19. Alfano, O.M., Negro, A.C., Cabrera, M.I., and Cassano, A.E. (1995) Scattering effects produced by inert particles in photochemical reactors. 1. Model and experimental verification. *Ind. Eng. Chem. Res.*, **34**, 488–499.
20. Siegel, R. and Howell, J.R. (2002) *Thermal Radiation Heat Transfer*, 6th edn, Hemisphere Publishing Corporation, Bristol, PA.
21. Van de Hulst, H.C. (1980) *Multiple Light Scattering*, Academic Press, New York.
22. Duderstadt, J.J. and Martin, W.R. (1979) *Transport Theory*, John Wiley & Sons, Inc., New York.
23. Theurich, J., Lindner, M., and Bahnemann, D.W. (1996) Photocatalytic degradation of 4-chlorophenol in aerated aqueous titanium dioxide suspensions: a kinetic and mechanistic study. *Langmuir*, **12**, 6368–6376.
24. Yang, H.G., Li, C.Z., Gu, H.C., and Fang, T.N. (2001) Rheological behavior of titanium dioxide suspensions. *J Colloid Interface Sci.*, **236**, 96–103.
25. Mills, A., Morris, S., and Davies, R. (1993) Photomineralisation of 4-chlorophenol sensitized by titanium dioxide: a study of the intermediates. *J. Photochem. Photobiol. A*, **70**, 183–191.
26. Satuf, M.L., Brandi, R.J., Cassano, A.E., and Alfano, O.M. (2007) Quantum efficiencies of 4-chlorophenol photocatalytic degradation and mineralization in a well-mixed slurry reactor. *Ind. Eng. Chem. Res.*, **46**, 43–51.
27. Özişik, M.N. (1973) *Radiative Transfer and Interactions with Conduction and Convection*, John Wiley & Sons, Inc., New York.
28. Braun, A.M., Maurette, M.T., and Oliveros, E. (1991) *Photochemical Technology*, John Wiley & Sons, Inc., New York.
29. Braslavsky, S.E., Braun, A.M., Cassano, A.E., Emeline, A.V., Litter, M.I., Palmisano, L., Parmon, V.N., and Serpone, N. (2011) Glossary of terms used in photocatalysis and radiation catalysis (IUPAC Recommendations 2011). *Pure Appl. Chem.*, **83**, 931–1014.
30. Doong, R.A., Chen, C.H., Maithreepala, R.A., and Chang, S.M. (2001) The influence of pH and cadmium sulfide on the photocatalytic degradation of 2-chlorophenol in titanium dioxide suspensions. *Water Res.*, **35**, 2873–2880.
31. Satuf, M.L., Brandi, R.J., Cassano, A.E., and Alfano, O.M. (2008) Photocatalytic degradation of 4-chlorophenol: a kinetic study. *Appl. Catal. B*, **82**, 37–49.
32. Al-Sayyed, G., D'Oliveira, J.C., and Pichat, P. (1991) Semiconductor-sensitized photodegradation of 4-chlorophenol in water. *J. Photochem. Photobiol. A*, **58**, 99–114.
33. Li, X., Cubbage, J.W., Tetzlaff, T.A., and Jenks, W.S. (1999) Photocatalytic degradation of 4-chlorophenol. 1. The hydroquinone pathway. *J. Org. Chem.*, **64**, 8509–8524.
34. Li, X., Cubbage, J.W., and Jenks, W.S. (1999) Photocatalytic degradation of 4-chlorophenol. 2. The 4-chlorocatechol pathway. *J. Org. Chem.*, **64**, 8525–8536.
35. Turchi, C.S. and Ollis, D.F. (1990) Photocatalytic degradation of organic water contaminants: mechanisms involving hydroxyl radical attack. *J. Catal.*, **122**, 178–192.
36. Almquist, C.B. and Biswas, P. (2001) A mechanistic approach to modeling the effect of dissolved oxygen in photo-oxidation reactions on titanium dioxide in aqueous systems. *Chem. Eng. Sci.*, **56**, 3421–3430.
37. Dijkstra, M.F.J., Panneman, H.J., Winkelman, J.G.M., and Kelly, J.J. (2002) Modeling the photocatalytic degradation of formic acid in a reactor with immobilized catalyst. *Chem. Eng. Sci.*, **57**, 4895–4907.
38. Terzian, R., Serpone, N., Minero, C., Pelizzetti, E., and Hidaka, H. (1990) Kinetic studies in heterogeneous photocatalysis 4. The photomineralization of a hydroquinone and a catechol. *J. Photochem. Photobiol. A*, **55**, 243–249.
39. Brandi, R.J., Alfano, O.M., and Cassano, A.E. (1996) Modeling of radiation absorption in flat plate photocatalytic reactor. *Chem. Eng. Sci.*, **54**, 3169–3174.
40. Brandi, R.J., Alfano, O.M., and Cassano, A.E. (1999) Rigorous model and experimental verification of the radiation field

in a flat plate solar collector simulator employed for photocatalytic reactions. *Chem. Eng. Sci.*, **54**, 2817–2827.

41. Satuf, M.L., José, S., Paggi, J.C., Brandi, R.J., Cassano, A.E., and Alfano, O.M. (2010) Reactor modeling in heterogeneous photocatalysis. Toxicity and biodegradability assessment. *Water Sci. Technol.*, **61**, 2491–2499.

14
Design and Optimization of Photocatalytic Water Purification Reactors

Tsuyoshi Ochiai and Akira Fujishima

14.1
Introduction

14.1.1
Market Transition of Industries Related to Photocatalysis

Environmental purification is one of the most important technologies for human life. Application of the strong oxidation ability of photoexcited TiO_2 for environmental purification has received growing attention [1–4]. Figure 14.1 shows the market transition of industries related to photocatalysis. This data represents the sales volume for companies which are members of the Photocatalysis Industry Association of Japan. Recently, the sales volume of "cleanup" application involved in photocatalytic environmental purifiers greatly increased. This trend indicates the increasing of the number of peoples who are interested in environmental issues such as food poisoning, sick house syndrome, swine influenza outbreak, and so on. Water pollution is a particularly serious problem for our life, however, the sales volume of photocatalytic water purification is actually smaller than that of photocatalytic air purification. Therefore, photocatalytic water purification is one of the most important applications of TiO_2 photocatalysis now.

14.1.2
Historical Overview

Since the discovery of water splitting using TiO_2 electrode by Fujishima and Honda, photocatalysis drew the attention of many people as a promising method for H_2 production [1]. However, although having a large surface-area-to-volume ratio, the aqueous suspension of TiO_2 powder shows very low reaction efficiency for H_2 production. Kawai and Sakata [5] concluded that the produced H_2 and O_2 gases might recombine to regenerate H_2O through the back reaction in the powder system, because the production sites of H_2 and O_2 gases are located close to each other. Thus, TiO_2 powder system is not very attractive from the viewpoint of H_2 production technology. On the other hand, Frank and Bard [6]

Photocatalysis and Water Purification: From Fundamentals to Recent Applications, First Edition. P. Pichat.

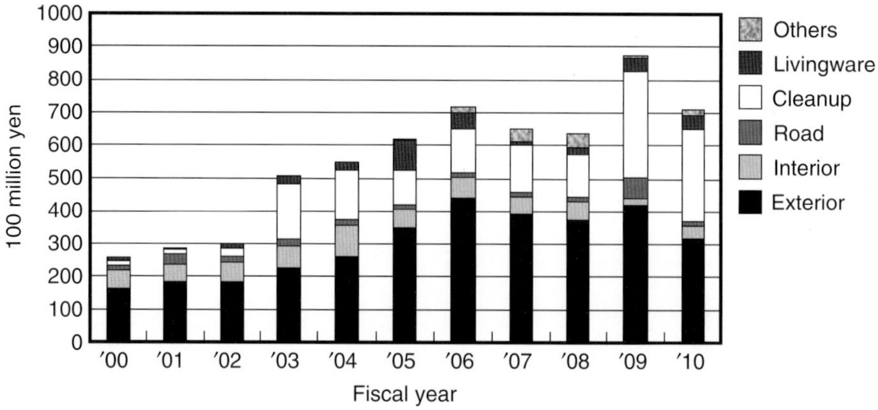

Figure 14.1 Market growth of industries related to photocatalysis. (Source: Photocatalysis Industry Association of Japan.)

showed the decomposition of cyanide in the presence of aqueous TiO_2 suspensions. After this report, many researchers demonstrated detoxification of various harmful compounds in water using powdered TiO_2. They found that the oxidation of harmful compounds was more interesting than the reduction reaction producing H_2. The reason is that various forms of active oxygen, such as $O_2^{\bullet-}$, $OH^{\bullet}$, and H_2O_2, produced by the following processes are responsible for the decomposition reactions [7]. In addition, the holes (h^+) generated in TiO_2 under UV irradiation were highly oxidizing, and most water contaminants were totally oxidized.

$$TiO_2 + h\nu \rightarrow h^+ \text{ (hole)} + e^- \text{ (electron)}$$

$$h^+ + OH^- \rightarrow OH^{\bullet}$$

$$h^+ + H_2O \rightarrow H^+ + OH^{\bullet}$$

$$e^- + O_2 \rightarrow O_2^{\bullet-}$$

$$2O_2^{\bullet-} + 2H^+ \rightarrow H_2O_2 + O_2$$

$$2OH \rightarrow H_2O_2$$

For the purpose of easy handling, the immobilization of TiO_2 powders on supports was carried out in the late 1980s [8]. Until now, although many researches carried on various reactors for photocatalytic water purification, TiO_2 photocatalysis could not be developed to the stage of effective real industrial technology. Chong *et al.* [9] summarized several key scientific and technical requirements for an effective process.

1) catalyst improvement for a high photoefficiency that can utilize a wider spectrum light at greater sensitivity;
2) catalyst immobilization strategy for a cost-effective solid–liquid separation;
3) improvement in the photocatalytic operation range (wide pH range and addition of small amount of oxidant additives);

4) integrated or coupling system for enhanced photomineralization or photodisinfection kinetics;
5) effective design of photocatalytic reactor system.

On the basis of these requirements (especially 2, 4, and 5), this chapter reviews several new technologies for photocatalytic water purification. Our recent studies and the future directions are also mentioned.

14.2 Catalyst Immobilization Strategy

14.2.1 Aqueous Suspension

Because of its simplicity, aqueous suspension systems are still investigated for fundamental analysis of reaction mechanism or optimization of the reaction condition [10–13]. Sun *et al.* [10] studied photocatalytic decomposition of diethyl phosphoramidate (DEPA) using the commercial TiO_2 powders, P25, and Hombikat UV 100 (HK). The photocatalytic decomposition was carried out in aqueous suspension containing a predetermined amount of TiO_2 powders, a certain amount of DEPA, and 800 ml of distilled water. The initial pH and temperature were adjusted further to study the effect of the pH and temperature. Interestingly, two ranges of initial pH were found where they achieved complete carbon oxidation, with one in the acid and one in near-neutral environments were observed. Moreover, the locations of these peaks of P25 are different from those of HK. Sun *et al.* discussed the reason of this result as follows. The charge of the compounds to be oxidized is determined by its pK_a and pH. In acidic pH both the TiO_2 surface and the amine substrate of DEPA were positively charged, repelled each other, and the reaction was retarded. On the other hand, ethyl phosphoramidate, the acidic hydrolysis product of DEPA, becomes positively charged and is repelled from the positively charged TiO_2 surface. Thus, the control of pH range for neutral condition of the compounds to be oxidized is important. The optimal DEPA concentration, temperature, and TiO_2 amount for efficient decomposition of DEPA were also investigated. All values were different between P25 and HK. It was affected by the different activation energy for total carbon decomposition of DEPA over HK and P25 (29.5 ± 1.0 and 24.3 ± 3.1 kJ mol^{-1}, respectively), which was calculated by the Arrhenius equation. In comparison, the reaction rate over HK increases more slowly with the concentration of the photocatalyst added. Therefore, TiO_2 type should be considered to achieve efficient decomposition. Intermediates and the mechanism of photocatalytic decomposition of contaminants are also investigated by using similar aqueous suspension systems. Sakkas *et al.* [11] reported the elimination of flufenacet (*N*-(4-fluorophenyl)-*N*-isopropyl-2-[5-(trifluoromethyl)-1,3,4-thiadiazol-2-yloxy]acetamide) from water mediated by photocatalysis. Response Surface Methodology, a collection of statistical and mathematical techniques, was employed for developing, improving, and optimizing

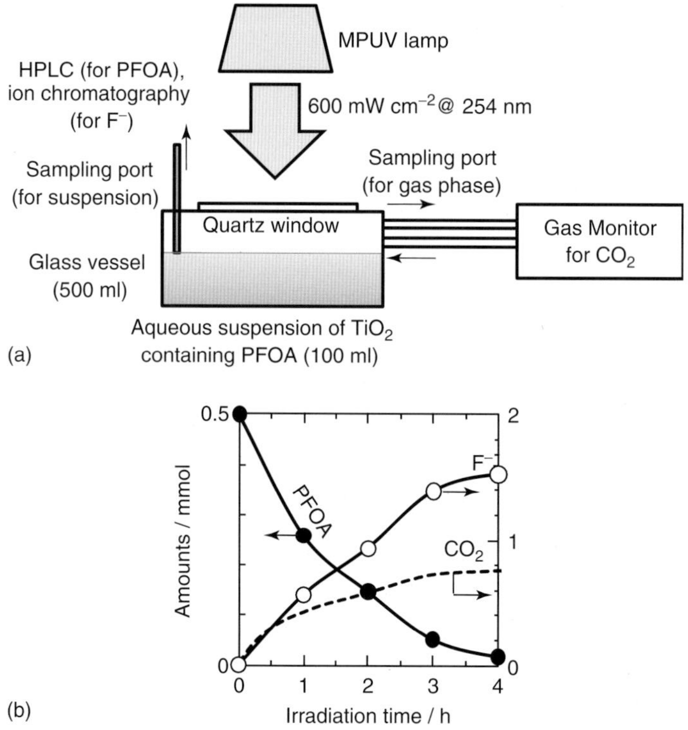

Figure 14.2 (a) Schematic illustration of the photocatalytic system. (b) Irradiation-time dependence of detected molar amounts of PFOA, CO_2, and F^- under irradiation with an MPUV lamp. (Source: Reprinted with permission from Ref. [13]. Copyright 2011 American Chemical Society.)

the processes. The results clearly showed the transformation pathways and the important role of the selection of the most appropriate reaction conditions in achieving highest decomposition efficiency for specific treatment cases.

We have reported the decomposition of environmentally persistent perfluorooctanoic acid (PFOA) in aqueous suspensions of P25 with the use of medium-pressure ultraviolet (MPUV) lamp irradiation under atmospheric pressure [13]. Figure 14.2a shows a schematic illustration of the system. A glass vessel equipped with a quartz window and sampling ports was used as a simple batch reactor. In a typical run, a 0.1 l of aqueous suspension containing 5 mmol l^{-1} of PFOA and 1.5 g l^{-1} of P25 was filled into the reactor. A MPUV lamp (U46C18, 600 mW cm^{-2} at 254 nm, Heraeus) was used as a light source for photocatalysis. During the reaction, the gas phase was analyzed by a photoacoustic field gas monitor and the aqueous phase was analyzed by ion chromatography and HPLC. In this condition, 5 mmol l^{-1} of PFOA was almost totally decomposed during 4 h of MPUV lamp irradiation under atmospheric pressure (Figure 14.2b). The result clearly shows that the PFOA decomposition followed pseudo-first-order kinetics with observed rate constant of 8.6×10^{-2} l^3 h^{-1} and stoichiometric production of CO_2 and F^-. In the proposed decomposition pathway, PFOA molecules adsorb onto the TiO_2 surface according to an adsorption equilibrium in aqueous suspension and could be easily decomposed by holes and radicals generated by the MPUV lamp irradiation. Under the present reaction conditions, a narrow region of TiO_2 concentrations around 1.5 g l^{-1} showed the maximum extent of PFOA decomposition, CO_2 formation, and F^- formation. The optimum rate can be attributed to a trade-off between an increase of photon absorption by the TiO_2 and a decrease in UV penetration below the surface of the TiO_2 suspension.

14.2.2 Immobilization of TiO_2 Particles onto Solid Supports

Aqueous suspension systems have some practical and economical disadvantages. The main problem is the separation of TiO_2 nanoparticles after treatment. Moreover, UV penetration depth in the suspended photocatalyst systems is limited because of the strong UV absorption by the TiO_2 nanoparticles. For a cost-effective TiO_2–water separation and easy handling of photocatalysts, the immobilization of TiO_2 nanoparticles on solid supports was investigated from the late 1980s [1–4]. Now various solid supports such as nonwoven fabric [14, 15], ceramic foam [16], and porous metal [17, 18] are available for fabrication of effective photocatalytic filters. Chapter 6 also mentions about this system.

Nonwoven fabric is one of the useful supports for fabrication of flexible, light weight, and cost-effective photocatalytic filters. TiO_2 nanoparticles could be immobilized onto the surface of nonwoven fabric by an inorganic binder (e.g., a suspension of colloidal SiO_2). It is easy to mass produce large-area filter with modest price. We have reported the photocatalytic inactivation and removal of algae with floated nonwoven fabric based photocatalytic filters [14]. On the other hand, porous ceramic foams have been chosen as solid supports for photocatalytic water and air purification, and disinfection systems [16, 19, 20]. Its high heat resistivity allows to immobilization of TiO_2 onto its surface by sintering via sol–gel technique without any binder. Moreover, the highly open porous structure with large surface area of these materials provides the excellent properties for gas and water passing while maintaining a high level of surface contact and low level of pressure drop. Thus, ceramic foam based photocatalytic filters are one of the most useful photocatalytic filter for commercial environmental purifier, especially air purifier for remediation of indoor air pollution caused by volatile organic compounds (VOCs) or disinfection of airborne pathogens.

Although numerous studies have proposed different application modes for photocatalyst-immobilized materials, further technological breakthroughs to realize high-quality products are still required for practical applications. To meet this requirement, we have fabricated the TiO_2 nanoparticles impregnated titanium mesh filter, TMiP™ [18]. The fabrication method of TMiP was shown in Figure 14.3. At first, both sides of a Ti-sheet (A4 size, 0.2 mm^t) were coated by different patterns of photoresistance. Then the Ti-sheet with resist pattern was immersed into acid solution and was converted to Ti-mesh by chemical etching. After etching, the

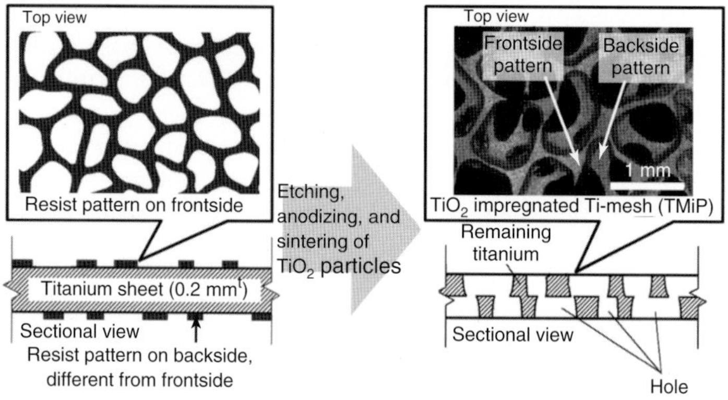

Figure 14.3 Schematic illustrations for fabrication of TMiP [18].

Ti-mesh was anodized at a voltage of 70 V in acid solution and was heated to prepare the TiO_2 layer on its surface. Finally, the anodized Ti-mesh was dip-coated in TiO_2 anatase sol and was heated at 550 °C for 3 h to sinter TiO_2 nanoparticles onto anodized Ti-mesh surface.

We have found many advantages of TMiP and its usefulness for air or water purification. TMiP could immobilize TiO_2 nanoparticles on its surface by sintering without any binder such as the ceramic foams, but TMiP does not break as easily as ceramic foams. Moreover, the high mechanical flexibility of Ti-mesh structure allows TMiP manipulation like a nonwoven fabric. Therefore, any geometry of modules for environmental purification could be designed by combination of UV sources and the other technologies. Now we are trying to fabricate various environmental purification units by using TMiP for application to various fields [18, 21–24].

14.3
Synergistic Effects of Photocatalysis and Other Methods

14.3.1
Deposition of Metallic Nanoparticles onto TiO_2 Surface for Disinfection

It is known that deposition of metallic nanoparticles such as Pt, Pd, and Ag onto TiO_2 surface could enhance the photocatalytic activity by removing electrons from TiO_2 and blocking electron–hole recombination. Metallic nanoparticles, especially Ag, are also known as nanodisinfectants already used in a number of commercial products. Therefore, many types of Ag–TiO_2 composite materials have been studied as efficient disinfectants [25–27].

Sunada *et al.* [28] studied the photocatalytic disinfection process of *Escherichia coli (E. coli)* on TiO_2 film. By atomic force microscope (AFM) measurement of *E. coli* cells on TiO_2-coated glass plate after different UV irradiation times, they concluded that the photokilling of bacteria on the TiO_2 surface could occur in two-step reaction mechanism: (i) partial decomposition of outer membranes of cells by the reactive species ($OH^\bullet$, H_2O_2, $O_2^{\bullet-}$) produced by the TiO_2 film and (ii) the structural and functional disordering of the cytoplasmic membrane because of lipid peroxidation and killing of the cell. In the first stage, cell viability was not lost very efficiently. However, the permeability of reactive species will be increased. Consequently, reactive species easily reach and attack the inner membrane. Wu *et al.* [26] investigated a more effective disinfection process of *E. coli* in water with Ag-deposited TiO_2 thin film. Compared with UV irradiation alone or UV/TiO_2, the inactivation of *E. coli* by the UV/Ag–TiO_2 process was enhanced and the photoreactivation of the bacteria was repressed. AFM measurements of *E. coli* indicated that the UV/Ag–TiO_2 process could expedite the destruction of cell wall and cell membrane. The formation of malondialdehyde and leakage of intracellular potassium ion (K^+) and protein also supported this result. In addition, Li *et al.* [27] reported synergistic bactericidal activity of Ag-deposited TiO_2 nanoparticles in both light and dark conditions. They found that Ag–TiO_2 particles exhibited different mechanisms of disinfection in light and dark conditions. Given the highly photoactive nature of the TiO_2 particles, the disinfection mechanism under UV was mainly due to the generation of reactive species as mentioned previously. In contrast, in the absence of UV, the mechanism was mainly due to direct contact with Ag surface and formation of toxic Ag species such as Ag^+, Ag(0), AgCl, and $AgCl_2$. This result indicates that Ag–TiO_2 system is useful not only in the light condition but also in the dark condition. On the other hand, Marugán *et al.* [29] noted that there are many differences between photocatalytic oxidation of chemicals and photocatalytic disinfection. The activation was based on the same physicochemical phenomena and consequently a good correlation between both processes was observed. However, osmotic stress, repairing mechanism, bacterial adhesion to the TiO_2 surface, and so on make disinfection kinetics significantly more complex than the kinetics observed for the oxidation of chemicals. Moreover, bacterial disinfection reactions are extremely sensitive to the composition of water and modifications of the TiO_2 in comparison with the oxidation of chemicals. For example, although *E. coli* inactivation efficiency by photocatalysis is enhanced by chloride, the decolorization rate of methylene blue by photocatalysis is negatively affected in the presence of chloride ions. This result may be caused by that chloride ions increased adsorption of the bacteria on the catalyst compared to other ionic species [30]. Thus, the activity observed for the photocatalytic oxidation of chemicals cannot be extrapolated to photocatalytic disinfection processes.

14.3.2
Combination with Advanced Oxidation Processes (AOPs)

Although TiO_2 surface can be improved by depositing noble metallic nanoparticles, this is a far from practical solution because of its high cost. Moreover, TiO_2 photocatalyst essentially has several limitations such as electron–hole recombination and difficulty in decomposition of large amount of pollutants or refractory chemicals. One of the solutions to these problems is combination with other processes such as electrolysis [31, 32], pulsed discharge processes [33, 34], photoelectrochemical cells [35, 36], and advanced oxidation processes (AOPs) [37, 38]. Especially, the combination with AOPs is useful for decomposition of contaminants at relatively low cost.

Beltrán *et al.* [38] reported effective decomposition of antibiotic sulfamethoxazole (SMT) in water by combination of TiO_2 photocatalysis and AOPs. They compared several processes to decompose SMT: ozone alone (O_3) and combined with UV-A radiation (O_3/UV-A, ozone photolysis) and photocatalytic oxidation (O_3/UV-A/TiO_2, photocatalytic ozonation), photocatalytic oxidation alone (UV-A/TiO_2). An aqueous solution of SMT (0.9 l) was charged in 1 l capacity glass photoreactor and was continuously bubbled by ozone–oxygen mixture gas. In photolytic experiments, the solution was irradiated with a UV lamp immersed in a glass well placed at the middle of the reactor. For experiments involving TiO_2, the solid was kept in suspension with a concentration of $1.5\ g\,l^{-1}$. Steadily, samples were withdrawn from the reactor and analyzed after filtration. Interestingly, the pseudo-first-order rate constant in total organic carbon (TOC) reduction kinetics of the O_3/UV-A/TiO_2 process is even higher than the sum of the rates due to UV-A/TiO_2 and O_3/UV-A processes under unbuffered condition. They concluded that the mechanism of photocatalytic ozonation mainly involves direct ozone reaction with SMT. Regarding TOC reduction, free radical oxidation and surface reactions are the main mechanisms of oxidation.

On the basis of this insight, we have developed a photocatalytic ozonation system for water purification by using TMiP (Section 14.2.2) [23]. Figure 14.4a shows the schematic illustration of the system. An aqueous solution of 1 l containing waterborne pathogens, *E. coli* or Qβ phage, or phenol was circulated through the units by a pump at a flow rate of $1\ l\,min^{-1}$ and was treated by the unit under several condition, that is, O_3 alone, UV-C alone, UV-C/TMiP, and O_3/UV-C/TMiP. The survival rate of *E. coli* was decreased for 10^5–10^7 by each condition. However, under

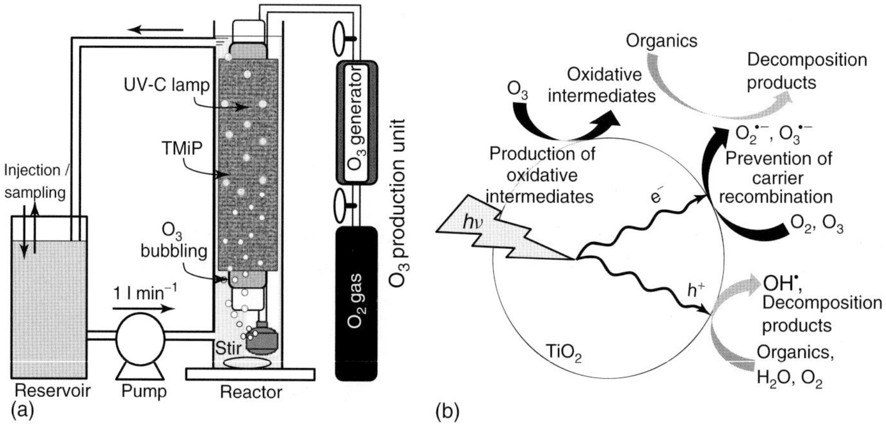

Figure 14.4 (a) Schematic illustration of the photocatalytic ozonation system. (b) The estimated reaction mechanism of photocatalytic ozonation on the TMiP surface [23].

UV-C alone or UV-C/TMiP conditions, more than 5 min of treatment time was required for total inactivation of *E. coli*. In contrast, O_3 alone or O_3/UV-C/TMiP conditions take a shorter treatment time (less than 1 min) for the inactivation. On the other hand, Qβ phage could not be inactivated efficiently under UV-C/TMiP condition. This difference may be due to the fact that it is difficult for the Qβ phage to contact the TiO_2 surface efficiently because its size (about 20 nm) is smaller than that of *E. coli* (about 1–2 μm). However, O_3 alone or O_3/UV-C/TMiP conditions could inactivate Qβ phage efficiently. Interestingly, inactivation efficiency under O_3/UV-C/TMiP condition is larger than under O_3 alone. Phenol decomposition shows this tendency more clearly. These results indicate that this photocatalytic ozonation unit could decompose waterborne pathogens and refractory chemicals more efficiently. Figure 14.4b shows the estimated reaction mechanism of photocatalytic ozonation on the TMiP surface. Ozone may play two roles in the system: preventing carrier recombination in photocatalysis and producing oxidative intermediates by photolytic and catalytic decomposition [39, 40]. These roles are important for efficient water purification. Moreover, as mentioned in Section 14.2.2, TMiP has more advantages than aqueous suspension systems or another photocatalytic filters. Therefore, it would be attractive to develop similar continuous-mode water purification systems for the practical treatment of highly contaminated water, such as sewage water.

14.4 Effective Design of Photocatalytic Reactor System

14.4.1 Two Main Strategies for the Effective Reactors

As mentioned in Section 14.2, immobilization of TiO_2 nanoparticles on the solid supports is useful for photocatalytic water purification. However, this time, the reaction efficiency is strongly limited by mass transfer from the bulk of the water to the TiO_2 surface because reaction could occur only at the liquid–solid interface. Moreover, UV irradiation onto the interface is necessary for photocatalysis. Although the operation conditions such as pH, temperature, and water quality also affect efficiency [9], reactive surface area and mass transfer in the system are more important in many cases. Therefore, the reactor design for effective photocatalytic water purification can be classified into two main strategies: (i) enlargement of reactive surface area and (ii) improvement of mass transfer.

On the basis of these strategies, Dionysiou *et al.* [41] developed the rotating disk photocatalytic reactor. The most recent work in this field was reported by Zhang *et al.* (Figure 14.5a) [42]. The rotating disk was made by following method. A circular titanium foil was anodized in an organic supporting electrolyte at 45 V for 6 h to prepare the one-dimensional TiO_2 nanotube arrays onto its surface. SEM image of the surface of the rotating disk is shown in the inset of Figure 14.5a. The

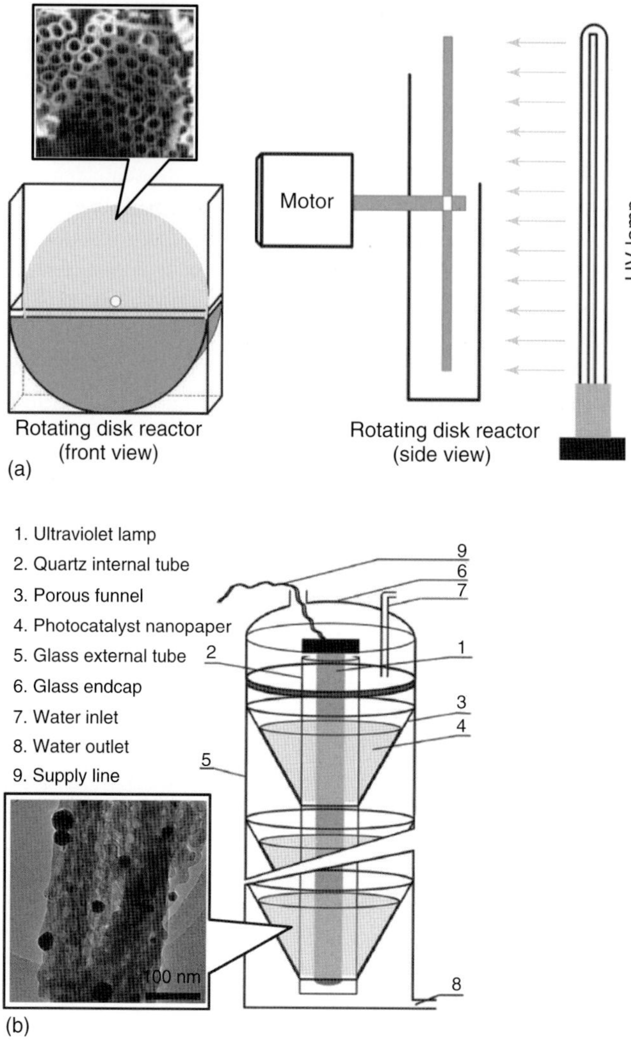

Figure 14.5 (a) Schematic illustration of rotating disk reactor system and SEM image of the surface of the disk [42]. (b) Schematic illustration of continuous-flow reactor system using Ag/TiO_2 nanopaper and SEM image of the surface of the nanopaper [43].

obtained nanotubes were 14.9 μm long with an average pore size of 75 nm and a wall thickness of 17 nm. The lower half of the disk was immersed into a rhodamine B aqueous solution of 36 ml. The disk was rotated at various speed and irradiated by UV-C lamp. Zhang *et al.* concluded that important parameters are rotating speed and high roughness of the disk. Rotation speed is related to the thickness of carried liquid film on the disk as well as the mass transfer coefficients of target contaminants, intermediates, and final products, all of which have influences on the photocatalysis. TiO_2 nanotube arrays realize relatively higher mass transfer coefficients and degradation efficiencies due to the more intense turbulence and large surface area. Indeed, the rotating disk reactor with TiO_2 nanotube arrays could fully realize the two main strategies for the effective reactors.

On the other hand, Zhou *et al.* [43] developed the continuous-flow reactor using Ag/TiO_2 paper. They fabricated the TiO_2 paper-like 2D freestanding membranes by a combination of hydrothermal process and a modified paper-making process. Then, Ag nanoparticles were deposited onto the surface of TiO_2 paper by photoreduction method (Figure 14.5b inset). Finally, Ag/TiO_2 papers placed on the inner wall of the reactor (Figure 14.5b, (4)). The sample solution is pumped into the upper inlet (7) of the reactor. UV light (1) radiates to the Ag/TiO_2 papers, and simultaneously sample solution flows through the Ag/TiO_2 papers, leading to the photocatalytic degradation reaction. The degradation products are discharged from the water outlet (8) of the reactor. Indeed, methyl orange removal rate of Ag/TiO_2 paper in the continuous-flow reactor was achieved to 100% in 40 min for single layer and only in 6 min for three layers. This reactor is also useful to remove gaseous contaminants such as toluene. Moreover, the Ag/TiO_2 papers still maintain a high-level activity even though used seven times in the same photocatalytic reactions and have a good antibacterial effect as mentioned in Section 14.3.1.

14.4.2
Design of Total System

For the purification of large amount of wastewater, we have to consider enlargement of reactor size and cost performance, reducing of energy consumption, and so on. Although the reactors mentioned in Section 14.4.1 are effective, they could not treat wastewater at a practical scale and low cost. Moreover, a typical UV source for photocatalysis, mercury lamps, should be avoided because of their environmental risk. Therefore, practical photocatalytic water purification technologies with mercury-free UV source such as UV-LED [44] or solar light [45, 46] are investigated.

We have developed an electrolysis–photocatalysis sequential water treatment system using a boron-doped diamond (BDD) electrode and a TiO_2 photocatalyst [47]. Figure 14.6 shows a photograph and a schematic illustration of the system, which consists mainly of an electrolysis unit, a photocatalysis unit, and photovoltaic cells. High-level wastewater was passed through a prefilter and stored in a tank. Then, the high-level wastewater was circulated through the electrochemical cell containing a BDD electrode and converted to low-level wastewater. The low-level wastewater was circulated through the TiO_2-coated quartz wool. Finally, the treated wastewater was passed through the final-filter and the ion-exchange-filter units. All electric power for the mechanical systems and the electrolysis was provided by photovoltaic cells. Furthermore, energy generated during daylight hours can charge a battery to provide electrical power during cloudy periods or at night. A chemical oxygen demand value of $106.1\,mg\,l^{-1}$ for the river water sample decreased to less than $1.0\,mg\,l^{-1}$ by this system. Moreover, total coliform and standard plate count values investigated by a bactericidal test were

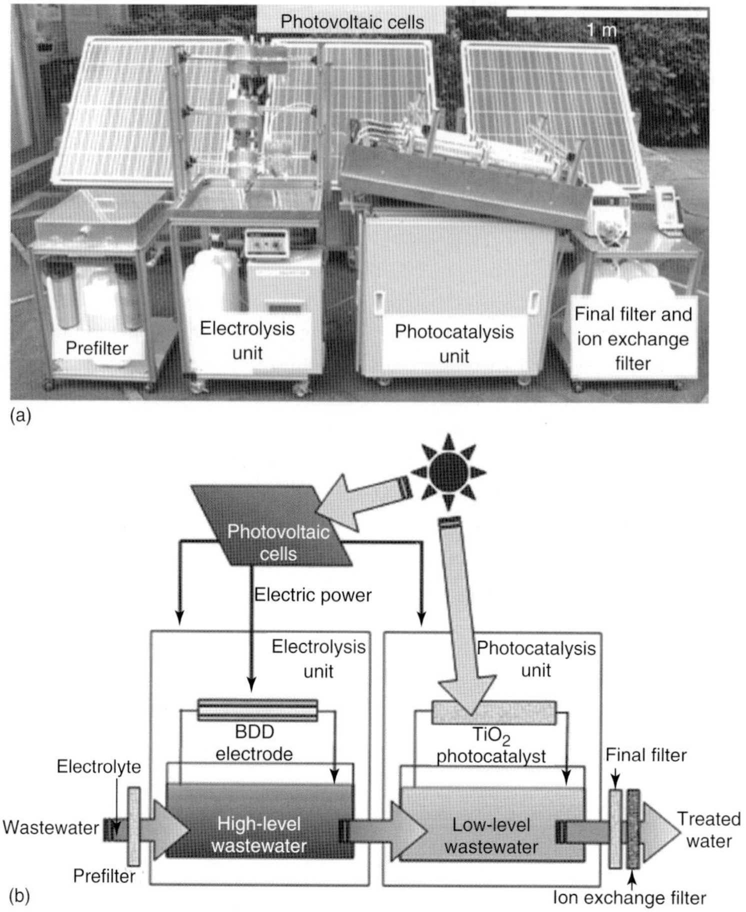

Figure 14.6 (a) A photograph of solar-driven electrochemical and photocatalytic water treatment system. (b) Schematic illustration of the system [47].

not detected in the treated water. These results indicate that the water treated by the system is suitable for drinking. On the basis of these results, this system could provide 12 l per day of drinking water from the river water using only solar energy. In addition, the estimated cost for drinking water is 26 yen l^{-1} in the system. This is extremely inexpensive compared with commercial waters. Therefore, this system may be useful for supplying drinking water during a disaster.

14.5 Future Directions and Concluding Remarks

One of the most serious limitations of TiO_2 photocatalysis is that TiO_2 can absorb only the UV. Thus UV lamp should be equipped near the TiO_2 photocatalyst surface inside the system. For solving this problem, The Project to Create Photocatalyst Industry for Recycling-oriented Society (New Energy and Industrial Technology Development Organization project) was conducted from 2007 to 2012 with an investment of 5.1 billion yen. This has given particular impetus to the research and development of high-sensitive visible-light-responsive photocatalysts. The project teams found that the visible light activity of WO_3 photocatalyst with cocatalysts such as Pt, Pd, and Cu(II) clusters drastically enhanced the oxygen reduction [48, 49]. Especially, Cu-modified WO_3 photocatalyst with a visible light reactivity was 10 times higher than that of the existing products which made from N-doped TiO_2. This project is expected to lead to the development of photocatalysis of a magnitude sufficient for deodorization, VOC elimination, and disinfection scenarios by interior applications. Thus, visible-light-responsive photocatalysts are mainly expected to be applicable for indoor self-cleaning and air purification; however, their practical application in water purification can also be expected. Because this is the most fundamental requirement of an effective reactor for water purification, catalyst improvement for a high photoefficiency that can utilize wider spectrum light at greater sensitivity, which is mentioned in Section 14.1.2. In conclusion, we can expect a large number of applications in photocatalytic water purification because of the expansion to new fields of research and development. In particular, we feel that since achieving a healthy and comfortable living environment is becoming an important issue, photocatalysis can fulfill an important role in water purification.

Acknowledgments

The work was supported in part by the Programs of Special Coordination Funds for Promoting Science and Technology of MEXT (Ministry of Education, Culture, Sports, Science and Technology), the Nippon Sheet Glass Foundation for Materials Science and Engineering, the Tokyo Ohka Foundation for the promotion of Science and Technology.

References

1. Fujishima, A. and Honda, K. (1972) Electrochemical photolysis of water at a semiconductor electrode. *Nature*, **238**(5358), 37–38.
2. Fujishima, A., Rao, T.N., and Tryk, D.A. (2000) Titanium dioxide photocatalysis. *J. Photochem. Photobiol., C*, **1**, 1–21.
3. Hashimoto, K., Irie, H., and Fujishima, A. (2005) TiO_2 photocatalysis: a historical overview and future prospects. *Jpn. J. Appl. Phys.*, **44**(12), 8269–8285.
4. Fujishima, A., Zhang, X., and Tryk, D.A. (2008) TiO_2 photocatalysis and related surface phenomena. *Surf. Sci. Rep.*, **63**, 515–582.
5. Kawai, T. and Sakata, T. (1980) Conversion of carbohydrate into hydrogen fuel by a photocatalytic process. *Nature*, **286**(5772), 474–476.
6. Frank, S.N. and Bard, A.J. (1977) Heterogeneous photocatalytic oxidation of cyanide ion in aqueous solutions at titanium dioxide powder. *J. Am. Chem. Soc.*, **99**(1), 303–304.
7. Min, L., Wu, X.-Z., Tetsuya, S., and Inoue, H. (2007) Time-resolved chemiluminescence study of the TiO_2

photocatalytic reaction and its induced active oxygen species. *Luminescence*, **22**(2), 105–112.
8. Matthews, R.W. (1987) Photooxidation of organic impurities in water using thin films of titanium dioxide. *J. Phys. Chem.*, **91**(12), 3328–3333.
9. Chong, M.N., Jin, B., Chow, C.W.K., and Saint, C. (2010) Recent developments in photocatalytic water treatment technology: a review. *Water Res.*, **44**(10), 2997–3027.
10. Sun, B., Vorontsov, A.V., and Smirniotis, P.G. (2011) Parametric studies of diethyl phosphoramidate photocatalytic decomposition over TiO_2. *J. Hazard. Mater.*, **186**, 1147–1153.
11. Sakkas, V.A., Calza, P., Vlachou, A.D., Medana, C., Minero, C., and Albanis, T. (2011) Photocatalytic transformation of flufenacet over TiO_2 aqueous suspensions: identification of intermediates and the mechanism involved. *Appl. Catal., B*, **110**, 238–250.
12. Szabó-Bárdos, E., Somogyi, K., Törö, N., Kiss, G., and Horváth, A. (2011) Photocatalytic decomposition of l-phenylalanine over TiO_2: identification of intermediates and the mechanism of photodegradation. *Appl. Catal., B*, **101**(3–4), 471–478.
13. Ochiai, T., Iizuka, Y., Nakata, K., Murakami, T., Tryk, D.A., Koide, Y., Morito, Y., and Fujishima, A. (2011) Efficient decomposition of perfluorocarboxylic acids in aqueous suspensions of TiO_2 photocatalyst with medium-pressure ultraviolet lamp irradiation under atmospheric pressure. *Ind. Eng. Chem. Res.*, **50**(19), 10943–10947.
14. Khataee, A.R., Fathinia, M., Aber, S., and Zarei, M. (2010) Optimization of photocatalytic treatment of dye solution on supported TiO_2 nanoparticles by central composite design: intermediates identification. *J. Hazard. Mater.*, **181**, 886–897.
15. Ochiai, T., Fukuda, T., Nakata, K., Murakami, T., Tryk, D., Koide, Y., and Fujishima, A. (2010) Photocatalytic inactivation and removal of algae with TiO_2-coated materials. *J. Appl. Electrochem.*, **40**(10), 1737–1742.
16. Yao, Y., Ochiai, T., Ishiguro, H., Nakano, R., and Kubota, Y. (2011) Antibacterial performance of a novel photocatalytic-coated cordierite foam for use in air cleaners. *Appl. Catal., B*, **106**(3–4), 592–599.
17. Hu, H., Xiao, W., Yuan, J., Shi, J., Chen, M., and Shang, G.W. (2007) Preparations of TiO_2 film coated on foam nickel substrate by sol–gel processes and its photocatalytic activity for degradation of acetaldehyde. *J. Environ. Sci.*, **19**(1), 80–85.
18. Ochiai, T., Hoshi, T., Slimen, H., Nakata, K., Murakami, T., Tatejima, H., Koide, Y., Houas, A., Horie, T., Morito, Y., and Fujishima, A. (2011) Fabrication of TiO_2 nanoparticles impregnated titanium mesh filter and its application for environmental purification unit. *Catal. Sci. Technol.*, **1**(8), 1324–1327.
19. Yazawa, T., Machida, F., Oki, K., Mineshige, A., and Kobune, M. (2009) Novel porous TiO_2 glass-ceramics with highly photocatalytic ability. *Ceram. Int.*, **35**(4), 1693–1697.
20. Plesch, G., Gorbar, M., Vogt, U.F., Jesenak, K., and Vargova, M. (2009) Reticulated macroporous ceramic foam supported TiO_2 for photocatalytic applications. *Mater. Lett.*, **63**(3–4), 461–463.
21. Ochiai, T., Nakata, K., Murakami, T., Morito, Y., Hosokawa, S., and Fujishima, A. (2011) Development of an air-purification unit using a photocatalysis-plasma hybrid reactor. *Electrochemistry*, **79**(10), 838–841.
22. Ochiai, T., Niitsu, Y., Kobayashi, G., Kurano, M., Serizawa, I., Horio, K., Nakata, K., Murakami, T., Morito, Y., and Fujishima, A. (2011) Compact and effective photocatalytic air-purification unit by using of mercury-free excimer lamps with TiO_2 coated titanium mesh filter. *Catal. Sci. Technol.*, **1**(8), 1328–1330.
23. Ochiai, T., Nanba, H., Nakagawa, T., Masuko, K., Nakata, K., Murakami, T., Nakano, R., Hara, M., Koide, Y., Suzuki, T., Ikekita, M., Morito, Y., and Fujishima, A. (2012) Development of an O_3-assisted photocatalytic water-purification unit by using a TiO_2

modified titanium mesh filter. *Catal. Sci. Technol.*, **2**(1), 76–78.

24. Slimen, H., Ochiai, T., Nakata, K., Murakami, T., Houas, A., Morito, Y., and Fujishima, A. (2012) Photocatalytic decomposition of cigarette smoke using a TiO_2-impregnated titanium mesh filter. *Ind. Eng. Chem. Res.*, **51**(1), 587–590.
25. Yao, Y., Ohko, Y., Sekiguchi, Y., Fujishima, A., and Kubota, Y. (2008) Self-sterilization using silicone catheters coated with Ag and TiO_2 nanocomposite thin film. *J. Biomed. Mater. Res. B: Appl. Biomater.*, **85B**(2), 453–460.
26. Wu, D., You, H., Jin, D., and Li, X. (2011) Enhanced inactivation of *Escherichia coli* with Ag-coated TiO_2 thin film under UV-C irradiation. *J. Photochem. Photobiol., A*, **217**(1), 177–183.
27. Li, M., Noriega-Trevino, M.E., Nino-Martinez, N., Marambio-Jones, C., Wang, J., Damoiseaux, R., Ruiz, F., and Hoek, E.M.V. (2011) Synergistic bactericidal activity of Ag-TiO_2 nanoparticles in both light and dark conditions. *Environ. Sci. Technol.*, **45**(20), 8989–8995.
28. Sunada, K., Watanabe, T., and Hashimoto, K. (2003) Studies on photokilling of bacteria on TiO_2 thin film. *J. Photochem. Photobiol., A*, **156**(1–3), 227–233.
29. Marugán, J., van Grieken, R., Pablos, C., and Sordo, C. (2010) Analogies and differences between photocatalytic oxidation of chemicals and photocatalytic inactivation of microorganisms. *Water Res.*, **44**(3), 789–796.
30. Gogniat, G., Thyssen, M., Denis, M., Pulgarin, C., and Dukan, S. (2006) The bactericidal effect of TiO_2 photocatalysis involves adsorption onto catalyst and the loss of membrane integrity. *FEMS Microbiol. Lett.*, **258**(1), 18–24.
31. Yu, H., Chen, S., Quan, X., Zhao, H., and Zhang, Y. (2008) Fabrication of a TiO_2-BDD heterojunction and its application as a photocatalyst for the simultaneous oxidation of an azo dye and reduction of Cr(VI). *Environ. Sci. Technol.*, **42**(10), 3791–3796.
32. Ochiai, T., Moriyama, H., Nakata, K., Murakami, T., Koide, Y., and Fujishima, A. (2011) Electrochemical and photocatalytic decomposition of perfluorooctanoic acid with a hybrid reactor using a boron-doped diamond electrode and TiO_2 photocatalyst. *Chem. Lett.*, **40**(7), 682–683.
33. Wang, T.C., Lu, N., Li, J., and Wu, Y. (2011) Plasma-TiO_2 catalytic method for high-efficiency remediation of p-nitrophenol contaminated soil in pulsed discharge. *Environ. Sci. Technol.*, **45**(21), 9301–9307.
34. Zhang, Y., Deng, S., Sun, B., Xiao, H., Li, L., Yang, G., Hui, Q., Wu, J., and Zheng, J. (2010) Preparation of TiO_2-loaded activated carbon fiber hybrids and application in a pulsed discharge reactor for decomposition of methyl orange. *J. Colloid Interface Sci.*, **347**(2), 260–266.
35. Liu, Y., Li, J., Zhou, B., Li, X., Chen, H., Chen, Q., Wang, Z., Li, L., Wang, J., and Cai, W. (2011) Efficient electricity production and simultaneously wastewater treatment via a high-performance photocatalytic fuel cell. *Water Res.*, **45**(13), 3991–3998.
36. Qin, G., Wu, Q., Sun, Z., Wang, Y., Luo, J., and Xue, S. (2012) Enhanced photoelectrocatalytic degradation of phenols with bifunctionalized dye-sensitized TiO_2 film. *J. Hazard. Mater.*, **199–200**, 226–232.
37. Michael, I., Hapeshi, E., Michael, C., and Fatta-Kassinos, D. (2010) Solar Fenton and solar TiO_2 catalytic treatment of ofloxacin in secondary treated effluents: evaluation of operational and kinetic parameters. *Water Res.*, **44**(18), 5450–5462.
38. Beltrán, F.J., Aguinaco, A., García-Araya, J.F., and Oropesa, A. (2008) Ozone and photocatalytic processes to remove the antibiotic sulfamethoxazole from water. *Water Res.*, **42**(14), 3799–3808.
39. Pierre, P. (2003) in *In Chemical Degradation Methods for Wastes and Pollutants* (ed M.A. Tarr), CRC Press, pp. 77–119.
40. Taranto, J., Frochot, D., and Pichat, P. (2007) Combining cold plasma and TiO_2 photocatalysis to purify gaseous

effluents: a preliminary study using methanol-contaminated air. *Ind. Eng. Chem. Res.*, **46**(23), 7611–7614.

41. Dionysiou, D.D., Balasubramanian, G., Suidan, M.T., Khodadoust, A.P., Baudin, I., and Laîné, J.-M. (2000) Rotating disk photocatalytic reactor: development, characterization, and evaluation for the destruction of organic pollutants in water. *Water Res.*, **34**(11), 2927–2940.
42. Zhang, A., Zhou, M., Han, L., and Zhou, Q. (2011) The combination of rotating disk photocatalytic reactor and TiO_2 nanotube arrays for environmental pollutants removal. *J. Hazard. Mater.*, **186**, 1374–1383.
43. Zhou, W., Du, G., Hu, P., Yin, Y., Li, J., Yu, J., Wang, G., Wang, J., Liu, H., Wang, J., and Zhang, H. (2011) Nanopaper based on Ag/TiO_2 nanobelts heterostructure for continuous-flow photocatalytic treatment of liquid and gas phase pollutants. *J. Hazard. Mater.*, **197**, 19–25.
44. Natarajan, K., Natarajan, T.S., Bajaj, H.C., and Tayade, R.J. (2011) Photocatalytic reactor based on UV-LED/TiO_2 coated quartz tube for degradation of dyes. *Chem. Eng. J.*, **178**, 40–49.
45. Plantard, G., Janin, T., Goetz, V., and Brosillon, S. (2012) Solar photocatalysis treatment of phytosanitary refuses: efficiency of industrial photocatalysts. *Appl. Catal., B*, **115–116**, 38–44.
46. Blanco, J., Malato, S., Fernández-Ibañez, P., Alarcón, D., Gernjak, W., and Maldonado, M.I. (2009) Review of feasible solar energy applications to water processes. *Renewable Sustainable Energy Rev.*, **13**(6–7), 1437–1445.
47. Ochiai, T., Nakata, K., Murakami, T., Fujishima, A., Yao, Y.Y., Tryk, D.A., and Kubota, Y. (2010) Development of solar-driven electrochemical and photocatalytic water treatment system using a boron-doped diamond electrode and TiO_2 photocatalyst. *Water Res.*, **44**(3), 904–910.
48. Arai, T., Horiguchi, M., Yanagida, M., Gunji, T., Sugihara, H., and Sayama, K. (2008) Complete oxidation of acetaldehyde and toluene over a Pd/WO_3 photocatalyst under fluorescent- or visible-light irradiation. *Chem. Commun.*, **43**, 5565–5567.
49. Abe, R., Takami, H., Murakami, N., and Ohtani, B. (2008) Pristine simple oxides as visible light driven photocatalysts: highly efficient decomposition of organic compounds over platinum-loaded tungsten oxide. *J. Am. Chem. Soc.*, **130**(25), 7780–7781.

15
Solar Photocatalytic Pilot Plants: Commercially Available Reactors

Sixto Malato, Pilar Fernández-Ibáñez, Maneil Ignacio Maldonado, Isabel Oller, and Maria Inmaculada Polo-López

15.1
Introduction

The specific hardware needed for solar photocatalytic applications have much in common with that used for thermal applications. As a result, both reactors and photocatalytic systems have followed conventional solar thermal collector designs, such as parabolic troughs and nonconcentrating collectors [1]. At this point, their designs begin to diverge [2], as (i) the fluid must be exposed to ultraviolet solar radiation, and, therefore, the absorber must be UV transparent and (ii) a temperature higher than ambient ($\sim$20 °C) does not play a significant role in the photocatalytic process, so no insulation is required. Contrary to solar thermal processes, which are based on the collection of large quantities of photons of all wavelengths to reach a specific temperature range, solar photocatalysis is based on the collection of only high-energy short-wavelength photons to promote photoreactions. Most of the solar photocatalytic processes use UV or near-UV sunlight (300–400 nm), but in some processes, up to 600 nm sunlight can be absorbed as in photo-Fenton or when new modified/doped photocatalysts are applied. Sunlight at wavelengths over 600 nm is normally not harvested nowadays in any photocatalytic process.

The original solar photoreactor designs for photochemical applications were based on line-focus parabolic-trough concentrators (PTCs; Figure 15.1). In part, this was a logical extension of the historical emphasis on trough units for solar thermal applications. Furthermore, PTC technology was relatively mature and existing hardware could be easily modified for photochemical processes. The first outdoors engineering-scale reactors developed were a converted solar thermal parabolic-trough collector in which the absorber/glazing-tube combination had been replaced by a simple Pyrex glass tube through which contaminated water could flow [3]. The main disadvantages of PTCs were described in late 1990s [4, 5]: (i) only direct radiation can be used, (ii) expensive, and (iii) have low optical (solar tracking/light reflection/concentration) and quantum efficiencies. It is well known that in the wavelength range of the solar spectrum that can be used

Photocatalysis and Water Purification: From Fundamentals to Recent Applications, First Edition. P. Pichat.

Figure 15.1 PTCs in operation. Note the brightness provoked by the irradiated photocatalyst (TiO_2) under concentrated solar light.

for the excitation of TiO_2, the diffuse and direct portion of the solar radiation reaching the surface of the earth are almost equal [6]. This means that a light concentrating system cannot employ much more than half of the solar radiation available for catalyst activation. Concerning quantum efficiency, it is well-known in photocatalysis since a long time ago [7, 8] that above a certain photon flux (q_p), reaction rate dependency on irradiance goes down from one ($r = kq_p^{1.0}$) to less than one order ($r = kq_p^{<1.0}$) until zero ($r = kq_p^0 = k$) because of the excess of photogenerated species (e^-, h^+, and $OH^{\bullet}$). Many articles on this aspect of photocatalysis provided information on irradiance at which the change of order is produced [2] and the main conclusion was that solar concentrating devices (as PTCs) should be disregarded for photocatalysis. In all cases, the described photonic efficiencies (moles of substrate degraded/incident moles of photons inside the reactor) were less than 1%.

Concentrating collectors focus only the direct sunlight and cannot collect the diffuse light. Thin clouds, dust, and haze reduce the direct-beam component of sunlight more than the diffuse component. As a result, nonconcentrating collectors can use a resource that is not only larger but also less variable than that available to concentrating collectors, permitting, in many locations, continual operation of the nonconcentrating detoxification system. Under cloudy conditions, nonconcentrating devices can continue operating (although at lower rates), whereas a trough unit would have to shut down. In middle 1990s, the first nonconcentrating collectors for photocatalysis were described as, in principle, they were cheaper than PTCs as they have no moving parts or solar tracking devices (see a few examples in Figure 15.2).

An extensive effort in the design of small nontracking collectors resulted in the design and testing of different nonconcentrating solar reactors [9]. However, the design of a robust nonconcentrating photoreactor is not trivial, as it must be weather resistant, chemically inert, and ultraviolet transmissive. In addition, flat plate nonconcentrating systems require significantly more photoreactor area than concentrating photoreactors and, as a consequence, full-scale systems (normally composed of hundreds of square meters of collectors) must be designed to withstand the operating pressures anticipated for fluid circulation through a large field. In uncovered, nonconcentrating systems exposed to the ambient, reactants, and catalyst could become contaminated. Very often the chemical inertness of the

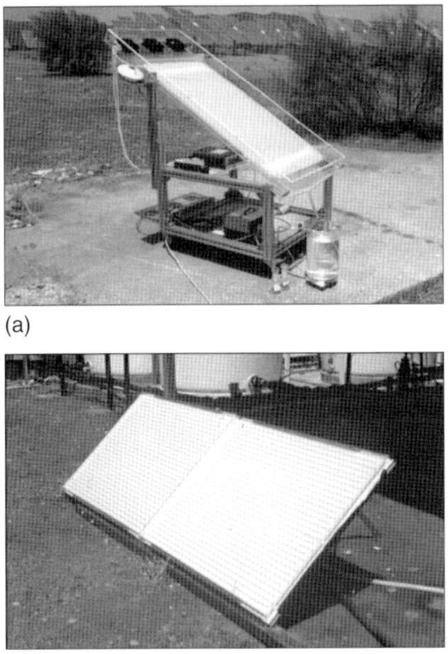

(a)

(b)

Figure 15.2 (a,b) Examples of nonconcentrating collectors developed during the 1990s [9].

materials used (to resist corrosion caused by outdoor operation and exposure to solar irradiation) for constructing the flat plate nonconcentrating collector would be difficult to guarantee. Consequently, the use of these photoreactors was abandoned by the main research groups.

To design a solar collector for photocatalytic purposes, there is a group of constraints that should be fulfilled: (i) the collection of UV radiation, (ii) working temperatures as close as possible to ambient temperature in order to avoid the loss of volatile organic compounds, (iii) quantum efficiency decreases with irradiance, and (iv) concentrators collect $1/R_C$ (concentration factor, R_C) of the available diffuse radiation. As a result of these considerations, the concentration for photocatalytic applications would be $R_C = 1$. Finally, its construction must be economical and should be easy and the reactor should correspond to a low pressure drop. As a consequence, the use of tubular photoreactors has a decisive advantage because of the inherent structural efficiency of tubing. Tubing is also available in a large variety of materials and sizes and is a natural choice for a pressurized fluid system.

15.2 Compound Parabolic Concentrators

There is a category of low-concentration collectors, called compound parabolic concentrators (CPCs), that are used in thermal applications. They are an interesting option, between parabolic concentrators and static flat systems, as they combine

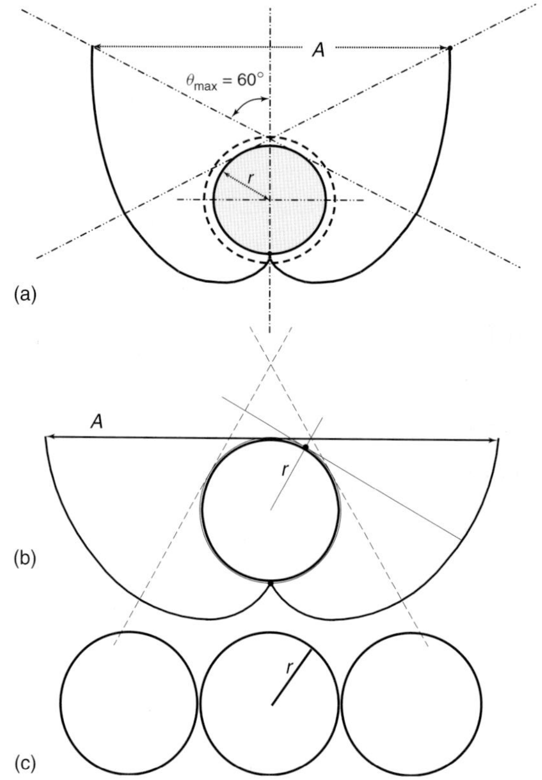

Figure 15.3 (a) Compound parabolic concentrator design, (b) compound parabolic concentrator $R_C = 1$, and (c) simple tubular photoreactors, note that $2\pi r = A$.

characteristics of each: they concentrate solar radiation, but they conserve the properties of the flat plate collectors, being static and collecting diffuse radiation. Thus they also constitute a good option for solar photochemical applications [10]. CPCs were invented in the 1960s [11] to achieve solar concentration with static devices. They do so irradiating the complete perimeter of the receiver, rather than just the "front" of it, as in conventional flat plates. R_C of a two-dimensional CPC collector is given by Eq. (15.1), where A is the aperture of the photoreactor (Figure 15.3a).

$$R_{C,CPC} = \frac{1}{\sin \theta_a} = \frac{A}{2\pi r} \tag{15.1}$$

The normal values for the semi-angle of acceptance (θ_a), for photocatalytic applications, are between 60° and 90° (Figure 15.3a). This wide angle of acceptance allows the receiver to collect both direct and a large part of the diffuse light ($1/R_C$ of it), with the additional advantage of decreasing errors of both the reflective surface and receiver tube alignment, which become important for achieving a low-cost photoreactor. A special case is one in which $\theta_a = 90°$ ($2\pi r = A$), whereby

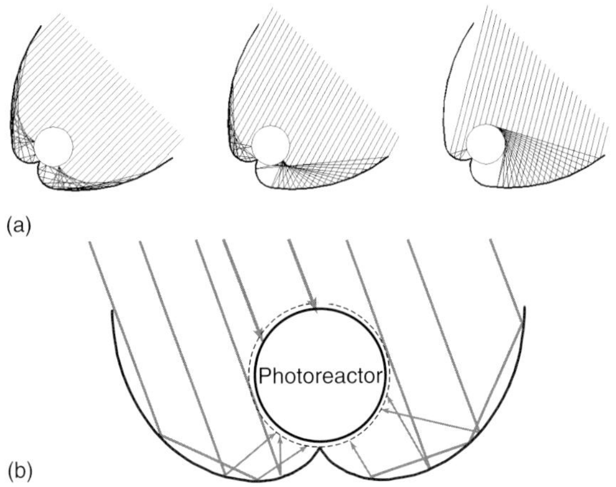

Figure 15.4 (a) Compound parabolic concentrator with $R_C > 1$. (b) Compound parabolic concentrator with $R_C = 1$.

$R_C = 1$ (nonconcentrating solar system; Figure 15.3b). When this occurs, all the UV radiation that reaches the aperture area of the CPC (direct and diffuse) can be collected and redirected to the reactor. If the CPC is designed for an acceptance angle of $+90^\circ$ to -90°, all incident solar diffuse radiation can be collected (Figure 15.4b). CPCs are static collectors with surface following an involute around a cylindrical reactor tube and designed with a $R_C = 1$, they have advantages of both PTCs and flat plate collectors. The light reflected by the CPC is distributed all around the tubular receiver so that almost the entire circumference of the receiver tube is irradiated and the light incident on the photoreactor is the same that would be impinging on a flat plate (Figure 15.4a). CPCs can make highly efficient use of both direct and diffuse solar radiation, without the need for solar tracking. There is no evaporation of possible volatile compounds because water does not heat up. They have high optical efficiency, because they make use of almost all the available radiation, and high quantum efficiency, as they do not receive a concentrated flow of photons. Flow also can be easily maintained turbulent inside the tube reactor. They are considered nowadays as the state of the art for solar collectors in photocatalysis [12].

The light reflected by the CPC is more or less distributed around the tubular photoreactor and as a result most of the reactor tube circumference is irradiated, but due to the ratio of CPC aperture to tube diameter, no single point on the tube receives much more than one-sun (not concentrated solar irradiation) of UV light. As a result, the UV light incident on the reactor is very similar to that of a one-sun photoreactor. As in the case of flat plate collectors, maximum yearly efficiency is normally obtained at the same collector angle inclination as the local latitude. Performance is very close to that of simple tubular photoreactors (see Figure 15.3c), but only about one-third (ratio between $2\pi r$ and diameter of the photoreactor) of the reactor tube material is required. As in a parabolic trough, the water is more easily piped and distributed than in many one-sun designs. CPC reflectors are usually made of polished aluminum and the structure can be a simple

Figure 15.5 Close view of CPC shape.

photoreactor support frame with connecting tubing (Figure 15.5). As this type of reflector is considerably less expensive than tubing, their use is very cost-effective compared to deploying nonconcentrating tubular photoreactors without use of any reflectors, but preserving the advantages of using tubing for the active photoreactor area. All these factors contribute to excellent CPC collector performance in solar photocatalytic applications.

15.3
Technical Issues: Reflective Surface and Photoreactor

In addition to the solar collector type, the most important technical issues related with solar photocatalysis hardware are the reflective surface and the reactor tube. The optical quality requirements of reflective surfaces for solar applications are usually related to the concentration required by the particular application under consideration. The higher the concentration desired, the stricter the requirements for quality of parameters. Light reflected off a polished or mirrored surface obeys the law of reflection: the angle between the incident ray and the normal to the surface is equal to the angle between the reflected ray and the normal. Even transparent glass specularly reflects a portion of incoming light. Diffuse reflection is typical of particulate substances such as powder. In the case of solar photocatalytic applications, the strictest requirements are those of PTCs, for example, UV-mirror materials need to have a specular reflectance between 300 and 400 nm in order to achieve concentration ratios of from 1 to 20. The greater the imperfections are, and particularly the reflective surface defects, the lower the effective concentration ratio is. So, the reverse is also true: the lower the effective concentration ratio, the higher the optical imperfections are and therefore the lower the quality of reflective surface required. This is an important additional factor in favor of nonconcentrating systems, as these lower quality requirements (lower specular reflectance) are directly translated into lower manufacturing cost, because the reflector element can represent 50% of the collector cost [13].

Another important factor is the reflective base material. For solar photocatalytic applications, the reflective surface must clearly be made of a highly reflective material for ultraviolet radiation. The reflectivity between 300 and 400 nm of traditional silver-coated mirrors is very low (reflected radiation/incident radiation) and aluminum-coated mirrors is the best option in this case. Aluminum is the only metal surface that is highly reflective throughout the ultraviolet spectrum. Reflectivities range from 92.3% at 280 nm to 92.5% at 385 nm. Comparable values for silver are 25.2 and 92.8%, respectively. A fresh deposited aluminum surface is fragile and needs to be protected from weathering and abrasion, but the conventional glass cover used for silver-backed mirrors has the drawback of significantly filtering UV light (an effect that is duplicated due to the light path two times through the glass). The surfaces currently available that best fit these requirements are (i) electropolished anodized aluminum and (ii) organic plastic films with an aluminum coating.

The requirements for the photocatalytic reactor are similar to other advanced oxidation processes, with the additional necessity of an irradiated photocatalyst. The photocatalytic reactor must contain the catalyst and be transparent to UV radiation providing good mass transfer of the contaminant to an irradiated photocatalyst surface with minimal pressure drop across the system. The square-root dependence on irradiance provides better photoefficiencies for one-sun designs. Nevertheless, for water treatment, the reactor must be hard enough to work under usable water pressure and tube configurations clearly seem to be the most appropriate for fluid containment and pumping. Adequate flow distribution inside the reactor must be assured. If the catalyst is used in suspensions (as usual in the case of TiO_2), the Reynolds number (Re) must always guarantee the turbulent flow. This is critical in avoiding catalyst settlement. Another important design issue is that reactor materials must not react with either the catalyst or the pollutants to be treated or their by-products. The choice of materials that are both transmissive to UV light and resistant to its destructive effects is limited. A one-sun reactor must be able to withstand summer temperatures of around 60–70 °C. Finally, resistance to low pH is needed because the production of inorganic and organic acids as reaction by-products or final products usually occurs. Common materials that meet these requirements are fluoropolymers and several types of glass (Figure 15.6). Quartz has excellent UV transmission as well as good temperature and chemical resistance, but the slight advantage in transmission in the terrestrial solar spectrum over other materials does not justify its high cost, which makes it completely unfeasible for photocatalytic applications.

Fluoropolymers are a good choice of plastic for photoreactors due to their good UV transmittance, excellent ultraviolet stability, and chemical inertness. Several different types, such as ETFE (ethylene-tetrafluoroethylene), PTFE (polytetrafluoroethylene), ECTFE (ethylenechloride-tetrafluroethylene), PVDF (polyvinylidene fluoride), FEP (fluorinated ethylenepropylene), PFA (Perfluoroalkoxy), and TFE (tetrafluoroethylene), can be extruded into tubing and used as a photoreactor. Tubular fluoropolymers are very strong, possess excellent tear resistance, and are flexible and lighter than glass. One of their greatest disadvantages is that, in order to achieve a desired pressure resistance, the wall thickness of a fluoropolymer tube may have to be increased, which in turn will lower its UV transmittance. In addition, due to the lack of rigidity, tube connections can withstand much lower

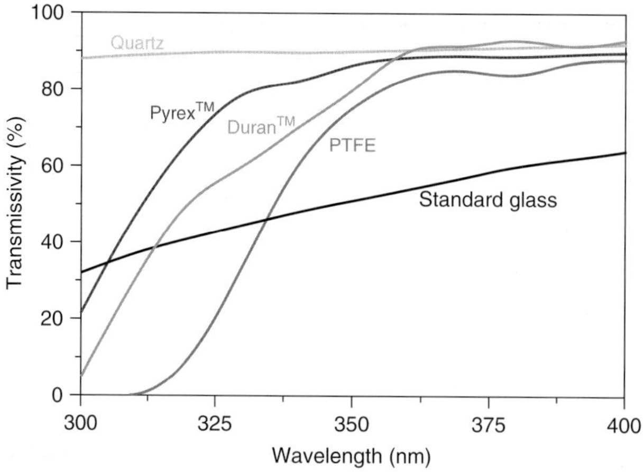

Figure 15.6 Transmittance of different materials suitable for the manufacture of photoreactor tubes.

pressures than rigid tubes. Glass is another alternative for photoreactors. Standard glass, used as protective surface, is not satisfactory because it absorbs part of the UV radiation that reaches it, due to its iron content. Borosilicate glass has good transmissive properties in the solar range with a cut-off of about 285 nm. Therefore, such a low-iron-content glass would seem to be the most adequate.

Therefore, as both fluoropolymers and glass are valid photoreactor materials, cost becomes an important issue. In large volumes, glass piping could be more expensive than fluoropolymer tubing, but from the perspective of performance, the choice is the material that has the best combination of tensile strength and UV transmittance. On this basis, if a large field is being designed, large collector area means also a considerable number of reactors and, as consequence, high pressure drop. Thus, fluoropolymer tubes are not the best choice of material because high pressure is linearly related to thickness and could result in higher cost. A detailed analysis is recommended for any specific design.

One of the most important parameters in a tubular photoreactor design is the diameter, as in both homogeneous or heterogeneous photocatalysis it must be guaranteed that all arriving useful photons are kept inside the reactor and do not go through it without intercepting a catalyst particle. The intensity of irradiation affects the relationship between reaction rate and catalyst concentration. The dispersion and absorption of light causes photon density to diminish over the length of the optical path within a catalyst suspension. It is also important to take into account that at higher irradiance, the catalyst concentration could be higher. The absorption of radiation in a photocatalytic tubular reactor is influenced by the geometry of the reactor and by the optical properties of the photocatalyst. The modeling of the radiation absorbed in solar photoreactors (tubular reactors and compound parabolic collectors) has been recently reported in [12] for the optical properties of TiO_2 P25 (formerly manufactured by Degussa, now by Evonik) averaged in the solar radiation spectrum.

In the case of TiO_2 heterogeneous photocatalysis, when catalyst concentration is very high, a "screening" effect produces excessive opacity in the solution, preventing the catalyst particles farthest in from being irradiated and reducing system efficiency. The lower the catalyst concentration, the less opaque the suspension. TiO_2 catalyst of $1\,g\,l^{-1}$ reduces transmittance to zero in a cylinder of 10 mm inner diameter with concentrated light in a parabolic-trough collector [14]. Therefore, in a wider diameter tube, only an outer layer is irradiated. This means that larger inner reactor diameter permits the use of lower optimum catalyst concentrations. Practical inner diameters for tubular photoreactor would be in the range of 25–50 mm. Diameters that are very small do not make sense because of the associated high pressure drop and very large diameters imply a considerable dark volume, thus reducing overall system efficiency.

Wavelengths where the catalyst does not absorb light are better for determining the optimum catalyst concentration as a function of light-path length. Under these conditions, measurement of photon losses is only affected by turbidity and it is easier to evaluate the effect of the photoreactor diameter. In Figure 15.7,

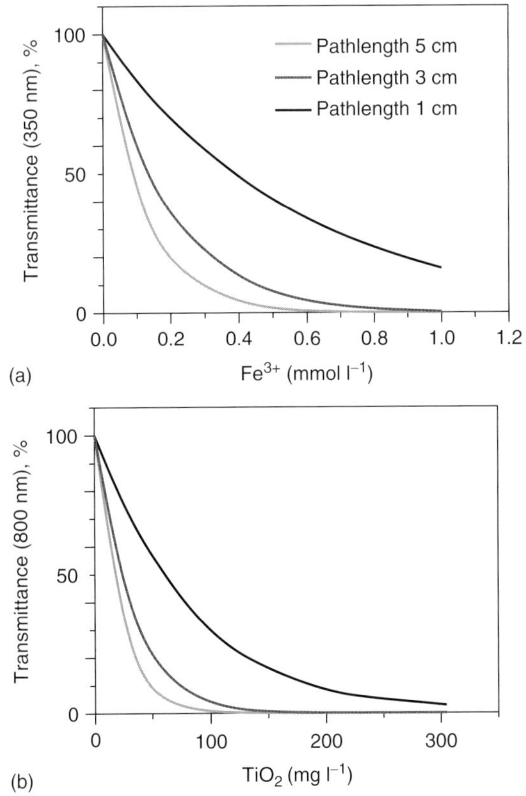

Figure 15.7 (a) Transmittance of Fe^{3+} dissolved in water. (b) Transmittance of TiO_2 suspension in water.

where these results are shown for different photoreactor tube diameters, it may be seen how larger photoreactor diameters result in a lower optimum catalyst concentration, and vice versa [15]. As these calculations are strongly affected by the experimental device, they had been experimentally demonstrated. After many experiments and simulations with different photoreactors, the optimum TiO_2 concentration obtained with sunlight is around a few hundreds of milligrams per liter with diameter of the photoreactor in the range of 25–50 mm [16, 17].

This chapter is focused in heterogeneous photocatalysis, but it is interesting to note than nowadays applications related with photo-Fenton are also very usual. As, from the point of view of the type and design of solar collectors both processes can be driven with very similar devices, it is important to explain the main reasons for this statement. The Fenton reactant consists, basically, of an aqueous solution of hydrogen peroxide and ferrous ions providing hydroxyl radicals. When the process is complemented with UV–vis radiation, it is called photo-Fenton. In this case, the process becomes catalytic. Fe^{3+} (related species and organic complexes) absorbs solar photons [18]. This effect must be considered when determining the optimum load as a function of light-path length in the photoreactor. It should also be remarked that the lowest wavelength in the solar spectrum at the Earth's surface is 300 nm. Therefore, absorption of light below this wavelength is irrelevant for the process. Figure 15.7a shows transmittance for iron(III) sulfate dissolved in water at pH 2.8 at 350 nm for different tube diameters. This figure shows that iron concentration is affected more by light-path length than the TiO_2 process, but it also indicates that the optimum diameter of the photoreactor is very similar in both cases (Figure 15.7b). As these calculations are strongly affected by different complexes formed by iron(III) during the photo-Fenton process, they had been experimentally demonstrated. The optimum concentrations of 0.2–0.5 mmol l^{-1} were obtained after many experiments with different photoreactors and different wastewaters under sunlight [19–21].

15.4 Suspended or Supported Photocatalyst

One of the major solar photoreactor system design issues is whether to use a suspended or supported catalyst. The majority of experiments to date have used small TiO_2 particles suspended in the contaminated water, which makes it necessary to recover them after treatment. Supported catalyst configurations eliminate the need for catalyst separation, but with the main objection of an important increase in treatment costs. The idea is to attach the catalyst to a support inside the reactor, which requires the catalyst to be anchored onto some type of inert support. Desirable characteristics of such a system would include being very active (comparable to slurries), have a low pressure drop, long lifetime, and reasonable cost. Other chapters of this book (Chapters 5 and 6) deal with this matter, but studies performed to date have not yet identified a fixed-catalyst system that performs as efficiently as slurry systems, at least for the solar treatment of water.

Fixed-catalyst designs must solve several problems. As the catalyst must be exposed to sunlight and in contact with the pollutant, the support must be configured to be efficiently irradiated and, at the same time, maintain a high flow rate in the water to ensure good mixing without significantly increasing system pressure. Also, the same criteria discussed for photoreactor materials must be kept in mind and applied when choosing a support (UV light transmission). Supports tested so far have included fiber glass beads, metal fibers, steel mesh, aluminum, many types of plastic, and ceramics in the most diverse shapes. Several important performance requirements are directly related to the process used for catalyst fixation, such as the durability of the coating, catalyst activity, and lifetime. Surface area of the catalyst coating must also be considered because a good contact of photons and target molecules with the catalyst is required for efficient performance.

In aqueous systems, compared to slurries, if immobilization of TiO_2 results in a reduction in performance an important direct consequence of this fact is the necessity of multiplying by the same factor the size of necessary solar collector field, making the overall system clearly less cost efficient and competitive. In addition to the above, a key question is how long supported catalysts will last in treatment of wastewater; a short period of activity would mean frequent replacement. To the contrary of fixed-catalyst configurations, slurry configurations have the advantage of a low pressure drop through the reactor and excellent fluid-to-catalyst mass transfer.

The need to remove the catalyst from the clean water after treatment was initially considered as the major disadvantage of a slurry system because ultrafiltration would be required. Although it is true that titanium dioxide P25 powder elementary particles are about 30 nm, or even smaller in other TiO_2, once in water, the particles always agglomerate [22] into larger ones (from 0.3 to 0.6 μm), facilitating the problem considerably, because microfiltration, which is much less expensive than ultrafiltration, can be used.

However, best performance can be obtained when microfiltration is combined with sedimentation. About 90% of the catalyst can easily be recovered by sedimentation and the rest by microfiltration [23]. The lifetime of membranes and the time between cleanings is very optimized nowadays [24]. This could be particularly important with high volumes of water. Titanium dioxide sedimentation is closely related to colloidal stability and TiO_2 aggregation conditions. The suspension can easily be destabilized adjusting the pH to point of zero charge (PZC) and the isoelectric point (IEP) on the surface of the catalyst particles, as both factors modify the surface charge. Progressive particle agglomeration and settlement is then obtained. In the case of TiO_2 (Degussa P25), at concentrations of 200 mg l^{-1}, the PZC is obtained at about pH 7 and only 5 h of storage is needed to recover 90–95%. This can be directly recovered in a conic-bottom tank and 5–10% of the remaining catalyst can be recovered by microfiltration. Catalyst recovered, usually in highly concentrated slurry, can be reused but not indefinitely. Slurry lifetime has been tested with satisfactory results under laboratory conditions but with real water treatments catalyst lifetime would be diminished due to poisoning by contaminants. However, in specific applications, tests should be performed to demonstrate how many times the catalyst can be reused without any problem. From this point of view, about a 5–10% addition of catalyst could be an interesting option to compensate possible loss and periodic replacement in case of catalyst poisoning.

15.5
Solar Photocatalytic Treatment Plants

The first outdoor engineering-scale reactor developed was a converted solar thermal parabolic-trough collector in which the absorber/glazing-tube combination had been replaced by a simple Pyrex glass tube through which contaminated water could flow [25]. The design procedure for a solar photocatalytic system requires the selection of a reactor, catalyst operating mode (slurry or fixed matrix), reactor-field configuration (series or parallel), treatment-system mode (once-through or batch), flow rate, pressure drop, pretreatment, catalyst and oxidant loading method, pH control, and so on. Usually, a photocatalytic plant is constructed with several solar collectors. All the modules are connected in series or parallel, but with valves that permit to bypass any number of them.

The most important sensors required for the system are temperature, pressure, and dissolved oxygen (at least in the reactor outlet). Other sensors, such as pH and selective electrodes, could be useful depending on the type of wastewater to be treated. A UV-radiation sensor must be placed in a position where the solar UV light reaching the photoreactor can be measured, permitting the evaluation of the incident radiation as a function of hour of the day, clouds, atmospheric or other environmental variations. Solar photocatalytic plants are frequently operated in a recirculating batch mode. The fluid is continuously pumped between a reactor zone and a tank in which no reaction occurs, until the desired degradation is achieved. The systems are operated in a discontinuous manner by recirculating the wastewater with an intermediate reservoir tank and centrifugal pump.

On the basis of accumulated experience in pilot plant design, construction, and testing [26], the first full-size demonstration plant was erected in 2000 for treating wastewater with slurries of TiO_2. This plant was designed to treat 1 m^3 of water contaminated with 100 m^2 of collector aperture area (Figure 15.8). The concentrated TiO_2 slurry and the air necessary for the reaction were injected in the circuit. Once the water was treated, the entire volume left to the sedimentation tank and the system was filled with more wastewater for another batch. Meanwhile, the treated water and the TiO_2 in the sedimentation tank underwent pH adjustment to achieve fast sedimentation of the catalyst, according to a patented procedure [27]. The concentrated catalyst slurry was transferred from the bottom of the sedimentation tank to another smaller tank from which the catalyst entered the photoreactor. The supernatant was removed through an outlet in the side almost at the bottom of the sedimentation tank, and enters another tank where the small fraction of the initial catalyst contained in the supernatant was removed by microfiltration with a

Figure 15.8 Full-size demonstration plant erected in 2000 for treating wastewater with slurries of TiO_2 [26].

suitable membrane before disposal of water. The membrane has two outlets, one for clean water and one for the concentrated slurry. The concentrated slurry was recirculated through the membrane until there are several grams of catalyst per liter. Then this was also reused.

The plant consisted of 2 parallel rows of 21 collectors and 31 m length each. East–west orientation was chosen with a small structural inclination (1%) to avoid the accumulation of rainwater on the CPC troughs. Collector inclination was equal to local latitude and the distance between rows was calculated to minimize shadowing of collectors. To this end, the angle of sunlight at noon on 21 December (lowest maximum sun elevation at north hemisphere) was used as a design parameter to define row separation (Figure 15.9). The final system design was modular with glass collector reactor tubes connected in series by high-density polyethylene (HDPE) quick connections. The hydraulic circuit was

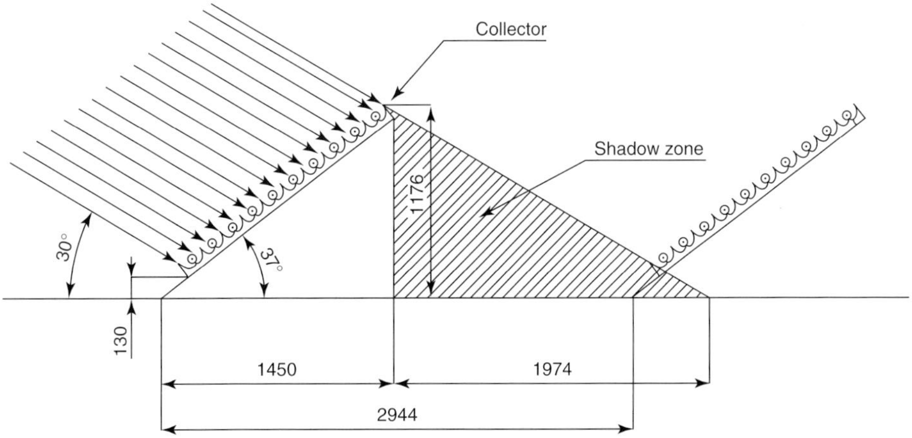

Figure 15.9 Design parameters to define row separation for minimize shadowing of collectors.

carefully designed to obtain the highest volumetric efficiency with minimum "dark zones." The different sections of the pipes must always be carefully calculated to guarantee similar flow rates in all the reactor tubes. Nominal flow was turbulent (Re between 10 000 and 20 000) to avoid catalyst settlement. All materials in contact with the water to be treated were carefully selected according to the nature of the contaminants and the required pH. Knowing the solar energy necessary to achieve certain treatment goal, a solar UVA sensor was incorporated within the electronic control devices, with the function of solar UV integration from the beginning of the treatment process. This sensor is connected to the control system and when the desired concentration was achieved the main pump transferred the water to the catalyst separation tank and advised the operator that the treatment has been completed. The control system was fed with design data (total energy needed for degrading contaminants until certain level) obtained at pilot plant scale with the same wastewater. But, in other cases online analytical devices (TOC (total organic carbon), COD (chemical oxygen demand), etc.) can be installed for controlling the plant. Other photocatalytic plants have been installed since 2000 and reviewed [9, 28, 29].

15.6
Specific Issues Related with Solar Photocatalytic Disinfection

There are a number of ways to enhance the disinfection of water using solar energy by photocatalytic inactivation of microorganisms. All these improvements have addressed the following criteria: (i) maximizing the collection of solar energy dose, (ii) enhancing the disinfecting efficacy especially against resistant waterborne pathogens, (iii) increasing the output of treated water in given solar exposure time, (iv) reducing the user dependence of the process, and (v) finding cheap and robust disinfection systems, which may also be constructed with local materials without sophisticated technological needs (this is especially important for developing countries), or (vi) optimizing photoreactor design taking into account the mechanism acting during disinfection and previous knowledge based on practical experiences on disinfection of real contaminated waters and wastewaters.

The first research on CPC prototype using TiO_2 solar photocatalysis for water disinfection objectives was done by Vidal *et al.* [30]. This solar photoreactor has $4.5\,m^2$ of CPC aperture and it was tilted at local latitude to maximize the available solar radiation for long exposure periods (few hours). They showed a 5 log decrease of *Escherichia coli* and *Enterococcus faecalis* in 30 min of solar treatment with TiO_2 suspensions ($0.5\,g\,l^{-1}$) and solar irradiation (mean UVA irradiance of $25\,W\,m^{-2}$). They also reported an economic analysis of this technology for future application not only to solar photocatalytic disinfection but also to decontamination of organic pollutants. Other contributions (e.g., [31]) demonstrated that three lab-scale solar photoreactors with aluminum reflectors consisting of compound parabolic, parabolic, and V-groove profiles all enhance the effect of the natural solar radiation, although the CPC is more efficient than the parabolic or V-groove profiles. Their results proved that bacterial deactivation rates using sunlight alone can be enhanced by low concentrations of titanium dioxide suspended in the water.

Although solar light or simulated solar light has a bactericidal effect, the addition of TiO_2 in the presence of radiation inactivates microorganisms (Chapter 11 deals with the basic concepts of the process) faster than in the absence of catalyst [32]. Most research on photocatalytic disinfection, such as in photocatalytic water decontamination, has been done with the commercial TiO_2 P25 photocatalyst. Nevertheless, some works have been carried out with pure anatase and doped titanium dioxide. For solar applications, visible-light-active materials are desirable (Chapters 5 and 14 deal with this matter). However, the smaller band gap, while absorbing a greater number of solar photons, gives a narrower voltage window to drive the redox reactions at the interface. Furthermore, metal sulfide semiconductors which absorb in the visible region of the spectrum, tend to undergo photo-anodic corrosion [33]. Considering cost, chemical and photochemical stability, availability, and lack of toxicity, the most suitable catalyst reported to date for the disinfection of water is TiO_2.

The configuration of the catalyst in the reactor can significantly alter in the disinfection results. The configuration of the catalyst in the photoreactor depends, among others factors, on the final application. If the system is designed for drinking water purification for human consumption, the use of TiO_2 in suspension, as a part of a routine intervention for improving the potability of water at household level (point-of-use water treatment) is not feasible. The photocatalyst particles would have to be removed after solar exposure and before consumption. Unfortunately, in photocatalytic disinfection it was even reported that immobilization of TiO_2 produces lower disinfection activity compared to slurry systems [34].

One example is TiO_2 deposited on glass rashing rings inside a tubular reactor of a CPC solar prototype that was compared with slurry systems in the work by Sordo *et al.* [35]. Maximum efficiency was shown by the slurry TiO_2 reactor due to the optimum contact between bacteria and catalyst. However, use of the fixed-bed reactor lead to inactivation rate quite close to that of the slurry. Not only the high titania surface area of this configuration is responsible for the bacteria inactivation but also the important contribution of the mechanical stress has to be considered. The main advantage of the fixed-bed TiO_2 catalyst is the outstanding stability, without deactivation effects after 10 reaction cycles, being readily applicable for continuous water treatment systems.

There are no available commercial systems based on solar photocatalytic disinfection, all we may find in the literature are prototypes developed with some particularities to work in the field with real contaminated water resources. Once a good disinfecting efficacy has been achieved, the most critical criterion for a good solar photoreactor performance is the increase of the output of treated water in the given solar exposure time. To address this objective, one must take into account the following limiting technical aspects.

1) The collection of solar light (using either CPC solar mirrors or other low-cost reflectors which increase the solar light collection) into the photoreactor must

be efficient for big volumes of water. Therefore, the water turbidity is critical; if the water is enough transparent (turbidity < 10 NTU (nephelometric turbidity units)) the optical reactor path length (i.e., diameter of the tubular photoreactor) can be increased up to 10 cm [36].

2) If the water is not very transparent, then the reactor tube must be reduced to few cm and the big volume requirement will be accomplished by the connection of several photoreactor modules [37].
3) The use of immobilized photocatalyst is desirable, but then the water flow rates must be as low as possible to increase the contact time between the disinfection target (bacterial cell, spore of bacteria or fungus, oocyst, virus, etc.) and the fixed catalyst. This limitation has been widely investigated [35, 38, 39] with the conclusion that the flow rate must be low enough (only a few liters per minute) to permit that the system delivers on the microorganism a minimum lethal dose of energy to reduce the microorganisms concentration as much as possible before the water circulates to dark regions of the reactor. In these dark bits of the reactor, bacteria may recover and regrow when the damage produced by the disinfecting agent is not strong enough to achieve total colonies abatement (i.e., complete disinfection). This situation leads to find residual bacterial concentration after the treatment regardless of the total solar energy delivered into the reactor system [39–41]. Therefore, the design of the reactor must be done with prior knowledge of that lethal solar energy dose, so that the residence time for complete disinfection is achieved in one run. Obviously, this energy requirement for complete killing depends strongly on several parameters such as water composition, turbidity, and nature of the biological contamination. Clear examples of this are the direct influence of ionic strength [42], natural organic matter [43], and natural consortium of bacteria in real waters or various ions [44].
4) There are limitations of photocatalytic solar disinfection when it is scaled up through the use of large batch volumes or continuous flow recirculation reactors [40]. Increasing flow rate has a negative effect on inactivation of bacteria as at a given time point there needs to be maximum exposure of bacteria to UV to ensure inactivation as compared to having bacteria exposed to sublethal doses over a long period of time. When the water with bacteria was kept static (Figure 15.10) under solar light, it was constantly irradiated and hence the needed uninterrupted UV dose was achieved and complete inactivation to the detection level took place. With continuous flow systems, the lethal dose was also deposited to the bacteria but in an intermittent manner and it did not produce complete inactivation. This statement has important implications for those attempting to scale-up solar systems through the use of pumped, recirculatory, continuous flow reactors. If the operational parameters are set such that the microbial pathogens are repeatedly exposed to sublethal doses of solar radiation followed by a period within which the cells have an opportunity to recover or repair, complete inactivation may not be achieved.
5) In photocatalytic disinfection, the electron acceptor is normally dissolved oxygen which is easily available from the air. With static batch systems,

Figure 15.10 Photograph of 25L-batch CPC static photoreactor located in PSA (Plataforma Solar de Almeria) facilities (Spain).

the concentration of dissolved oxygen will be rapidly depleted and must be replenished to maintain photocatalytic activity. Furthermore, the solubility of oxygen in water is reduced by temperature. This must be taken into consideration as the temperature within solar irradiated reactors can reach 55°C. New designs of reactor must address the need for replenishment of dissolved oxygen in photocatalytic disinfection systems. An alternative may be to introduce other oxidants, for example, H_2O_2; however, this would give rise to a dependence on consumable chemicals that may be affordable or undesirable depending on the final application of the system.

6) Other aspects such as reducing the user dependence of the process or making the system as cheap and robust as possible are worthy to consider so that the reactor can be used worldwide for solving a number of issues related with water safety, mainly for human consumption purposes in developing countries.

Recently, Polo-López *et al.* [41] have addressed the practical problems associated with increasing the treated water output using a continuous flow concept. A novel sequential batch photoreactor was designed and constructed with the aim of decreasing the treatment time required, increasing the total volume of water treated per day and reducing user dependency. The new photoreactor incorporated

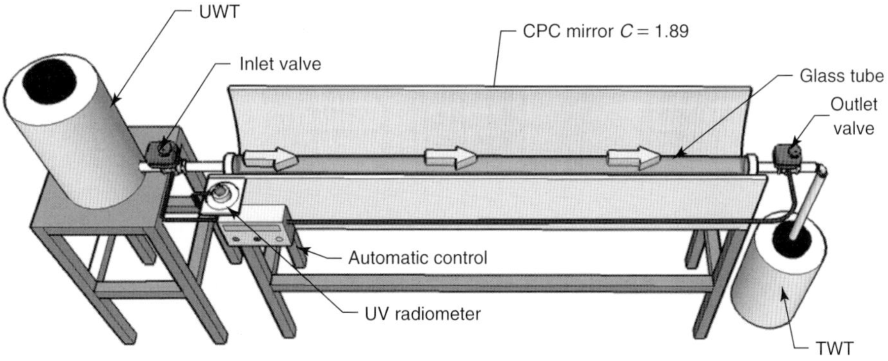

Figure 15.11 Schematic of the sequential batch system. UWT, untreated water tank and TWT, treated water tank. (Courtesy of Dr. Christian Navntoft, Argentina.)

a CPC of concentration factor of 1.89 (Figure 15.11) and the treatment time was automatically controlled by an electronic UVA sensor. The feedback sensor system controlled the gravity-filling of the reactor from an untreated water reservoir, and controlled the discharge of the treated water into a clean reservoir tank following receipt of the predefined UVA dose. The reactor was tested using *E. coli* from well water under real sunny conditions in Southern Spain. They found that this system would permit processing of six sequential batches of 2.5 l each day (i.e., 15 l of solar purified water each per day). The system is modular; therefore, it may be scaled up to allow several CPC photoreactors to be used under the control of a single UVA sensor. For example, six systems like this could produce around 90 l of potable water per day (for several households), and it could produce approximately 31 500 l during a typical year [41].

15.7 Conclusions

The current lack of data for comparison of solar photocatalysis with other technologies definitely presents an obstacle toward an industrial application. One issue would be to give sound examples of technoeconomic studies. Another aspect should be the assessment of the environmental impact in its broadest sense. These studies should assess if solar photocatalysis is a promising technology compared to others. To lead to industrial application, it will be critical that the photocatalytic processes can be developed up to a stage, where the process is cost efficient compared to other processes, robust (small to moderate changes to the wastewater stream do not affect the plant's efficiency and operability strongly), predictable (process design and up-scaling can be done reliably), easy to implement (suppliers and engineering companies can start marketing the process without huge initial investment costs, which could only be recovered by high turnovers), and safe regarding the environment (minimize risks of leakage, discharge of not sufficiently treated effluent).

Little by little, TiO_2 disinfection research went from basic laboratory studies to the first trials with real disinfection applications. Microorganisms that are very resistant to UVA irradiation have been successfully inactivated by TiO_2 photocatalysis and therefore recent TiO_2 disinfection research focuses more on disinfection applied to more resistant microorganisms. The irradiation way on the photoreactor has a high influence on the disinfection behavior. When light exposure is made continuously (without temporal interruptions) the disinfection effect of the process is more fast and efficient than when the light is applied intermittently. Therefore, there are aspects that are essential to generate a technology: (i) optimization of photoreactors taking into account the processes specific requirements, (ii) development of viable process schemes (batch, continuous, semicontinuous), (iii) development of process control strategies, (iv) assessment of the influence of the water chemical parameters, and (v) finding out applications different to potable water disinfection.

Acknowledgments

The authors wish to thank the Spanish Ministry of Science and Innovation for its financial assistance under the "FOTOREG" Project (Ref. CTQ2010-20740-C03-02).

References

1. Dillert, R., Cassano, A.E., Goslich, R., and Bahnemann, D. (1999) Large scale studies in solar catalytic wastewater treatment. *Catal. Today*, **54**, 267–282.
2. Malato, S., Blanco, J., Vidal, A., and Richter, C. (2002) Photocatalysis with solar energy at a pilot-plant scale: an overview. *Appl. Catal. B: Environ.*, **37**, 1–15.
3. Goswami, D.Y. (1997) A review of engineering developments of aqueous phase solar photocatalytic detoxification and disinfection processes. *J. Sol. Energy Eng. Trans. ASME*, **119**(2), 101–107.
4. Alfano, O.M., Bahnemann, D., Cassano, A.E., Dillert, D., and Goslich, R. (2000) Photocatalysis in water environments using artificial and solar light. *Catal. Today*, **58**, 199–230.
5. Malato, S., Gimenez, J., Richter, C., Curco, D., and Blanco, J. (1997) Low concentrating CPC collectors for photocatalytic water detoxification. Comparison with a medium concentrating solar collector. *Wat. Sci. Tech.*, **35**(4), 157–164.
6. Bird, R.E., Hulstrom, R.L., and Lewis, L.J. (1983) Terrestrial solar spectral data sets. *Sol. Energy*, **30**(6), 563–573.
7. D'Oliveira, J.-C., Al-Sayyed, G., and Pichat, P. (1990) Photodegradation of 2- and 3-chlorophenol in titanium dioxide aqueous suspensions. *Env. Sci. Tech.*, **24**, 990–996.
8. Al-Sayyed, G., D'Oliveira, J.-C., and Pichat, P. (1991) Semiconductor-sensitized photodegradation of 4-chlorophenol in water. *J. Photochem. Photobiol. A*, **58**, 99–114.
9. Bahnemann, D. (2004) Photocatalytic water treatment: solar energy applications. *Sol. Energy*, **77**(5), 445–459.
10. Ajona, J.A. and Vidal, A. (2000) The use of CPC collectors for detoxification of contaminated water: design, construction and preliminary results. *Sol. Energy*, **68**(1), 109–120.
11. Rabl, A. (1976) Optical and thermal properties of compound parabolic concentrators. *Sol. Energy*, **18**(6), 497–511.
12. Colina-Márquez, J., MacHuca-Martínez, F., and Puma, G.L. (2010) Radiation absorption and optimization of solar photocatalytic reactors for environmental applications. *Env. Sci. Tech.*, **44**(13), 5112–5120.
13. Blanco-Galvez, J. and Malato-Rodríguez, S. (2003) *Solar Detoxification*, UNESCO Publishing ISBN:. ISBN: 92-3-103916-4
14. Ollis, D.F. (1991) *Photochemical Conversion and Storage of Solar Energy*, Kluwer Academic Publishers, pp. 593–622.
15. Fernández-Ibáñez, P., Malato, S., and De Las Nieves, F.J. (1999) Relationship between TiO_2 particle size and reactor diameter in solar photodegradation efficiency. *Catal. Today*, **54**, 195–204.
16. Giménez, J., Curcó, D., and Queral, M.A. (1999) Photocatalytic treatment of phenol and 2,4-dichlorophenol in a solar plant in the way to scaling-up. *Catal. Today*, **54**, 229–244.
17. Colina-Márquez, J., Machuca-Martínez, F., and Li Puma, G. (2009) Photocatalytic mineralization of commercial herbicides in a pilot-scale solar CPC reactor: photoreactor modeling and reaction kinetics constants independent of radiation field. *Env. Sci.. Tech.*, **43**, 8953–8960.
18. Pignatello, J.J., Oliveros, E., and MacKay, A. (2006) Advanced oxidation processes for organic contaminant destruction based on the Fenton reaction and related chemistry. *Critical Rev. Environ. Sci. Technol.*, **36**, 1–84.
19. Fallmann, H., Krutzler, T., Bauer, R., Malato, S., and Blanco, J. (1999) Applicability of the photo-Fenton method

for treating water containing pesticides. *Catal. Today*, **54**, 309–319.

20. Gernjak, W., Krutzler, T., Glaser, A., Malato, S., Cáceres, J., Bauer, R., Fernández-Alba, A. R. (2003) Photo-Fenton treatment of water containing natural phenolic pollutants. *Chemosphere*, **50**, 71–78, 130.

21. Zapata, A., Velegraki, T., Sánchez-Pérez, J.A., Mantzavinos, D., Maldonado, M.I., and Malato, S. (2009) Solar photo-Fenton treatment of pesticides in water: effect of iron concentration on degradation and assessment of ecotoxicity and biodegradability. *Appl. Catal. B: Environ.*, **88**, 448–454.

22. Fernández-Ibáñez, P., de las Nieves, F.J., and Malato, S. (2000) Titanium dioxide/electrolyte solution interface: electron transfer phenomena. *J. Colloid Interf. Sci.*, **227**, 510–516.

23. Fernández-Ibáñez, P., Blanco, J., Malato, S., and de las Nieves, F.J. (2003) Application of the colloidal stability of TiO_2 particles for recovery and reuse in solar photocatalysis. *Wat. Res.*, **37**, 3180–3188.

24. Ulbricht, M. (2006) Advanced functional polymer membranes. *Polymer*, **47**(7), 2217–2262.

25. Alpert, D.J., Sprung, J.L., Pacheco, J.E., Prairie, M.R., Reilly, H.E., Milne, T.A., and Nimlos, M.R. (1991) Sandia national laboratories' work in solar detoxification of hazardous wastes. *Sol. Energy Mater.*, **24**(1–4), 594–607.

26. Blanco, J., Malato, S., Fernández, P., Vidal, A., Morales, A., Trincado, P., Oliveira, J.C., Minero, C., Musci, M., Casalle, C., Brunotte, M., Tratzky, S., Dischinger, N., Funken, K.-H., Sattler, C., Vincent, M., Collares-Pereira, M., Mendes, J.F., and Rangel, C.M. (2000) Compound parabolic concentrator technology development to commercial solar detoxification applications. *Sol. Energy*, **67**(4–6), 317–330.

27. Blanco, J., Malato, S., de las Nieves, J., and Fernández P. (2001) Method of sedimentation of colloidal semiconductor particles. European Patent Application EP-1-101-737-A1. Applicant: CIEMAT. 23-05-2001, European Patent Office bulletin 2001/21.

28. Malato, S., Fernández-Ibáñez, P., Maldonado, M.I., Blanco, J., and Gernjak, W. (2009) Decontamination and disinfection of water by solar photocatalysis: recent overview and trends. *Catal. Today*, **147**, 1–59.

29. Chong, M.N., Jin, B., Chow, C.W.K., and Saint, C. (2010) Recent developments in photocatalytic water treatment technology: a review. *Wat. Res.*, **44**, 2997–3027.

30. Vidal, A.I., Dićaz, A., El Hraiki, Romero, M., Muguruza, I., Senhaji, F., and Gonzaćlez, J. (1999) *Catal. Today*, **54**, 183.

31. McLoughlin, O.A., Fernández-Ibañez, P., Gernjak, W., Malato Rodríguez, S., and Gill, L.W. (2004) Photocatalytic disinfection of water using low cost compound parabolic collectors. *Sol. Energy*, **77**, 625.

32. Markowska-Szczupak, A., Ulfigb, K., and Morawskia, A.W. (2011) The application of titanium dioxide for deactivation of bioparticulates: an overview. *Catal. Today*, **169**, 249–257.

33. Byrne, J.A., Fernandez-Ibañez, P., Dunlop, P.S.M., Alrousan, D.M.A., and Hamilton, J.W.J. (2011) Photocatalytic enhancement for solar disinfection of water: a review. *Int. J.. Photoenergy*, 1–12.

34. Fernández, P., Blanco, J., Sichel, C., and Malato, S. (2005) Water disinfection by solar photocatalysis using compound parabolic collectors. *Catal. Today*, **101**, 345–352.

35. Sordo, C., Van Grieken, R., Marugan, J., and Fernandez-Ibañez, P. (2010) Solar photocatalytic disinfection with immobilised TiO_2 at pilot-plant scale. *Wat. Sci. Technol.*, **61**, 507–512.

36. Ubomba-Jaswa, E., Navntoft, C., Polo-López, I., Fernández-Ibáñez, P., and McGuigan, K.G. (2010) Investigating the microbial inactivation efficiency of a 25 L batch solar disinfection (SODIS) reactor enhanced with a compound parabolic collector (CPC) for household use. *J. Chem. Techn. Biotech.*, **85**, 1028–1037.

37. Polo-López, M.I., Fernández-Ibáñez, P., García-Fernández, I., Oller, I., Salgado-Tránsito, I., and Sichel, C. (2010) Resistance of Fusarium sp spores to solar TiO_2 photocatalysis: influence of spore type and water (scaling-up results). *J. Chem. Technol. Biotech.*, **85**, 1038–1048.

38. Gumy, D., Rincon, A.G., Hajdu, R., and Pulgarin, C. (2006) Solar photocatalysis for detoxification and disinfection of water: different types of suspended and fixed TiO_2 catalysts study. *Sol. Energy*, **80**, 1376–1381.
39. Sichel, C., Tello, C., de Cara, M., and Fernández-Ibáñez, P. (2007) Effect of UV solar intensity and dose on the photocatalytic disinfection of bacteria and fungi. *Catal. Today*, **129**, 152–160.
40. Ubomba-Jaswa, E., Navntoft, C., Polo-López, I., Fernández-Ibáñez, P., and McGuigan, K.G. (2009) Solar disinfection of drinking water (SODIS): an investigation of the effect of UVA dose on inactivation efficiency. *Photochem. Photobiol. Sci.*, **8**(5), 587–595.
41. Polo-López, M.I., Fernández-Ibáñez, P., Ubomba-Jaswa, E., Navntoft, C., Garcia-Fernandez, I., Dunlop, P.S.M., Schmidt, M., Byrne, J.A., and McGuigan, K.G. (2011) Elimination of water pathogens with solar radiation using and automated sequential batch CPC Reactor. *J. Haz. Ma.*, **196**, 16–21.
42. Sichel, C., Fernández-Ibáñez, P., Blanco, J., and Malato, S. (2007) Effects of experimental conditions on E. Coli survival during solar photocatalytic water disinfection. *J. Photochem. Photobiol. A: Chem.*, **189**, 239–246.
43. Rincón, A.G. and Pulgarin, C. (2004) Effect of pH, inorganic ions, organic matter and H_2O_2 on E. coli K_{12} photocatalytic inactivation by TiO_2 implications in solar water disinfection. *App. Catal. B: Environ.*, **51**, 283–302.
44. Rincón, A.G. and Pulgarin, C. (2004) Bactericidal action of illuminated TiO_2 on pure Escherichia coli and natural bacterial consortia: post-irradiation events in the dark and assessment of the effective disinfection time. *App. Catal. B: Environ.*, **49**, 99–112.

Index

Photocatalysis and Water Purification: From Fundamentals to Recent Applications, First Edition. P. Pichat.